ENRICHED NUMERICAL TECHNIQUES

ENRICHED NUMERICAL TECHNIQUES

Implementation and Applications

Edited by

AZHER JAMEEL

National Institute of Technology Srinagar, Hazratbal, Srinagar, Jammu and Kashmir, India

GHULAM ASHRAF UL HARMAIN

National Institute of Technology Srinagar, Hazratbal, Srinagar, Jammu and Kashmir, India

INDRA VIR SINGH

Indian Institute of Technology Roorkee, Roorkee, Uttarakhand, India

MAGD ABDEL WAHAB

Ghent University, Ghent, Belgium

ELSEVIER

ACADEMIC PRESS

An imprint of Elsevier

Academic Press is an imprint of Elsevier
125 London Wall, London EC2Y 5AS, United Kingdom
525 B Street, Suite 1650, San Diego, CA 92101, United States
50 Hampshire Street, 5th Floor, Cambridge, MA 02139, United States

ISBN: 978-0-443-15362-4 (print)
ISBN: 978-0-443-15361-7 (online)

For Information on all Academic Press publications
visit our website at https://www.elsevier.com/books-and-journals

Publisher: Matthew Deans
Acquisitions Editor: Sophie Harrison
Editorial Project Manager: Mason Malloy
Production Project Manager: Sharmila Kirouchenadassou
Cover Designer: Greg Harris

Typeset by MPS Limited, Chennai, India

Contents

List of contributors

Mahvash Afzal Department of Mechanical Engineering, Islamic University of Science and Technology, Awantipora, Jammu and Kashmir, India

Mumtaz Ahmad Department of Mechanical Engineering, National Institute of Technology Srinagar, Hazratbal, Srinagar, Jammu and Kashmir, India

Danish Ali Department of Mechanical Engineering, Jamia Millia Islamia, New Delhi, India

Sanjeev Anand School of Energy Management, Shri Mata Vaishno Devi University, Kakryal, Katra, Jammu and Kashmir, India

Yatheshth Anand School of Mechanical Engineering, Shri Mata Vaishno Devi University, Kakryal, Katra, Jammu and Kashmir, India; School of Energy Management, Shri Mata Vaishno Devi University, Kakryal, Katra, Jammu and Kashmir, India

Qazi Junaid Ashraf Department of Mechanical Engineering, University of Kashmir, Hazratbal, Srinagar, Jammu and Kashmir, India

Gagandeep Bhardwaj Mechanical Engineering Department, Thapar Institute of Engineering and Technology, Patiala, Punjab, India

Mohd Afzal Bhat Department of Mechanical Engineering, Islamic University of Science and Technology, Awantipora, Jammu and Kashmir, India

Kapil Chopra School of Mechanical Engineering, Shri Mata Vaishno Devi University, Kakryal, Katra, Jammu and Kashmir, India

Neha Duhan Department of Mechanical and Industrial Engineering, Indian Institute of Technology Roorkee, Roorkee, Uttarakhand, India

Kishan Dwivedi Design Against Failure and Fracture Group, School of Mechanical and Materials Engineering, Indian Institute of Technology Mandi, Mandi, India

R.K. Godara Mechanical Engineering Department, Thapar Institute of Engineering and Technology, Patiala, Punjab, India

Ovais Gulzar Department of Mechanical Engineering, Islamic University of Science and Technology, Awantipora, Jammu and Kashmir, India

Vibhushit Gupta School of Mechanical Engineering, Shri Mata Vaishno Devi University, Kakryal, Katra, Jammu and Kashmir, India; School of Energy Management, Shri Mata Vaishno Devi University, Kakryal, Katra, Jammu and Kashmir, India

G.A. Harmain Department of Mechanical Engineering, National Institute of Technology Srinagar, Hazratbal, Srinagar, Jammu and Kashmir, India

Azher Jameel Department of Mechanical Engineering, National Institute of Technology Srinagar, Hazratbal, Srinagar, Jammu and Kashmir, India

Showkat Ahmad Kanth Department of Mechanical Engineering, National Institute of Technology Srinagar, Hazratbal, Srinagar, Jammu and Kashmir, India

Majid Hameed Koul Department of Mechanical Engineering, Islamic University of Science and Technology, Awantipora, Jammu and Kashmir, India

Amit Kumar School of Mechanical Engineering, Shri Mata Vaishno Devi University, Kakryal, Katra, Jammu and Kashmir, India

Avnish Kumar MPKI Packaging Pvt-Ltd, Chennai, Tamil Nadu, India

Aazim Shafi Lone Department of Mechanical Engineering, National Institute of Technology Srinagar, Hazratbal, Srinagar, Jammu and Kashmir, India

Ishfaq Amin Maekai Department of Mechanical Engineering, National Institute of Technology Srinagar, Hazratbal, Srinagar, Jammu and Kashmir, India

Mohd Junaid Mir Department of Mechanical Engineering, Islamic University of Science and Technology, Awantipora, Jammu and Kashmir, India

B.K. Mishra Department of Mechanical and Industrial Engineering, Indian Institute of Technology Roorkee, Roorkee, Uttarakhand, India

Shuhaib Mushtaq Department of Mechanical Engineering, Islamic University of Science and Technology, Awantipora, Jammu and Kashmir, India

S.M. Muzakkir Department of Mechanical Engineering, Jamia Millia Islamia, New Delhi, India

Himanshu Pathak Design Against Failure and Fracture Group, School of Mechanical and Materials Engineering, Indian Institute of Technology Mandi, Mandi, India

Mehnaz Rasool Department of Mechanical Engineering, Islamic University of Science and Technology, Awantipora, Jammu and Kashmir, India

Ahmed Raza Design Against Failure and Fracture Group, School of Mechanical and Materials Engineering, Indian Institute of Technology Mandi, Mandi, India

Sanjay Sharma School of Energy Management, Shri Mata Vaishno Devi University, Kakryal, Katra, Jammu and Kashmir, India

Ummer Amin Sheikh Department of Mechanical Engineering, Islamic University of Science and Technology, Awantipora, Jammu and Kashmir, India

I.V. Singh Department of Mechanical and Industrial Engineering, Indian Institute of Technology Roorkee, Roorkee, Uttarakhand, India

Mohammad Talha Design Against Failure and Fracture Group, School of Mechanical and Materials Engineering, Indian Institute of Technology Mandi, Mandi, India

Sahil Thappa Department of Mechanical Engineering, Maharishi Markandeshwar (Deemed to be University), Ambala, Haryana, India; School of Energy Management, Shri Mata Vaishno Devi University, Kakryal, Katra, Jammu and Kashmir, India

Shubham Kumar Verma School of Energy Management, Shri Mata Vaishno Devi University, Kakryal, Katra, Jammu and Kashmir, India; Built Environment Lab, Indian Institute of Technology, Gandhinagar, Gujarat, India

Aanchal Yadav Mechanical Engineering Department, Thapar Institute of Engineering and Technology, Patiala, Punjab, India

About the editors

Dr. Azher Jameel is currently working as an assistant professor in the Department of Mechanical Engineering, National Institute of Technology Srinagar, India. Before that, he worked as an assistant professor at the Shri Mata Vaishno Devi University, Katra and Islamic University of Science and Technology, Awantipora for more than 8 years. He has obtained a Bachelor of Technology in mechanical engineering from the National Institute of Technology Srinagar, India and a Master of Technology in CAD, CAM, and robotics from the Indian Institute of Technology Roorkee, India. He pursued his PhD in the areas of advanced computational mechanics at the National Institute of Technology Srinagar, India. Dr. Jameel has been actively involved in teaching and research for the last 10 years in the areas of advanced computational mechanics, fracture and fatigue in structures, large elastoplastic deformations, and experimental solid mechanics. His research interests include enriched numerical techniques, meshfree methods, isogeometric analysis, and coupled computational techniques. He has published more than 70 research papers in reputed journals and conferences. He has attended numerous conferences and workshops both in India and abroad and has been an active reviewer of various international journals. Dr. Jameel has supervised two PhD scholars and is currently supervising four PhD research scholars who are working in the areas of experimental and computational mechanics. He has supervised several postgraduate and undergraduate dissertations and projects. He has also delivered expert lectures on computational mechanics at various universities in India.

Dr. Ghulam Ashraf Ul Harmain is currently working as a professor in the Department of Mechanical Engineering at the National Institute of Technology Srinagar, India. He has served as the head of the department and dean of research in addition to various important positions at NIT Srinagar. Prof. Harmain obtained his PhD degree from the University of Victoria, Canada and Master of Technology from the Indian Institute of Technology Delhi, India. He has been actively involved in teaching and research for the last 35 years in the areas of advanced computational mechanics, fracture and fatigue in structures, thermoelasticity, tribology and maintenance, large elastoplastic deformations, and experimental solid mechanics. He has published more than 100 research papers in reputed journals and conferences. He has been an active reviewer of various international journals and conferences. Dr. Harmain has supervised more than 10 PhD students and is currently supervising six PhD research scholars who are working in the areas of experimental and computational mechanics. He has supervised several postgraduate and undergraduate dissertations.

Dr. Indra Vir Singh is currently working as a professor in the Department of Mechanical Engineering at the Indian Institute of Technology Roorkee, India.

Dr. Singh carried out his postdoctoral research at Shinshu University Nagano, Japan and has obtained his PhD degree from BITS Pilani, India. He has been actively involved in teaching and research for the last 20 years in the areas of advanced computational mechanics, fracture and fatigue in structures, large elasto-plastic deformations, experimental fracture mechanics, machine learning, multiscale modeling, gradient damage, and phase field modeling. He has published more than 250 research papers in reputed journals and conferences. Dr. Singh has supervised more than 16 PhD students and is currently supervising about 10 PhD research scholars who are working in the areas of experimental and computational mechanics. He has supervised more than 50 postgraduate dissertations. Dr. Indra Vir Singh is a well-known figure in the areas of computational mechanics because of his vast contributions in this area.

Dr. Magd Abdel Wahab is currently working as a professor and chair of the Department of Applied Mechanics at the Ghent University, Belgium. Dr. Wahab is also an adjunct professor at several universities, including Ton Duc Thang University, Vietnam, Duy Tan University, Vietnam, HUTECH University, Vietnam, and Nanjing Tech University, China. He has obtained PhD from KU Leuven, Belgium and a Doctor of Science from the University of Surrey, United Kingdom. He has published more than 550 research papers in highly reputed journals and conferences. He has been an active reviewer of various international journals and conferences. Dr. Wahab has authored several books, including *Mechanics of Adhesives in Composite and Metal Joints*, *Finite Elements in Fracture Mechanics*, and *Dynamics and Vibration: An Introduction*. More than 30 students have completed their doctoral programs under the guidance of Dr. Wahab. He has been actively involved in teaching and research in the areas of advanced finite element analysis, isogeometric analysis, multi-scale analysis, fretting fatigue and wear, automotive and aerospace applications, mechanical and vibrating systems, and civil engineering constructions. Dr. Wahab is the editor-in-chief of the *Journal of Applied Mechanics* and also serves as a member of the editorial boards of several reputed journals.

Preface

Enriched numerical techniques have gained tremendous importance and have found wide applications in different areas of engineering in the last few decades. Researchers working in the areas of solid mechanics, fracture mechanics, large elastoplastic deformations, and other nonlinear structural problems have made important contributions in the domain of enriched numerical techniques. This book aims to cover the basic fundamentals and applications of different enriched numerical techniques such as the extended finite element method (XFEM), element-free Galerkin method (EFGM), extended isogeometric analysis (XIGA), and coupled numerical techniques. The level-set method has found immense importance in representing different discontinuities such as cracks, holes, inclusions, and other bimaterial irregularities. This book covers the level-set methodologies adopted for tracking different geometries of material irregularities present in the domain. The implementation of various enrichment strategies for modeling different types of discontinuities is also presented in detail.

In Chapter 1, level-set methodologies for different geometries of material interfaces have been discussed. The level-set functions for circular, elliptical, rectangular, hexagonal, and octagonal-shaped discontinuities are presented. This is followed by Chapter 2, which discusses the higher order enrichment functions for fracture simulation in isotropic and orthotropic material mediums. Chapter 3 includes the analysis of cracks in three-dimensional engineering materials by employing the XFEM. Three-dimensional cracked domains containing plane edge cracks, horizontal and inclined penny cracks, and lens-shaped cracks are presented in this chapter. Chapter 4 presents the XFEM based on higher order shear deformation theory for free vibration analysis of cracked plates followed by stability analysis of stiffened trapezoidal composite panels in Chapter 5. Different enrichment strategies employed in XFEM for modeling different engineering problems are reported.

The implementation and mathematical foundations of the EFGM are discussed in Chapter 6 which is followed by the investigation of three-dimensional fracture mechanics problems by the EFGM in Chapter 7. Modeling and simulation of different types of cracks such as penny shaped, lens type, and plane crack problems are reported in this chapter. Elastoplastic crack growth in steel alloys has been investigated by the EFGM in Chapter 8 which is followed by the modeling and simulation of large elastoplastic deformations in bimaterial engineering components in Chapter 9. Modeling and simulation of Hertzian contact problems by the enriched EFGM have been reported in Chapter 10. The basic mathematical foundations for investigating large elastoplastic deformations by the enriched EFGM are presented in Chapter 11.

The implementation of the XIGA is reported in Chapter 12, where the fundamentals of various enrichment strategies and mathematical modeling of different engineering problems are discussed in detail. The discussion is followed by Chapter 13 which presents the applications of XIGA in modeling the strong discontinuities produced by the presence of cracks in engineering structures. Chapter 14 discusses the implementation of XIGA for modeling the behavior of engineering components subjected to thermomechanical loads. The application of XIGA in investigating thermal absorber coatings is presented in Chapter 15. A detailed review of the applications of enriched numerical techniques in modeling the behavior of structural wood is presented in Chapter 16 which is followed by the analysis of crack growth in parabolic trough collectors in Chapter 17. The implementation of XIGA for investigating the behavior of cracks in reinforced composite materials subjected to thermomechanical loads is discussed in Chapter 18. The book concludes with Chapter 19 which presents the implementation and applications of the XFEM in investigating the behavior of edge dislocations in engineering materials.

Azher Jameel
Ghulam Ashraf Ul Harmain
Indra Vir Singh
Magd Abdel Wahab

CHAPTER

1

Development of level set methodologies for different geometric discontinuities

Danish Ali[1], Avnish Kumar[2], S.M. Muzakkir[1] and Mahvash Afzal[3]

[1]Department of Mechanical Engineering, Jamia Millia Islamia, New Delhi, India [2]MPKI Packaging Pvt-Ltd, Chennai, Tamil Nadu, India [3]Department of Mechanical Engineering, Islamic University of Science and Technology, Awantipora, Jammu and Kashmir, India

1.1 Introduction

The modeling of complex-shaped discontinuities has always been a challenge in the field of enriched numerical techniques, which model different material irregularities irrespective of the mesh selected for analysis. The classical finite element method (FEM), no doubt, has always provided a potential framework for finding the approximate solution of partial differential equations, but it is very difficult to model discontinuities present in structural components such as cracks, contact surfaces, holes, and inclusions of different shapes, sizes, and orientations within the framework of conventional FEM because of the requirements of mesh conformation and element distortions. In the explicit microstructure modeling, the standard FEM faces the difficulties of generating the mesh for complex geometries [1]. The FEM also encounters many difficulties while modeling evolving discontinuities because, during the analysis, re-meshing of the domain is needed. One method to eliminate such difficulties is to make use of digital image-based FEM, as discussed in [2], but such an approach is computationally expensive. Another approach is to represent interfaces independently of element boundaries using regular meshes. In this approach, heterogeneity is modeled only at the integration points [3] for those elements that are cut by internal interfaces. The Voronoi cell FEM [4] is another technique in which the influence of embedded inclusions is incorporated in the stiffness matrix. In this technique, discretization of both

Enriched Numerical Techniques
DOI: https://doi.org/10.1016/B978-0-443-15362-4.00019-X

displacement and stress fields is needed. Developing level set functions (LSFs) and their coupling with the enriched numerical techniques such as the extended FEM (XFEM), the element-free Galerkin method (EFGM), and coupled FE-EFGM [5] provides another way to eliminate all the mesh-related difficulties faced by FEM, including conformal meshing, mesh adaption, and mesh refinement. The numerically enriched techniques use partition of unity and are used to represent discontinuities of all types, independent of the mesh or the grid [5–7]. The technique of partitioning unity is discussed in [8]. The identification and application of the partition of unity are explained by Oden and coworkers [9–12] through a numerical technique known as the generalized FEM.

1.2 Level set method

The level set method (LSM), introduced by Osher and Sethian [13], is a strong and effective computational tool used to represent and track static and evolving discontinuities present in structural components [13,14]. This method has the potential to capture changes in topology in contrast to the methods [15–19], which require extra work to depict the changes in topology. After the introduction of LSM [13], the method is being widely used in the areas of computer graphics, computational geometry, image processing, computational physics [20–22], and many other problems in engineering [23]. LSMs define the material interface clearly and unambiguously [24,25]. The LSM has found wide application in various fields, most importantly in the modeling of cracks, voids, and inclusions present in the structural components using enriched numerical techniques [26–30].

In this chapter, level set theory is employed for the representation of a nonevolving or static interface using a function known as the LSF $\emptyset(x)$. The key point here is to represent the interface with a LSF, which vanishes at the interfaces. The value of LSF is negative on one side and positive on the other side of material interfaces. Based on the LSF update procedure, Dijk et al. [31] classified LSMs into implicit methods and explicit methods. In implicit methods, LSF is evolved with the help of the Hamilton–Jacobi (HJ) equation [32–35], while in explicit methods, mathematical programming techniques are used to update parameterized LSF [31,36–41]. Various types of discontinuities are inherently present in structural components, which may develop during their manufacturing or may occur during operational stages. These discontinuities or irregularities present in the domain are usually in the form of cracks, voids, inclusions, contact interfaces, or bi-material interface, as shown in Fig. 1.1. Such discontinuities are categorized into two kinds. The first kind is the one that separates the domain (Ω) into two distinct subdomains (Ω_A and Ω_B). This type of discontinuity may be open, such as contact surface, or closed like a hole or an inclusion, as can be seen in Fig. 1.2. Single-LSF, named normal LSF ($\emptyset_N$), is required to represent such discontinuities completely. Normal level set ($\emptyset_N$) for any point in structural domain can be defined as its normal distance from the interface. Another type of discontinuity is the one that does not divide the structural domain into two distinct regions such as a crack interface. Such material irregularities require two LSFs, that is the normal level set and tangential LSFs, for their complete representation. For such cases, normal level set ($\emptyset_N$) represents normal distance of any point from the

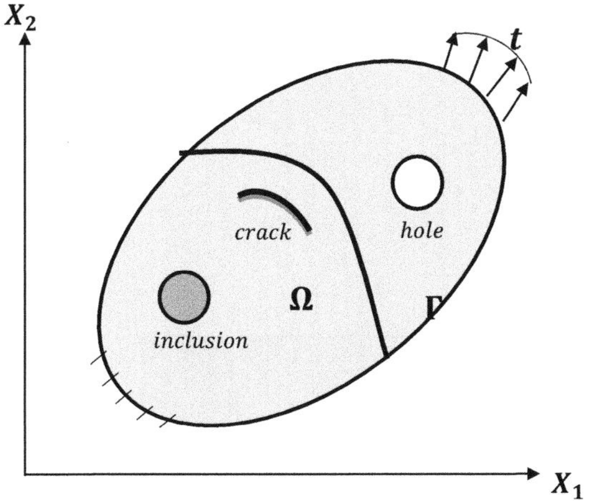

FIGURE 1.1 Domain with various types of discontinuities.

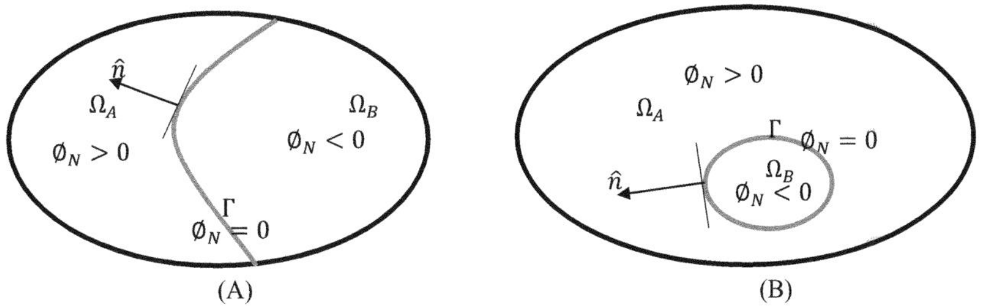

FIGURE 1.2 Representation of level set functions: (A) open interface and (B) closed interface.

crack surface, and the tangential level set ($\emptyset_T$) represents the tangential distance of the point from the tip of the crack.

1.3 Normal level set function

Suppose Γ represents any open or closed interface that separates the structural domain into two different regions given by Ω_A and Ω_B (Fig. 1.2). Normal LSF '$\emptyset_N(x)$' for the interface can be defined as

$$\emptyset_N(x) > 0 \ \forall \ x \in \Omega_A; \ \emptyset_N(x) < 0 \ \forall \ x \in \Omega_B; \ \emptyset_N(x) = 0 \ \forall \ x \in \Gamma. \tag{1.1}$$

The LSF, described above, can be easily defined by signed distance functions, which take positive values on one side and negative values on the other side of material irregularities [42]. For defining the signed distance function, a point in the structural domain lying closest to the discontinuity is chosen such that $\|\mathbf{x} - \mathbf{x}_\Gamma\|$ is minimum. Here, $\mathbf{x}$ represents the corresponding point, and $\mathbf{x}_\Gamma$ is the closest point on the discontinuity as represented in Fig. 1.3.

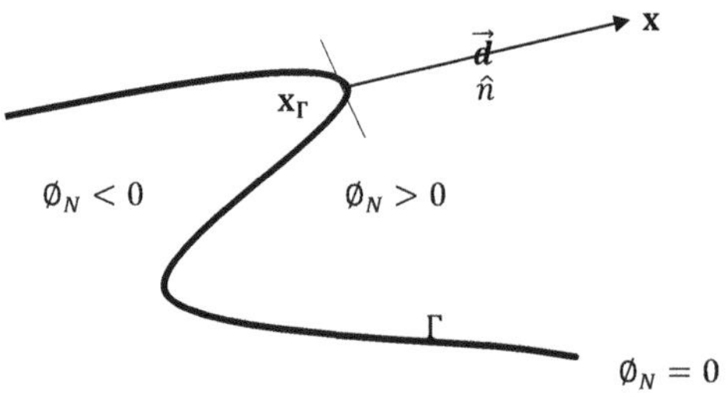

FIGURE 1.3 Signed distance function.

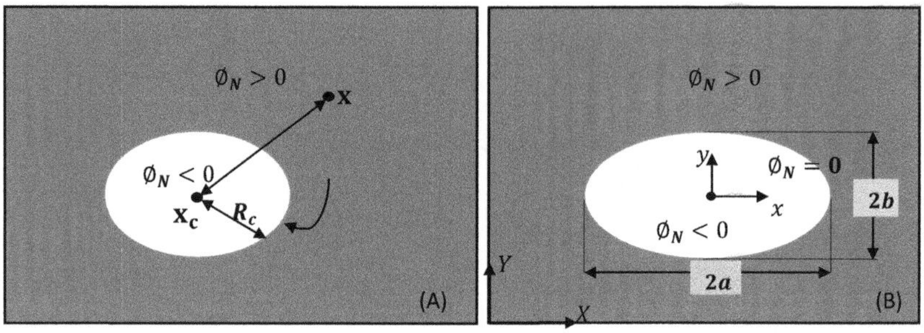

FIGURE 1.4 Description of level sets: (A) circular discontinuity and (B) elliptical discontinuity.

$(\mathbf{x}-\mathbf{x}_\Gamma)$ is, therefore, a vector $\boldsymbol{d}$ which remains orthogonal to the discontinuity Γ at a point $\mathbf{x}_\Gamma$. Thus we can define the normal LSF as

$$\emptyset_N(x) = \boldsymbol{d}\cdot\hat{n} = (\mathbf{x}-\mathbf{x}_\Gamma)\cdot\hat{n} \tag{1.2}$$

1.4 Level set representation for closed discontinuities

Closed discontinuities separate the domain under consideration into two distinct subdomains and, therefore, are represented completely with the help of normal level sets only. These discontinuities may be circular (with holes and inclusions), elliptical, triangular, rectangular, or may resemble any closed curve. The circular discontinuities include holes and inclusions. For representing the discontinuities of circular geometries, normal LSF can be defined as

$$\emptyset_N(x) = \|\mathbf{x}-\mathbf{x}_\mathbf{c}\| - R_c \tag{1.3}$$

where x is any point in the structural domain, $\mathbf{x}_\mathbf{c}$ denotes the center of circle and R_c is the radius of circle, as shown in Fig. 1.4. It is clear that the level set values will remain negative inside the circle and positive outside. The function will have a zero value for all data points lying on the boundary of the circle. In the similar fashion, normal LSF for the elliptical discontinuity can be described as

$$\emptyset_N(x) = \left\| \left(\frac{x}{a}\right)^2 + \left(\frac{y}{b}\right)^2 \right\| - 1 \tag{1.4}$$

LSF of an ellipse is in the local coordinate system, due to which some transformation is needed to define the level sets in global coordinate system. Here, x and y denote local coordinates of the elliptical interface with origin at the center of the given ellipse. Suppose $\mathbf{x} = [x, y, 1]^T$ and $\mathbf{X} = [X, Y, 1]^T$ represent the local and global systems of coordinates, respectively, then mapping between the local and global coordinates can be described as $\mathbf{X} = \mathbf{T}\mathbf{x}$, where $\mathbf{T}$ is the corresponding transformation matrix, which includes rotation as well as translation. The transformation matrix can be defined as

$$T = \begin{bmatrix} \cos\theta & -\sin\theta & 0 \\ \sin\theta & \cos\theta & 0 \\ 0 & 0 & 1 \end{bmatrix} \begin{bmatrix} 1 & 0 & x_c \\ 0 & 1 & y_c \\ 0 & 0 & 1 \end{bmatrix} \tag{1.5}$$

With the incorporation of a transformation matrix $\mathbf{T}$ during the mapping between local and global coordinates, discontinuities present in the domain of interest at different orientations and if are of different sizes can be represented easily. The LSF for elliptical discontinuity in terms of global coordinates is written as

$$\emptyset_N(x) = \left\| \left(\frac{X}{a}\right)^2 + \left(\frac{Y}{b}\right)^2 \right\| - 1 \tag{1.6}$$

Description of level sets for an elliptical discontinuity can be seen in Fig. 1.4. The radial distance of any point from elliptical interface in local coordinate system is given by $(\boldsymbol{x} - \boldsymbol{x}_E)$, where $\boldsymbol{x}_E$ is assumed to be the locus of an elliptical discontinuity in the local coordinate system. In terms of global coordinates, the radial vector can be obtained as $\mathbf{T}(\boldsymbol{x} - \boldsymbol{x}_E)$.

Triangular discontinuity is a closed type of discontinuity which divides the domain into two separate regions. Therefore a triangular discontinuity is also represented completely by a single-LSF called the normal LSF, which is generated from the equation of triangle. Let the equations of the three sides of the triangle be given by $A_1x + B_1y = 1$, $A_2x + B_2y = 1$ and $A_3x + B_3y = 1$. Then, the normal LSF for the triangle can be written as

$$\begin{aligned} \emptyset_N(X) = {} & Q(a)Q(b)(A_1X + B_1Y) + q(a)q(c)(A_2X + B_3Y) + q(a)Q(c)(A_3X + B_3Y) \\ & + q(b)Q(a)(A_3X + B_3Y) - 1 \end{aligned} \tag{1.7}$$

where

$$Q(a) = \frac{1}{2}[1 + \text{sign}(y - ax)]; \quad q(a) = \frac{1}{2}[1 - \text{sign}(y - ax)] \tag{1.8}$$

$$Q(b) = \frac{1}{2}[1 + \text{sign}(y - bx)]; \quad q(b) = \frac{1}{2}[1 - \text{sign}(y - bx)] \tag{1.9}$$

$$Q(c) = \frac{1}{2}[1 + \text{sign}(y - cx)]; \quad q(c) = \frac{1}{2}[1 - \text{sign}(y - cx)] \tag{1.10}$$

In the above equations, a–c are the slopes of the lines joining the vertices of the triangular discontinuity to the origin of the local coordinate system. Normal LSF for triangular discontinuity is in the coordinate system, where x and y represent the local coordinates of

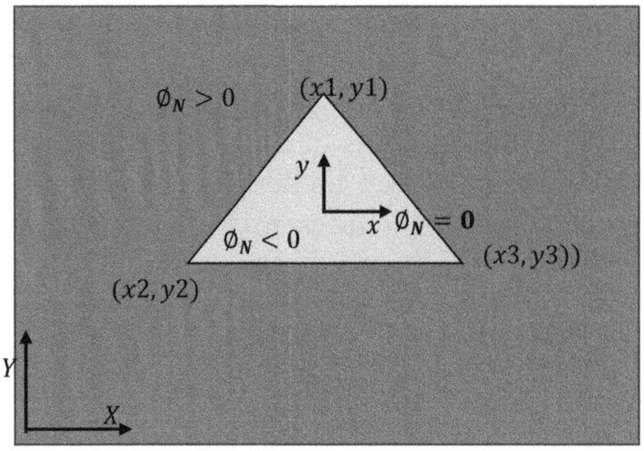

FIGURE 1.5 Level set description of triangular discontinuity.

the triangle. Therefore LSF for triangular discontinuity in terms of global coordinates, therefore, is written as

$$\emptyset_N(X) = Q(a)Q(b)(A_1X + B_1Y) + q(a)q(c)(A_2X + B_3Y) + q(a)Q(c)(A_3X + B_3Y) + q(b)Q(a)(A_3X + B_3Y) - 1 \quad (1.11)$$

The level set description of the triangular discontinuity can be seen in Fig. 1.5. All points lying within the triangular discontinuity exhibit negative values of LSFs, whereas the points lying outside the material irregularity show positive level set values. The level set values remain zero along the triangular interface.

Rectangular discontinuity is again a closed-type discontinuity; therefore it is represented completely by the normal LSF only, which is generated from the mathematical equation of rectangle as

$$\emptyset_N(x) = \left|\frac{x}{a} + \frac{y}{b}\right| + \left|\frac{x}{a} - \frac{y}{b}\right| - 1 \quad (1.12)$$

where a and b are the sides of the rectangular discontinuity. For $a = b$, the LSF given above represents a discontinuity of the shape of square. Normal LSF given by Eq. (1.2) for rectangular discontinuity is in the local coordinate system. Here, x and y are local coordinates of rectangle, with the origin located at the center of the rectangular discontinuity. The LSF for rectangular discontinuity in global coordinates is stated as

$$\emptyset_N(x) = \left|\frac{X}{a} + \frac{Y}{b}\right| + \left|\frac{X}{a} - \frac{Y}{b}\right| - 1 \quad (1.13)$$

The level set description of level set for rectangular discontinuity is shown in Fig. 1.6. The region lying inside the rectangle will have negative level set values, whereas the region lying outside the triangle will have positive level set values.

Regular hexagonal and octagonal discontinuities are yet another type of closed discontinuity that divides the domain into two subdomains (Fig. 1.7). The normal LSF, which represents the regular hexagonal and octagonal discontinuities completely, is given by

$$\emptyset_N(x) = 2|x| + |x - \sqrt{3}y| + |x + \sqrt{3}y| - a \quad (1.14)$$

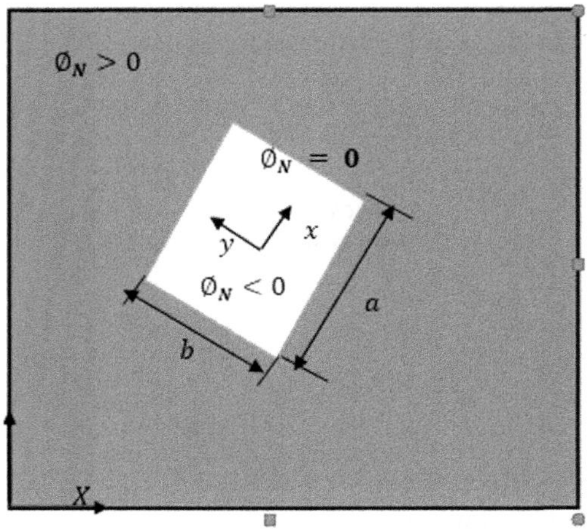

FIGURE 1.6 Level set representation of rectangular discontinuity.

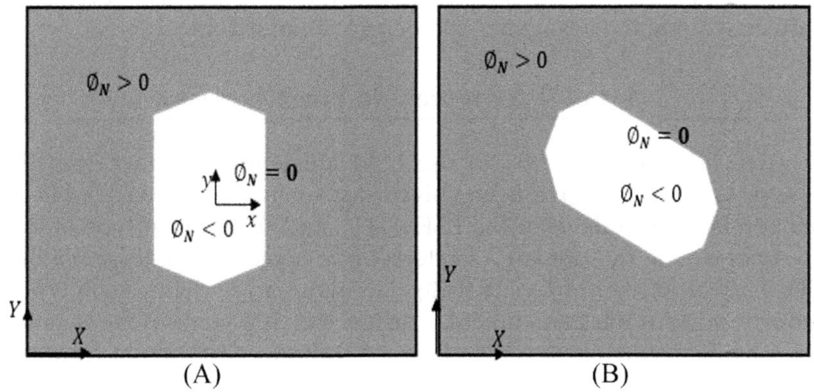

FIGURE 1.7 Description of level sets: (A) hexagonal discontinuity and (B) octagonal discontinuity.

$$\emptyset_N(x) = 2(|x| + |y|) + \sqrt{2}(|x - y| + |x + y|) - a \tag{1.15}$$

where a is the in radius of the regular polygon. The above-given equations are in the local coordinate system. The following equations represent the LSFs for regular hexagonal and octagonal discontinuities, respectively, in the global coordinate system:

$$\emptyset_N(x) = 2|X| + |X - \sqrt{3}Y| + |X + \sqrt{3}Y| - a \tag{1.16}$$

$$\emptyset_N(x) = 2(|X| + |Y|) + \sqrt{2}(|X - Y| + |X + Y|) - a \tag{1.17}$$

where **X** and Y represent the global coordinates of the regular hexagon and octagon. The level set description of level sets for regular hexagonal and octagonal discontinuities is

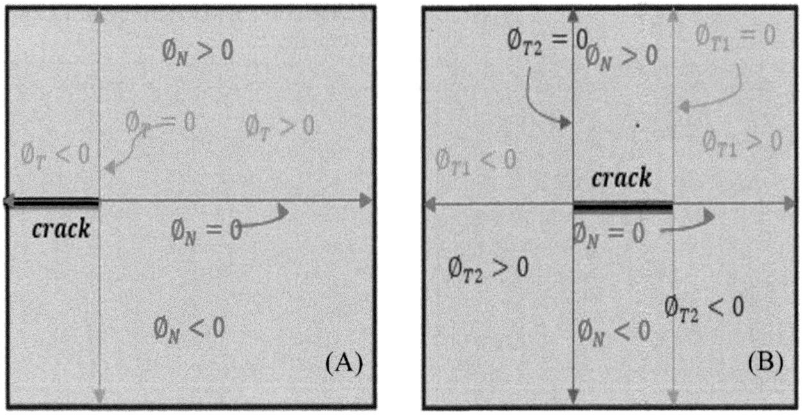

FIGURE 1.8 Level set methodology of cracks: (A) edge crack and (B) center crack.

shown in Fig. 1.7. All points located inside the hexagonal and octagonal interfaces show negative level set values, whereas the points outside have positive level sets. Level set values remain zero along the hexagonal and octagonal interfaces.

1.5 Level set representation for cracks

Stolarska et al. [43] were the first to use LSM efficiently for crack growth problems. Since then, several problems have been solved successfully using LSMs [44–46]. Signed distance function is used to develop the LSFs [41]. Crack is an open type of discontinuity, that is it does not divide the domain of interest into separate subdomains; therefore two types of LSFs, normal ($\emptyset_N$) and tangential ($\emptyset_T$), are required to define such types of discontinuities completely. $\emptyset_N$ is the perpendicular distance of any point in the structural domain from the crack boundary, while $\emptyset_T$ is the tangential distance of a point from the crack tip. Cracks present in structural components may be edge or center cracks. The representation of an edge crack is possible with the help of one normal LSF and one tangential LSF. On the contrary, a center crack is represented completely by one normal and two tangential LSFs for each crack tip, as every center crack has two crack tips. The level set methodology for center and edge cracked specimens can be seen in Fig. 1.8. In order to construct the LSFs, consider a crack with two crack tips. Suppose two crack tips of an embedded crack are given by $E(x_1, y_1)$ and $F(x_2, y_2)$, as shown in Fig. 1.9. Let us consider a point $P(x_p, y_p)$ lying inside the structural domain considered here. Using a simple two-point formula, the general equation of the crack interface can be obtained as

$$(y_1 - y_2)x + (x_2 - x_1)y + x_1y_2 - x_2y_1 = 0 \tag{1.18}$$

From the above equation, the normal level set of point $P(x_p, y_p)$ can be obtained as [47]

$$\emptyset_N = PQ = \frac{(y_1 - y_2)x_p + (x_2 - x_1)y_p + x_1y_2 - x_2y_1}{\sqrt{(x_2 - x_1)^2 + (y_2 - y_1)^2}} \tag{1.19}$$

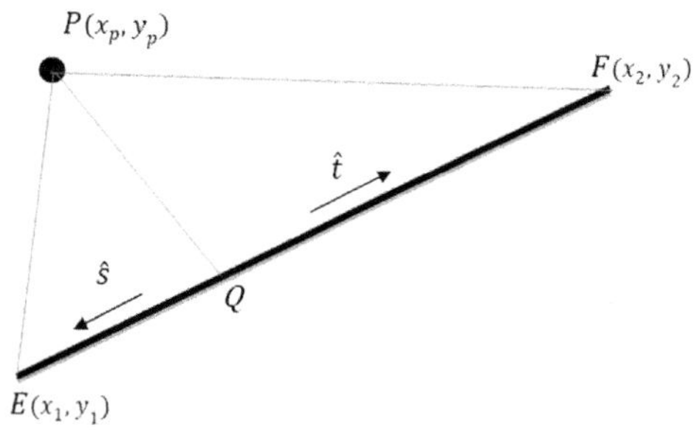

FIGURE 1.9 Level set formulations for cracks.

Similarly, two tangential level sets defined for two crack tips are represented by QF and QE, which are obtained as $\emptyset_{T1} = QF = \overrightarrow{PF} \cdot \hat{t}$ and $\emptyset_{T2} = QE = \overrightarrow{PE} \cdot \hat{s}$, where $\hat{t}$ and $\hat{s}$ denote unit vectors along the boundary of crack, which are further expressed as $\hat{t} = \frac{\overrightarrow{EF}}{|\overrightarrow{EF}|}$ and $\hat{s} = \frac{\overrightarrow{FE}}{|\overrightarrow{FE}|}$. The crack boundary inside the structural domain is completely defined by $\emptyset_N = 0$ and $\emptyset_T < 0$. Similarly, crack tips have both normal and tangential level sets equal to zero. The complete description of crack can be given by

$$\left\{ \begin{array}{c} \text{For } P \in \Gamma_{crack},\ \emptyset_N = 0 \text{ and } \emptyset_T < 0 \\ \text{For } P \in \Gamma_{tip},\ \emptyset_N = 0 \text{ and } \emptyset_T = 0 \end{array} \right\} \tag{1.20}$$

1.6 Evaluation of split and tip elements

LSFs used by the enriched techniques like XFEM do help in representing the discontinuities independently of nodal distributions selected for investigation. While modeling discontinuities with the help of these enriched techniques, we come across two types of elements in addition to the classical four-noded finite elements. These additional two types of elements are split and tip elements. The split elements are cut by the discontinuity completely, while the tip element contains the tip and is not cut completely by the discontinuity. While modeling open discontinuities, both split and tip elements are present, but when closed discontinuities are modeled, only split elements are present. Calculation of split and tip elements from the level set values can easily be done using an appropriate algorithm. The elements that are not cut by material irregularities have level sets of all four nodes either positive or all level sets are negative. On the contrary, the elements that are cut by material irregularities have few nodes having positive level set values and some nodes showing negative level set values. Figs. 1.10 and 1.11 show the description of normal and tangential level sets for circular and crack-type material interfaces. Level sets of split elements for closed discontinuities follow the condition $\emptyset_{N,\max} \times \emptyset_{N,\min} < 0$, where $\emptyset_{N,\max}$ and $\emptyset_{N,\min}$ are the maximum and minimum values of normal level sets of four nodes of the element. The criterion given above is valid only for split elements.

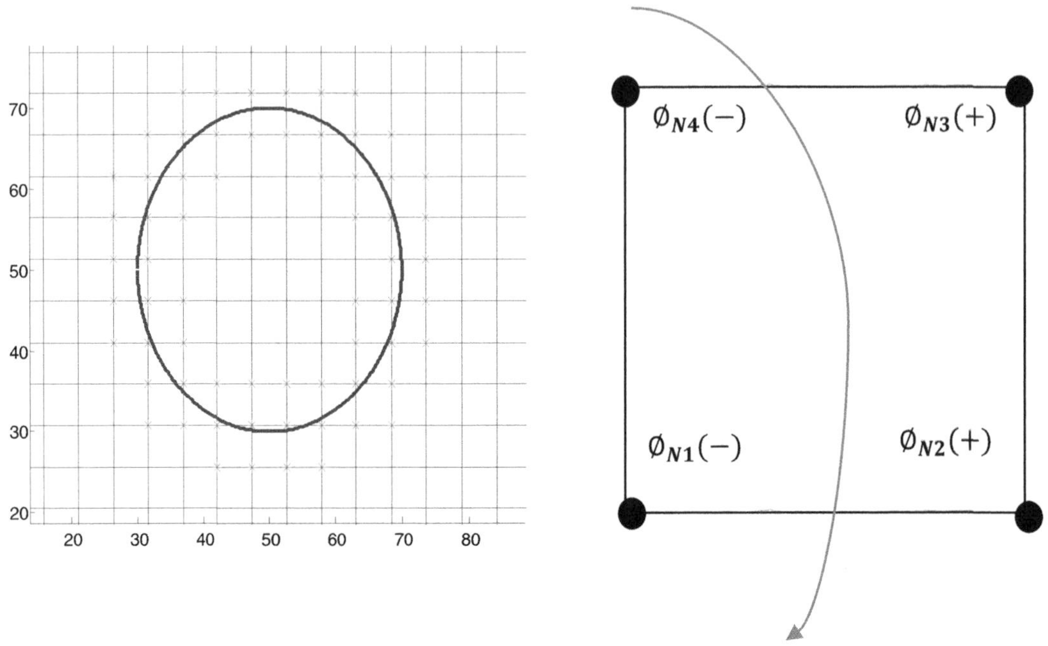

FIGURE 1.10 Level set values for split elements of a circular discontinuity.

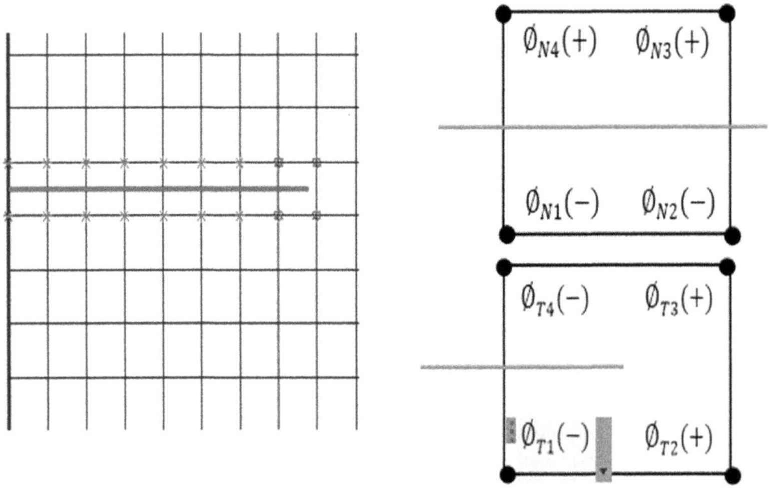

FIGURE 1.11 Level set description for split element and tip element of a crack.

Other elements of the FE mesh that are not cut by internal interfaces satisfy the condition $\emptyset_{N,\max} \times \emptyset_{N,\min} > 0$. Open discontinuities like cracks are completely represented by both normal and tangential LSFs. The split elements of the finite element mesh satisfy the relation $\emptyset_{N,\max} \times \emptyset_{N,\min} < 0$ and $\emptyset_{T,\max} < 0$. And the tip elements satisfy the condition given as $\emptyset_{N,\max} \times \emptyset_{N,\min} < 0$ and $\emptyset_{T,\max} \times \emptyset_{T,\min} < 0$.

1.7 Effect of material irregularities on cracks

The presence of inclusions and holes in a cracked structure has a drastic effect on its load-carrying capacity. It is very important to characterize the behavior of cracks in structural components containing different material irregularities, such as inclusions, holes, and other defects, in order to design the safe and reliable engineering structures. A large number of computational techniques have been employed for modeling the behavior of cracks in engineering specimens. The techniques that have been extensively used to model the phenomenon of crack growth in engineering components in the past few decades include the standard FEM [48], XFEM [3,49], boundary element method [50–52], and EFGM [30,47,53,54]. Among all the tools mentioned above, FEM is still the most dominant and widely applied computational tool used for finding the solutions of various engineering problems. In three-dimensional configurations, it is very challenging for any numerical technique to evaluate various fracture mechanics parameters under mixed-mode loadings, such as SIFs and J-integrals. The techniques that are used for the evaluation of fracture mechanics parameters include the J-integral technique, the virtual crack extension method, and singularity techniques. The most general and useful way to investigate the behavior of cracked structures is to use the J-integral approach. The J-integral technique uses the energy approach, and methods based on energy approach have higher convergence rates compared to other techniques.

1.7.1 Displacement approximations

Equilibrium equation for a deformable body having an external surface "Γ", enclosing a three-dimensional domain "Ω", can be written as

$$\int_{\Omega} \boldsymbol{\sigma}{:}\boldsymbol{\varepsilon} d\Omega - \int_{\Omega} \mathbf{b}{:}\mathbf{u} d\Omega - \int_{\Gamma_t} \mathbf{t}{:}\mathbf{u} d\Gamma_t = 0 \tag{1.21}$$

where the external surface "Γ" is divided into traction (Γ_t), displacement (Γ_u), and traction-free crack surfaces (Γ_c). $\boldsymbol{\sigma}$ and $\boldsymbol{\varepsilon}$ are the stresses and strains, respectively. Stresses can be defined in terms of strains as $\boldsymbol{\sigma} = \mathbf{D}\boldsymbol{\varepsilon}$, where $\mathbf{D}$ denotes the Hooke's matrix. Body and external force vectors are shown as $\mathbf{b}$ and $\mathbf{t}$, respectively. For the given domain, displacement field and geometry can be approximated in FEM as

$$u^h(\mathbf{x}) = \sum_{i=1}^{n} N_i^e(\mathbf{x}) u_i ; \mathbf{x} = \sum_{i}^{n} \mathbf{x}_i \check{N}_i(\xi, \eta, \zeta) \tag{1.22}$$

where $\check{N}_i(\xi, \eta, \zeta) = \frac{1}{8}(1 + \xi_i \xi)(1 + \eta_i \eta)(1 + \zeta_i \zeta)$ and (ξ, η, ζ) are the coordinates of master element. Substitution of the displacement approximation in the equilibrium equation yields the final system of discrete equations as $[K]\{u\} = \{f\}$, such that

$$K = \int_{\Omega} B^T D B \, d\Omega \tag{1.23}$$

$$f = \int_{\Omega} \mathbf{N}^T \mathbf{b} d\Omega + \int_{\Gamma} \mathbf{N}^T \mathbf{t} d\Gamma \tag{1.24}$$

$$\boldsymbol{B} = \begin{bmatrix} N_{i,x} & 0 & 0 \\ 0 & N_{i,y} & 0 \\ 0 & 0 & N_{i,z} \\ N_{i,y} & N_{i,x} & 0 \\ 0 & N_{i,z} & N_{i,y} \\ N_{i,z} & 0 & N_{i,x} \end{bmatrix} \tag{1.25}$$

1.7.2 Evaluation of fracture parameters

The J-integral for cracked bodies can be defined in standard form as

$$J = \int_{\Gamma} \left(W dy - T \frac{\partial u}{\partial x} ds \right) \tag{1.26}$$

where Γ represents any arbitrary closed contour surrounding the crack tip, ds is the coordinate measured along the closed contour, W denotes strain energy density, and the applied force vector is given by T. Domain-based interaction integral is written as

$$M^{(1,2)} = \int_{\Omega} \left[\sigma_{ij}^{(1)} \frac{\partial u_i^{(2)}}{\partial x_1} + \sigma_{ij}^{(2)} \frac{\partial u_i^{(1)}}{\partial x_1} - W^{(1,2)} \delta_{ij} \right] \frac{\partial q}{\partial x_j} d\Omega \tag{1.27}$$

where "q" is equal to 0 and 1 along the contour and crack tip, respectively. Actual and auxiliary states of the loaded body are represented by "1" and "2", respectively. Mutual strain energy for any two equilibrium states of the given structural domain is given as

$$W^{(1,2)} = \frac{1}{2} \left(\sigma_{ij}^{(1)} \varepsilon_{ij}^{(2)} + \sigma_{ij}^{(2)} \varepsilon_{ij}^{(1)} \right) = \sigma_{ij}^{(1)} \varepsilon_{ij}^{(2)} = \sigma_{ij}^{(2)} \varepsilon_{ij}^{(1)} \tag{1.28}$$

The relationship between the interaction integral and mixed-mode SIFs (K_I, K_{II}, K_{III}) in three-dimensional cracked components can be written as

$$M^{(1,2)} = \frac{2}{E^*} \left(K_I^{(1)} K_I^{(2)} + K_{II}^{(1)} K_{II}^{(2)} \right) + \frac{2(1+\nu)}{E} \left(K_{III}^{(1)} K_{IIII}^{(2)} \right) \tag{1.29}$$

where $E^* = \frac{E}{1-\mu^2}$ for plane strain and $E^* = E$ for plane stress. For obtaining the mode-I SIFs, we use $K_I^{(2)} = 1$ and $K_{II}^{(2)} = K_{III}^{(2)} = 0$, which gives $K_I^{(1)} = \frac{M^{(1,I)} E^*}{2}$. When $K_I^{(2)} = K_{III}^{(2)} = 0$ and $K_{II}^{(2)} = 1$, mode-II SIF is obtained as $K_{II}^{(1)} = \frac{M^{(1,II)} E^*}{2}$. Similarly, when $K_I^{(2)} = K_{II}^{(2)} = 0$ and $K_{III}^{(2)} = 1$, the mode-III SIFs can be obtained as $K_{III}^{(1)} = \frac{M^{(1,III)} E^*}{2(1+\nu)}$.

1.8 Numerical results and discussions

The current book chapter considers various numerical problems involving engineering specimens containing material irregularities of circular, elliptical, rectangular, hexagonal, and octagonal geometries. Engineering components containing different edges and embedded cracks are also considered for analysis. The numerical models and algorithms on level set methodologies developed in the current work have been coded in MATLAB® to obtain level sets and identify the corresponding split and tip elements in the mesh. At the end of this chapter, several numerical problems are presented to investigate the behavior of three-dimensional cracked plates in the presence of inclusions and holes by invoking the standard FEM. The effect of the position and size of inclusions and holes on the three-dimensional cracked structures by using the conventional FEM is presented at the end of this chapter.

1.8.1 Circular discontinuities

A square plate with dimensions 100×100mm containing a hole of diameter 30mm at the center of the plate is considered for analysis, as can be seen in Fig. 1.12. Normal level set contours of circular hole are presented in Fig. 1.13. It is clearly visible that the normal LSF remains zero along the boundary of circular hole. The level set values are positive outside and negative inside of the circular hole. Split elements for the circular discontinuity have been shown in Fig. 1.14. The results derived in the present study clearly demonstrate that the level set methodology presented in the current work has a strong potential to represent different geometries of material irregularities present in the domain.

1.8.2 Elliptical discontinuity

A square plate with dimensions $200 \times 200\text{mm}^2$ containing elliptical discontinuity at the center of the plate is considered here. The length of the major axis of the elliptical

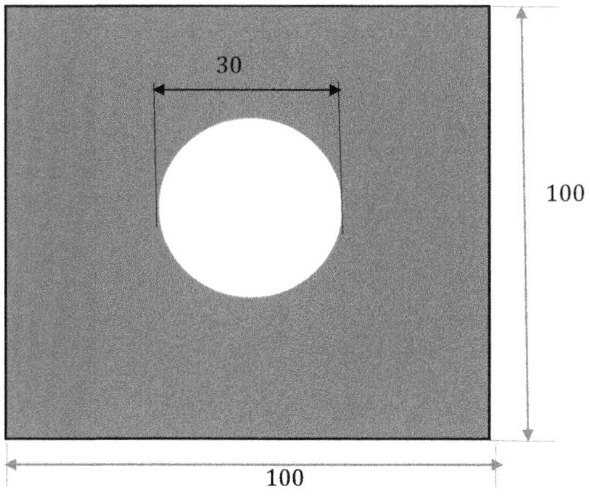

FIGURE 1.12 Square plate with a circular discontinuity.

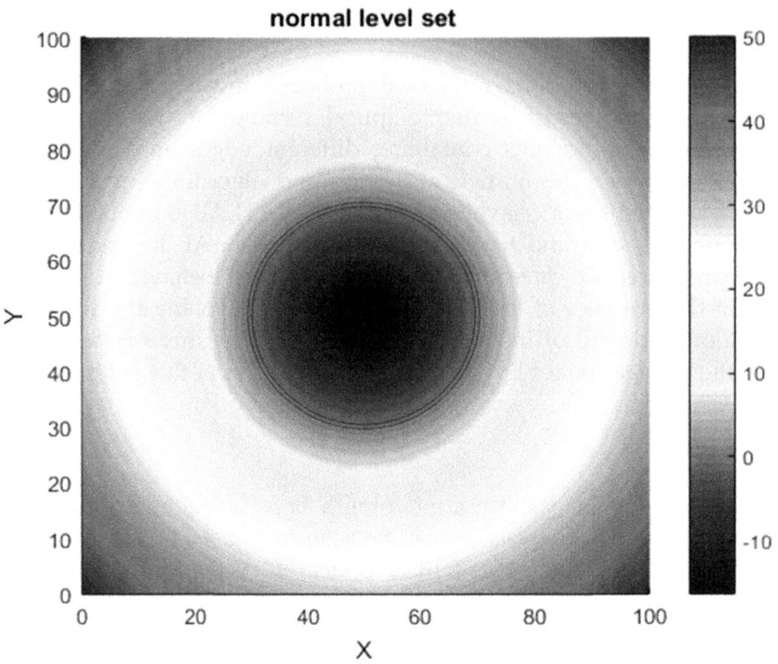

FIGURE 1.13 Normal level set contours for circular discontinuity. Please see the online version to view the color image of the figure.

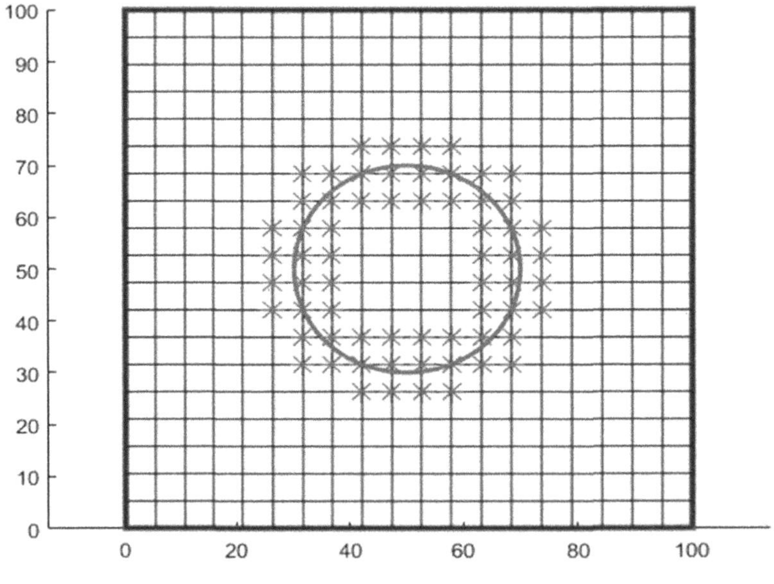

FIGURE 1.14 Split elements of circular discontinuity.

discontinuity is $2a$ and that of minor axis is $2b$. Normal level set contours for the elliptical discontinuity are shown in Figs. 1.15 and 1.16. It is easily concluded that the normal LSF remains zero along the boundary of elliptical hole. Values of the level

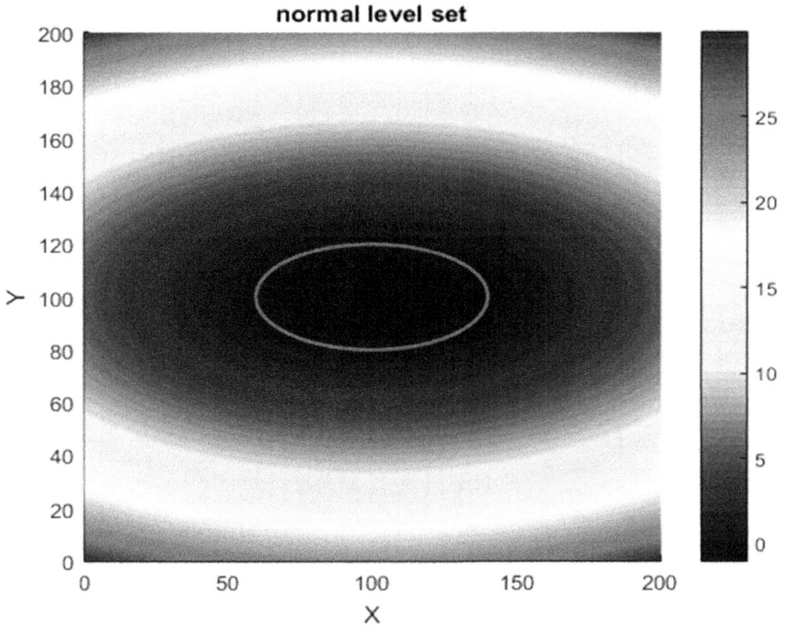

FIGURE 1.15 Normal level set contours of elliptical discontinuity at 0 degree. Please see the online version to view the color image of the figure.

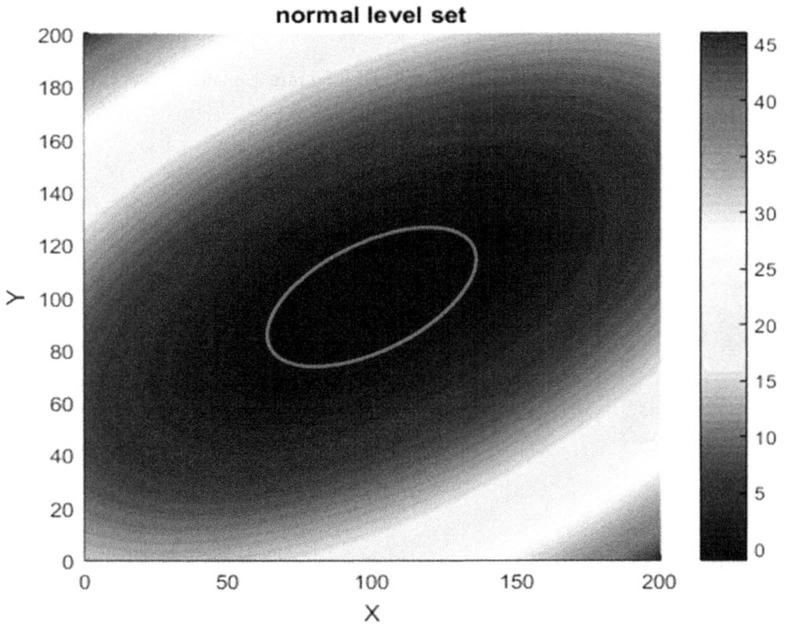

FIGURE 1.16 Normal level set contours of elliptical discontinuity at 30 degrees. Please see the online version to view the color image of the figure.

sets are positive outside and negative inside the elliptical discontinuity. Split elements for the elliptical discontinuity at different orientations of the ellipse have been shown in Fig. 1.17.

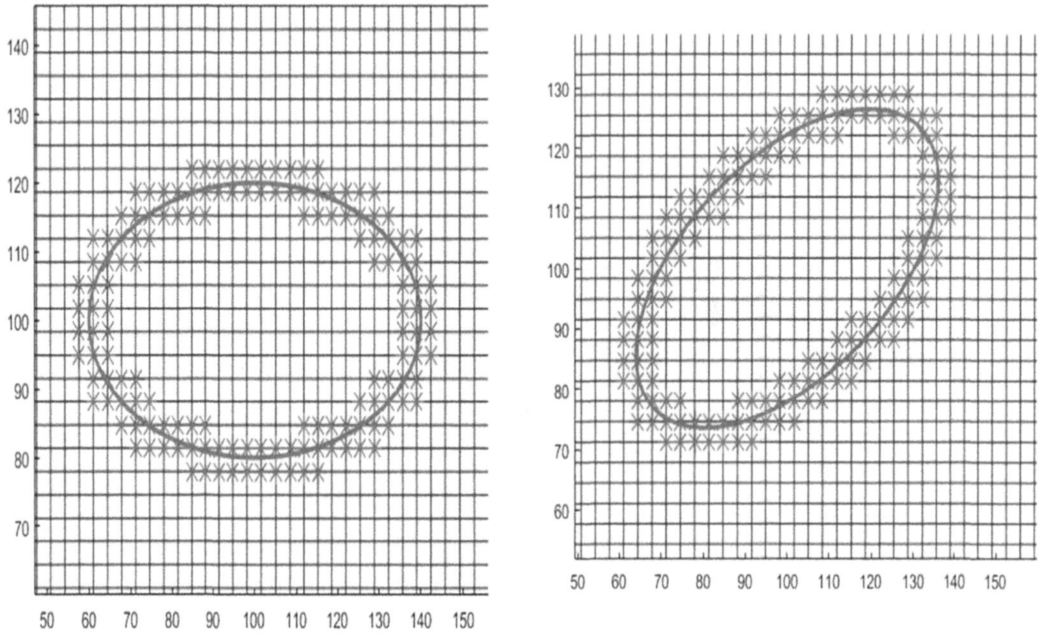

FIGURE 1.17 Split elements of elliptical discontinuity at 0 and 30 degrees.

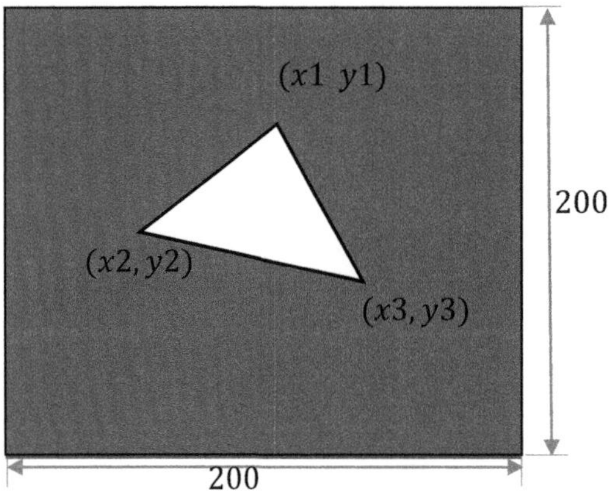

FIGURE 1.18 Square plate with triangular discontinuity.

1.8.3 Triangular discontinuity

A square plate with dimensions $200 \times 200\text{mm}^2$ containing triangular discontinuity at the center of the plate is shown in Fig. 1.18. The methodology employed to obtain the level sets for a triangular discontinuity has been discussed earlier. Normal level set contours and split elements for the triangular discontinuity can be seen in Fig. 1.19. It can be easily observed that the normal LSF remains zero along the boundary of triangle.

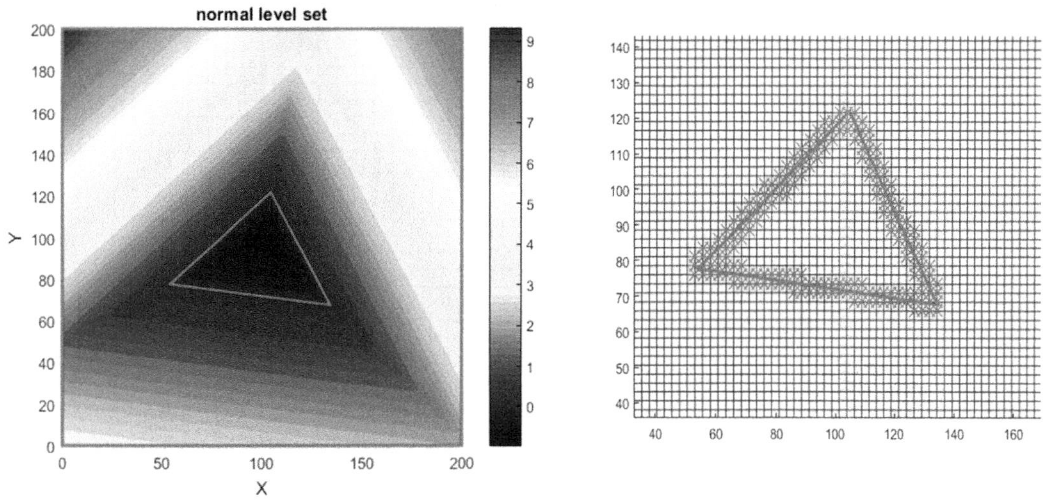

FIGURE 1.19 Triangular discontinuity level set contours and split elements. Please see the online version to view the color image of the figure.

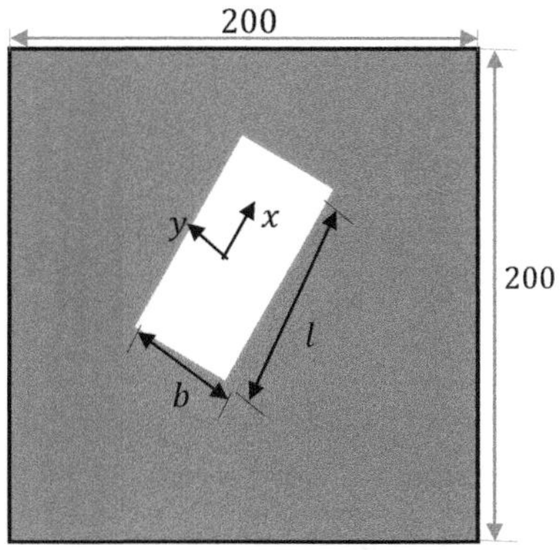

FIGURE 1.20 Square plate with rectangular discontinuity.

1.8.4 Rectangular discontinuity

A square plate with dimensions $200 \times 200\text{mm}^2$ and containing a rectangular discontinuity at the center of the plate is shown in Fig. 1.20. The split elements for the rectangular discontinuity have been shown in Fig. 1.21. Normal level set contours for the rectangular discontinuity at different orientations can be seen in Fig. 1.22. The normal level set values remain zero along the boundary of rectangular hole. The level set values are positive outside and negative inside the rectangular discontinuity.

1.8.5 Hexagonal discontinuity

A square plate with dimensions $200 \times 200\text{mm}^2$ and containing a regular hexagonal discontinuity is considered here for analysis. The normal level set contours for the regular

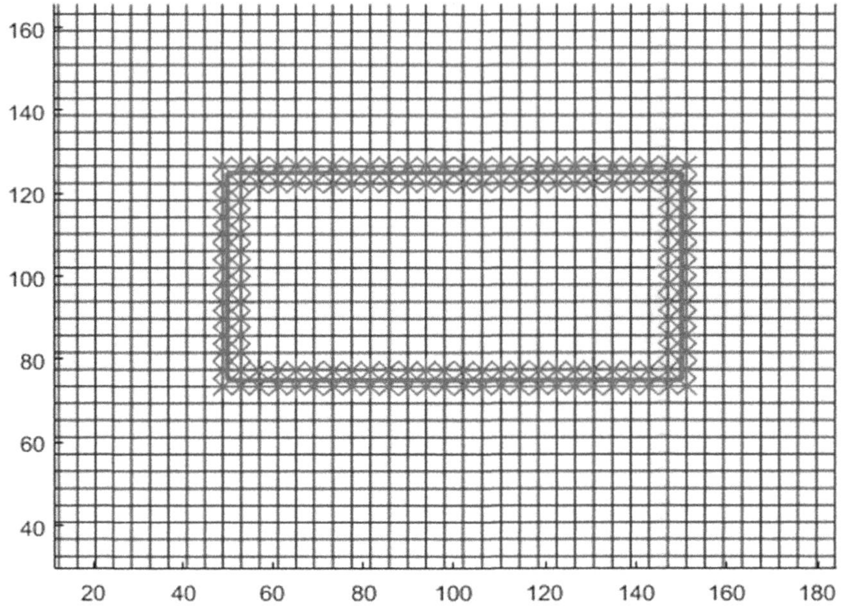

FIGURE 1.21 Split elements for rectangular discontinuity.

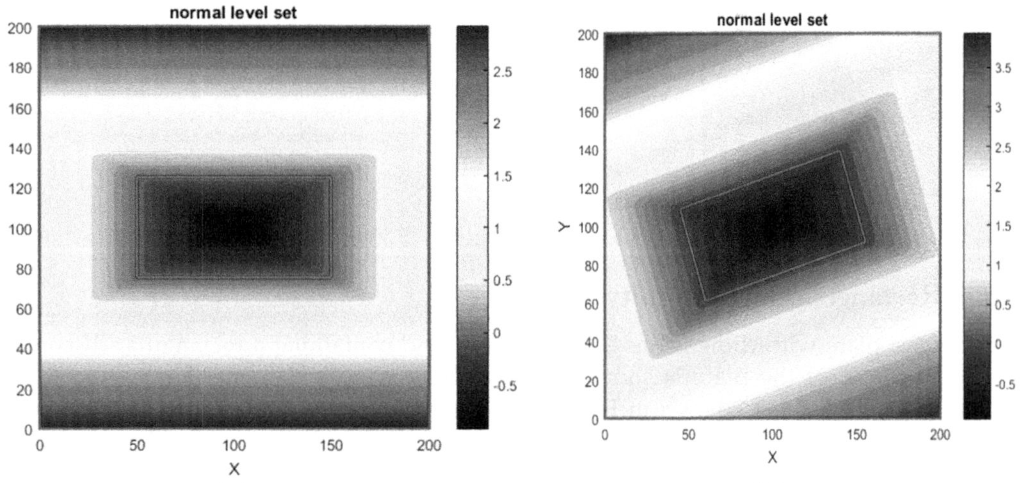

(A) (B)

FIGURE 1.22 Normal level set contours of rectangular discontinuity (A) at 0 degree and (b) at 45 degrees. Please see the online version to view the color image of the figure.

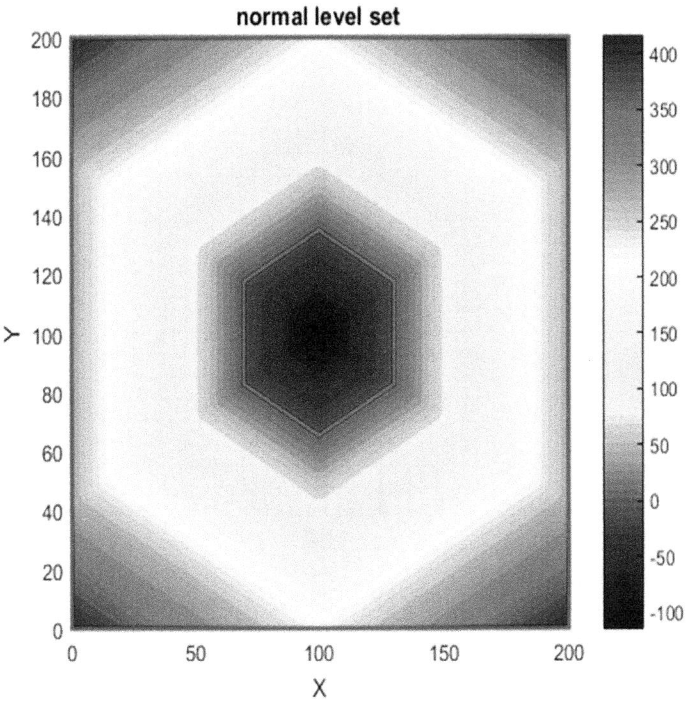

FIGURE 1.23 Normal level set contours of hexagonal discontinuities. Please see the online version to view the color image of the figure.

hexagonal discontinuities are shown in Fig. 1.23. The split elements for the regular hexagonal discontinuities can be seen in Fig. 1.24.

1.8.6 Octagonal discontinuity

A square plate with dimensions $200 \times 200mm^2$ and containing a regular octagonal discontinuity is considered here for analysis. The normal level set contours for the regular octagonal discontinuities are depicted in Fig. 1.25. The split elements for the regular octagonal discontinuities are shown in Fig. 1.26.

1.8.7 A rectangular plate carrying cracks

A rectangular plate with dimensions $200 \times 200mm^2$ carrying edge and center cracks has been considered for analysis. Normal and tangential level set contours obtained for an edge crack component are shown Fig. 1.27. The normal level set contours of an inclined center crack are presented in Fig. 1.28. Tangential level set contours for an inclined center

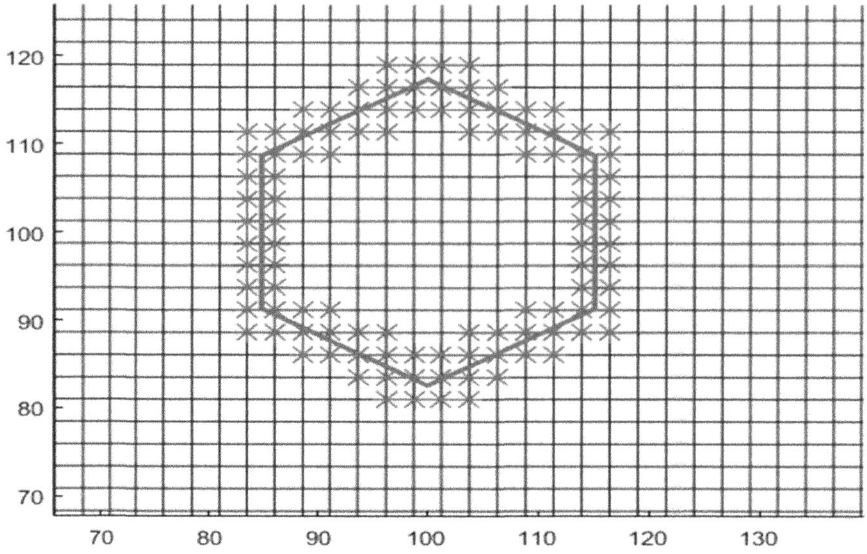

FIGURE 1.24 Split elements of hexagonal discontinuity.

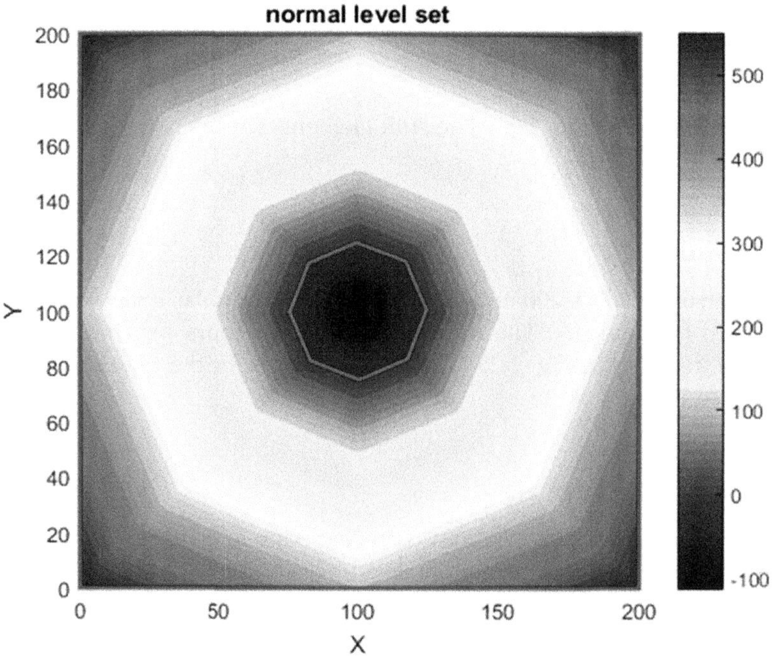

FIGURE 1.25 Normal level set contours of octagonal discontinuities. Please see the online version to view the color image of the figure.

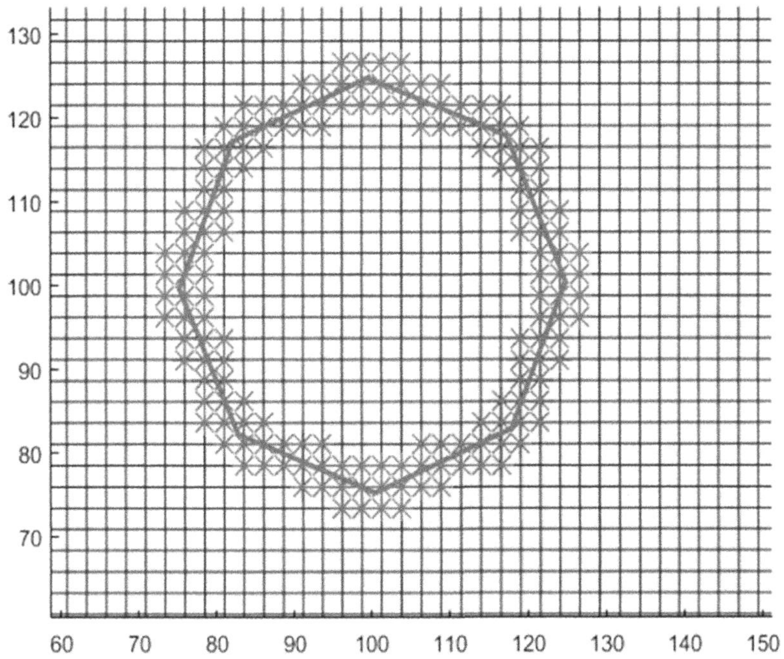

FIGURE 1.26 Split elements of octagonal discontinuity.

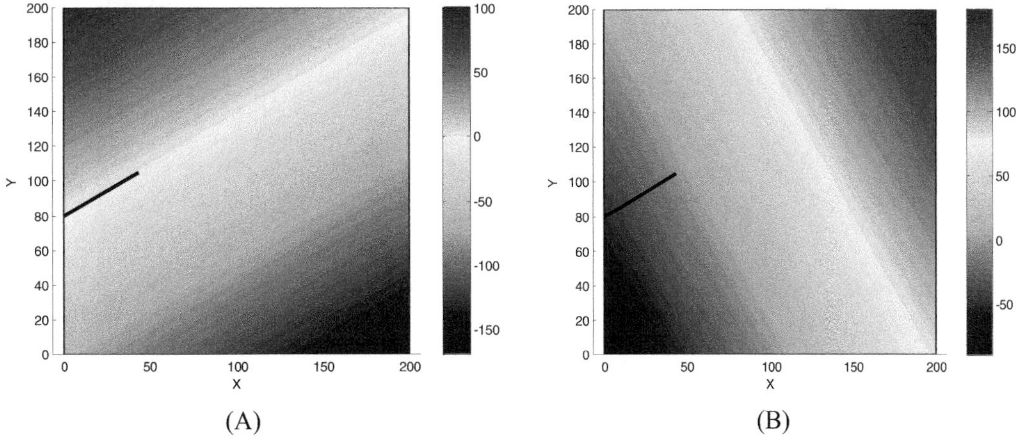

FIGURE 1.27 Level set contours for an inclined edge crack: (A) normal level sets and (B) tangential level sets. Please see the online version to view the color image of the figure.

crack corresponding to two crack tips are shown in Figure Fig. 1.29. It can be seen that the edge cracks are completely represented by one tangential LSF, whereas the center cracks are fully represented by two tangential LSFs corresponding to two crack tips.

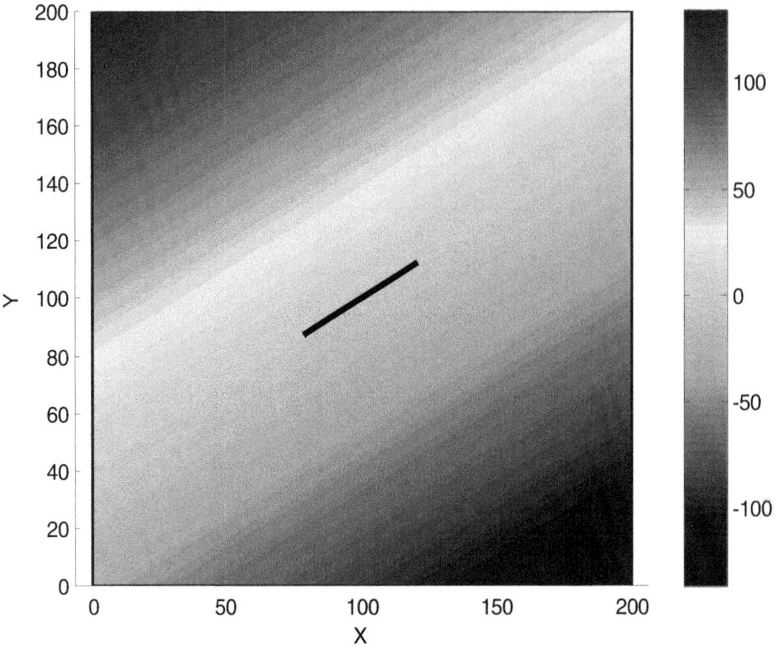

FIGURE 1.28 Normal level set contours for an inclined center crack. Please see the online version to view the color image of the figure.

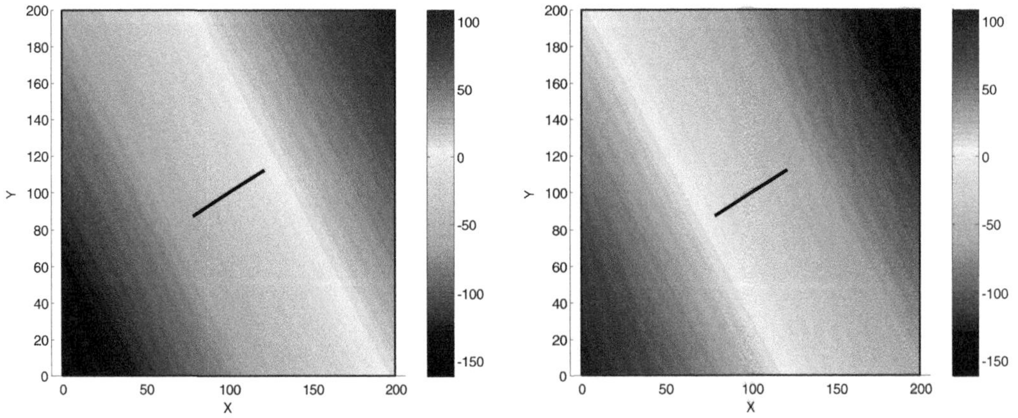

(A) (B)

FIGURE 1.29 Two tangential level set contours for an inclined center crack corresponding to two crack tips: (A) represents the tangential level set contours corresponding to left crack tip and (B) shows the tangential level set contours for right crack tip. Please see the online version to view the color image of the figure.

1.8.8 Effect of size of inclusion

A center-cracked three-dimensional structural steel plate, containing an inclusion of Young's modulus of 30 GPa, and dimensions $200 \times 100 \times 30\ mm^3$, is considered for analysis

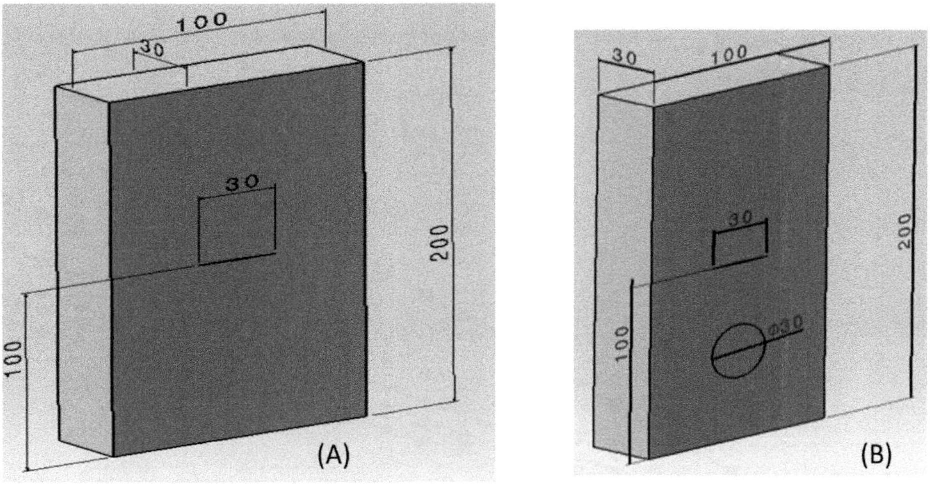

FIGURE 1.30 Three-dimensional center-cracked component: (A) without inclusion and (B) with inclusion.

as presented in Fig. 1.30. The center crack is assumed to have a length of 30 mm. The cracked specimen is subjected to a static load of 40 MPa at the top surface, whereas the bottom surface of the plate remains fixed during simulation. The position of the inclusion is kept fixed, whereas its radii have been changed from 5 to 25 mm. The variations of mode-I SIFs and J-integrals with load for various radii of inclusion can be seen in Figs. 1.31 and 1.32, respectively. It is observed that the presence of weaker inclusions in center-cracked components decreases the load-bearing capacity of the component, as expected. As the size of inclusion increases, the effect of the inclusion becomes more pronounced.

1.8.9 Effect of position of inclusion

The same structural steel component, having a center crack length of 30 mm as shown in Fig. 1.30, is taken for analysis. The effect of the position of inclusion on a three-dimensional center-cracked specimen is presented here. The position of inclusion keeps on changing while its radius is assumed to be 15 mm, which is kept fixed during analysis. The cracked specimen is subjected to a tensile load of 40 MPa at upper surface, whereas the bottom edge of lower surface plate remains fixed during simulation. Mode-I SIFs and J-integrals vary with load for different positions of the inclusion, as shown in Figs. 1.33 and 1.34, respectively. It has been observed that, as the inclusion approaches the crack, the influence of inclusion on the load-bearing capacity of the engineering three-dimensional structures becomes more pronounced.

1.8.10 Effect of size of hole

A three-dimensional edge-cracked plate, made of structural steel, with dimensions $200 \times 100 \times 30\text{mm}^3$ is considered for investigation, as shown in Fig. 1.35. The crack is

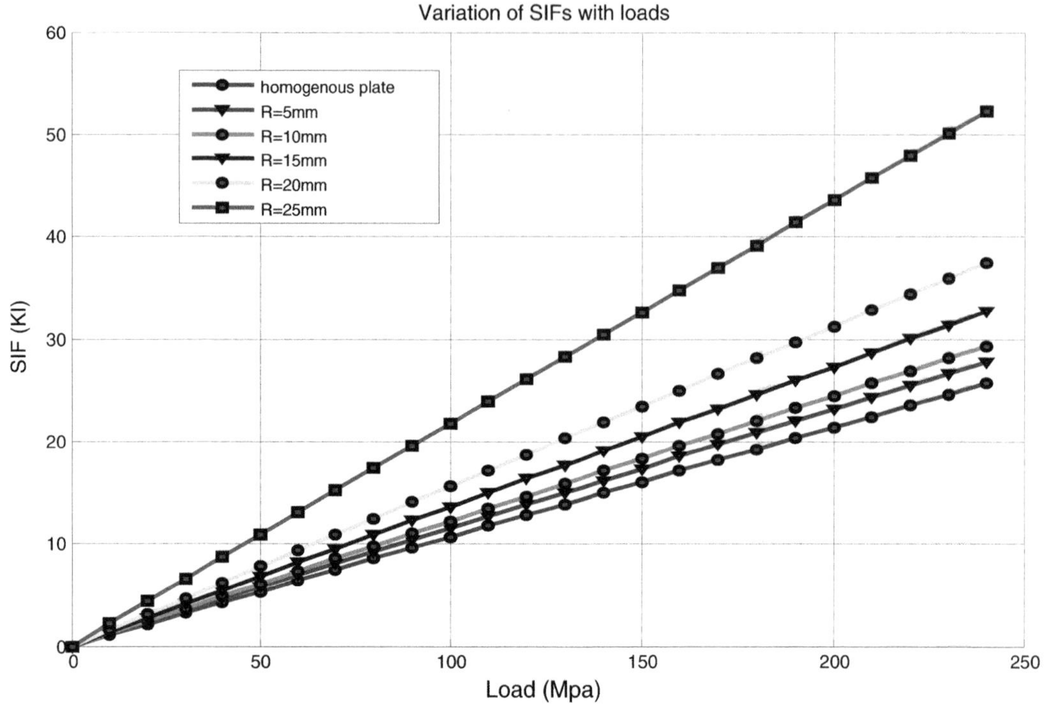

FIGURE 1.31 Variation of mode-I stress intensity factors (SIFs) with load. Please see the online version to view the color image of the figure.

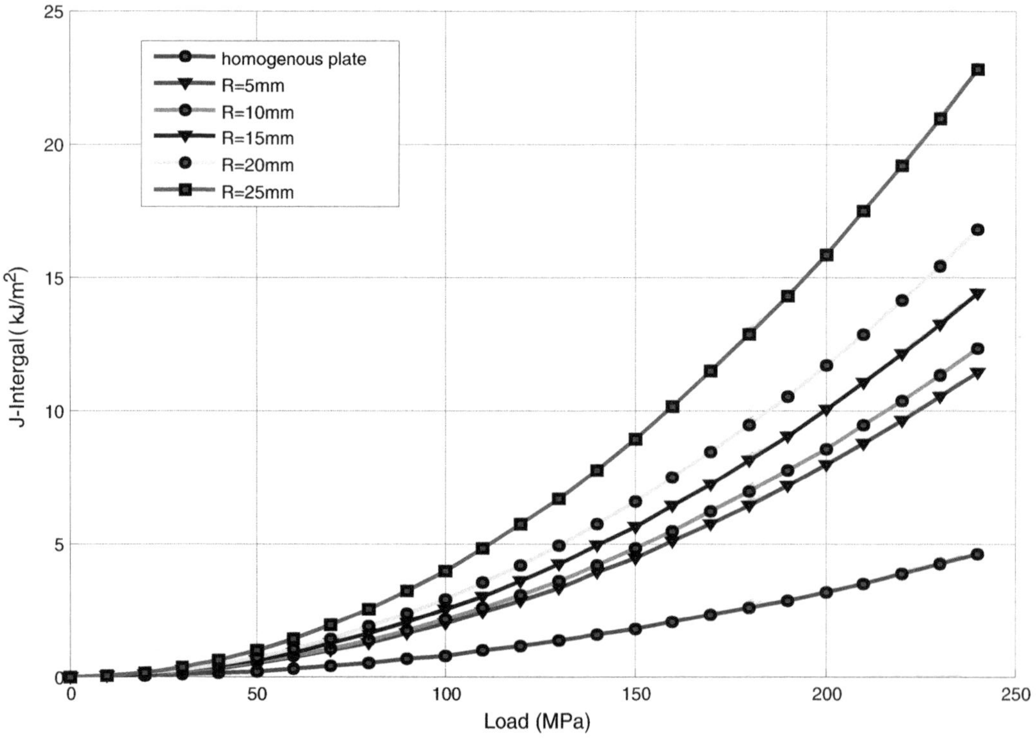

FIGURE 1.32 Variation of J-integral with load. Please see the online version to view the color image of the figure.

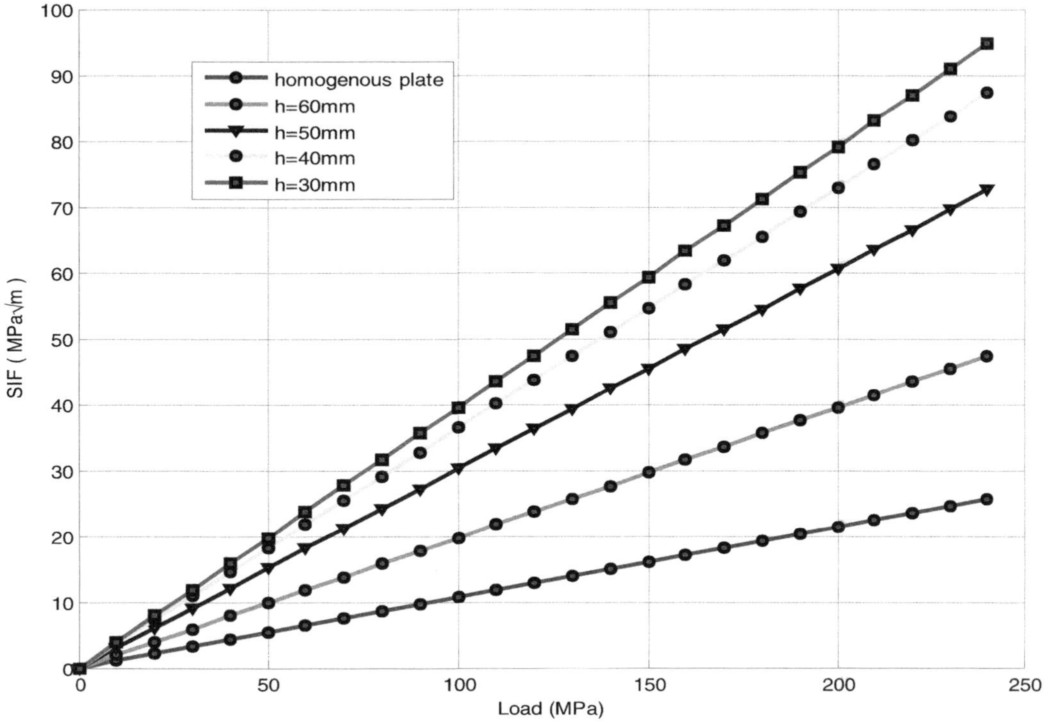

FIGURE 1.33 Variation of mode-I stress intensity factors (SIFs) with load for different positions of inclusion. Please see the online version to view the color image of the figure.

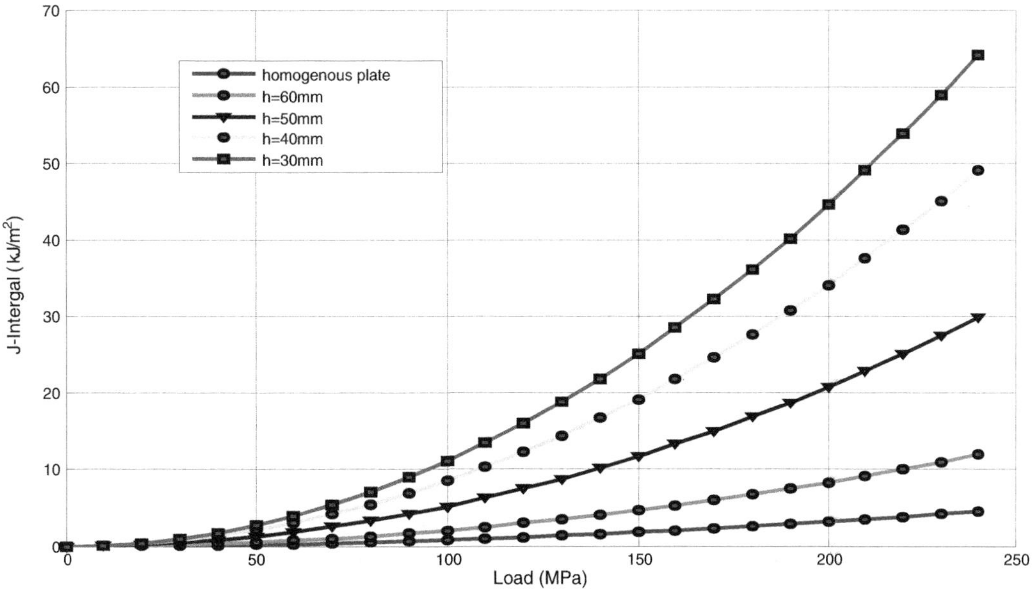

FIGURE 1.34 Variation of J-integral with load for different inclusion positions. Please see the online version to view the color image of the figure.

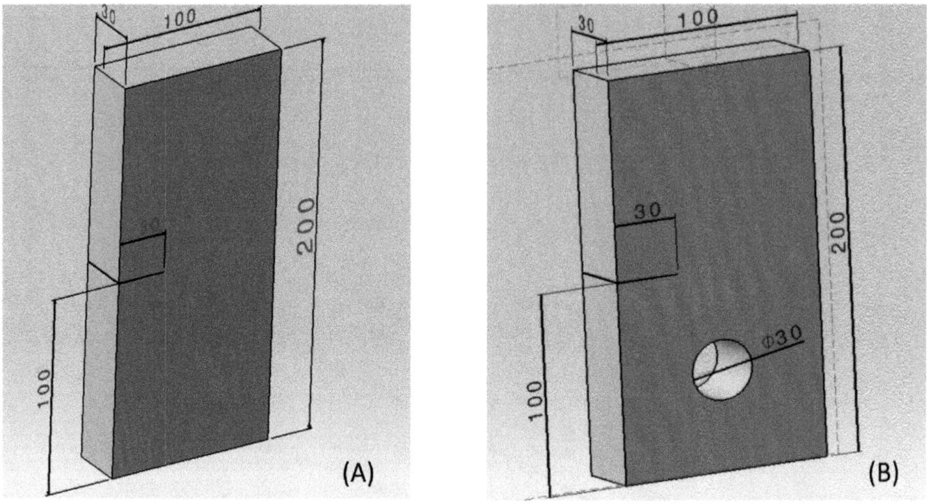

FIGURE 1.35 Three-dimensional edge-cracked component: (A) without hole and (B) with hole.

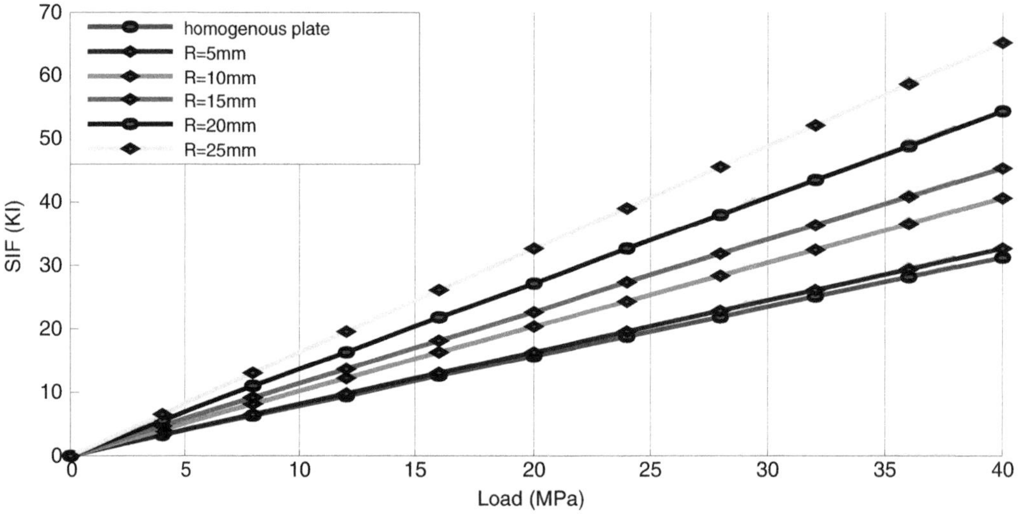

FIGURE 1.36 Variation of mode-I stress intensity factors (SIFs) with load. Please see the online version to view the color image of the figure.

assumed to have a length of 30mm. The cracked specimen is subjected to a tensile load of 40 MPa at upper surface, whereas the bottom edge of lower surface plate remains fixed during simulation. The position of the hole does not change during simulation, whereas the radii of holes have been changed from 5 to 30 mm. Variations of mode-I SIFs and J-integral with load for hole sizes are shown in Figs. 1.36 and 1.37, respectively. As expected, it has been observed that the presence of holes in cracked structures reduces the load-bearing capacity of the structure.

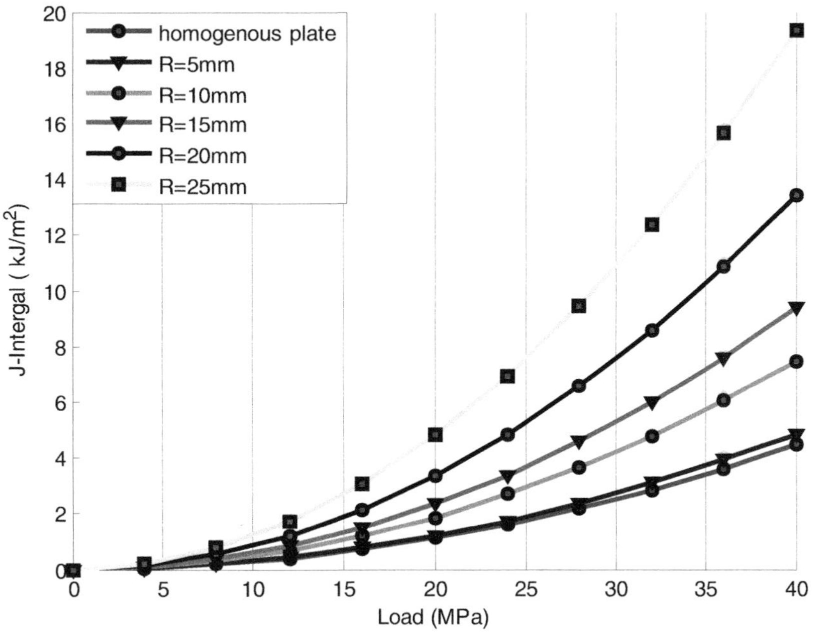

FIGURE 1.37 Variation of J-integral with load. Please see the online version to view the color image of the figure.

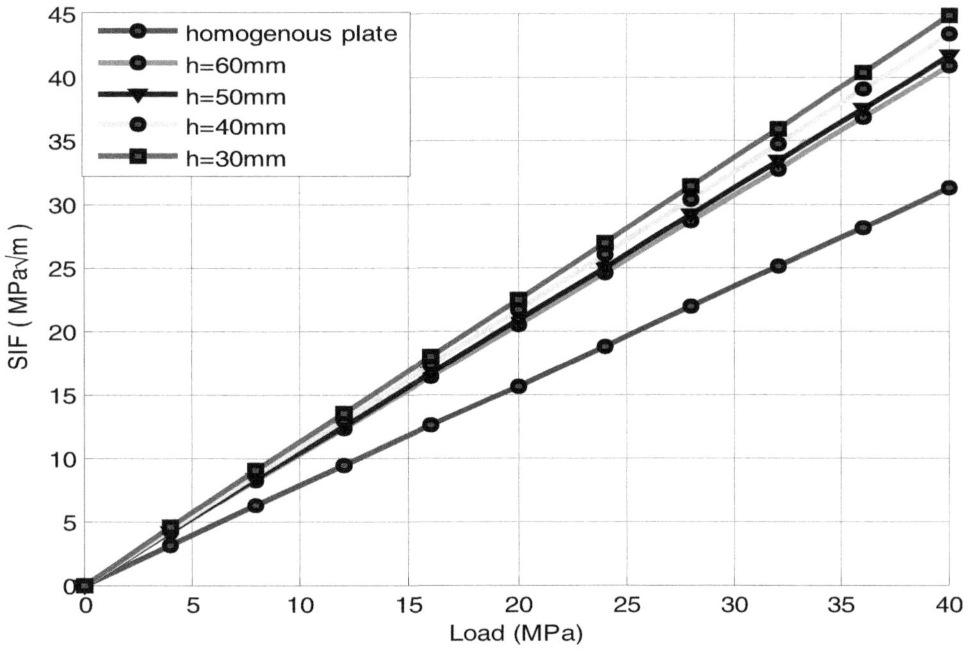

FIGURE 1.38 Variation of mode-I stress intensity factors (SIFs) with load for different positions of holes. Please see the online version to view the color image of the figure.

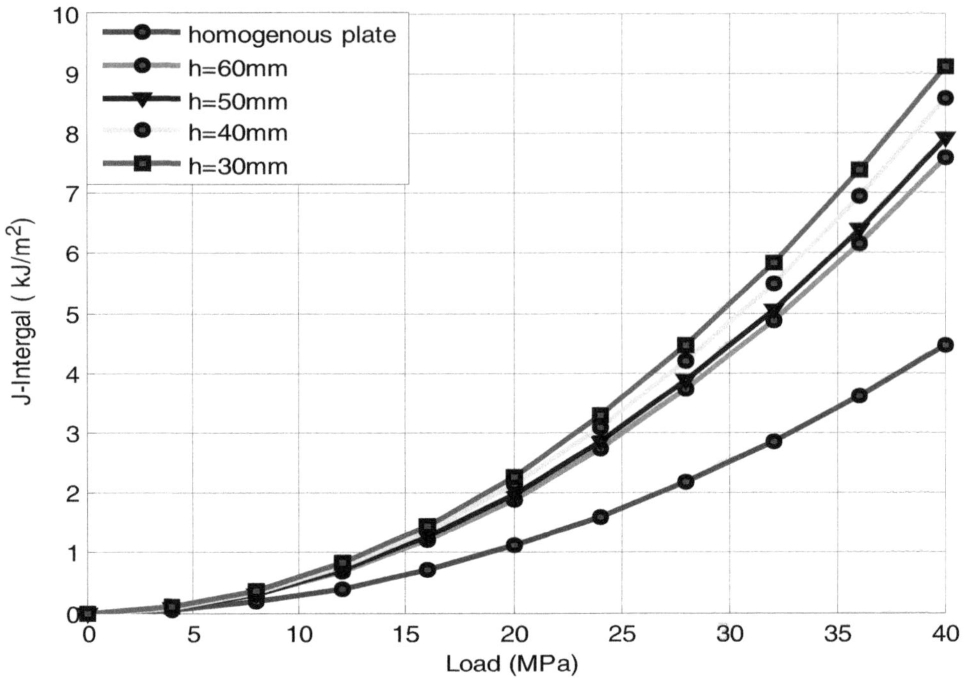

FIGURE 1.39 Variation of J-integral with load for different hole position. Please see the online version to view the color image of the figure.

1.8.11 Effect of position of hole

The same three-dimensional structural steel component, as shown in Fig. 1.35, having an edge crack of length 30mm, is selected for analysis. The effect of position of holes on three-dimensional edge-cracked components is discussed here. The radius of hole is assumed to be 15mm, whereas its position keeps on changing during analysis. The cracked specimen is subjected to a tensile load of 40 MPa at upper surface, whereas the bottom edge of lower surface plate remains fixed during simulation. The variations of mode-I SIFs and the J-integral with load for different hole positions are presented in Figs. 1.38 and 1.39, respectively. It is observed that the holes reduce the load-bearing capacity of the cracked structure. The effect of holes becomes more pronounced as the hole approaches the crack present in engineering components.

1.9 Conclusion

Level set algorithms for different types of complex-shaped material irregularities have been developed in the current study. Several numerical problems involving engineering specimens containing different geometries of discontinuities, such as circular, elliptical, rectangular, hexagonal, and octagonal discontinuities, have been considered for analysis.

Engineering components containing different edges and embedded cracks are also considered for analysis. The numerical models and algorithms on level set methodologies developed in the current work have been coded in MATLAB to obtain level sets and identify the corresponding split and tip elements in the mesh. The normal LSFs vanish along the boundary of interfaces. Level set values remain positive outside the material discontinuities and negative inside. It has also been observed that the edge cracks are fully described by one tangential LSF, whereas the center cracks require two tangential LSFs for their complete description. Also in this chapter, the behavior of three-dimensional cracked specimens in the presence of inclusions and holes has been analyzed. The influence of the size and position of inclusions and holes on different fracture mechanics parameters has been presented. It is observed that the load-carrying capacity of cracked structures reduces due to the presence of weaker inclusions and holes in them. The effect of these weak material irregularities becomes more pronounced when their size increases. The position of inclusions and holes has a drastic effect on the fracture mechanics parameters and load-carrying capacity of the cracked structures.

References

[1] Wentorf R, Collar R, Shepard MS, Fish J. Automated modelling for complex woven microstructures. Computer Methods in Applied Mechanics and Engineering 1999;172(1–4):273–91.

[2] Hollister SJ, Kikuchi N. Homogenization theory and digital imaging: a basis for studying the mechanics and design principles of bone tissue. Biotechnology & Bioengineering 1994;94(43):586–96.

[3] Jameel A, Harmain GA. Modeling and numerical simulation of fatigue crack growth in cracked specimens containing material discontinuities. Strength of Materials 2016;48(2):294–307.

[4] Lone AS, Jameel A, Harmain GA. A coupled finite element-element free Galerkin approach for modeling frictional contact in engineering components. Materials Today: Proceedings 2018;5:18745–54.

[5] Jameel A, Harmain GA. Effect of material irregularities on fatigue crack growth by enriched techniques. International Journal for Computational Methods in Engineering Science and Mechanics 2020;21:109–33.

[6] Kanth SA, Harmain GA, Jameel A. Modeling of nonlinear crack growth in steel and aluminum alloys by the element free Galerkin method. Materials Today: Proceedings 2018;5:18805–14.

[7] Sukumar N, Chopp DL, Moes N, Belyschko T. Modeling holes end inclusions by level sets in the extended finite-element method. Computer Methods in Applied Mechanics and Engineering 2001;190:6183–200.

[8] Jameel A, Harmain GA. Fatigue crack growth analysis of cracked specimens by the coupled finite element-element free Galerkin method. Mechanics of Advanced Materials and Structures 2019;26:1343–56.

[9] Jameel A, Harmain GA. Extended iso-geometric analysis for modeling three dimensional cracks. Mechanics of Advanced Materials and Structures 2019;26:915–23.

[10] Duarte CA, Babuska I, Oden JT. Generalized finite element methods for three dimensional structural mechanics problems. Computers & Structures 2002;77:215–32.

[11] Jameel A, Harmain GA. A coupled FE-IGA technique for modeling fatigue crack growth in engineering materials. Mechanics of Advanced Materials and Structures 2019;26:1764–75.

[12] Oden JT, Duarte CA, Zienkiewicz OC. A new cloud-based *hp* finite element method. Computer Methods in Applied Mechanics and Engineering 1998;153(1–2):117–26.

[13] Osher S, Setheian JA. Fronts propagating with curvature dependent speed: algorithm based on Hamilton–Jacobi formulations. Journal of Computational Physics 1988;79(1):12–49.

[14] Moes N, Cloirec M, Cartraud P, Remacle JF. A computational approach to handle complex microstructure geometries. Computer Methods in Applied Mechanics and Engineering 2003;192(28–30):3163–77.

[15] Juric D, Tryggvason G. A front tracking method for dendritic solidification. Journal of Computational Physics 1996;123:127–48.

[16] Lone AS, Jameel A, Harmain GA. Enriched element free Galerkin method for solving frictional contact between solid bodies. Mechanics of Advanced Materials and Structures 2022;0:1–13.

[17] Singh K, Jameel A, Harmain GA. Investigations on crack tip plastic zones by the extended iso-geometric analysis. Materials Today: Proceedings 2018;5:19284–93.
[18] Tryggvason G, Bunner B, Esmaeeli A, Juric D, Al-Rawahi N, Tauber W, et al. A front-tracking method for the computations of multiphase flow. Journal of Computational Physics 2001;169:708–59.
[19] Unverdi SO, Tryggvason G. A front-tracking method for viscous, incompressible, multifluid flows. Journal of Computational Physics 1992;100:25–37.
[20] Kanth SA, Lone AS, Harmain GA, Jameel A. Modelling of embedded and edge cracks in steel alloys by XFEM. Materials Today: Proceedings 2020;26:814–18.
[21] Lone AS, Kanth SA, Jameel A, Harmain GA. XFEM modelling of frictional contact between elliptical inclusions and solid bodies. Materials Today: Proceedings 2020;26:819–24.
[22] Gibou F, Fedkiw R, Osher S. A review of level-set methods and some recent applications. Journal of Computational Physics 2017;353:82–109.
[23] Sethian JA. Level set methods and fast marching methods: evolving interfaces in computational geometry, fluid mechanics, computer vision, and materials science. Cambridge, UK: Cambridge University Press; 1999.
[24] Gupta V, Jameel A, Anand S, Anand Y. Analysis of composite plates using isogeometric analysis: a discussion. Materials Today: Proceedings 2021;44:1190–4.
[25] Kanth SA, Jameel A, Harmain GA. Investigation of fatigue crack growth in engineering components containing different types of material irregularities by XFEM. Mechanics of Advanced Materials and Structures 2021;0:1–13.
[26] Chopp DL, Sukumar N. Fatigue crack propagation of multiple coplanar cracks with the coupled extended finite element/fast marching method. International Journal of Engineering Science 2003;41(8):845–69.
[27] Jameel A, Harmain GA. Large deformation in bi-material components by XIGA and coupled FE-IGA techniques. Mechanics of Advanced Materials and Structures 2022;29:850–72.
[28] Sukumar N, Chopp DL, Moran B. Extended finite element method and fast marching method for three -dimensional fatigue crack propagation. Engineering Fracture Mechanics 2003;70(1):29–48.
[29] Lone AS, Jameel A, Harmain GA. Modelling of contact interfaces by penalty based enriched finite element method. Mechanics of Advanced Materials and Structures 2022;0:1–13.
[30] Harmain GA, Jameel A, Najar FA, Masoodi JH. Large elasto-plastic deformations in bi-material components by coupled FE-EFGM. IOP Conference Series: Materials Science and Engineering 2017;225(012295):1–7.
[31] van Dijk NP, Maute, Maute K, et al. Level-set methods for structural topology optimization: a review. Structural and Multidisciplinary Optimization 2013;48(3):437–72.
[32] Osher SJ, Santosa F. Level set methods for optimization problems involving geometry and constraints I. Frequencies of a two-density inhomogeneous drum. Journal of Computational Physics 2001;171(1):272–88.
[33] Allaire G, Jouve F, Toader AM. A level-set method for shape optimization. Comptes Rendus Mathématique 2002;334(12):1125–30.
[34] Wang MY, Wang X, Wang D, Guo. A level set method for structural topology optimization. Computer Methods in Applied Mechanics and Engineering 2003;192(1–2):227–46.
[35] Allaire G, Jouve F, Toader AM. Structural optimization using sensitivity analysis and a level-set method. Journal of Computational Physics 2004;194(1):363–93.
[36] Wang S, Wang MY. Radial basis functions and level set method for structural topology optimization. International Journal for Numerical Methods in Engineering 2006;65(12):2060–90.
[37] Luo Z, Wang MY, et al. A level set-based parameterization method for structural shape and topology optimization. International Journal for Numerical Methods in Engineering 2008;76(1):1–26.
[38] Kreissl S, Pingen G, Maute K. An explicit level set approach for generalized shape optimization of fluids with the lattice Boltzmann method. International Journal for Numerical Methods in Fluids 2011;65 (5):496–519.
[39] van Dijk NP, Langelaar M, van Keulen F. Explicit level-set-based topology optimization using an exact Heaviside function and consistent sensitivity analysis. International Journal for Numerical Methods in Engineering 2012;91(1):67–97.
[40] de Ruiter MJ, van Keulen F. Topology optimization using a topology description function. Structural and Multidisciplinary Optimization 2004;26(6):406–16.
[41] Belytschko T, Black T, Moes N, Sukumar N, Usui S. Structured extended finite element methods of solids defined by implicit surfaces. International Journal for Numerical Methods in Engineering 2003;56:609–35.

[42] Sethian JA. A marching level set method for monotonically advancing fronts. Proceedings of the National Academy of Sciences of the United States of America 1996;93(4):1591–5.

[43] Stolarska M, Chopp DL, Moes N, Belyschko T. Modelling crack growth by level sets in the extended finite element method. International Journal for Numerical Methods in Engineering 2001;51:943–60.

[44] Sheikh UA, Jameel A. Elasto-plastic large deformation analysis of bi-material components by FEM. Materials Today: Proceedings 2020;26:1795–802.

[45] Khoei R, Nikbakht M. An enriched finite element algorithm for numerical computation of contact friction problems. International Journal of Mechanical Sciences 2007;49:183–99.

[46] H. Chen, C. Gerlach, T. Belytschko, Dynamic crack growth with XFEM, in: Proc. of 6th USACM, Dearborn, 2001.

[47] Jameel A, Harmain GA. Fatigue crack growth in presence of material discontinuities by EFGM. International Journal of Fatigue 2015;81:105–16.

[48] Cheung S, Luxmoore AR. A finite element analysis of stable crack growth in an aluminium alloy. Engineering Fracture Mechanics 2003;70:1153–69.

[49] Daux C, Moes N, Dolbow J. Arbitrary branched and intersecting cracks with the extended finite element method. International Journal for Numerical Methods in Engineering 2000;48:1741–60.

[50] Portela A, Aliabadi M, Rooke D. The dual boundary element method: effective implementation for crack problem. International Journal for Numerical Methods in Engineering 1991;33:1269–87.

[51] Yan AM, Nguyen-Dang H. Multiple-cracked fatigue crack growth by BEM. Computational Mechanics 1995;16:273–80.

[52] Yan X. A boundary element modeling of fatigue crack growth in a plane elastic plate. Mechanics Research Communications 2006;33:470–81.

[53] Belytschko T, Gu L, Lu YY. Fracture and crack growth by element-free Galerkin methods. Modelling and Simulation in Materials Science and Engineering 1994;2:519–34.

[54] Duflot M, Nguyen-Dang H. Fatigue crack growth analysis by an enriched meshless method. Journal of Computational and Applied Mathematics 2004;168:155–64.

CHAPTER

2

Higher order enrichment functions for fracture simulation in isotropic and orthotropic material medium

Kishan Dwivedi and Himanshu Pathak

Design Against Failure and Fracture Group, School of Mechanical and Materials Engineering, Indian Institute of Technology Mandi, Mandi, India

2.1 Introduction

The existence of cracks, which may spread and ultimately cause structural failure, is crucial to the safety of engineering structures. In practically all engineering components, cracks inherently exist or may develop during the operational stage. Crack propagation reduces the stiffness and life of components. Therefore the study of cracks in engineering components has always been an area of interest for engineers. Various fracture mechanics parameters provide detailed information regarding the behavior of cracks present in the domain, such as the stress intensity factor (SIF), J-integrals, and crack tip opening displacement (CTOD). SIFs are crucial parameters that quantify the intensity of stress singularity at crack tips of a cracked material. This quantity is used in linear elastic fracture mechanics (SIF) to determine crack propagation and stability. Therefore a precise evaluation of SIFs is important for structural integrity investigations. Even though several analytical solutions [1–9] for SIFs have been developed, they are all obtained for basic geometries and common loading scenarios. SIFs are often extracted from finite element (FE) analysis using numerical techniques for more widespread fracture situations. Many computational techniques have been developed to model static and moving cracks in engineering materials, such as the standard FE method, extended FE method (XFEM), element-free Galerkin (EFG) method (EFGM), boundary element method, iso-geometric analysis, and other coupled numerical techniques. Although the conventional FE method has always remained the dominant numerical tool for different engineering problems, it uses conformal elements for simulating discontinuity in fracture analysis. In this method, to capture the

Enriched Numerical Techniques
DOI: https://doi.org/10.1016/B978-0-443-15362-4.00020-6

stress field, an element must be refined near the crack tip, and remeshing is necessary to depict the progress of the crack tip/front as the crack grows. Due to remeshing, solution accuracy decreases, and this method is computationally expensive. FEM modeling of crack propagation is tedious because the updated mesh contains the surface of cracks. To simulate the propagation of cracks without remesh, many approaches have been created. The most effective among them is the XFEM approach, which employs an enrichment strategy. Compared to the FE method, this method provides a more accurate solution and no need for remeshing as the crack grows. This method is considerably simpler to tackle various material modeling problems, including crack growth, dislocations' creation, grain boundaries' modeling, and the evolution of phase boundaries. Belytschko and Black et al. [10] initially introduced this XFEM method. Then Mos, Dolbow, and Belytschko et al. [11] enhanced the XFEM approach. XFEM for fracture mechanics problems has reached its full potential. Displacement approximation in XFEM is divided into regular and enriching parts. These enriched parts are associated with an extra degree of freedom that carries the information about the strong discontinuity and singularity. We begin by reviewing a few of the most significant developments in the XFEM method. Daux et al. [12] used XFEM to model crack problems with multiple branches and holes. XFEM was used in three-dimensional (3D) fracture mechanics by Sukumar et al. [13]. Stolarska et al. [14] investigated an algorithm combining the level set technique with the crack growth problem in XFEM. This level set technique captures the crack position, including the crack tip in domain. Mos, Gravouil, and Belytschko et al. [15] modified XFEM with the signed distance functions to analyze three-dimensional nonplanar cracks. Ayhan and Nied et al. [16] developed an enhanced FE technique and estimated SIFs for three-dimensional fracture problems. Prevost and Sukumar et al. [17] simulate 2D crack modeling in biomaterials and isotropic media using the XFEM method. Huang, Sukumar, and Prevost et al. [18] further show how to simulate SIFs in crack increment simulation numerically. Lee et al. [19] introduced mesh superposition approach and combined it with XFEM for simulating moving and stationary cracks. Budyn et al. [20] performed XFEM to simulate multiple crack growth problems in a brittle material. Zi et al. [21] performed crack growth XFEM simulation on brittle material with multiple cracks. Researchers have performed recent investigations on complex LEFM problems. Legrain, Mos, and Verron et al. [22] apply XFEM approach to massively strained materials like rubber. Mos, Bechet, and Tourbier et al. [23] achieved the optimal convergence rate by using XFEM with essential boundary conditions. XFEM approach to simulating cracked orthotropic media was proposed by Asadpoure, Mohammadi, and Vafai et al. [24]. To replicate the orthotropic cracked media, Asadpoure and Mohammadi et al. [25] updated their prior model by including additional enrichment functions. The impact of crack shielding and different configurations of microcracks on the SIFs was investigated by Loehnert and Belytschko et al. [26] with XFEM approach. Sukumar et al. [27], integrating the fast marching approach with the XFEM method, devised a numerical approach for simulating the nonplanar crack propagation. To simulate the 2D crack growth, Tabarraei and Sukumar et al. [28] used XFEM on quadtree and polygonal FE meshes. Significant effort has been put into enhancing the accuracy of the XFEM approach. The conventional XFEM uses a local partition of unity (PU); hence, only some nodes are enhanced. As a result, some blending elements, which include both standard and enhanced nodes, do not fill the PU. Accuracy and convergence rates are reduced by the presence of these components. The blending elements are one of the key problems with the XFEM approach. To improve the performance, an improved strain approach is proposed in blending elements by Belytschko, Chessa, and Wang et al. [29] to improve the performance. Legay, Wang, and Belytschko et al. [30] used the XFEM in the spectral FEs there; we do not need to implement this blending element. Fries and Belytschko et al. [31] created an intrinsic XFEM approach to treat arbitrary discontinuities without blending elements. A corrected XFEM technique based on weight function was developed by Fries et al. [32] for blending elements. A Galerkin formulation without blending elements was created by Gracie et al. [33]. Benvenuti, Tralli, and Ventura et al. [34] developed an XFEM model that developed stress and strain fields under certain assumptions to handle displacement discontinuity. To provide a seamless transition between the unenriched and enriched domains, Ventura, Gracie, and Belytschko et al. [35] devised a weighted blending method where a smooth weight function premultiplied the enrichment function. Higher order shape functions were used by Tarancon et al. [36] to minimize the negative impacts of the blending elements. Shibanuma and Utsunomiya et al. [37] introduced a PU-based XFEM formulation to address the issue of blending elements. With finite deformation theory, Wriggers et al. [38] expanded the initially corrected XFEM approach for three-dimensional situations. Using a domain decomposition methodology, Menk and Bordas et al. [39] described a method for obtaining stiffness matrices without any enrichment. To improve LEFM's performance over the conventional XFEM, Chen et al. [40] introduced the strain smoothing method in XFEM approach. The main goals of XFEM implementation in dynamic fracture have been to simulate crack propagation and estimate SIFs for any dimensional fracture mechanics problems. In the XFEM framework, Combescure, Rethore, Gravouil et al. [41] suggested an energy-conserving strategy to describe the time-dependent fracture. Menouillard et al. [42,43] described a time-stepping strategy for enriched elements based on a lumping mass matrix method. Elguedj, Gravouil, and Maigre et al. [44] proposed an expanded lumping mass strategy for time-dependent modeling with XFEM. Fries and Zilian et al. [45] obtained convergence of several time integration techniques in the context of the XFEM for moving interfaces. For dynamic fracture problems, Belytschko and Menouillard et al. [46] investigated a method for enriching the XFEM that employed meshless approximation. For releasing tip elements, Belytschko and Menouillard et al. [47] presented a solution to ensure the continuity of forces with enhanced degree of freedom (DOF). A novel enrichment approach was introduced by Menouillard et al. [48] and investigated the impact of several directional criteria on the crack path using the XFEM method. Motamedi and Mohammadi et al. [49,50] evaluate SIFs in time-dependent moving cracks for composite materials. Esna Ashari and Mohammadi et al. [51] introduced the XFEM, in which the stress singularities around the debonding crack tip were modeled using orthotropic bi-material enrichment functions. Liu, Menouillard, and Belytschko et al. [52] created a higher order XFEM approach based on the spectral element method for modeling dynamic fracture. This method significantly reduced numerical oscillations and increased the accuracy of SIFs. Motamedi and Mohammadi [53] first introduced the orthotropic tip enrichment function and performed an XFEM simulation for crack propagation in composites. The enrichment functions were derived from the analytical solutions. A convergence investigation for an XFEM in cracked domains was performed by Chahine, Laborde, and Renard et al. [54], utilizing a cut-off function. A stress recovery approach for LEFM problems is presented by Rodenas et al. [55] and produces accurate discretization errors using XFEM. In the domain of fracture mechanics, Panetier, Ladeveze, and Chamoin et al. [56] used XFEM to describe a technique for determining the local error boundaries by assessing the discretization error. Rodenas et al. [57] introduced an error estimator using the XFEM to produce energy norm errors for LEFM problems. Lew et al. [58,59] developed a discontinuous Galerkin-based XFEM method, and optimal convergence was achieved compared to the existing XFEM method. Chahine, Nicaise, and Renard et al. [60] estimate error on SIFs with respect to a priori error estimates for the XFEM with a cut-off function and enrichment function. Prange, Loehnert, and Wriggers et al. [61] offered an error estimate in crack analysis using XFEM. Byfut and Schroder et al. [62] introduced a higher order XFEM approach by combining the conventional XFEM approach with a higher order FE method. Gonzalez-Albuixech et al. [63] looked at the convergence of SIF derived from domain energy integral for 2D crack problems utilizing XFEM. Rodenas et al. [64] introduced the moving least squares-based method and the estimated energy norm error for the XFEM method. Ruter, Gerasimov, and Stein et al. [65] introduced a posteriori error estimator in the XFEM method to estimate errors in fracture parameters in LEFM domain. After this, researchers have looked into more advanced XFEM analysis and LEFM fracture mechanics problems. Park et al. [66] created a mapping method that may be used for two- and three-dimensional problems to introduce the effect of weak singularities that come from XFEM approximation. Mousavi and Sukumar et al. [67] used the Gaussian integration method that evaluated weak form integrals effectively and constructed the Gauss quadrature rule in two dimensions. By using strain smoothing on higher order elements, Bordas et al. [68,69] investigate the accuracy of the enhanced XFEM. Legrain, Allais, and Cartraud et al. [70] used the XFEM, paid special attention to hanging node enrichment, and developed a method for maintaining displacement continuity. Baydoun et al. [71] used a hybrid explicit–implicit technique and provided propagation criteria for 3D fracture problems. Minnebo et al. [72] compute SIF and stiffness matrix using the interaction integral approach. In the three-dimensional XFEM analysis, Benvenuti et al. [73] presented the Gauss quadrature integrals with unique and discontinuous functions. A curvilinear gradient correction utilized is given by Gonzalez-Albuixech et al. [74]. Pathak et al. [75] made three-dimensional crack modeling simple and effective, in which many piecewise curve segments are used in crack front and used higher order shape functions to approximate the level set functions. Many researchers have implemented XFEM for real-world engineering problems. To detect and identify cracks in components, Rabinovich, Givoli, and Vigdergauz et al. [76,77] proposed a numerical tool based on a combination of genetic algorithms with XFEM. Nistor, Pantale, and Caperaa et al. [78] used the XFEM code to model crack propagation under dynamic loads. Holl et al. [79] introduced a method based on multiscale to examine moving cracks in three-dimensional domain. Pathak et al. [80] used XFEM and the EFG approach (EFGM) for bi-material interfacial fracture problems under mode I and mixed mode loading problems. Singh IV et al. [81] used the XFEM method to analyze the effect of multiple discontinuities like small cracks, holes, and inclusions on the fatigue life of homogeneous cracked plates. He observed that the existence of many discontinuities has a substantial impact on the fatigue life of the materials. Jameel et al. [82] used the XFEM method to estimate the bi-material interface specimen's fatigue life. He observes that a weak bi-material discontinuity decreases the specimen's critical crack length and fatigue life. He concludes that as we decrease the distance among the discontinuities, the fatigue life of the component also decreases. Jameel et al. [83] used the extended iso-geometric approach (XIGA) to analyze three-dimensional crack behavior. He estimated the SIF of the three-dimensional crack specimen under a mixed mode loading condition based on interaction integral approach. The obtained result from XIGA is validated with the XFEM and EFGM methods. Kanth et al. [84] used the EFG technique to describe and simulate the nonlinear fatigue fracture propagation phenomena in steel and aluminum alloys. He considers moving least square shape function and the geometric and material nonlinearity in the formulation part. Jameel et al. [85] proposed a unique method for modeling the formation of fatigue cracks in two dimensional specimens with various types of material discontinuities, such as holes and bi-material interfaces, based on the coupled FE and EFGMs. Dwivedi et al. [86] estimate fatigue life for fiber-reinforced polymer composite material at different volume fractions of fiber using the XFEM. In several numerical examples, he considers multiple discontinuities like cracks and holes. Suman et al. [87] proposed a novel XFEM–artificial neural network approach to predict the fatigue performance of single-side patched aluminum panels. Raza et al. [88] performed in-plane free vibration analysis of the cracked functionally graded plates using XFEM and reported the influence of cracks and their orientation on linear frequency. Raza et al. [89] investigated the in-plane free vibration of bi-material FGM plate to examine the influence of interfacial cracks in deterministic and stochastic environments by implementing XFEM. Raza et al. [90] implemented extended the FE method to simulate the free flexural vibration of cracked porous functionally graded plates. Raza et al. [91] incorporated higher order shear deformation theory in extended FE formulation to solve the cracked porous plate in thermal environment. Raza et al. [92] implemented the stochastic extended FE approach to examine the influence of material uncertainty on free flexural vibration response. Sachin et al. [93] performed stable crack growth analysis in ductile materials using a new enrichment technique in conjunction with XFEM. In this technique, he used the ramp function with the Heaviside function for complete crack modeling. Deng et al. [94] performed fatigue crack growth analysis with variable node element concept using the XFEM approach. He used variable node elements to reduce the computational cost of the simulation. For the future scope of the study, he suggested that this proposed approach is also suitable for fatigue crack growth analysis in functionally graded material under a thermal loading environment.

In the framework of XFEM, Liu et al. [95] introduced enrichment by higher order terms for the first time in 2004. He modified the original XFEM tip enrichment, replacing the asymptotic fields with first M terms ($r^{1/2}$, r, $r^{3/2}$....) for the branch functions. Several benchmark instances are examined, and the findings suggest that high-order enrichment terms may attain great accuracy. Duarte et al. [96] investigate higher order components of the asymptotic field using h–p clouds. This technique encouraged Zamani et al. [97] and Rethore et al. [98] to obtain higher accuracy using suitable enrichment scheme adjustments. Zamani et al. [99] got higher accuracy in SIF by using higher order terms at the crack tip to enrich displacement and temperature fields in thermo-elastic crack problems. The four branch functions in the original XFEM were changed by Xiao and Karihaloo [100] to a predefined number of crack tip higher order terms. Higher order XFEM was used by Cheng et al. [101] to address curved discontinuity problems. Saxby et al. [102] implement a higher order XFEM approach for straight and curved discontinuous domain issues. To attain the best convergence rate for Poisson problems and linear elasticity problems with curved discontinuities, an improved modified XFEM was constructed. Mousavi et al. [103] employed higher order XFEM approach to investigate the many intersecting and branching fractures in the elastic domain using the harmonic enrichment function.

The literature review clearly shows that the XFEM technique is often used for geometrical discontinuity simulation. The XFEM models different types of discontinuities more effectively than the conventional FE method. The problems of conformal meshing do not arise in the XFEM. Instead, the displacement-based approximations are enriched with additional enrichment functions that introduce the effect of these discontinuities in the formulations. XFEM models all types of discontinuities independent of the mesh chosen for analysis. XFEM has been used to explore fracture behavior in practically all structural materials. A higher order enrichment function at the crack tip has been included to improve the computational accuracy of the XFEM approach. However, there is a need for accuracy improvement in the XFEM approach for fracture modeling. This chapter implements crack tip higher order enrichment functions for fracture modeling in an isotropic and orthotropic material medium.

2.2 Displacement approximation with extended finite element method

The basic difference between the XFEM and the conventional FE methods lies in the modeling of different types of discontinuities. The conventional FE method considers all types of discontinuities during mesh generation, whereas XFEM does not consider the discontinuities during meshing. Instead, the standard displacement-based approximations are modified for the elements lying near the crack interface. As seen in Fig. 2.1, a body (Ω) with a geometrical discontinuity (crack) has been considered. On Γ_t and, Γ_u boundary conditions for traction and displacement are specified. The crack

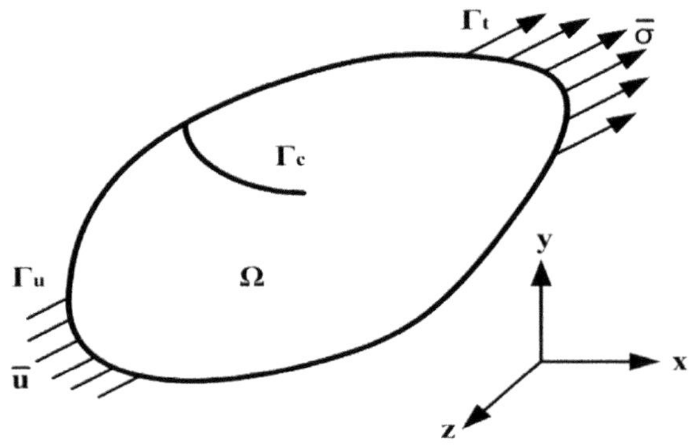

FIGURE 2.1 Domain with discontinuity.

surfaces are traction-free. The linear elastic problem's equilibrium equation and boundary conditions may be summed up as [104]

$$\nabla.\boldsymbol{\sigma} + \boldsymbol{b} = 0 \ \ in \ \Omega \tag{2.1}$$

$$\boldsymbol{\sigma}.\boldsymbol{n} = \overline{\sigma} \ \ on \ \ \Gamma_t \tag{2.2a}$$

$$\boldsymbol{\sigma}.\boldsymbol{n} = 0 \ \ on \ \ \Gamma_{c^+} \tag{2.2b}$$

$$\boldsymbol{\sigma}.\boldsymbol{n} = 0 \ \ on \ \ \Gamma_{c^-} \tag{2.2c}$$

where the Cauchy stress tensor is represented by σ, the displacement field by u, the boundary unit outward normal by n, and the body forces per unit volume b.

The XFEM framework models different discontinuity types by enriching the primary variable approximation with additional mathematical functions. These mathematical functions are enrichment and are added to the mathematical models by employing the PU approach. Level set method (LSM) is used to identify the enriched nodes. The enriched displacement-based approximation in the XFEM can be expressed as

$$\boldsymbol{u}^h(\mathbf{x}) = \boldsymbol{u}(\mathbf{x})_{FEM} + a(\mathbf{x})_{Split \ \ enrichment} + b(\mathbf{x})_{Tip \ \ enrichment} \tag{2.3}$$

Additionally, the enriched displacement-based approximation may be written as [105]

$$u^h(x) = \sum_{(i=1)}^{n} \Psi_i(x)u_i + \underbrace{\sum\nolimits_{j=1}^{n} \Psi_j(x)\left[\Xi(x) - \Xi(x_j)\right]a_j}_{j=\text{Split element}} + \underbrace{\Sigma_{k=1}^{n}\Psi_k(x)[\gamma(x) - \gamma(x_k)]b_k}_{k=\text{Tip element}} \tag{2.4}$$

where $\Psi(\boldsymbol{x})$ represents the Lagrange basis FE interpolation functions. $\Xi(\boldsymbol{x})$ represents the Heaviside jump function used to enrich the displacement approximations of split elements, and $\gamma(\boldsymbol{x})$ denotes the crack tip enrichment function used to enrich the displacement approximations of crack tip elements. $\boldsymbol{a}_j$ and $\boldsymbol{b}_k$ are used here as an extra degree of freedom for enriched nodes. Domain discretization with enriched elements/nodes can be seen in Fig. 2.2. In this figure, the green nodes represent the tip nodes, and the yellow nodes represent the split nodes.

2.2.1 Partition of unity

The extended FE framework describes the PU as a group of shape functions, the values of which add up to unity at every location in the domain of the boundary value problem. Local enrichment functions used in the XFEM method are incorporated into the approximation space through this PU property. At each point x in the FE domain, the sum of all the shape function ψ_i that contain unity can be seen in the following expression:

$$\sum_{i \in I} \psi_i(x) = 1 \tag{2.5}$$

By choosing any other function $f(x)$, the following properties will be satisfied:

$$\sum_{i \in I} \psi_i(x) f(x) = f(x) \tag{2.6}$$

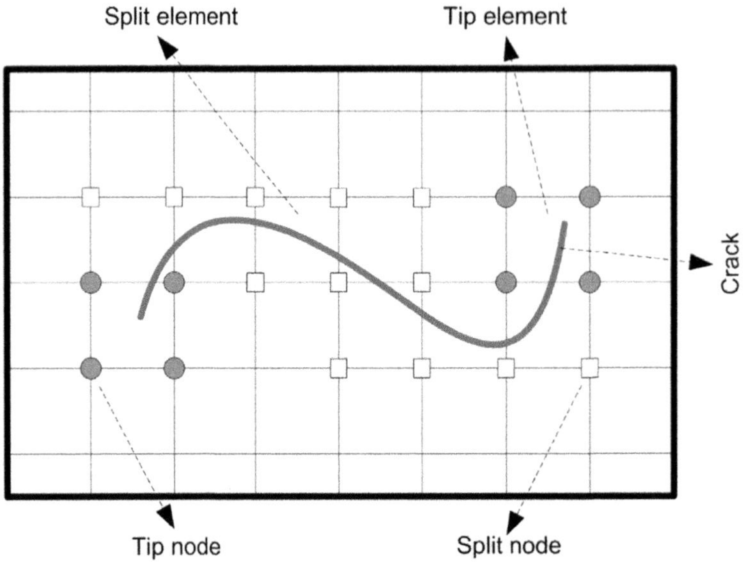

FIGURE 2.2 Domain discretization with XFEM. *XFEM*, Extended finite element method.

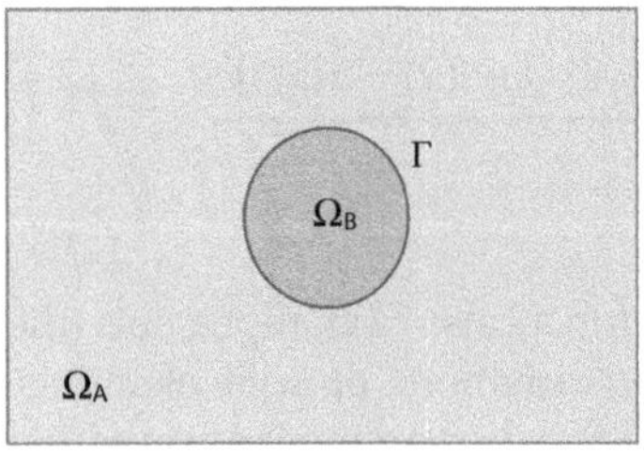

FIGURE 2.3 Level set function.

2.2.2 Level set method with extended finite element method

In the framework of the XFEM method, the basic idea behind the LSM is to define the function at each node in the FE domain to capture the discontinuity. Osher and Sethian et al. [106] developed this method as a straightforward and adaptable way to compute and analyze the motion of an interface in two or three dimensions. This approach has curve, surface, and volume functions for problems with one, two, and three dimensions in the domain. Combining LSM with XFEM techniques provides an efficient capacity for numerical modeling. These techniques used in crack growth simulation make it easier to choose enriched nodes and define enrichment functions. The only drawback of this method is that if we use it for higher dimensional domain problems, it requires higher storage and increases the computational cost. The circular hole geometry in the domain is captured in the below figure using the level set function (φ) (Fig. 2.3):

$$\begin{aligned} \varphi(x) &> 0 \quad \text{if } x \in \Omega_A \\ \varphi(x) &= 0 \quad \text{if } x \in \Gamma \\ \varphi(x) &< 0 \quad \text{if } x \in \Omega_B \end{aligned} \tag{2.7}$$

2.2.3 Heaviside enrichment function

The Heaviside jump function enriches elements through which the crack entirely passes, and this function effectively captures significant discontinuity because of the crack surface. At each node, this function is defined with a value of −1 below the crack surface and +1 above the crack surface. Fig. 2.4 shows a 3D plot of the Heaviside function. A mathematical model of the Heaviside function is shown in the following equation:

$$\Xi(x) = \begin{cases} +1 & \text{if} \quad \varphi(x) \geq 0 \\ -1 & \text{if} \quad \varphi(x) \leq 0 \end{cases} \tag{2.8}$$

where $\varphi(x)$ is the level set function, and it is defined at each node. The level set function value represents the minimum displacement of node with respect to crack surface. On one side of the crack, its value is positive; on the other side of the crack, its value is negative; and on the crack surface, its value is zero. The LSM is a strong numerical tool for representing different discontinuities present in the domain.

2.2.4 Branch enrichment functions for isotropic material

The purpose of branch enrichment $\gamma(x)$ is to describe elements with singularities in the stress field. The branch enrichment function is defined only for those elements that the crack has partially sliced. Such elements are known as the tip elements, and they carry the crack tip. Severe stress gradients are present in the tip elements. A displacement field

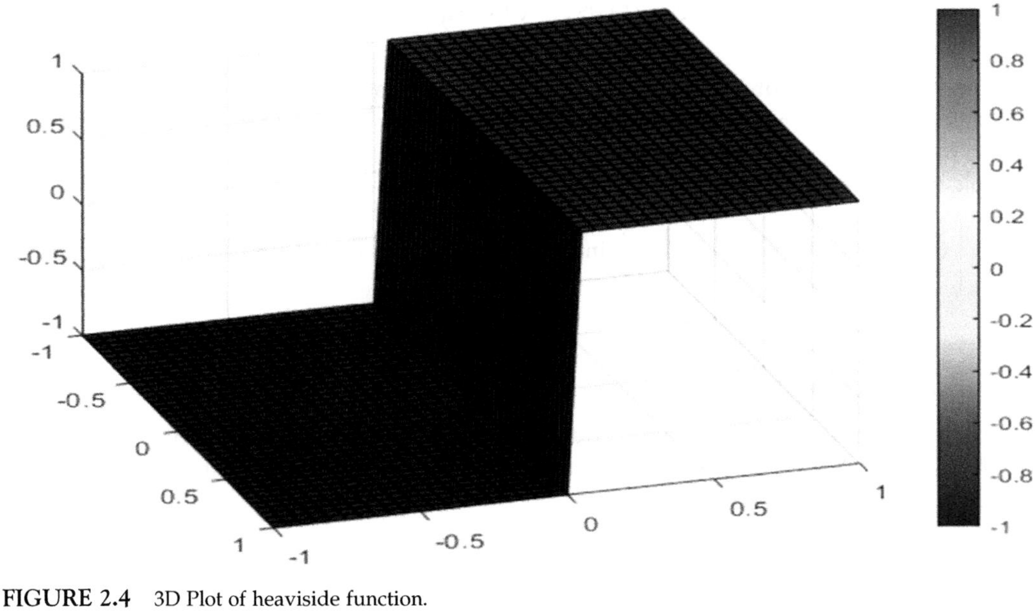

FIGURE 2.4 3D Plot of heaviside function.

close to the crack tip is used to obtain this branch enrichment function. This displacement field for isotropic material was obtained analytically [107] and can be written in the following equation:

$$u = \frac{K_I}{2\overline{\mu}}\sqrt{\frac{r}{2\pi}}\left\{\cos\frac{\phi}{2}\left(-1+\overline{\kappa}+2\sin^2\frac{\phi}{2}\right)\right\} + \frac{K_{II}}{2\overline{\mu}}\sqrt{\frac{r}{2\pi}}\left\{\left(1+\overline{\kappa}+2\cos^2\frac{\phi}{2}\right)\left(\sin\frac{\phi}{2}\right)\right\} \quad (2.9a)$$

$$v = \frac{K_I}{2\overline{\mu}}\sqrt{\frac{r}{2\pi}}\left\{\left(1+\overline{\kappa}-2\cos^2\frac{\phi}{2}\right)\left(\sin\frac{\phi}{2}\right)\right\} + \frac{K_{II}}{2\overline{\mu}}\sqrt{\frac{r}{2\pi}}\left\{\left(-1+\overline{\kappa}-2\sin^2\frac{\phi}{2}\right)\left(-\cos\frac{\phi}{2}\right)\right\} \quad (2.9b)$$

$$\text{Here, } \overline{\mu} = \frac{E}{2(1+\nu)} \quad (2.10)$$

$$\overline{\kappa} = \begin{cases} \dfrac{3-\nu}{1+\nu} & \text{plane stress} \\ 3-4\nu & \text{plane strain} \end{cases} \quad (2.11)$$

The displacements in the x and y directions are denoted by the variables u and v, respectively, in the equation above. K_I and K_{II}, respectively, represent SIFs in mode I and mode II situations. r and ϕ are the polar coordinates system near the crack tip. $\overline{\mu}$ is the shear modulus, while $\overline{\kappa}$ is the Kolovos constant. ν is the Poisson ratio, and E is the elastic constant. The isotropic crack tip enrichment functions obtained from displacement fields are shown in the following equation:

$$\gamma(x) = \left[\sqrt{r}\sin\frac{\phi}{2} \quad \sqrt{r}\cos\frac{\phi}{2} \quad \sqrt{r}\cos\frac{\phi}{2}\sin\phi \quad \sqrt{r}\sin\frac{\phi}{2}\sin\phi\right] \quad (2.12)$$

To increase the solution accuracy, the isotropic crack tip enrichment function is augmented with higher order terms like r^2, $r^2\sin2\phi$, and $r^2\cos2\phi$ as enrichment functions, as seen in Eq. (2.13). Fig. 2.5 displays the crack tip higher order enrichment functions for isotropic materials:

$$\gamma(x) = \left[\sqrt{r}\sin\frac{\phi}{2} \quad \sqrt{r}\cos\frac{\phi}{2} \quad \sqrt{r}\cos\frac{\phi}{2}\sin\phi \quad \sqrt{r}\sin\frac{\phi}{2}\sin\phi \quad r^2 \quad r^2\sin2\phi \quad r^2\cos2\phi\right] \quad (2.13)$$

2.2.5 Branch enrichment functions for orthotropic material

Different enrichment functions are used to enrich the displacement field in the case of orthotropic materials. The enrichment functions for the crack tip elements in the orthotropic material medium are defined as [104]

$$\gamma(x) = \left[\sqrt{r}\cos\frac{\phi_1}{2}\sqrt{g_1(\phi)} \quad \sqrt{r}\cos\frac{\phi_2}{2}\sqrt{g_2(\phi)} \quad \sqrt{r}\sin\frac{\phi_1}{2}\sqrt{g_1(\phi)} \quad \sqrt{r}\sin\frac{\phi_2}{2}\sqrt{g_2(\phi)}\right] \quad (2.14)$$

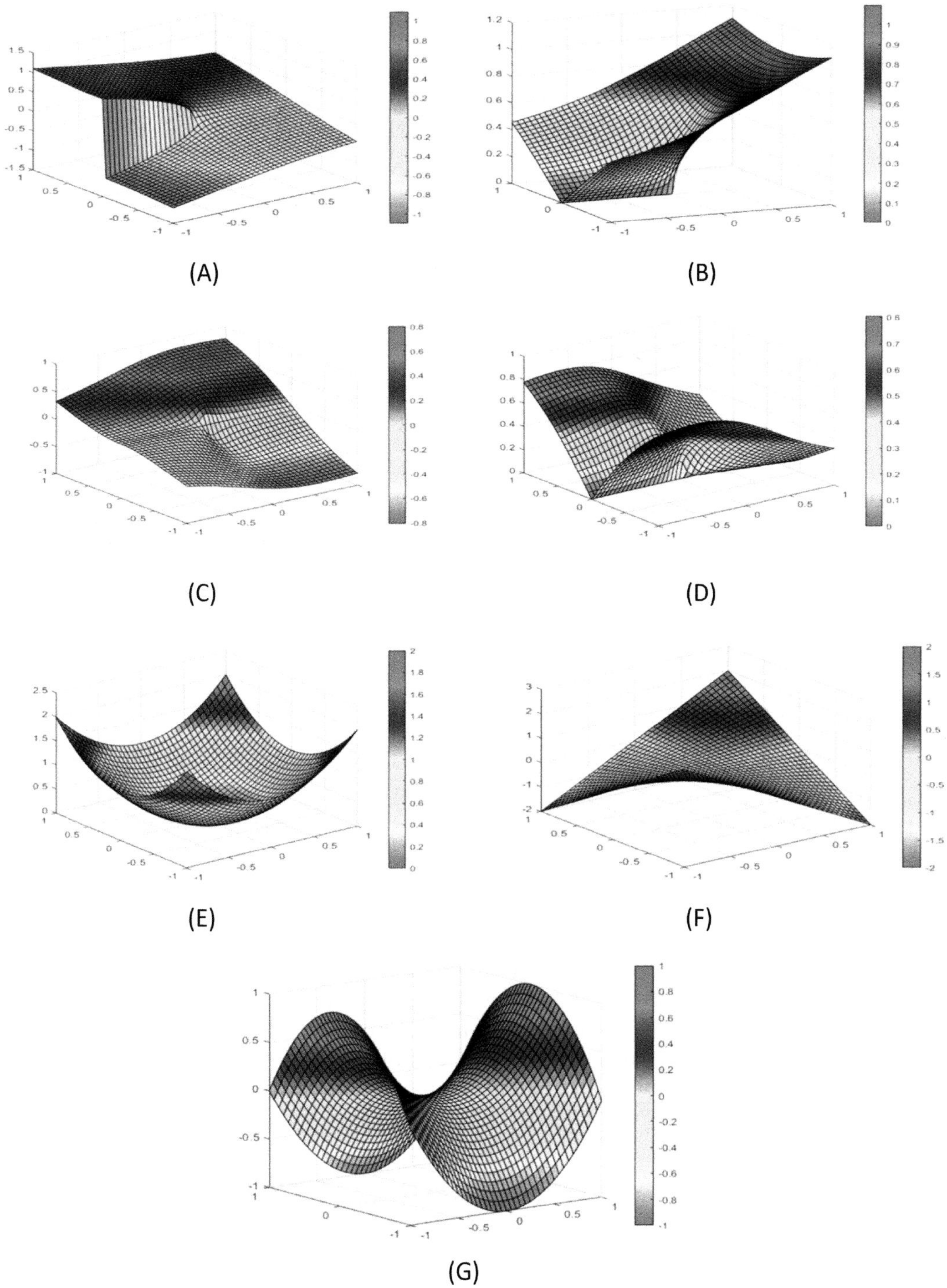

FIGURE 2.5 Three-dimensional plot of each terms of isotropic branch enrichment functions (A)–(G) in sequence as given in Eq. (2.13).

Furthermore, higher order terms are added to above written tip enrichment function:

$$\begin{bmatrix} \sqrt{r}\cos\frac{\phi_1}{2}\sqrt{g_1(\phi)} & \sqrt{r}\cos\frac{\phi_2}{2}\sqrt{g_2(\phi)} & \sqrt{r}\sin\frac{\phi_1}{2}\sqrt{g_1(\phi)} & \sqrt{r}\sin\frac{\phi_2}{2}\sqrt{g_2(\phi)} \\ & r^2\sqrt{g_1(\phi)} & r^2\sqrt{g_2(\phi)} & \\ r^2\sin 2\phi_1\sqrt{g_1(\phi)} & r^2\cos 2\phi_1\sqrt{g_1(\phi)} & r^2\sin 2\phi_2\sqrt{g_2(\phi)} & r^2\cos 2\phi_2\sqrt{g_2(\phi)} \end{bmatrix} \tag{2.15}$$

The methods used to create these higher order enrichment functions consider all conceivable stress–strain states and displacements close to the crack tip. Additionally, [25] finds independent crack tip function terms:

$$\phi_j = \arctan\left(\frac{\mu_{jy}\sin(\phi)}{\mu_{jx}\sin(\phi) + \cos(\phi)}\right) \tag{2.16}$$

$$g_j(\phi) = \sqrt{\left(\mu_{jy}\sin(\phi)\right)^2 + \left(\cos(\phi) + \mu_{jx}\sin(\phi)\right)^2} \tag{2.17}$$

In the previous equation, μ_{jx} and μ_{jy} stand in for the real and imaginary roots of the material's characteristic equation, where $j = 1$ and 2. After applying the compatibility and equilibrium conditions, this characteristic equation [25] is produced:

$$\overline{C}_{11}\mu^4 - 2\overline{C}_{16}\mu^3 + \left(2\overline{C}_{12} + \overline{C}_{66}\right)\mu^2 - 2\overline{C}_{26}\mu + \overline{C}_{22} = 0 \tag{2.18}$$

Crack tip higher order enrichment functions are plotted in Fig. 2.6. The plotted enrichment functions show singularity and discontinuity.

2.2.6 Numerical integration

Gauss quadrature technique is used for numerical integration across different elements in the XFEM. To capture the effect of the discontinuities more accurately, Gauss quadrature rules are modified in the XFEM. If the element is discontinuous, ordinary Gauss quadrature cannot be applied to model the discontinuous displacement fields efficiently. Higher order Gauss quadrature is required for the numerical integration of such elements. To integrate these elements, subtriangulation is performed with several smaller triangular elements using the PU approach. The subtriangulation of an enriching element can be seen in Fig. 2.7. The actual objective of subtriangulation is to generate Gauss points around the crack. Once these Gauss points are created around the crack, they can be utilized directly for integration without considering the subtriangles. The distribution of Gauss points over the different elements in domain can be seen in Fig. 2.7.

2.2.7 Fracture parameter (stress intensity factor) calculation

Domain-based interaction integral technique [105,108–114] was used to calculate SIF. This technique accurately and efficiently extracts individual SIFs under mixed mode loading conditions. This interaction integral is computed around the crack tip using J-integral

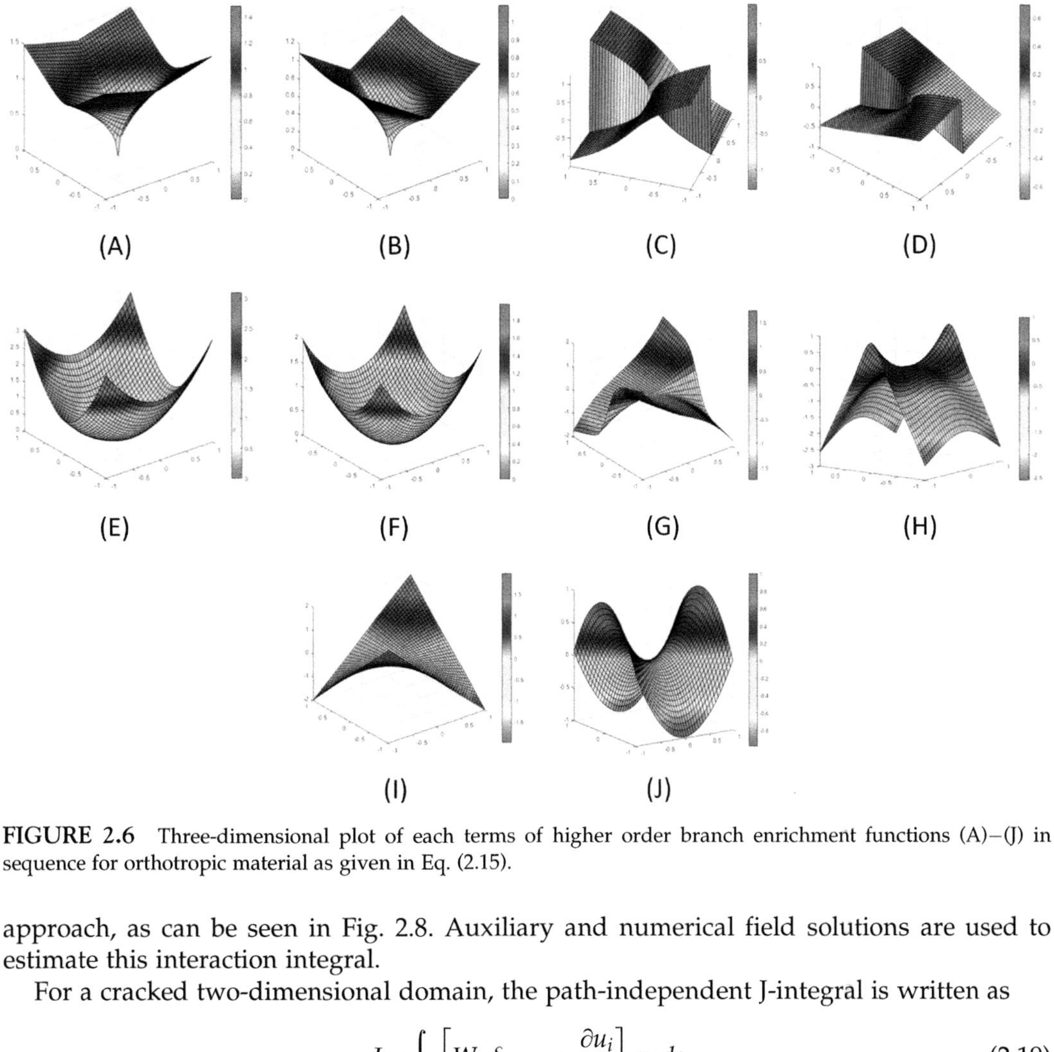

FIGURE 2.6 Three-dimensional plot of each terms of higher order branch enrichment functions (A)–(J) in sequence for orthotropic material as given in Eq. (2.15).

approach, as can be seen in Fig. 2.8. Auxiliary and numerical field solutions are used to estimate this interaction integral.

For a cracked two-dimensional domain, the path-independent J-integral is written as

$$J = \int_{\Gamma} \left[W \ \delta_{1j} - \sigma_{ij} \frac{\partial u_i}{\partial x_1} \right] n_j \, ds \tag{2.19}$$

where W is the strain energy density, n_j is the unit normal to the contour, and δ_{1j} is the Kronecker delta function. Γ is an arbitrary contour for J-integral calculations around the crack tip. To evaluate domain-based interaction integrals, two independent equilibrium states have also been imposed in the J-integral; state "a" corresponds to the real state, while state "b" is taken as an auxiliary state. The real state has been extracted from the numerical solution of the fracture domain, whereas the auxiliary state has been generated using analytical solutions of asymptotic stress and displacement fields. It may be stated as [105] after imposing auxiliary and real field solutions in the J-integral:

$$J^{(a+b)} = I^{(a,b)} + J^{(a)} + J^{(b)} \tag{2.20}$$

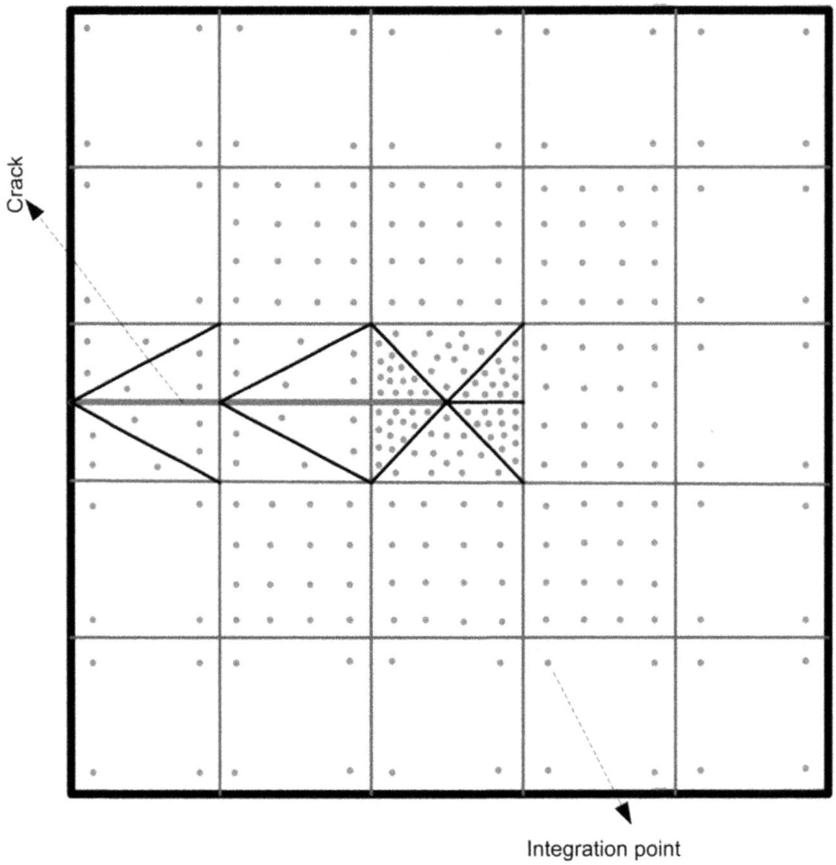

FIGURE 2.7 Gauss point distribution over the different element.

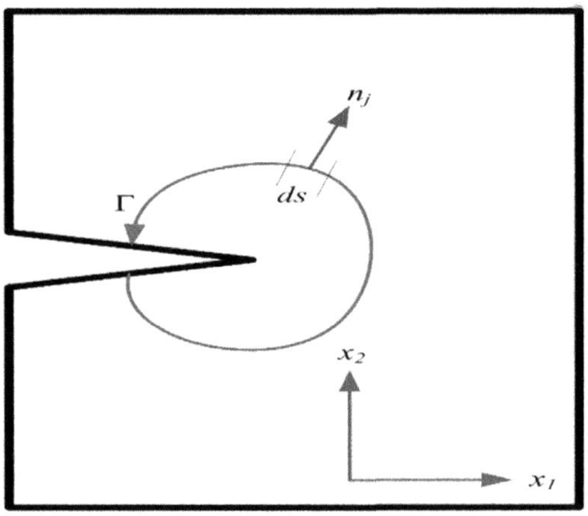

FIGURE 2.8 Path-independent integral over the crack.

where $I^{(a,b)}$ is the interaction integral term, and $J^{(a+b)}$ is the J-integral of the superimposed state. The interaction integral under mechanical loading has the following domain form [105]:

$$I^{(a,b)} = \int_A \left[\sigma_{ij}^{(a)} \frac{\partial u_i^{(b)}}{\partial x_1} + \sigma_{ij}^{(b)} \frac{\partial u_i^{(a)}}{\partial x_1} - \frac{1}{2} \left(\sigma_{ij}^{(a)} \varepsilon_{ij}^{(b)} + \sigma_{ij}^{(b)} \varepsilon_{ij}^{(a)} \right) \delta_{ij} \right] \frac{\partial q}{\partial x_j} dA \tag{2.21}$$

where $\sigma_{ij}{}^{(a)}$ and $\varepsilon_{ij}^{(a)}$ are the numerically derived Cauchy stress and engineering strain, respectively, whereas $\sigma_{ij}{}^{(b)}$ and $\varepsilon_{ij}^{(b)}$ are the Cauchy stress and engineering strain obtained via auxiliary field equations. q is a scalar weight function inside the contour. The interaction integral can be expressed in terms of the mixed mode stress intensity component that has been observed [104]:

$$I = 2m_{11}K_I^a K_I^b + m_{12}\left(K_I^a K_{II}^b + K_I^b K_{II}^a\right) + 2m_{22}K_{II}^a K_{II}^b \tag{2.22}$$

$$m_{11} = -\frac{\overline{C}_{22}}{2}\mathrm{Im}\left(\frac{\mu_1 + \mu_2}{\mu_1\mu_2}\right) \tag{2.23a}$$

$$m_{12} = -\frac{\overline{C}_{22}}{2}\mathrm{Im}\left(\frac{1}{\mu_1\mu_2}\right) + \frac{\overline{C}_{11}}{2}\mathrm{Im}(\mu_1\mu_2) \tag{2.23b}$$

$$m_{22} = \frac{\overline{C}_{11}}{2}\mathrm{Im}(\mu_1 + \mu_2) \tag{2.23c}$$

Mixed mode SIFs (K_I and K_{II}) can be computed after imposing $K_I^b = 1$, $K_{II}^b = 0$, and vice versa . A complete solution algorithm is presented in Fig. 2.9.

2.3 Result and discussions

The two-dimensional isotropic/orthotropic domain with discontinuity under traction and fatigue load conditions is analyzed using the suggested higher order XFEM technique. A comparative study between the higher order XFEM technique and the XFEM approach is performed in each numerical example. The numerical results are presented as stress contours and comparative SIFs.

2.3.1 Edge crack plate with an isotropic material medium

The suggested higher order XFEM technique is tested for accuracy by modeling an edge crack plate of isotropic material with dimensions $a = 50$ mm, $b = 100$ mm, $c = 100$ mm, and $d = 200$ mm, as shown in Fig. 2.10A. The material's Poisson ratio is 0.3 and has a Young modulus of 200×10^3 MPa. The boundary condition of the plate is shown in Fig. 2.10A, and it comprises the application of a 1 MPa tensile traction stress to the plate's top edge. The crack in the model is colored red and is 50 mm in length. This model's SIF is assessed using the higher order XFEM and the XFEM technique shown in Table 2.1. With a total of 4800 DOF, the domain is discretized. Fig. 2.10B and Fig. 2.10C depict the stress contour obtained by different computational methods. Compared to

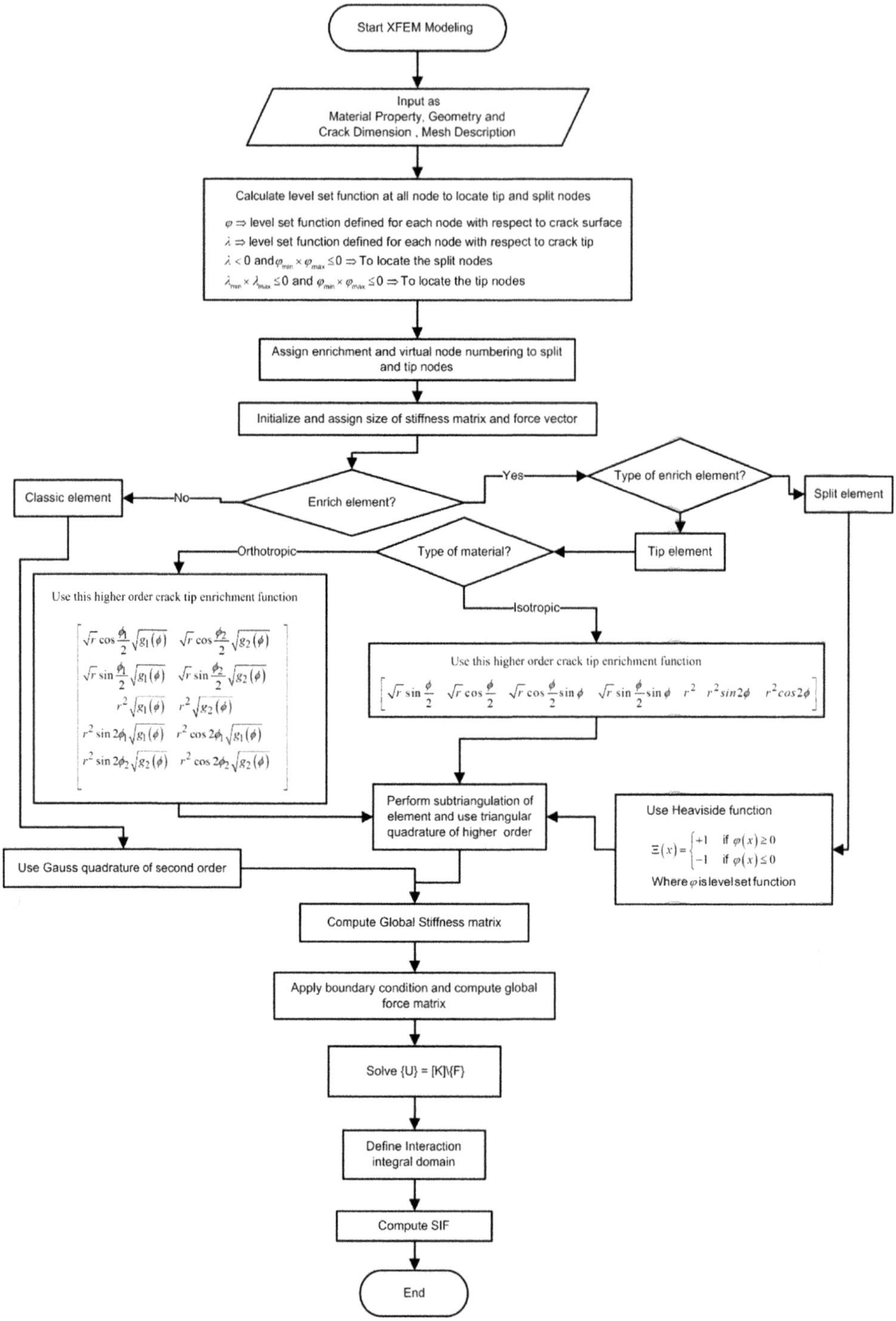

FIGURE 2.9 Flow chart for complete solution algorithm.

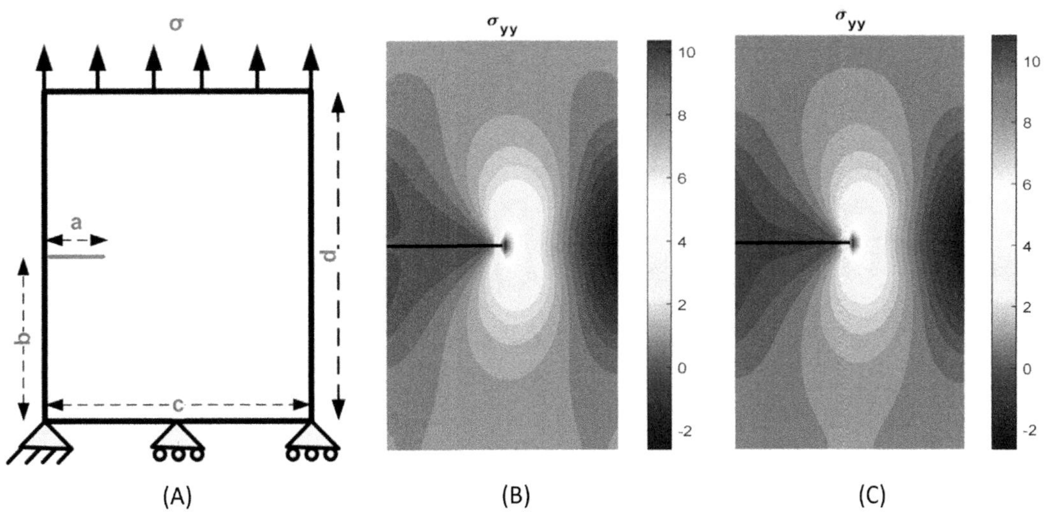

FIGURE 2.10 Edge crack isotropic plate with **(A)** boundary condition, **(B)** stress contour (MPa) obtained from XFEM approach, and **(C)** stress contour (MPa) obtained from higher order XFEM approach. *XFEM*, Extended finite element method.

TABLE 2.1 Stress intensity factor (SIF) (K_1) comparison for different computational approaches.

Computational approach	SIF (MPa $\sqrt{mm}$) at crack tip	Degree of freedom (total)	Exact SIF [104] (MPa $\sqrt{mm}$) at crack tip	SIFs deviation (%)
XFEM	34.7845	4800	35.4234	1.8
Higher order XFEM	35.0171	4800	--	1.1

XFEM, Extended finite element method.

earlier standard XFEM methods, the proposed higher order tip enrichment strategy has a high convergence rate, as illustrated in Fig. 2.11.

2.3.2 Center crack plate with isotropic material medium

To test the precision of the suggested computational technique, a center crack isotropic plate of dimensions $a = 50$ mm, $b = 100$ mm, $c = 100$ mm, and $d = 200$ mm is simulated, as shown in Fig. 2.12A. The plate's lower edge is restrained along the y-axis, while the plate's upper edge is mechanically tensed at a maximum of 1 MPa. $\nu = 0.3$ and $E = 200 \times 10^3$ MPa are the Poisson ratio and Young modulus of an isotropic domain, respectively. The stress contour obtained by different computational techniques with a crack length of 50 mm is shown in Fig. 2.12B and C. The discretized domain has a total of 40×60 nodes. Fig. 2.13 illustrates the SIF error in terms of L-2 norms based on the number of nodes. This figure shows that the higher order XFEM approach has a higher rate of convergence than the traditional XFEM method. Table 2.2 compares the analytical SIF for a 50 mm crack length and the SIF derived from different computational methodologies.

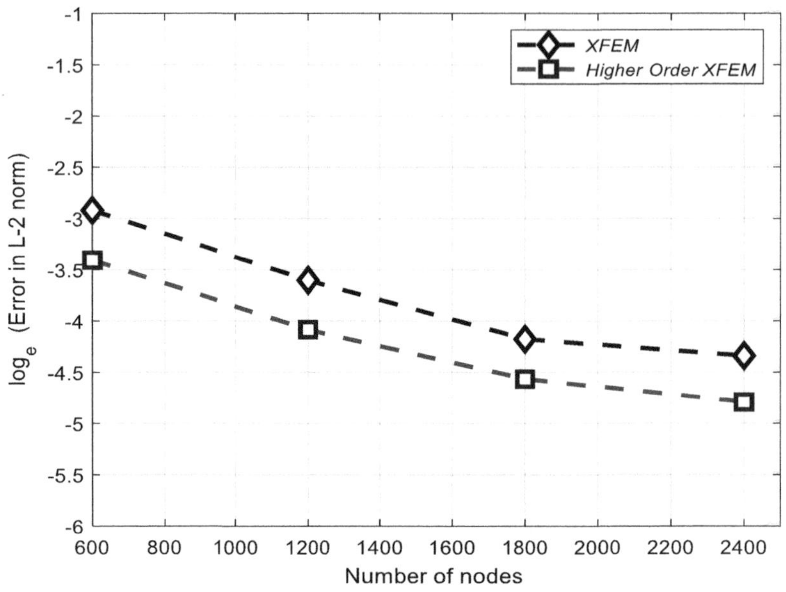

FIGURE 2.11 SIFs error in L2 norms for isotropic edge crack plate. *SIF*, Stress intensity factor.

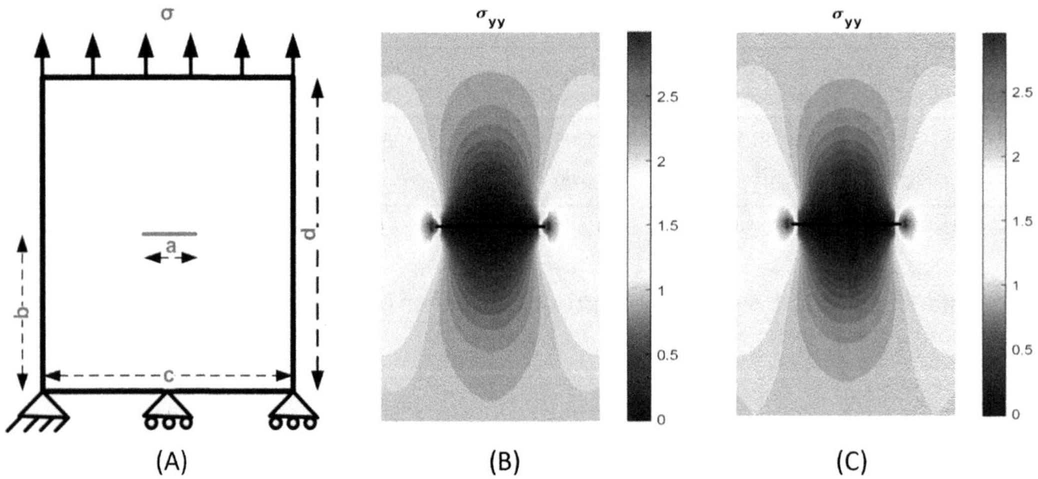

FIGURE 2.12 Isotropic central crack plate with **(A)** boundary condition, **(B)** stress contour (MPa) obtained from XFEM **(C)** stress contour (MPa) obtained from higher order XFEM. *XFEM*, Extended finite element method.

2.3.3 Edge crack plate with an orthotropic material medium

The suggested computational method simulates a two-dimensional edge crack orthotropic medium. Fig. 2.14A illustrates the domain dimensions as $a = 50$ mm, $b = 100$ mm, $c = 100$ mm, and $d = 200$ mm. Table 2.3 lists the orthotropic material properties of the plate. Fig. 2.14A shows the boundary condition for a plate with a 1 MPa tensile traction stress applied to the plate's upper edge. The entire domain is discretized using 6400 DOF.

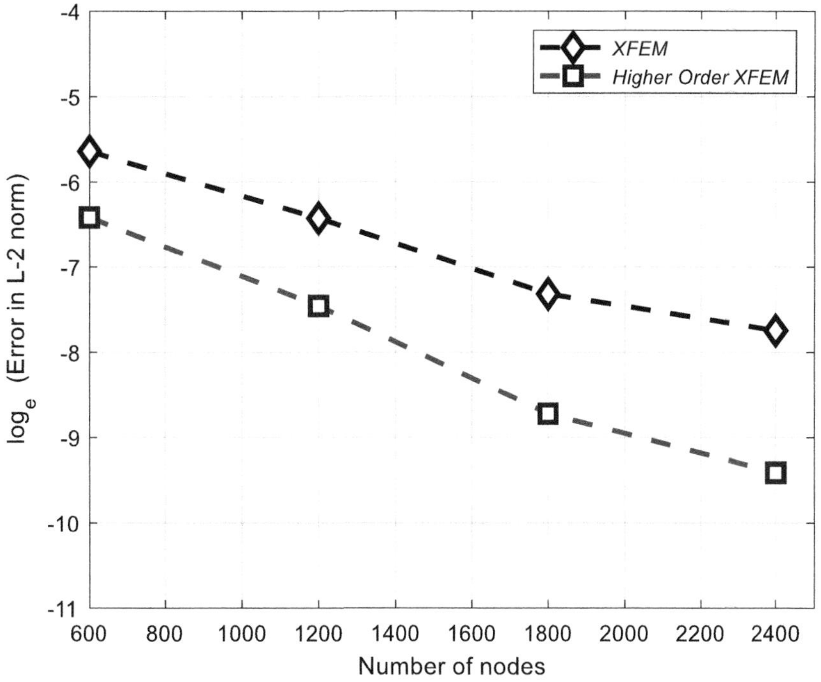

FIGURE 2.13 SIFs error in term of L-2 norms for isotropic central crack plate. *SIF*, Stress intensity factor.

TABLE 2.2 Comparisons in stress intensity factor (SIF) (K_I) for various computational techniques.

	XFEM		Higher order XFEM	
Number of nodes	**SIF (MPa $\sqrt{mm}$)**	**Deviation (%) from [115]**	**SIF (MPa $\sqrt{mm}$)**	**Deviation (%) from [115]**
600	10.3940	0.83	10.4407	0.38
1200	10.4246	0.54	10.4607	0.19
1800	10.4525	0.27	10.4738	0.066
2400	10.4595	0.2	10.4847	0.038

XFEM, Extended finite element method.

Fig. 2.14B and C represents the stress contour of an edge crack orthotropic plate with a 0 degree laminar orientation obtained using XFEM and higher order XFEM techniques, respectively. Fig. 2.15 shows the numerical SIFs that were anticipated during the postprocessing stage for various laminar orientations (0, 30, 60, and 90 degrees). Fig. 2.15 compares the acquired SIF to previous research and displays the results. At every depicted lamina orientation, the obtained SIFs strongly accord with the literature. Fig. 2.15 shows SIF variation for the two distinct computational methods used at various laminar orientation angles. Fig. 2.15 illustrates how the higher order tip enrichment function has improved the computational accuracy for calculating fracture parameters (SIFs). The higher order XFEM technique improves computational accuracy in fracture modeling issues; it can be observed.

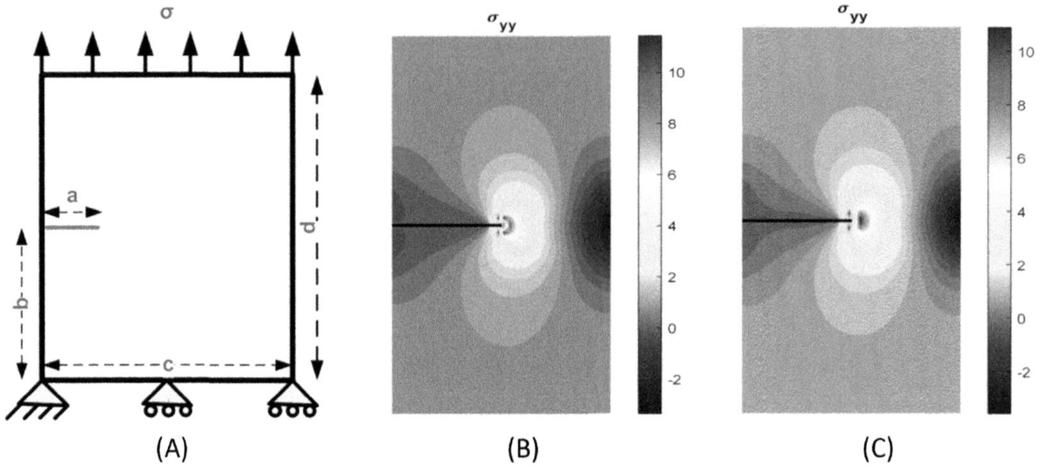

FIGURE 2.14 Orthotropic edge crack plate with **(A)** boundary conditions, **(B)** stress contour (MPa) at 0° laminar orientation using XFEM, and **(C)** stress contour (MPa) at 0 degree laminar orientation using higher order XFEM. *XFEM*, Extended finite element method.

TABLE 2.3 The simulation examples used orthotropic material properties [104].

E_1	E_2	ν_{12}	G_{12}
114.8 GPa	11.7 GPa	0.21	9.66 GPa

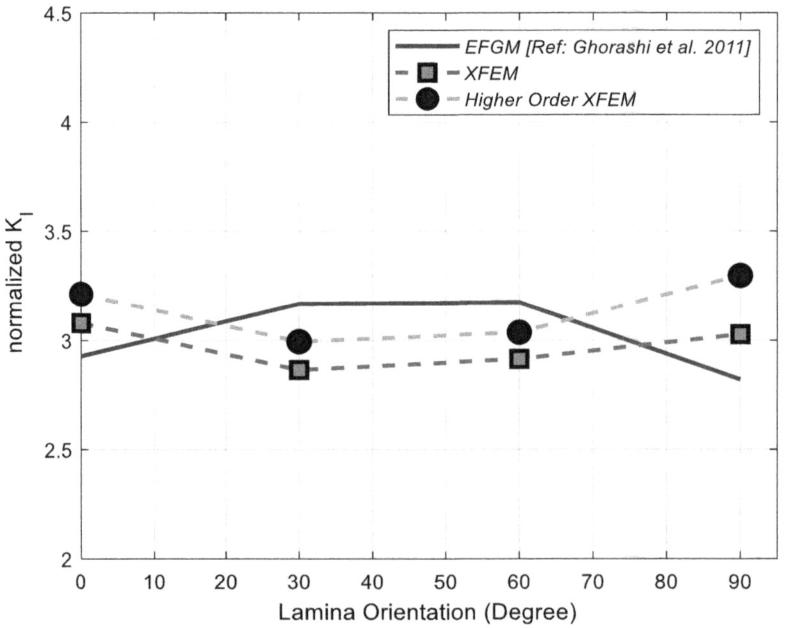

FIGURE 2.15 Numerical SIFs for edge crack in orthotropic plate problem obtained by two computational approaches. *SIF*, Stress intensity factor.

2.3.4 Center crack plate with an orthotropic material medium

The center crack orthotropic plate has been simulated to confirm the accuracy of the recommended numerical technique. The crack plate in Fig. 2.16A has the following measurements: $a = 50$ mm, $b = 100$ mm, $c = 100$ mm, $d = 200$ mm, and $e = 25$ mm. This analysis uses the material property given in Table 2.3. Fig. 2.16A represents the boundary condition of the plate, and the plate's upper edge is subject to a 1 MPa tensile traction stress. A total of 6400 DOFs are utilized to discretize the problem domain. The stress field was obtained in the postprocessing stage using two distinct computational methods and is shown in Fig. 2.16B and C. Additionally, SIFs at various lamina orientations (0, 30, 45, 60, 75, and 90 degrees) at the tip have been computed from different computational approaches and are shown in Fig. 2.17. The SIF values are raised to 60 degrees laminar orientation on the SIF plot and then dropped to 90 degrees. This figure provides a graphic representation of the effect of lamina orientation on SIFs. As can be observed, the accuracy of the produced SIFs is improved by the higher order enrichment function.

2.3.5 Edge crack with multiple holes orthotropic plate under fatigue loading

Further, we performed fatigue life analysis on an edge crack with multiple holes orthotropic plate, the dimensions of which are given as $a = 25$ mm, $b = 100$ mm, $d = 200$ mm, $k = 25$ mm, $m = 25$ mm, $e = 25$ mm, $f = 40$ mm, $g = 35$ mm, $h = 35$ mm, and $i = 40$ mm as shown in Fig. 2.18A. Multiple holes of radius 10 mm are distributed over the domain of the plate. Discontinuity in the domain is represented by red. The bottom edge of plate is restricted to move in y direction, and top edge of the plate is subjected to cyclic mechanical loading of a maximum stress value is 1 MPa. Fatigue analysis is performed for stress ratios 0 and 0.5 at different laminar angles of orthotropy like 0 and 60 degrees using higher order

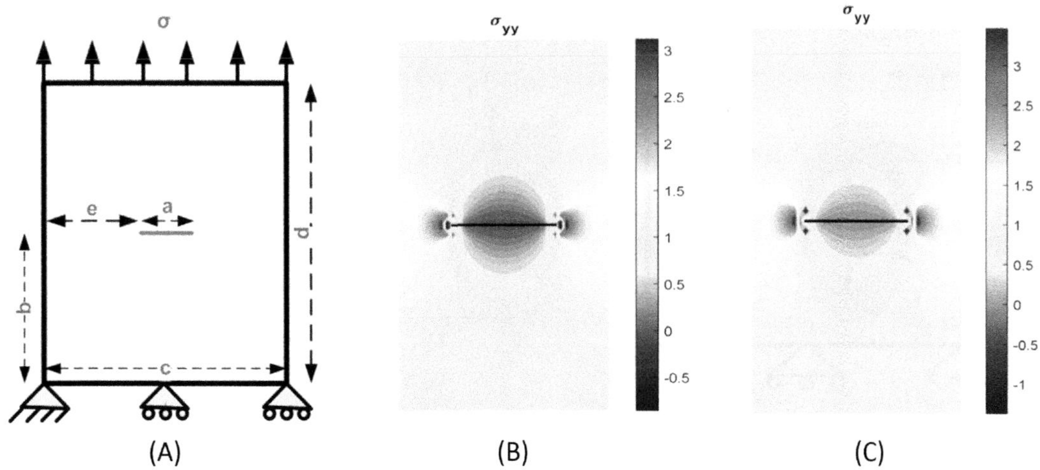

FIGURE 2.16 Center crack orthotropic plate with **(A)** boundary conditions, **(B)** stress contour (MPa) at 0 degrees laminar orientation using XFEM **(C)** stress contour (MPa) at 0 degree laminar orientation using higher order XFEM. *XFEM*, Extended finite element method.

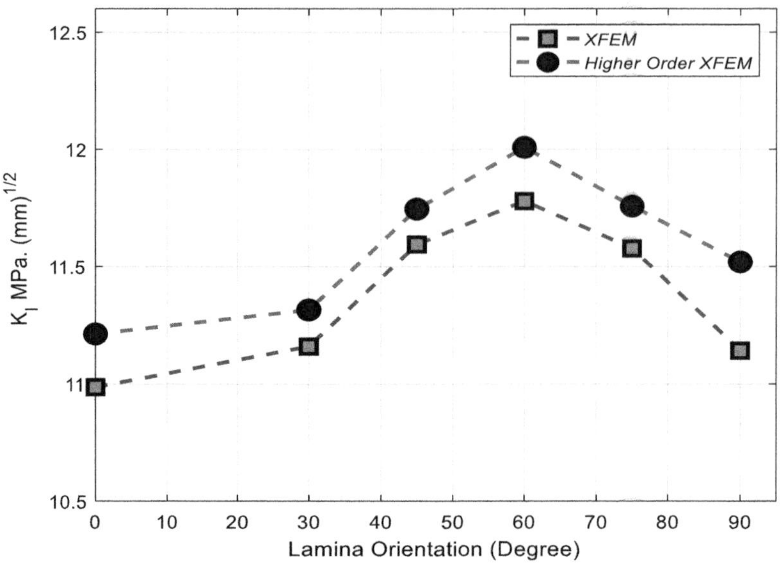

FIGURE 2.17 Comparative plot at different lamina orientations for SIF (K_1). *SIF*, Stress intensity factor.

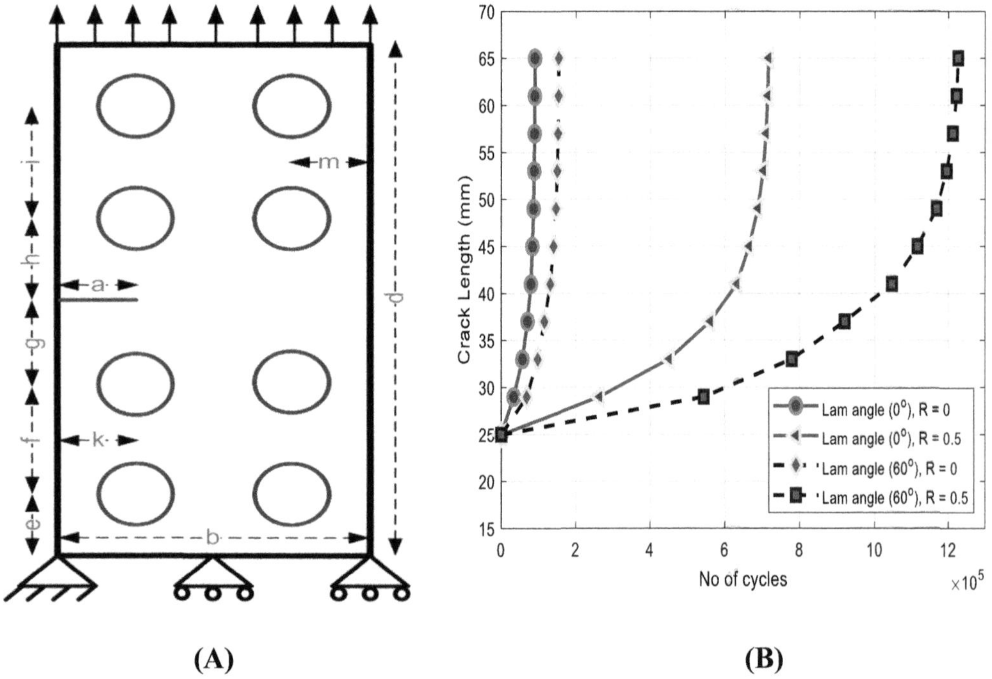

FIGURE 2.18 Multiple-hole orthotropic plate with **(A)** boundary conditions and **(B)** fatigue life curve at different laminar orientation angles and stress ratios using higher order XFEM approach. *XFEM*, Extended finite element method.

XFEM approach. The material property of the orthotropic plate is taken from Table 2.3. Paris law is used to perform the fatigue life estimation with, Paris constant $C=2\times10^{-8}$, and $m=2.997$ [116]. Maximum circumferential tensile stress criteria [117] are used to estimate the fatigue crack growth direction. Each step of the simulation is considered with a 4 mm increment. A total 3200 number of nodes are used to discretize the domain. Fatigue life curve is compared for edge crack with multiple holes on an orthotropic plate at different stress ratios and laminar orientation angles, as shown in Fig. 2.18B. From this figure, we can observe that fatigue life is maximum for 60 degrees laminar angle with 0.5 stress ratio, and fatigue life is minimum for 0 degree laminar angle with 0 stress ratio.

2.3.6 Edge crack with multiple holes/cracks orthotropic plate under fatigue loading

In this numerical example, we increase discontinuity in domain compared to previous case and analyze the fatigue life of plate using the higher order XFEM approach. Edge crack plate dimensions are given as $a=25$ mm, $b=100$ mm, $d=200$ mm, $k=25$ mm, $m=25$ mm, $e=25$ mm, $f=40$ mm, $g=35$ mm, $h=35$ mm, $i=40$ mm as shown in Fig. 2.19A. Multiple holes of 10 mm radius and multiple cracks of 10 mm are spread over the domain of

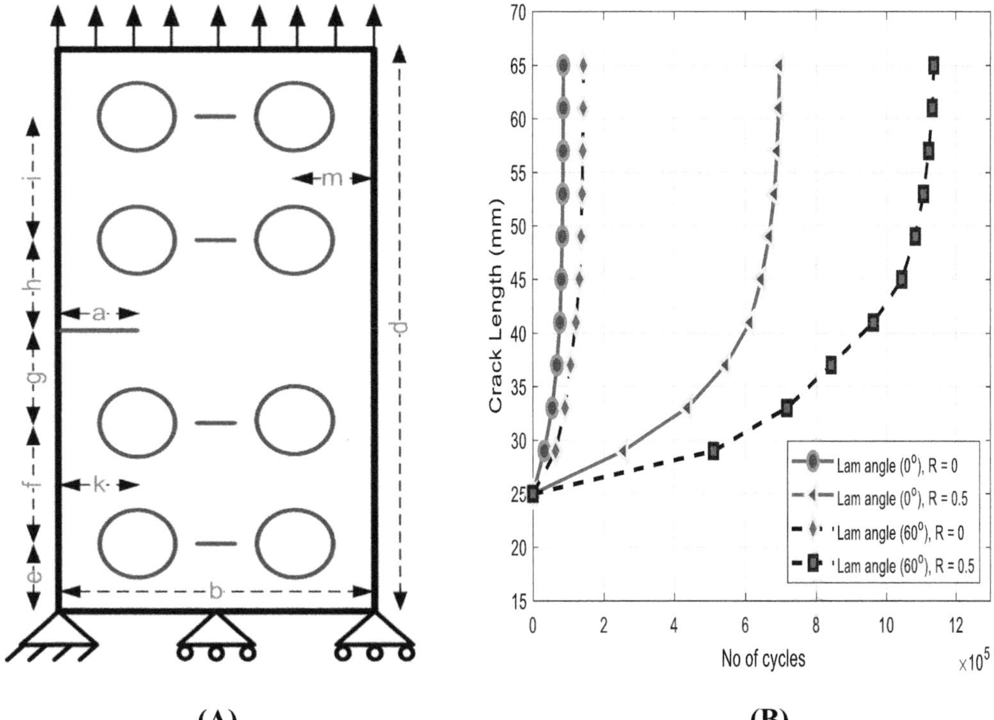

FIGURE 2.19 Multiple holes and cracks orthotropic plate with (A) boundary conditions, (B) fatigue life curve at different laminar orientation angles and stress ratios using higher order XFEM approach. *XFEM*, Extended finite element method.

orthotropic plate. All discontinuities in domain are represented by red. Material property of orthotropic lamina is taken from Table 2.3. All discontinuities are represented by red can be seen in Fig. 2.19A. Mechanical fatigue loading of maximum value 1 MPa with stress ratios 0 and 0.5 are applied to the upper edge of the plate, and lower edge is restricted to move in y direction. Simulation is performed with a total of 3200 nodes. Paris law is used to perform the fatigue life estimation with, Paris constant $C = 2 \times 10^{-8}$, and $m = 2.997$. A crack increment of 4 mm is used in each step of simulation. Fatigue life curve is also compared for orthotropic plate at different stress ratios, and different laminar angles of orientation are presented in Fig. 2.19B. From this figure, we also obtain the maximum fatigue life for 60 degrees laminar orientation angle with 0.5 stress ratio and minimum fatigue life is obtained for 0 degree laminar orientation angle with 0 stress ratio.

2.4 Conclusion

This chapter presents and applies an accurate computational method to simulate geometrical discontinuities in different materials. The accuracy of higher order tip enrichment functions has been addressed in several numerical examples. From the numerical result, we observe the following conclusions which are illustrated below.

- The accuracy of the solution in terms of SIF is improved by higher order crack tip functions.
- The stress contours show that the suggested enrichment approach successfully captures the domain discontinuities.
- A higher order XFEM technique is the greatest option for precisely modeling geometrical discontinuities, as can be seen.
- As we increase the stress ratio (0–0.5) and laminar angle (0–60 degrees), the fatigue life of the orthotropic plate is increasing.
- As we increase the discontinuity in the domain of the orthotropic plate, the fatigue life decreases.
- To explore fatigue crack development issues in the composite laminate material under a thermo-mechanical loading environment, the proposed computational technique (higher order XFEM) may be extended.

References

[1] Sih GC, Paris PC, Irwin GR. On crack in rectilinearly anisotropic bodies. International Journal of Fracture Mechanics 1965;1:189–203. Available from: https://doi.org/10.1007/BF00186854.

[2] Viola A, Piva A, Radi E. Crack propagation in an orthotropic medium under general loading. Engineering Fracture Mechanics 1989;34:1155–74. Available from: https://doi.org/10.1016/0013-7944(89)90277-4.

[3] Bowie OL, Freese CE. Central crack in-plane orthotropic rectangular sheet. International Journal of Fracture Mechanics 1972;8:49–58. Available from: https://doi.org/10.1007/BF00185197.

[4] Barnett DM, Asaro RJ. The fracture mechanics of slit-like cracks in anisotropic elastic media. Journal of the Mechanics and Physics of Solids 1972;20:353–66. Available from: https://doi.org/10.1016/0022-5096(72)90013-0.

[5] Bogy. The plane solution for anisotropic elastic wedges under normal and shear traction. Journal of Applied Mechanics 1972;39:1103–9. Available from: https://doi.org/10.1115/1.3422837.

[6] Kuo MC, Bogy DB. Plane solutions for the displacement and traction displacement problem for anisotropic elastic wedges. Journal of Applied Mechanics 1974;41:197–203. Available from: https://doi.org/10.1115/1.3423223.

[7] Nobile L, Carloni C. Fracture analysis for orthotropic cracked plates. Composite Structure 2005;68:285–93. Available from: https://doi.org/10.1016/j.compstruct.2004.03.020.

[8] Carloni C, Nobile L. Crack initiation behaviour of orthotropic solids as predicted by the strain energy density theory. Theoretical Applied Fracture Mechanics 2002;38:109–19. Available from: https://doi.org/10.1016/S0167-8442(02)00089-7.

[9] Carloni C, Piva A, Viola E. An alternative complex variable formulation for an inclined crack in an orthotropic medium. Engineering Fracture Mechanics 2003;70:2033–58. Available from: https://doi.org/10.1016/S0013-7944(02)00258-8.

[10] Belytschko T, Black T. Elastic crack growth in finite elements with minimal remeshing. International Journal of Numerical Method in Engineering 1999;45:601–20. Available from: https://doi.org/10.1002/(SICI)1097-0207(19990620)45:5 < 601::AID-NME598 > 3.0.CO;2-S.

[11] Moes N, Dolbow J, Belytschko T. A finite element method for crack growth without remeshing. International Journal for Numerical Method in Engineering 1999;46:131–50. Available from: https://doi.org/10.1002/(SICI)1097-0207(19990910)46:1 < 131::AID-NME726 > 3.0.CO;2-J.

[12] Daux C, Moes N, Dolbow J. Arbitrary branched and intersecting cracks with the extended finite element method. International Journal for Numerical Method in Engineering 2000;48:1741–60. Available from: https://doi.org/10.1002/1097-0207(20000830)48:12 < 1741::AID-NME956 > 3.0.CO;2-L.

[13] Sukumar N, Moes N, Moran B, Belytschko T. Extended finite element method for three-dimensional crack modelling. International Journal for Numerical Method in Engineering 2000;48:1549–70. Available from: https://doi.org/10.1002/1097-0207(20000820)48:11 < 1549::AID-NME955 > 3.0.CO;2-A.

[14] Stolarska M, Chopp DL, Moes N, Belytschko T. Modelling of crack growth by level set in the extended finite element method. International Journal for Numerical Method in Engineering 2001;51:943–60. Available from: https://doi.org/10.1002/nme.201.

[15] Moes N, Gravouil A, Belytschko T. Non-planer 3D crack growth by the extended finite element and level set, Part I: Mechanical model. International Journal for Numerical Method in Engineering 2002;53:2549–68. Available from: https://doi.org/10.1002/nme.429.

[16] Ayhan AO, Nied HF. Stress intensity factors for three-dimensional surface cracks using enriched finite elements. International Journal for Numerical Method in Engineering 2002;54:899–921. Available from: https://doi.org/10.1002/nme.459.

[17] Sukumar N, Prevost JH. Modeling quasi-static crack growth with the extended finite element method. Part I: computer implementation. International Journal of Solid and Structures 2003;40:7513–37. Available from: https://doi.org/10.1016/j.ijsolstr.2003.08.002.

[18] Huang R, Sukumar N, Prevost JH. Modeling quasi-static crack growth with the extended finite element method Part II: numerical applications. International Journal of Solid and Structures 2003;40:7539–52. Available from: https://doi.org/10.1016/j.ijsolstr.2003.08.001.

[19] Lee SH, Song JH, Yoon YC. Combined extended and superimposed finite element method for cracks. International Journal for Numerical Method in Engineering 2004;59:1119–36. Available from: https://doi.org/10.1002/nme.908.

[20] Budyn E, Zi G, Moes N, Belytschko T. A method for multiple crack growth in brittle materials without remeshing. International Journal for Numerical Method in Engineering 2004;61:1741–70. Available from: https://doi.org/10.1002/nme.1130.

[21] Zi G, Song JH, Budyn E, Lee SH, Belytschko T. A method for growing multiple cracks without remeshing and its application to fatigue crack growth. Modelling and Simulation in Materials Science and Engineering 2004;12:901–15. Available from: https://doi.org/10.1088/0965-0393/12/5/009.

[22] Legrain G, Moes N, Verron E. Stress analysis around crack crack tips in finite strain problems using the extended finite element method. International Journal for Numerical Method in Engineering 2005;63:290–314. Available from: https://doi.org/10.1002/nme.1291.

[23] Moes N, Bechet E, Tourbier M. Imposing Dirichlet boundary conditions in the extended finite element method. International Journal for Numerical Method in Engineering 2006;67:1641–69. Available from: https://doi.org/10.1002/nme.1675.

[24] Asadpoure A, Mohammadi S, Vafai A. Crack analysis in orthotropic media using the extended finite element method. Thin Walled Structures 2006;44:1031–8. Available from: https://doi.org/10.1016/j.tws.2006.07.007.
[25] Asadpoure A, Mohammadi S. Developing new enrichment functions for crack simulation in orthotropic media by the extended finite element method. International Journal for Numerical Method in Engineering 2007;69:2150–72. Available from: https://doi.org/10.1002/nme.1839.
[26] Loehnert S, Belytschko T. Crack shielding and amplification due to multiple microcracks interacting with a macrocrack. International Journal of Fracture 2007;145:1–8. Available from: https://doi.org/10.1007/s10704-007-9094-1.
[27] Sukumar N, Chopp DL, Bechet EB, Moes N. Three-dimensional non-planar crack growth by a coupled extended finite element and fast marching method. International Journal for Numerical Method in Engineering 2008;76:727–48. Available from: https://doi.org/10.1002/nme.2344.
[28] Tabarraei A, Sukumar N. Extended finite element method on polygonal and quadtree meshes. Computer Methods in Applied Mechanics and Engineering 2008;197:425–38. Available from: https://doi.org/10.1016/j.cma.2007.08.013.
[29] Chessa J, Wang H, Belytschko T. On the construction of blending elements for local partition of unity enriched finite elements. International Journal for Numerical Method in Engineering 2003;57:1015–38. Available from: https://doi.org/10.1002/nme.777.
[30] Legay A, Wang HW, Belytschko T. Strong and week arbitrary discontinuities in spectral finite elements. International Journal for Numerical Method in Engineering 2005;64:991–1008. Available from: https://doi.org/10.1002/nme.1388.
[31] Fries TP, Belytschko T. The intrinsic XFEM: a method for arbitrary discontinuities without additional unknowns. International Journal for Numerical Method in Engineering 2006;68:1358–85. Available from: https://doi.org/10.1002/nme.1761.
[32] Fries TP. A corrected XFEM approximation without problems in blending elements. International Journal for Numerical Method in Engineering 2008;75:503–32. Available from: https://doi.org/10.1002/nme.2259.
[33] Gracie R, Wang H, Belytschko T. Blending in the extended finite element method by discontinuous Galerkin and assumed strain method. International Journal for Numerical Method in Engineering 2008;74:1645–69. Available from: https://doi.org/10.1002/nme.2217.
[34] Benvenuti E, Tralli A, Ventura G. A regularized XFEM model for the transition from continuous to discontinuous displacements. International Journal for Numerical Method in Engineering 2008;74:911–44. Available from: https://doi.org/10.1002/nme.2196.
[35] Ventura G, Gracie R, Belytschko T. Fast integration and weight function blending in the extended finite element method. International Journal for Numerical Method in Engineering 2009;77:1–29. Available from: https://doi.org/10.1002/nme.2387.
[36] Tarancon JE, Vercher A, Giner E, Fuenmayor FJ. Enhanced blending elements for XFEM applied to linear elastic fracture mechanics. International Journal for Numerical Method in Engineering 2009;77:126–48. Available from: https://doi.org/10.1002/nme.2402.
[37] Shibanuma K, Utsunomiya T. Reformulation of XFEM based on PUFEM for solving problem caused by blending elements. Finite Element in Analysis and Design 2009;45:806–16. Available from: https://doi.org/10.1016/j.finel.2009.06.007.
[38] Loehnert S, Mueller-Hoeppe DS, Wriggers P. 3D corrected XFEM approach and extension to finite deformation theory. International Journal for Numerical Method in Engineering 2011;86:431–52. Available from: https://doi.org/10.1002/nme.3045.
[39] Menk A, Bordas SPA. A robust preconditioning technique for the extended finite element method. International Journal for Numerical Method in Engineering 2011;85:1609–32. Available from: https://doi.org/10.1002/nme.3032.
[40] Chen L, Rabczuk T, Bordas SPA, Liu GR, Zeng KY, Kerfriden P. Extended finite element method with edge-based strain smoothing (ESm-XFEM) for linear elastic crack growth. Computer Methods in Applied Mechanics and Engineering 2012;209:250–65. Available from: https://doi.org/10.1016/j.cma.2011.08.013.
[41] Rethore J, Gravouil A, Combescure A. An energy-conserving scheme for dynamic crack growth using the extended finite element method. International Journal for Numerical Method in Engineering 2005;63:631–59. Available from: https://doi.org/10.1002/nme.1283.
[42] Menouillard T, Rethore J, Combescure A, Bung H. Efficient explicit time stepping for the extended finite element method. International Journal for Numerical Method in Engineering 2006;68:911–39. Available from: https://doi.org/10.1002/nme.1718.

[43] Menouillard T, Rethore J, Moes N. Mass lumping strategies for XFEM explicit dynamics: application to crack propagation. International Journal for Numerical Method in Engineering 2008;74:447–74. Available from: https://doi.org/10.1002/nme.2180.
[44] Elguedj T, Gravouil A, Maigre H. An explicit dynamics extended finite element method. Part I: mass lumping for arbitrary enrichment functions. Computer Method in Applied Mechanics and Engineering 2009;198:2297–317. Available from: https://doi.org/10.1016/j.cma.2009.02.019.
[45] Fries TP, Zilian A. On time integration in the XFEM. International Journal for Numerical Methods in Engineering 2009;79:69–93. Available from: https://doi.org/10.1002/nme.2558.
[46] Menouillard T, Belytschko T. Dynamic fracture with meshfree enriched XFEM. Acta Mechanica 2010;213:53–69. Available from: https://doi.org/10.1007/s00707-009-0275-z.
[47] Menouillard T, Belytschko T. Smoothed nodal forces for improved dynamic crack propagation modeling in XFEM. International Journal for Numerical Methods in Engineering 2010;84:47–72. Available from: https://doi.org/10.1002/nme.2882.
[48] Menouillard T, Song JH, Duan Q, Belytschko T. Time dependent crack tip enrichment for dynamic crack propagation. International Journal of Fracture 2010;162:33–49. Available from: https://doi.org/10.1007/s10704-009-9405-9.
[49] Motamedi D, Mohammadi S. Dynamic analysis of fixed cracks in composites by the extended finite element method. Engineering Fracture Mechanics 2010;77:3373–93. Available from: https://doi.org/10.1016/j.engfracmech.2010.08.011.
[50] Motamedi D, Mohammadi S. Dynamic crack propagation analysis of orthotropic media by the extended finite element method. International Journal of Fracture 2010;161:21–39. Available from: https://doi.org/10.1007/s10704-009-9423-7.
[51] Esna Ashari S, Mohammadi S. Fracture analysis of FRP-reinforced beams by orthotropic XFEM. Journal of Composite Materials 2012;46:1367–89. Available from: https://doi.org/10.1177/0021998311418702.
[52] Liu ZL, Menouillard T, Belytschko T. An XFEM/spectral element method for dynamic crack propagation. International Journal of Fracture 2011;169:183–98. Available from: https://doi.org/10.1007/s10704-011-9593-y.
[53] Motamedi D, Mohammadi S. Fracture analysis of composites by time independent moving-crack orthotropic XFEM. International Journal of Mechanical Sciences 2012;54:20–37. Available from: https://doi.org/10.1016/j.ijmecsci.2011.09.004.
[54] Chahine E, Laborde P, Renard Y. A quasi-optimal convergence result for fracture mechanics with XFEM. Comptes Rendus Mathematique 2006;342:527–32. Available from: https://doi.org/10.1016/j.crma.2006.02.002.
[55] Rodenas JJ, Gonzalez-Estrada OA, Tarancon JE, Fuenmayor FJ. A recovery-type error estimator for the extended finite element method based on singular + smooth stress field splitting. International Journal for Numerical Methods in Engineering 2008;76:545–71. Available from: https://doi.org/10.1002/nme.2313.
[56] Panetier J, Ladeveze P, Chamoin L. Strict and effective bounds in goal-oriented error estimation applied to fracture mechanics problems solved with XFEM. International Journal for Numerical Methods in Engineering 2010;81:671–700. Available from: https://doi.org/10.1002/nme.2705.
[57] Rodenas JJ, Gonzalez-Estrada OA, Dez P, Fuenmayor P. Accurate recovery-based upper error bounds for the extended finite element framework. Computer Methods in Applied Mechanics and Engineering 2010;199:2607–21. Available from: https://doi.org/10.1016/j.cma.2010.04.010.
[58] Shen Y, Lew A. An optimally convergent discontinuous Galerkin-based extended finite element method for fracture mechanics. International Journal for Numerical Methods in Engineering 2010;82:716–55. Available from: https://doi.org/10.1002/nme.2781.
[59] Shen Y, Lew A. Stability and convergence proofs for a discontinuous-Galerkin-based extended finite element method for fracture mechanics. Computer Methods in Applied Mechanics and Engineering 2010;199:2360–82. Available from: https://doi.org/10.1016/j.cma.2010.03.008.
[60] Nicaise S, Renard Y, Chahine E. Optimal convergence analysis for the extended finite element method. International Journal for Numerical Methods in Engineering 2011;86:528–48. Available from: https://doi.org/10.1002/nme.3092.
[61] Prange C, Loehnert S, Wriggers P. Error estimation for crack simulations using the XFEM. International Journal for Numerical Methods in Engineering 2012;91:1459–74. Available from: https://doi.org/10.1002/nme.4331.
[62] Byfut A, Schroder A. Hp-adaptive extended finite element method. International Journal for Numerical Methods in Engineering 2012;89:1392–418. Available from: https://doi.org/10.1002/nme.3293.

[63] Gonzalez-Albuixech VF, Giner E, Tarancon JE, Fuenmayor FJ, Gravouil A. Convergence of domain integrals for stress intensity factor extraction in 2-D curved cracks problems with the extended finite element method. International Journal for Numerical Methods in Engineering 2013;94:740–57. Available from: https://doi.org/10.1002/nme.4478.
[64] Rodenas JJ, Gonzalez-Estrada OA, Fuenmayor FJ, Chinesta F. Enhanced error estimator based on a nearly equilibrated moving least squares recovery technique for FEM and XFEM. Computational Mechanics 2013;52:321–44. Available from: https://doi.org/10.1007/s00466-012-0814-7.
[65] Ruter M, Gerasimov T, Stein E. Goal-oriented explicit residual-type error estimates in XFEM. Computational Mechanics 2013;52:361–76. Available from: https://doi.org/10.1007/s00466-012-0816-5.
[66] Park K, Pereira JP, Duarte CA, Paulino GH. Integration of singular enrichment functions in the generalized/extended finite element method for three-dimensional problems. International Journal for Numerical Methods in Engineering 2009;78:1220–57. Available from: https://doi.org/10.1002/nme.2530.
[67] Mousavi SE, Sukumar N. Generalized Gaussian quadrature rules for discontinuities and crack singularities in the extended finite element method. Computer Methods in Applied Mechanics and Engineering 2010;199:3237–49. Available from: https://doi.org/10.1016/j.cma.2010.06.031.
[68] Bordas SPA, Rabczuk T, Hung NX, Nguyen VP, Natarajan S, Bog T, et al. Strain smoothing in FEM and XFEM. Computers and Structures 2010;88:1419–43. Available from: https://doi.org/10.1016/j.compstruc.2008.07.006.
[69] Bordas SPA, Natarajan S, Kerfriden P, Augarde CE, Mahapatra DR, Rabczuk T, et al. On the performance of strain smoothing for quadratic and enriched finite element approximations (XFEM/GFEM/PUFEM). International Journal for Numerical Methods in Engineering 2011;86:637–66. Available from: https://doi.org/10.1002/nme.3156.
[70] Legrain G, Allais R, Cartraud P. On the use of the extended finite element method with quadtree/octree meshes. International Journal for Numerical Methods in Engineering 2011;86:717–43. Available from: https://doi.org/10.1002/nme.3070.
[71] Baydoun M, Fries TP. Crack propagation criteria in three dimensions using the XFEM and an explicit-implicit crack description. International Journal of Fracture 2012;178:51–70. Available from: https://doi.org/10.1007/s10704-012-9762-7.
[72] Minnebo H. Three-dimensional integration strategies of singular functions introduced by the XFEM in the LEFM. International Journal for Numerical Methods in Engineering 2012;92:1117–38. Available from: https://doi.org/10.1002/nme.4378.
[73] Benvenuti E, Ventura G, Ponara N. Finite element quadrature of regularized discontinuous and singular level set functions in 3D problems. Algorithms 2012;5:529–44. Available from: https://doi.org/10.3390/a5040529.
[74] Gonzalez-Albuixech VF, Giner E, Tarancon JE, Fuenmayor FJ, Gravouil A. Domain integral formulation for 3-D curved and non-planar cracks with the extended finite element method. Computer Methods in Applied Mechanics and Engineering 2013;264:129–44. Available from: https://doi.org/10.1016/j.cma.2013.05.016.
[75] Pathak H, Singh A, Singh IV, Yadav SK. A simple and efficient XFEM approach for 3D cracks simulations. International Journal of Fracture 2013;181:189–208. Available from: https://doi.org/10.1007/s10704-013-9835-2.
[76] Rabinovich D, Givoli D, Vigdergauz S. XFEM-based crack detection scheme using a genetic algorithm. International Journal for Numerical Methods in Engineering 2007;71:1051–80. Available from: https://doi.org/10.1002/nme.1975.
[77] Rabinovich D, Givoli D, Vigdergauz S. Crack identification by "arrival time" using XFEM and a genetic algorithm. International Journal for Numerical Methods in Engineering 2009;77:337–59. Available from: https://doi.org/10.1002/nme.2416.
[78] Nistor I, Pantale O, Caperaa S. Numerical implementation of the extended finite element method for dynamic crack analysis. Advances in Engineering Software 2008;39:573–87. Available from: https://doi.org/10.1016/j.advengsoft.2007.06.003.
[79] Holl M, Rogge T, Loehnert S, et al. 3D multiscale crack propagation using the XFEM applied to a gas turbine blade. Computational Mechanics 2014;53:173–88. Available from: https://doi.org/10.1007/s00466-013-0900-5.
[80] Pathak H, Singh A, Singh IV. Numerical simulation of bi-material interfacial cracks using EFGM and XFEM. International Journal of Mechanics and Materials in Design 2012;8:9–36. Available from: https://doi.org/10.1007/s10999-011-9173-3.

[81] Singh IV, Mishra BK, Bhattacharya S, Patil RU. The numerical simulation of fatigue crack growth using extended finite element method. International Journal of Fatigue 2012;36:109–19. Available from: https://doi.org/10.1016/j.ijfatigue.2011.08.010.
[82] Jameel A, Harmain GA. Modeling and numerical simulation of fatigue crack growth in cracked specimens containing material discontinuities. Strength of Materials 2016;48(2):294–307. Available from: https://doi.org/10.1007/s11223-016-9765-0.
[83] Jameel A, Harmain GA. Extended iso-geometric analysis for modeling three dimensional cracks. Mechanics of Advanced Materials and Structures 2019;26:915–23. Available from: https://doi.org/10.1080/15376494.2018.1430275.
[84] Kanth SA, Harmain GA, Jameel A. Modeling of nonlinear crack growth in steel and aluminum alloys by the element free Galerkin method. Materials Today: Proceedings 2018;5(9):18805–14. Available from: https://doi.org/10.1016/j.matpr.2018.06.227.
[85] Jameel A, Harmain GA. Fatigue crack growth analysis of cracked specimens by the coupled finite element-element free Galerkin method. Mechanics of Advanced Materials and Structures 2019;26:1343–56. Available from: https://doi.org/10.1080/15376494.2018.1432800.
[86] Dwivedi K, Arora G, Pathak H. Fatigue crack growth in CNT-reinforced polymer composite. Journal of Micromechanics and Molecular Physics 2022;7:173–4. Available from: https://doi.org/10.1142/S242491302241003X.
[87] Suman S, Dwivedi K, Anand S, Pathak H. XFEM-ANN approach to predict the fatigue performance of a composite patch repaired aluminum panel. Composite Part C: Open Access 2022;9100326. Available from: https://doi.org/10.1016/j.jcomc.2022.100326.
[88] Raza A, Pathak H, Talha M. Vibration characteristics of cracked functionally graded structures using XFEM. Journal of Physics: Conference Series 2019;1240:012028. Available from: https://doi.org/10.1088/1742-6596/1240/1/012028.
[89] Raza A, Pathak H, Talha M. Stochastic extended finite element implementation for natural frequency of cracked functionally gradient and bi-material structures. International Journal of Structural Stability and Dynamics 2021;21:2150044. Available from: https://doi.org/10.1142/S0219455421500449.
[90] Raza A, Pathak H, Talha M. Computational investigation of porosity effect on free vibration of cracked functionally graded plates using XFEM. Materialstoday Proceedings 2022;61:96–102. Available from: https://doi.org/10.1016/j.matpr.2022.03.654.
[91] Raza A, Pathak H, Talha M. Influence of microstructural defects on free flexural vibration of cracked functionally graded plates in thermal medium using XFEM. Mechanics Based Design of Structures and Machines 2022;1–24. Available from: https://doi.org/10.1080/15397734.2022.2066544.
[92] Raza A, Talha M, Pathak H. Influence of material uncertainty on vibration characteristics of higher order cracked functionally gradient plates using XFEM. International Journal of Applied Mechanics 2021;132150062. Available from: https://doi.org/10.1142/S1758825121500629.
[93] Kumar S, Bhardwaj G. A new enrichment scheme in XFEM to model crack growth behaviour in ductile materials. Theoretical and Applied Fracture Mechanics 2018;96:296–307. Available from: https://doi.org/10.1016/j.tafmec.2018.05.008.
[94] Deng H, Yan B, Okabe T. Fatigue crack propagation simulation method using XFEM with variable node element. Engineering Fracture Mechanics 2022;269108533. Available from: https://doi.org/10.1016/j.engfracmech.2022.108533.
[95] Liu X, Xiao Q, Karihaloo BL. XFEM for direct evaluation of mixed mode SIFs in homogeneous and bi-materials. International Journal of Numerical Methods in Engineering 2004;59:1103–18. Available from: https://doi.org/10.1002/nme.906.
[96] Duarte C, Oden J. An h-p adaptive method using clouds. Computer method in Applied Mechanics and Engineering 1996;139(14):263–88. Available from: https://doi.org/10.1016/S0045-7825(96)01085-7.
[97] Zamani A, Gracie R, Reza Eslami M. Cohesive and non-cohesive fracture by higher order enrichment of XFEM. International Journal for numerical method in Engineering 2012;90(4):452–83. Available from: https://doi.org/10.1002/nme.3329.
[98] Rethore J, Roux S, Hild F. Hybrid analytical and extended finite element method (HAX-FEM): a new enrichment procedure for cracked solids. International Journal for numerical method in Engineering 2010;81(3):269–85. Available from: https://doi.org/10.1002/nme.2691.

[99] Zamani A, Gracie R, Eslami M. Higher order tip enrichment of extended finite element method in thermoelasticity. Computational Mechanics 2010;46(6):851–66. Available from: https://doi.org/10.1007/s00466-010-0520-2.

[100] Xiao QZ, Karihaloo BL, Liu XY. Direct determination of SIF and higher order terms of mixed mode cracks by a hybrid crack element. International Journal of Fracture 2004;125:207–25. Available from: https://doi.org/10.1023/B:FRAC.0000022229.54422.13.

[101] Cheng KW, Fries TP. Higher order XFEM for curved strong and week discontinuities. International Journal for Numerical Method in Engineering 2010;82:564–90. Available from: https://doi.org/10.1002/nme.2768.

[102] Saxby BA, Hazel AL. Improving the modified XFEM for optimal higher-order approximation. International Journal of Numerical Method in Engineering 2018;0:1–23. Available from: https://doi.org/10.1002/nme.6214.

[103] Mousavi SE, Grinspun E, Sukumar N. Higher-order extended finite elements with harmonic enrichment functions for complex crack problems. International Journal for Numerical Methods in Engineering 2010;0:1–29. Available from: https://doi.org/10.1002/nme.3098.

[104] Ghorashi SS, Mohammadi S, Sabbagh-Yazdi S-R. Orthotropic enriched element free Galerkin method for fracture analysis of composites. Engineering Fracture Mechanics 2011;78:1906–27. Available from: https://doi.org/10.1016/j.engfracmech.2011.03.011.

[105] Moran B, Shih CF. A general treatment of crack tip contour integrals. International Journal of Fracture 1987;35(4):295–310. Available from: https://doi.org/10.1007/BF00276359.

[106] Osher S, Sethian JA. Fronts propagating with curvature dependent speed: algorithms based on Hamilton-Jacobi formulations. Journal of Computational Physics 1988;79:12–49.

[107] Jameel A, Harmain GA. A coupled FE-IGA technique for modeling fatigue crack growth in engineering materials. Mechanics of Advanced Materials and Structures 2019;26:1764–75. Available from: https://doi.org/10.1080/15376494.2018.1446571.

[108] Jameel A, Harmain GA. Large deformation in Bi-material components by XIGA and coupled FE-IGA techniques. Mechanics of Advanced Materials and Structures 2022;29:850–72. Available from: https://doi.org/10.1080/15376494.2020.1799120.

[109] Kanth SA, Jameel Azher, Harmain GA. Investigation of fatigue crack growth in engineering components containing different types of material irregularities by XFEM. Mechanics of advanced materials and structures, Vol. 0. Taylor and Francis; 2021. p. 1–13.

[110] A. Jameel, G.A. Harmain, Effect of material irregularities on fatigue crack growth by enriched techniques, International Journal for Computational Methods in Engineering Science and Mechanics (Taylor and Francis), 21, pp. 109–133, 2020.

[111] Jameel A, Harmain GA. Fatigue crack growth in presence of material discontinuities by EFGM. International Journal of Fatigue (Elsevier) 2015;81:105–16.

[112] Jameel Azher, Harmain GA. Fatigue crack growth analysis of cracked specimens by the coupled finite element-element free Galerkin method. Mechanics of advanced materials and structures, 26. Taylor and Francis; 2019. p. 1343–56.

[113] Kanth SA, Lone AS, Harmain GA, Jameel Azher. Modelling of embedded and edge cracks in steel alloys by XFEM. Materials Today: Proceedings, 26. Elsevier; 2020. p. 814–8.

[114] Kanth SA, Lone AS, Harmain GA, Jameel Azher. Elasto plastic crack growth by XFEM: a review. Materials Today: Proceedings, 18. Elsevier; 2019. p. 3472–81.

[115] Nassar AA. Evaluation of critical stress intensity factor (K_{ic}) for plates using new crack extension technique. Journal of Engineering Technology 2012;31:730–40. Available from: https://uotechnology.edu.iq/english/tec_magaz/2013/volum312013/No.04.A.2013/Text%20(11).pdf.

[116] Nicholas T., Chen S.E., Boyajian D. Mode I fatigue of the carbon fiber-reinforced plastic-concrete interface bond. Society for Experimental Mechanics; 2010. Available from: https://doi.org/10.1111/j.1747-1567.2011.00739.x.

[117] Erdogan F, Sih GC. On the crack extension in plates under plane loading and transverse shear. Journal of Basic Engineering 1963;85:519–25. Available from: https://doi.org/10.1115/1.3656897.

CHAPTER

3

Extended finite element method for three-dimensional cracks

Azher Jameel[1], *Qazi Junaid Ashraf*[2], *Mumtaz Ahmad*[1] *and Majid Hameed Koul*[3]

[1]Department of Mechanical Engineering, National Institute of Technology Srinagar, Hazratbal, Srinagar, Jammu and Kashmir, India [2]Department of Mechanical Engineering, University of Kashmir, Hazratbal, Srinagar, Jammu and Kashmir, India [3]Department of Mechanical Engineering, Islamic University of Science and Technology, Awantipora, Jammu and Kashmir, India

3.1 Introduction

The presence of different defects, such as cracks, cannot be eliminated in structural components. Different types of cracks may develop during operational or manufacturing stages. The initiation and propagation of cracks cannot be eliminated completely but can be controlled to increase the residual life of cracked components, which invokes the investigation of cracks in three-dimensional specimens to enhance reliability and the residual life of engineering structures. There are various fracture mechanics parameters that give the complete description of the behavior of cracks in engineering components, such as the stress intensity factors (SIFs), J-integral, and crack tip opening displacements (CTODs). SIFs serve as an important parameter in investigating the behavior of static and propagating cracks. J-integral plays a dominant role in describing the behavior of cracks in elasto-plastic fracture mechanics (EPFMs). Different numerical techniques are available for determining the behavior of cracks in engineering specimens. The conventional finite element method (FEM) has always proved to be a potential computational tool for modeling different engineering problems. Some of the important computational techniques that have been used in the past include the boundary element technique [1–3], extended FEM (XFEM) [4–6], mesh-free techniques [7–11], methods based on peridynamics [12–14], phase field models [15–17], and conventional FEM [18–21]. Developing a conformal mesh in FEM for irregularly shaped discontinuities is computationally more demanding and costly, which limits the application of this approach for modeling different types of material irregularities. The

Enriched Numerical Techniques
DOI: https://doi.org/10.1016/B978-0-443-15362-4.00014-0

problem gets more complicated if the discontinuity evolves with time, as is the case with propagating cracks and large deformation problems. FEM suffers extreme mesh distortion in modeling large deformation problems and hence demands remeshing of the domain throughout the simulation. Remeshing is once again computationally more costly, especially in the case of three-dimensional problems where remeshing, conformal meshing, and mesh refinements are very cumbersome.

The limitations of conformal meshing, element refinements, and various other grid-related issues do not arise in XFEM, which is an extended version of classical FEM. All types of cracks and other material irregularities are simulated independent of the nodal distribution selected for analysis [22–24]. In XFEM, the classical variable approximations are modified by adding suitable enrichment functions to model various material irregularities occurring in the domain. Due to the elimination of all mesh-related drawbacks, XFEM provides a computationally less expensive framework as compared to conventional FEM [25–34]. The enrichment functions are included in the formulations with the help of the partition of unity approaches [35–39]. The level set method serves as a strong tool for keeping track of moving discontinuities in the domain. The level set method represents the discontinuity by assigning a zero value to the interface. The level set function changes its sign from positive to negative across the interface. The work reported in the chapter used XFEM to simulate the behavior of discontinuities created by cracks in three-dimensional cracked components subjected to monotonic tensile loads. The interaction integral has been for the evaluation of the mixed mode SIFs in three-dimensional cracked components subjected to mixed mode loadings. The mathematical models on XFEM and other related algorithms have been coded in MATLAB to perform different numerical tests on three-dimensional cracked specimens. The analytical solutions available in literature serve as the benchmark or reference solutions for validating the MATLAB codes developed in the current work. Several numerical problems on cracked specimens are simulated in the current work to demonstrate the capabilities of the XFEM in investigating the behavior of different types of cracks in engineering specimens. Different types of plane edge cracks, horizontal and inclined penny cracks, and lens-shaped cracks in three-dimensional engineering components have been considered for analysis in the current work.

3.2 Extended finite element method for cracks

As discussed earlier, XFEM has evolved as a potential numerical tool for investigating different types of cracks in structural materials, as it models material irregularities irrespective of the nodal distribution selected for investigation. This section provides the detailed mathematical foundations for the modeling of cracks in 3D engineering components by the numerical framework based on XFEM. In order to derive the mathematical models for modeling the behavior of cracks in engineering components, consider an engineering domain "Ω", which is surrounded by the outer boundary "Γ", as can be seen in Fig. 3.1. The body forces are represented by }b," and the applied tractions are given by }t." The crack surfaces are traction-free and are shown as Γ_c, and the traction boundaries are

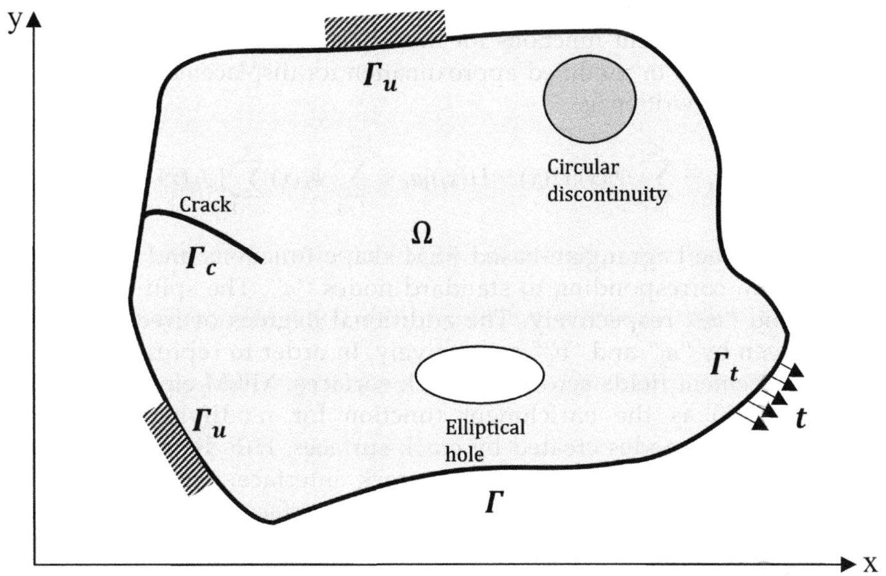

FIGURE 3.1 Different material irregularities present in structural components.

represented by Γ_t. The prescribed displacements are imposed at the displacement boundary given by Γ_u.

For the given engineering domain subjected to the applied loads and the imposed boundary conditions, the equilibrium equation can be written in standard form as

$$\nabla.\sigma + b = 0 \text{ in } \Omega \tag{3.1}$$

$$\sigma.n = t \text{ on } \Gamma_t \tag{3.2a}$$

$$\sigma.n = o \text{ on } \Gamma_c^+ \tag{3.2b}$$

$$\sigma.n = \mathrm{O} \text{ on } \Gamma_c^- \tag{3.2c}$$

$$u = U \text{ on } \Gamma_u \tag{3.2d}$$

The strains are related to the displacements as $\varepsilon = \boldsymbol{Bu}$, where "$\boldsymbol{B}$" represents the strain–displacement matrix, which is constructed from the derivatives of XFEM-based shape functions, and "$\boldsymbol{u}$" denotes the displacement field. In linear elastic materials, the stresses can be defined in terms of strains by Hooke's law, which can be put in standard form as $\boldsymbol{\sigma} = \boldsymbol{D}\varepsilon$, where $\boldsymbol{\sigma}$ represents the stresses induced and $\boldsymbol{D}$ is Hooke's tensor. The equilibrium equation defined above can be expressed in variational form as

$$\int_\Omega \delta\varepsilon^T \sigma d\Omega - \int_\Omega \delta \boldsymbol{u}^T \boldsymbol{b} d\Omega - \int_{\Gamma_t} \delta \boldsymbol{u}^T \boldsymbol{t} d\Gamma_t = 0 \tag{3.3}$$

As discussed earlier, XFEM modifies the classical FEM-based Lagrangain approximations with suitable enrichment functions for modeling material irregularities occurring in the domain. The enriched or modified approximation for displacement fields in XFEM for modeling cracks can be written as

$$\boldsymbol{u}^h(\boldsymbol{x}) = \sum_{i=1}^{n} \Psi_i(\boldsymbol{x})\boldsymbol{u}_i + \sum_{j=1}^{n_s} \Psi_j(\boldsymbol{x})[H(\boldsymbol{x}) - H(\boldsymbol{x}_j)]\boldsymbol{a}_j + \sum_{k=1}^{n_T} \Psi_k(\boldsymbol{x}) \sum_{\alpha=1}^{4} [\beta_\alpha(\boldsymbol{x}) - \beta_\alpha(\boldsymbol{x}_k)]\boldsymbol{b}_k^\alpha \tag{3.4}$$

where Ψ_i represent the Lagrangian-based FEM shape functions and "$\boldsymbol{u}_i$" are the nodal degrees of freedom corresponding to standard nodes "n". The split and tip nodes are given by "n_s" and "n_T" respectively. The additional degrees of freedom for split and tip nodes are given by "$\boldsymbol{a}_j$" and "$\boldsymbol{b}_k^\alpha$" respectively. In order to represent the discontinuities in the displacement fields across the crack surfaces, XFEM employs the Heaviside jump function (HJF) as the enrichment function for modifying the displacement approximations of split nodes created by crack surfaces. HJF defines the crack surface by maintaining the zero value across the crack interface, and the function value changes from $+1$ to -1 across the two sides of the interface. In polar coordinates (r, θ), the enrichment functions for modifying the displacement fields at the crack tip can be defined as

$$\left[\beta_\alpha(\boldsymbol{x}) = \sqrt{r}\cos\frac{\theta}{2}, \sqrt{r}\sin\frac{\theta}{2}, \sqrt{r}\cos\frac{\theta}{2}\sin\theta, \sqrt{r}\sin\frac{\theta}{2}\cos\theta\right] \tag{3.5}$$

The final numerical model for simulating three-dimensional cracks in engineering materials can be derived by substituting the XFEM-based variable approximations in the equilibrium equation. By substituting the modified XFEM approximation in the appropriate form of equilibrium equation, the final discrete system of algebraic equations can be obtained as $[\boldsymbol{K}^e]\{\boldsymbol{d}^e\} = \{\boldsymbol{f}^e\}$, where $\{\boldsymbol{d}^e\} = \{\boldsymbol{u}\boldsymbol{a}\boldsymbol{b}_1 \quad \boldsymbol{b}_2 \quad \boldsymbol{b}_3 \quad \boldsymbol{b}_4\}^T$. Different matrices in the final XFEM model can be defined as

$$[\boldsymbol{K}^e] = \begin{bmatrix} \boldsymbol{K}^{uu} & \boldsymbol{K}^{ua} & \boldsymbol{K}^{ub} \\ \boldsymbol{K}^{au} & \boldsymbol{K}^{aa} & \boldsymbol{K}^{ab} \\ \boldsymbol{K}^{bu} & \boldsymbol{K}^{ba} & \boldsymbol{K}^{bb} \end{bmatrix}; \{\boldsymbol{f}^e\} = \left\{\boldsymbol{f}^u \quad \boldsymbol{f}^a \quad \boldsymbol{f}^{b1} \quad \boldsymbol{f}^{b2} \quad \boldsymbol{f}^{b3} \quad \boldsymbol{f}^{b4}\right\}^T \tag{3.6}$$

$$\boldsymbol{K}^{rs} = \int_{\Omega^e} (\mathbf{B}^r)^T \boldsymbol{D}\boldsymbol{B}^s d\Omega \quad ;(r, s = u, a, b) \tag{3.7}$$

$$\boldsymbol{f}^u = \int_{\Omega^e} \Psi^T \boldsymbol{b} d\Omega + \int_{\Gamma^e} \Psi^T \boldsymbol{t} d\Gamma \tag{3.8}$$

$$\boldsymbol{f}^a = \int_{\Omega^e} \Psi^T (H(\boldsymbol{x}) - H(\boldsymbol{x}_j))\boldsymbol{b} d\Omega + \int_{\Gamma^e} \Psi^T (H(\boldsymbol{x}) - H(\boldsymbol{x}_j))\boldsymbol{t} d\Gamma \tag{3.9}$$

$$\boldsymbol{f}^{b\alpha} = \int_{\Omega^e} \Psi^T(\beta_\alpha(\boldsymbol{x}) - \beta_\alpha(\boldsymbol{x}_k))\boldsymbol{b} d\Omega + \int_{\Gamma^e} \Psi^T(\beta_\alpha(\boldsymbol{x}) - \beta_\alpha(\boldsymbol{x}_k))\boldsymbol{t} d\Gamma; (\alpha = 1, 2, 3, 4) \tag{3.10}$$

$$B^u = \begin{bmatrix} \Psi_{i,x} & 0 & 0 \\ 0 & \Psi_{i,y} & 0 \\ 0 & 0 & \Psi_{i,z} \\ \Psi_{i,y} & \Psi_{i,x} & 0 \\ 0 & \Psi_{i,z} & \Psi_{i,y} \\ \Psi_{i,z} & 0 & \Psi_{i,x} \end{bmatrix} \tag{3.11}$$

$$B^a = \begin{bmatrix} (\Psi_j(H(x)-H(x_j)))_{\ x} & 0 & 0 \\ 0 & (\Psi_j(H(x)-H(x_j)))_{\ y} & 0 \\ 0 & 0 & (\Psi_j(H(x)-H(x_j)))_{\ z} \\ (\Psi_j(H(x)-H(x_j)))_{\ y} & (\Psi_j(H(x)-H(x_j)))_{\ x} & 0 \\ 0 & (\Psi_j(H(x)-H(x_j)))_{\ z} & (\Psi_j(H(x)-H(x_j)))_{\ y} \\ (\Psi_j(H(x)-H(x_j)))_{\ z} & 0 & (\Psi_j(H(x)-H(x_j)))_{\ x} \end{bmatrix} \tag{3.12}$$

$$B^{b\alpha} = \begin{bmatrix} (\Psi_k\ (\beta_\alpha(x)-\beta_\alpha(x_k)))_{\ x} & 0 & 0 \\ 0 & (\Psi_k\ (\beta_\alpha(x)-\beta_\alpha(x_k)))_{\ y} & 0 \\ 0 & 0 & (\Psi_k\ (\beta_\alpha(x)-\beta_\alpha(x_k)))_{\ z} \\ (\Psi_k\ (\beta_\alpha(x)-\beta_\alpha(x_k)))_{\ y} & (\Psi_k\ (\beta_\alpha(x)-\beta_\alpha(x_k)))_{\ x} & 0 \\ 0 & (\Psi_k\ (\beta_\alpha(x)-\beta_\alpha(x_k)))_{\ z} & (\Psi_k\ (\beta_\alpha(x)-\beta_\alpha(x_k)))_{\ y} \\ (\Psi_k\ (\beta_\alpha(x)-\beta_\alpha(x_k)))_{\ z} & 0 & (\Psi_k\ (\beta_\alpha(x)-\beta_\alpha(x_k)))_{\ x} \end{bmatrix} \tag{3.13}$$

3.2.1 Choice of enrichment functions

The selection of enrichment functions is very crucial in XFEM. Enrichment functions are defined to enrich the variable approximations for the elements lying near crack interfaces. Enrichment functions should be expressed in such a way that they capture stress singularities developed ahead of the crack tip accurately and effectively. Unlike classical FEM, we encounter three types of elements in XFEM. There are two classical four-noded finite elements, the elements completely cut by the interfaces, known as split elements, and the elements containing the crack tips, known as tip elements. Since different stress gradients can be seen at split elements and tip elements, there is a need to define separate enrichment functions for modifying the variable approximations of these elements. In XFEM, the enrichment procedure is carried out by adding one additional node at the split nodes, and for two-dimensional problems, each additional node will add two additional degrees of freedom to the system. There is a discontinuity in the primary variable across the split elements, which can be easily modeled by the HJF that has been discussed earlier. If there is a point $\mathbf{x}$ in the domain and x_Γ represents the point close to $\mathbf{x}$

on the crack surface and **n** denotes the unit normal at the crack surface, the Heaviside function H(**x**) can be defined as

$$H(\mathbf{x}) = \begin{Bmatrix} +1 & \text{if}(\mathbf{x} - x_\Gamma)\mathbf{n} \geq 0 \\ -1 & \text{otherwise} \end{Bmatrix} \tag{3.14}$$

With the inclusion of HJF, the discontinuities produced at the split elements can be modeled easily. In the case of tip elements, it can be seen that the entire element is not cut by the crack surface. Hence, HJF cannot be used to modify the variable approximations for tip elements. The additional enrichment functions for modeling the stress singularities near the crack tip can be derived from the analytical solutions of the displacement fields available in LEFM, which can be written in standard form as

$$u(r,\theta) = \frac{1}{2\nu}\sqrt{\frac{r}{2\pi}}\left\{K_I\cos\frac{\theta}{2}(\kappa - \cos\theta) + K_{II}\sin\frac{\theta}{2}(\kappa + 2 + \cos\theta)\right\} \tag{3.15}$$

$$u(r,\theta) = \frac{1}{2\nu}\sqrt{\frac{r}{2\pi}}\left\{K_I\cos\frac{\theta}{2}(\kappa - \cos\theta) + K_{II}\sin\frac{\theta}{2}(\kappa + 2 + \cos\theta)\right\} \tag{3.16}$$

where K_I, K_{II} are the SIFs, r and θ are the crack tip coordinates expressed in polar form, and κrepresents the Kolsov constant, which is defined as

$$\kappa_{\text{planestress}} = \frac{3-\nu}{1+\nu} \tag{3.17}$$

$$\kappa_{\text{planestrain}} = 3 - 4\nu \tag{3.18}$$

From the expressions of the displacement fields, as given above, it can be shown that the analytical solution for the displacement field at the crack tip lies within the span of four functions, as given in the following equations:

$$\beta_\alpha(r,\theta) = \left[\sqrt{r}cos\frac{\theta}{2}, \sqrt{r}sin\frac{\theta}{2}, \sqrt{r}cos\frac{\theta}{2}sin\theta, \sqrt{r}sin\frac{\theta}{2}sin\theta\right] \tag{3.19}$$

Thus these four functions provide a strong background for carrying out the enrichment of the displacement fields for the tip elements in LEFM. Since four additional functions are required to be added to the classical variable approximations, four additional enriched nodes have to be placed on the standard nodes, giving rise to eight extra degrees of freedom at each tip node. The second enrichment function $\sqrt{r}sin\frac{\theta}{2}$ helps to impart the required discontinuity across the crack surface, and remaining three enrichment functions help to improve the quality of approximations in the domain.

For EPFM, the analytical or exact solutions for displacement fields are not known. One of the most commonly used solutions describing the singularities for power-hardening materials is the Hutchinson–Rice–Rosengren (HRR) solution [15]. The Fourier analysis of HRR fields helps to develop the suitable enrichment functions for the elasto-plastic crack growth problems. For the areas lying very close to crack tips, elastic strain rates are negligible in comparison to the plastic strain rates. The HRR solutions for the asymptotic fields near the crack tip can be expressed as

$$\sigma_{ij} = \sigma_0 \left(\frac{J}{\alpha \sigma_0 \varepsilon_0 I_n r} \right)^{\frac{1}{n+1}} \tilde{\sigma}_{ij}(\theta, n) \tag{3.20}$$

$$\varepsilon_{ij} = \alpha \varepsilon_0 \left(\frac{J}{\alpha \sigma_0 \varepsilon_0 I_n r} \right)^{\frac{n}{n+1}} \tilde{\varepsilon}_{ij}(\theta, n) \tag{3.21}$$

$$u_i = \alpha \varepsilon_0 r \left(\frac{J}{\alpha \sigma_0 \varepsilon_0 I_n r} \right)^{\frac{n}{n+1}} \tilde{u}_i(\theta, n) \tag{3.22}$$

where $I_n$is a dimensionless constant, $\alpha$and n are the Ramberg–Osgood coefficients, and $\tilde{\sigma}_{ij}$, $\tilde{\varepsilon}_{ij}$, and $\tilde{u}_i$ are dimensionless angular functions of θ and n. J represents the J-integral. σ_{ij} and ε_{ij} are the stresses and strains, and u_i denotes the displacement field. The Ramberg–Osgood model for representing the power-law hardening in the material for uni-axial stress–strain relationship can be written as

$$\frac{\varepsilon}{\varepsilon_0} = \frac{\sigma}{\sigma_0} + \alpha \left(\frac{\sigma}{\sigma_0} \right)^n \tag{3.23}$$

where σ_0 is the virgin yield stress, $\varepsilon_0 = \frac{\sigma_0}{E}$ denotes elastic strain at yield, α is the material parameter and n is the exponent of hardening. If hardening exponent n equals 1, then we have a linear elastic material. On the other hand, if $n = \infty$, then we have a rigid-perfectly plastic material. For obtaining the elasto-plastic fields around the crack tip, the Fourier decomposition of HRR displacement fields is to be carried out. Elguedj et al. have shown that the displacement field in cracked specimens subjected to pure mode-I and mode-II loadings can be represented by expressing the HRR functions in the form of the following basis functions:

$$r^{\frac{n}{n+1}} \left\{ \left(cos\frac{k\theta}{2}, sin\frac{k\theta}{2} \right); k\epsilon[1, 3, 5, 7] \right\} \tag{3.24}$$

Elguedj et al. proposed three different types of enrichment schemes for modeling crack tip singularities in elasto-plastic fracture analysis. The enrichment functions used in elasto-plastic analysis are shown as follows:

$$\beta_1(r, \theta) = r^{\frac{n}{n+1}} \left\{ \sin\frac{\theta}{2}, \cos\frac{\theta}{2}, \sin\frac{\theta}{2}\sin\theta, \cos\frac{\theta}{2}\sin\theta, \sin\frac{\theta}{2}\sin2\theta, \cos\frac{\theta}{2}\sin2\theta \right\} \tag{3.25}$$

$$\beta_2(r, \theta) = r^{\frac{n}{n+1}} \left\{ \sin\frac{\theta}{2}, \cos\frac{\theta}{2}, \sin\frac{\theta}{2}\sin\theta, \cos\frac{\theta}{2}\sin\theta, \sin\frac{\theta}{2}\sin3\theta, \cos\frac{\theta}{2}\sin3\theta \right\} \tag{3.26}$$

$$\beta_3(r, \theta) = r^{\frac{n}{n+1}} \left\{ \sin\frac{\theta}{2}, \cos\frac{\theta}{2}, \sin\frac{\theta}{2}\sin\theta, \cos\frac{\theta}{2}\sin\theta, \sin\frac{\theta}{2}\sin2\theta, \cos\frac{\theta}{2}\sin2\theta, \sin\frac{\theta}{2}\sin3\theta, \cos\frac{\theta}{2}\sin3\theta \right\} \tag{3.27}$$

It was observed that the addition of the enrichment terms $\sin\frac{\theta}{2}\sin2\theta$ and $\cos\frac{\theta}{2}\sin2\theta$ improves the quality of the approximation, but they result in an ill-conditioned stiffness matrix and, hence, create problems in inverting the stiffness matrix. However, the addition of the two extra enrichment functions $\sin\frac{\theta}{2}\sin3\theta$ and $\cos\frac{\theta}{2}\sin3\theta$ also increases the quality of the approximation, but they also result in a well-conditioned stiffness matrix, even though both additions result in the same amount of accuracy in the approximation. Consequently, they proposed the following two types of enrichment functions for modeling elasto-plastic crack growth problems:

$$\beta_1(r,\theta) = r^{n/(n+1)}\left\{\sin\frac{\theta}{2}, \cos\frac{\theta}{2}, \sin\frac{\theta}{2}\sin\theta, \cos\frac{\theta}{2}\sin\theta\right\} \tag{3.28}$$

$$\beta_2(r,\theta) = r^{n/(n+1)}\left\{\sin\frac{\theta}{2}, \cos\frac{\theta}{2}, \sin\frac{\theta}{2}\sin\theta, \cos\frac{\theta}{2}\sin\theta, \sin\frac{\theta}{2}\sin3\theta, \cos\frac{\theta}{2}\sin3\theta\right\} \tag{3.29}$$

3.3 Basic fracture mechanics principles

The conventional theory of elasticity cannot completely define the displacement and stress fields, if the bodies contain different flaws or cracks, because no consideration is given to stress singularities present at the crack tip. Therefore the application of fracture mechanics principles is very important to find out the mechanical behavior of the bodies in the presence of cracks. It has been seen that the extent of plastic zones and other nonlinear effects in most of the engineering components is extremely small in comparison to the size of the cracks present in structural components. Due to this reason, the application of linear theories is sufficient and adequate to govern the behavior of the cracked bodies. On the contrary, if the extent of plastic zone is significant, then linear principles cannot define govern the behavior of the crack bodies accurately, and the application of elasto-plastic theories finds extreme importance.

3.3.1 Linear elastic fracture mechanics principles

Linear elastic principles (LEFM) are valid if the extent of the plastic zone developed ahead of the advancing crack is negligible as compared to other geometrical dimensions of cracked components. LEFM principles describe the stress–strain fields in the area lying close to the crack tips in terms of applied loads, geometrical parameters like crack size and crack orientation, and the material properties. The three types of loadings (modes I–III) develop three different types of stresses in the cracked structural component. In the case of mixed loading, the stress fields developed at the crack tip can be obtained as the combination of the stress fields under different fracture modes. Using LEFM principles, the stresses induced near the crack tips can be written in standard form as

$$\sigma_{ij} = \frac{1}{\sqrt{2\pi r}}\left(K_I f_{ij}^{I}(\theta) + K_{II} f_{ij}^{II}(\theta) + K_{III} f_{ij}^{III}(\theta)\right) \tag{3.30}$$

where K_I, K_{II}, K_{III} are the SIFs for different fracture modes, respectively, and $f_{ij}^{i}(\theta)$ are the corresponding trigonometric functions used to express the stresses induced ahead of the crack tip. The analytical solutions for the displacements and stresses at the crack tip have been obtained by Irwin and Westergaard for three modes of fracture as

Mode-I:

$$\sigma_{xx} = \frac{K_I}{\sqrt{2\pi r}} cos\frac{\theta}{2}\left(1 - sin\frac{\theta}{2}cos\frac{3\theta}{2}\right) \tag{3.31}$$

$$\sigma_{yy} = \frac{K_I}{\sqrt{2\pi r}} cos\frac{\theta}{2}\left(1 - sin\frac{\theta}{2}sin\frac{3\theta}{2}\right) \tag{3.32}$$

$$\sigma_{xy} = \frac{K_I}{\sqrt{2\pi r}} sin\frac{\theta}{2}cos\frac{\theta}{2}cos\frac{3\theta}{2} \tag{3.33}$$

$$\sigma_{zz} = \nu\left(\sigma_{xx} + \sigma_{yy}\right); \tau_{xz} = \tau_{yz} = 0 \tag{3.34}$$

$$u_x = \frac{K_I}{2G}\sqrt{\frac{r}{2\pi}} cos\frac{\theta}{2}(\kappa - cos\theta) \tag{3.35}$$

$$u_y = \frac{K_I}{2G}\sqrt{\frac{r}{2\pi}} sin\frac{\theta}{2}(\kappa - cos\theta) \tag{3.36}$$

Mode-II:

$$\sigma_{xx} = -\frac{K_{II}}{\sqrt{2\pi r}} sin\frac{\theta}{2}\left(2 + cos\frac{\theta}{2}cos\frac{3\theta}{2}\right) \tag{3.37}$$

$$\sigma_{yy} = \frac{K_{II}}{\sqrt{2\pi r}} sin\frac{\theta}{2}cos\frac{\theta}{2}cos\frac{3\theta}{2} \tag{3.38}$$

$$\sigma_{xy} = \frac{K_{II}}{\sqrt{2\pi r}} cos\frac{\theta}{2}\left(1 - sin\frac{\theta}{2}cos\frac{3\theta}{2}\right) \tag{3.39}$$

$$\sigma_{zz} = \nu\left(\sigma_{xx} + \sigma_{yy}\right); \tau_{xz} = \tau_{yz} = 0 \tag{3.40}$$

$$u_x = \frac{K_{II}}{2G}\sqrt{\frac{r}{2\pi}} sin\frac{\theta}{2}(2 + \kappa + cos\theta) \tag{3.41}$$

$$u_y = \frac{K_{II}}{2G}\sqrt{\frac{r}{2\pi}} cos\frac{\theta}{2}(2 - \kappa - cos\theta) \tag{3.42}$$

Mode-III:

$$\tau_{xz} = -\frac{K_{III}}{\sqrt{2\pi r}} sin\frac{\theta}{2} \tag{3.43}$$

$$\tau_{yz} = \frac{K_{III}}{\sqrt{2\pi r}} cos\frac{\theta}{2} \tag{3.44}$$

$$\sigma_{xx} = \sigma_{yy} = \sigma_{zz} = \tau_{xy} = 0 \tag{3.45}$$

$$\sigma_{zz} = \nu(\sigma_{xx} + \sigma_{yy}); \tau_{xz} = \tau_{yz} = 0 \tag{3.46}$$

$$u_z = \frac{K_{III}}{G}\sqrt{\frac{2r}{\pi}} sin\frac{\theta}{2}; u_x = u_y = 0 \tag{3.47}$$

3.3.2 The stress intensity factor

The SIF plays an important role and acts as an important parameter in linear elastic fracture mechanics, describing the intensity of crack tip stresses developed at the crack tip. SIF defines an important property of the material known as fracture toughness (K_{IC}), which gives a measure of the resistance offered by the material to the extension of cracks through the domain. The crack in any structural component will not grow as long as the SIF is less than the fracture toughness of the material. The SIF equations for various common geometries have already been obtained and can be found in any fracture mechanics text book. However, it should be noted that these equations are valid for linear elastic fracture mechanics only. Mode-I SIFs for any cracked body can be written in standard form as

$$K_I = f\left(\frac{a}{w}\right)\sigma\sqrt{\pi a} \tag{3.48}$$

where $2a$ is the crack length, w is the thickness of the specimen and $f\left(\frac{a}{w}\right)$ defines the geometry correction factor that can be determined by stress analysis. For an infinite center-cracked plate for which $a \ll w$, we have $K_I = \sigma\sqrt{\pi a}$. For a center-cracked specimen with finite width, the geometry correction factor $f\left(\frac{a}{w}\right)$ can be written as

Another geometrical correction factor was proposed by Fedderson, for obtaining the SIF for a center-cracked plate of finite width, as

$$f\left(\frac{a}{w}\right) = \sqrt{\frac{w}{\pi a}\tan\left(\frac{\pi a}{w}\right)}; \left(\text{accuracy of } 0.5\% \text{ for } \frac{a}{w} \leq 0.35\right) \tag{3.49}$$

$$f\left(\frac{a}{w}\right) = \sqrt{\sec\left(\frac{\pi a}{w}\right)}; \left(\text{accuracy of } 0.3\% \text{ for } \frac{a}{w} \leq 0.35\right) \tag{3.50}$$

Mode-I SIFs for an infinite plate with a single edge crack, such that $a \ll w$, can be given by $K_I = 1.12\sigma\sqrt{\pi a}$. However, appropriate geometric correction factors are applied to take the effect of finite geometries into consideration. The geometric correction factor for a single-edge cracked specimen can be written as

$$f\left(\frac{a}{w}\right) = 1.122 - 0.231\left(\frac{a}{w}\right) + 10.55\left(\frac{a}{w}\right)^2 - 21.71\left(\frac{a}{w}\right)^3 + 30.382\left(\frac{a}{w}\right)^4 \tag{3.51}$$

which has an accuracy of $0.5\% for \frac{a}{w} \leq 0.6$

Similarly, the geometric correction factor for a double-edge cracked specimen can be expressed as

$$f\left(\frac{a}{w}\right) = \frac{1.122 - 1.122(a/w) - 0.82(a/w)^2 + 3.768(a/w)^3 - 3.04(a/w)^4}{\sqrt{1-(2a/w)}} \tag{3.52}$$

which has an accuracy of 0.5%for any value of a/w. The geometric correction factors for all other geometries are available in literature easily. The crack will grow when the value of the SIF becomes equal to the fracture toughness of the material, i.e. fracture occurs when $K_I = K_{IC}$. Griffith introduced a fracture parameter in terms of the elastic energy release rate$G$, which is related to SIFs for linear elastic problems as

$$G = \frac{K_I^2}{E'} \tag{3.53}$$

such that $E' = E$ for plane stress and $E' = E/(1-\nu^2)$ for plane strain cases. The crack growth occurs when $G = G_{IC} = K_{IC}^2/E'$, where G_{IC} denotes the critical energy release rate. It can be seen that the Griffith energy and SIF-based criteria are equivalent in linear elastic fracture mechanics. Griffith's energy theory was originally derived for brittle materials, where the resistance to crack growth comes mainly from the surface energy of the material. It has been found that the energy theory did not provide accurate results in the case of ductile materials, where significant energy is absorbed during the plastic deformation at the crack tip. Later on, Griffith's energy criterion was modified by several researchers so that it could be applied to ductile materials also with sufficient accuracy.

3.3.3 The J-integral

LEFM principles cannot characterize the behavior of cracked bodies in the presence of large plasticity effects. In order to characterize the fracture behavior accurately, Rice proposed a new fracture mechanics parameter, known as J-integral, which is path-independent and can be defined for two-dimensional problems as

$$J = \int_\Gamma \left(W dy - T\frac{\partial u}{\partial x} ds \right) \tag{3.54}$$

where Γ is a closed contour surrounding the crack tip. W is the strain energy density and T represents the applied tractions. x and y are the local crack tip Cartesian coordinates and ds denotes the length measured along the contour. For linear elastic problems, the J-integral represents the elastic energy release rate G, but the J-integral finds more applications for nonlinear material behavior. The J-integral for a pure bending specimen, containing deep cracks, was obtained as

$$J = \frac{2}{b}\int_0^{\Omega} M d\Omega \tag{3.55}$$

where $b = w - a$ is the left out ligament, and w represents width of the specimen. M denotes bending moment, and Ω represents the rotation of the ends of the specimen. In terms of the applied load, the J-integral for triple-point bend or compact specimens can be written as

$$J = \frac{2}{b}\int_0^{u} P du = \frac{2A_{total}}{b} \tag{3.56}$$

where A_{total} represents the total area under the load-displacement curve. Concept of J-integral defines another important facture mechanics parameter called the fracture instability toughnessJ_{RC}, which gives the resistance to fracture at the onset of fracture instability, which is expected to occur in EPFM when the crack driving force (J curve) becomes tangent to the $J - R$ curve at J_{RC}. Thus fracture instability occurs when

$$J = J_R \text{ or } \frac{dJ}{da} = \frac{dJ_R}{da} \tag{3.57}$$

The materials showing steep R curves do not usually experience unstable crack propagation. Paris et al. [7] defined another fracture mechanics parameter called the tearing modulus, which describes the onset of tearing instability during crack growth. We have

$$T_{app} = \frac{E}{\sigma_{ys}^2}\frac{dJ}{da} \text{ and } T_R = \frac{E}{\sigma_{ys}^2}\frac{dJ_R}{da} \tag{3.58}$$

where dJ/da represents the driving tearing force and dJ_R/da is the tearing resistance of the material. Thus fracture instability is expected to occur when $J_{app} = J_R$ and $T_{app} = T_R$.

3.3.4 The crack tip opening displacement

The concept of CTOD was first presented by Wells to extend linear elastic fracture mechanics approach to elasto-plastic fracture conditions. The $CTOD$ for a center-cracked plate, using Irwin's plastic zone estimate and crack-tip displacement fields, was obtained as

$$CTOD = \frac{4}{\pi}\frac{K_I^2}{E'\sigma_{ys}} \tag{3.59}$$

However, Wells found out that the factor $\frac{4}{\pi}$ in the above equation is inconsistent with the energy balance approach and subsequently omitted this factor. A more accurate expression of $CTOD$ for perfectly plastic materials was obtained by employing the strip yield model, which was proposed by Dugdale. The expression for $CTOD$ was obtained as

$$CTOD = \frac{8\sigma_{ys}a}{\pi E'} \ln sec\left(\frac{\pi\sigma}{2\sigma_{ys}}\right) \tag{3.60}$$

If the applied stresses are very low, i.e.$\sigma \ll \sigma_{ys}$, the above equation reduces to

$$CTOD = \frac{K_I^2}{E'\sigma_{ys}} \tag{3.61}$$

Rice and Rosengren applied the J-integral to the Dugdale's yield strip model and obtained the relation for $CTOD$ as

$$CTOD = \frac{J}{m\sigma_{ys}} \tag{3.62}$$

where m is the constraint factor, the value of which lies between 0 and 2. For plane stress conditions, $m = 1$. The $CTOD$ criterion states that fracture occurs when $CTOD$ exceeds the critical value of$CTOD$, i.e. fracture occurs when$CTOD \geq (CTOD)_C$. During elasto-plastic analysis, the$CTOD$ is split into elastic and plastic components. The elastic portion of $CTOD$ is calculated from the SIF values, whereas the plastic component is evaluated from the plastic $CMOD$ using the plastic hinge model. The general expression for $CTOD$ was obtained as

$$CTOD = \frac{K_I^2}{2\sigma_{ys}E'} + \frac{\left[r_p(w-a) + \Delta a\right](CMOD)_{pl}}{\left[r_p(w-a) + a + Z\right]} \tag{3.63}$$

where a denotes the length of the crack, Δa represents the crack extension, r_p denotes plastic rotation factor, Z measures the distance from measure point to the front face of the specimen and w is width of the component. The plastic rotation factor r_p depends on a/w ratio and other related material properties.

ASTM E1820-11 revised $CTOD$ evaluation methods in 2005 and proposed a new expression for evaluating $CTOD$ as

$$CTOD = \frac{J}{m\sigma_Y} = \frac{1}{m\sigma_Y}\left(\frac{K_I^2}{E'} + J_{pl}\right) \tag{3.64}$$

where J_{pl} is evaluated in the similar way as discussed in previous sections. The factor m is a function of crack size, geometry of specimen and material properties and can be evaluated as

$$m = A_0 - A_1\left(\frac{\sigma_{ys}}{\sigma_{ts}}\right) + A_2\left(\frac{\sigma_{ys}}{\sigma_{ts}}\right)^2 - A_3\left(\frac{\sigma_{ys}}{\sigma_{ts}}\right)^3 \tag{3.65}$$

For SE(B) specimens, we have

$$A_0 = 3.18 - 0.22\left(\frac{a_0}{w}\right); A_1 = 4.32 - 2.23\left(\frac{a_0}{w}\right) \tag{3.66}$$

$$A_2 = 4.44 - 2.29\left(\frac{a_0}{w}\right); A_3 = 2.05 - 1.06\left(\frac{a_0}{w}\right) \tag{3.67}$$

For C(T) specimens, we have

$$A_0 = 3.62; A_1 = 4.21; A_2 = 4.33; A_3 = 2.00 \tag{3.68}$$

ASTM E1290-08e1 adopted the similar procedure and obtained the expression for evaluating the $CTOD$ as

$$CTOD = \frac{J}{m\sigma_Y} = \frac{1}{m\sigma_Y}\left(\frac{K_I^2}{E'} + \frac{\eta_{CMOD}A_{CMOD}^{Pl}}{(w-a_0)\{1+Z/(0.8a_0+0.2w)\}}\right) \tag{3.69}$$

where A_{CMOD}^{Pl} represents the plastic area under $load - CMOD$ curve and η_{CMOD}denotes $CMOD$-based plastic geometry correction factor, which has been defined as

$$\eta_{CMOD} = 3.667 - 2.199\left(\frac{a}{w}\right) + 0.437\left(\frac{a}{w}\right)^2; 0.25 \le \frac{a}{w} \le 0.7 \tag{3.70}$$

3.3.5 The interaction integral

The interaction integral proves to be a very effective tool for calculating the SIFs in cracked components subjected to mixed mode loading conditions. This integral uses auxiliary stress, strain or displacement fields ($\sigma^{aux}, \varepsilon^{aux}, u^{aux}$) for evaluation of mixed mode SIFs. The choice of auxiliary fields remains arbitrary. However, auxiliary fields are chosen in such a manner that they ease the evaluation of SIFs. The J-integral for a homogenous cracked specimen can be written as

$$J = \int_\Gamma \left(W\delta_{1j} - \sigma_{ij}\frac{\partial u_i}{\partial x_1}\right)n_j d\Gamma \tag{3.71}$$

where W denotes strain energy density and n_j represents jth component of unit normal vector to the contour Γ around the tip of crack. The application of divergence theorem to the above integral expresses the J-integral as

$$J = \int_A \left(\sigma_{ij}\frac{\partial u_i}{\partial x_1} - W\delta_{1j}\right)\frac{\partial q}{\partial x_j}dA + \int_A \frac{\partial}{\partial x_j}\left(\sigma_{ij}\frac{\partial u_i}{\partial x_1} - W\delta_{1j}\right)q dA \tag{3.72}$$

where q represents a weight function that remains unity at the crack tip and the weight function vanishes along the contour Γ. A represents area inside the contour surrounding the crack tip. The J-integral has been expressed in the equivalent domain form that is suitable for numerical treatment. Second integrand of the above equation vanishes for homogenous materials, which gives

$$J = \int_A \left(\sigma_{ij}\frac{\partial u_i}{\partial x_1} - W\delta_{1j}\right)\frac{\partial q}{\partial x_j}dA \tag{3.73}$$

Further, let us consider two independent equilibrium states "1" and "2" of a cracked domain, where "1" represents actual state of the body and "2" represents the auxiliary state of the cracked body. These two states can be superimposed, which produces another equilibrium state "3", for which we can write

$$J^{(3)} = \int_A \left[\left(\sigma_{ij}^{(1)} + \sigma_{ij}^{(2)} \right) \frac{\partial \left(u_i^{(1)} + u_i^{(2)} \right)}{\partial x_1} - W^{(3)} \delta_{1j} \right] \frac{\partial q}{\partial x_j} dA \tag{3.74}$$

where $W^{(3)}$ is the strain energy for state "3", which is expressed as

$$W^{(3)} = W^{(1)} + W^{(2)} + W^{(1,2)} = \frac{1}{2} \left(\sigma_{ij}^{(1)} + \sigma_{ij}^{(2)} \right) \left(\varepsilon_{ij}^{(1)} + \varepsilon_{ij}^{(2)} \right) \tag{3.75}$$

Now, the J-integral for state "3" can be further written as

$$J^{(3)} = J^{(1)} + J^{(2)} + M^{(1,2)} \tag{3.76}$$

where $J^{(1)}$ and $J^{(2)}$ denote J-integrals for equilibrium states "1" and "3", and $M^{(1,2)}$ represents the interaction integral presenting the interaction between the two given states. We have

$$J^{(1)} = \int_A \left(\sigma_{ij}^{(1)} \frac{\partial u_i^{(1)}}{\partial x_1} - W^{(1)} \delta_{1j} \right) \frac{\partial q}{\partial x_j} dA \tag{3.77}$$

$$J^{(2)} = \int_A \left(\sigma_{ij}^{(2)} \frac{\partial u_i^{(2)}}{\partial x_1} - W^{(2)} \delta_{1j} \right) \frac{\partial q}{\partial x_j} dA \tag{3.78}$$

$$M^{(1,2)} = \int_A \left(\sigma_{ij}^{(1)} \frac{\partial u_i^{(2)}}{\partial x_1} + \sigma_{ij}^{(2)} \frac{\partial u_i^{(1)}}{\partial x_1} - W^{(1,2)} \delta_{1j} \right) \frac{\partial q}{\partial x_j} dA \tag{3.79}$$

such that $W^{(1,2)}$ represents mutual strain energy density of equilibrium states "1" and "2" defined by

$$W^{(1,2)} = \frac{1}{2} \sigma_{ij}^{(1)} \varepsilon_{ij}^{(2)} + \sigma_{ij}^{(2)} \varepsilon_{ij}^{(1)} = \sigma_{ij}^{(1)} \varepsilon_{ij}^{(2)} = \sigma_{ij}^{(2)} \varepsilon_{ij}^{(1)} \tag{3.80}$$

In linear elastic fracture mechanics, the J-integral defines the elastic energy release rate, and it can be related to SIFs as

$$J = \frac{1}{E'} \left(K_I^2 + K_{II}^2 \right) \tag{3.81}$$

where $E' = E$ for plane stress and $E' = E / \left(1 - \nu^2 \right)$ for plane strain. Thus J-integral for the superimposed state "3" becomes

$$J^{(3)} = \frac{1}{E'} \left[\left(K_I^{(1)} + K_I^{(2)} \right)^2 + \left(K_{II}^{(1)} + K_{II}^{(2)} \right)^2 \right] \tag{3.82}$$

$$J^{(3)} = \frac{1}{E'}\left[\left(K_I^{(1)^2} + K_{II}^{(1)^2}\right) + \left(K_I^{(2)^2} + K_{II}^{(2)^2}\right) + 2K_I^{(1)}K_I^{(2)} + 2K_{II}^{(1)}K_{II}^{(2)}\right] \tag{3.83}$$

which can be further written as $J^{(3)} = J^{(1)} + J^{(2)} + M^{(1,2)}$ such that

$$J^{(1)} = \frac{1}{E'}\left[\left(K_I^{(1)} + K_I^{(2)}\right)^2\right] \tag{3.84}$$

$$J^{(2)} = \frac{1}{E'}\left[\left(K_{II}^{(1)} + K_{II}^{(2)}\right)^2\right] \tag{3.85}$$

$$M^{(1,2)} = \frac{2}{E'}\left[K_I^{(1)}K_I^{(2)} + K_{II}^{(1)}K_{II}^{(2)}\right] \tag{3.85}$$

The individual SIFs in mixed mode loading can be derived by making an appropriate choice for the auxiliary state ("2"). For obtainingK_I, the auxiliary state of the cracked body is chosen to be pure mode-I state where $K_I^{(2)} = 1$ and $K_{II}^{(2)} = 0$. Similarly, K_{II} can be derived by selecting the auxiliary state as pure mode-II state such that $K_I^{(2)} = 0$ and $K_{II}^{(2)} = 1$. In view of this, the mixed mode SIFs can be obtained as

$$K_I^{(1)} = \frac{M^{(1,I)}E'}{2}; K_{II}^{(1)} = \frac{M^{(1,II)}E'}{2} \tag{3.86}$$

3.4 Numerical results and discussions

This section discusses several fracture mechanics problems that have been solved to establish the potential and accuracy of the proposed technique in simulating the effect of the presence of cracks in engineering specimens subjected to monotonic loads. The mathematical models on XFEM and other related algorithms have been coded in MATLAB to perform different numerical tests on three-dimensional cracked specimens. The analytical or benchmark solutions reported in literature serve as the benchmark or reference solutions for validating the MATLAB codes developed in the current work. The numerical problems considered here simulate different types of cracks, such as plane edge cracks, horizontal and inclined penny cracks and lens-type cracks in 3D structural components.

3.4.1 Investigation of a plane edge crack

This example considers a cuboidal domain, containing a plane edge crack with length 50 mm and having dimensions $100 \times 175 \times 150$ mm^3, for analysis by the XFEM. The problem description is shown in Fig. 3.2. The Young's modulus of the cracked structural component is taken as 200 GPa and Poisson's ratio is 0.3. The boundary conditions are also shown in Fig. 3.2. A monotonic tensile load equal to 100 MPa is imposed at the top surface. The domain discretization of the cracked component in XFEM is shown in Fig. 3.3. XFEM develops a uniform mesh to represent any type of discontinuity present in the domain. Remeshing and conformal meshing problems are not seen in XFEM, which makes

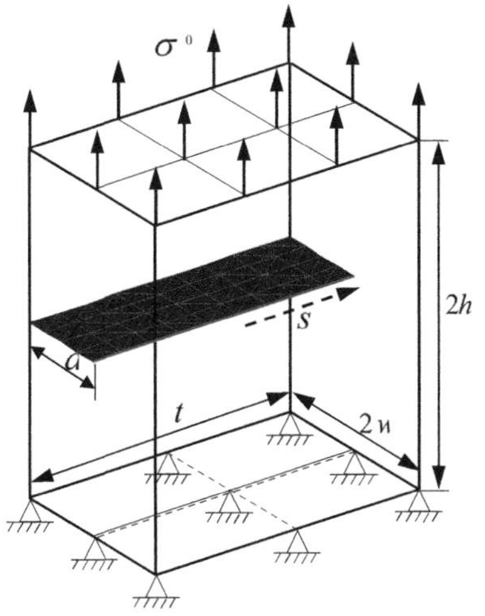

FIGURE 3.2 Problem description for a cuboidal domain with a plane edge crack.

it computationally less expensive as compared to conventional FEM. The variation of SIFs with normalized length (s/t) can be seen in Fig. 3.4. In this figure, "s" denotes the arc length from the crack center measured along the crack front. It has been found that results obtained in the current study are in close agreement with the analytical solutions available in literature [40]. It is quite evident from the numerical simulations that the proposed approach has a strong potential in modeling the behavior of static and propagating cracks in three-dimensional structural components.

3.4.2 Cuboidal domain with a horizontal penny crack

The behavior of a horizontal penny crack placed at the center of a cuboidal structural component is presented here. The cuboidal domain is considered to be cubical with each side equal to 200 mm, and it contains a penny crack at the center having a radius of 10 mm, as shown in Fig. 3.5. Same properties, as in the previous case, have been considered for analysis. A tensile load of 100 MPa is applied at the top surface, and the bottom surface remains fixed during the application of the load. The XFEM mesh for the 3D domain containing a penny crack can be seen in Fig. 3.6. The distribution of mode-I SIFs along the crack surface is shown in Fig. 3.7. The numerical simulations carried out in the current study show a close agreement with the analytical or benchmark solutions published in literature [40]. The numerical simulations also show the accuracy, and potential of the proposed approach in modeling three-dimensional crack growth in engineering materials.

3.4.3 Cuboidal domain with inclined penny crack

A cubical domain containing an inclined penny crack (Fig. 3.8) is investigated here. The geometry of the cubical specimen and crack length are chosen to be the same as in the previous case. Same properties, as in the previous case, have been considered for analysis. A tensile load of 100 MPa acts at the top surface, whereas bottom area remains fixed during

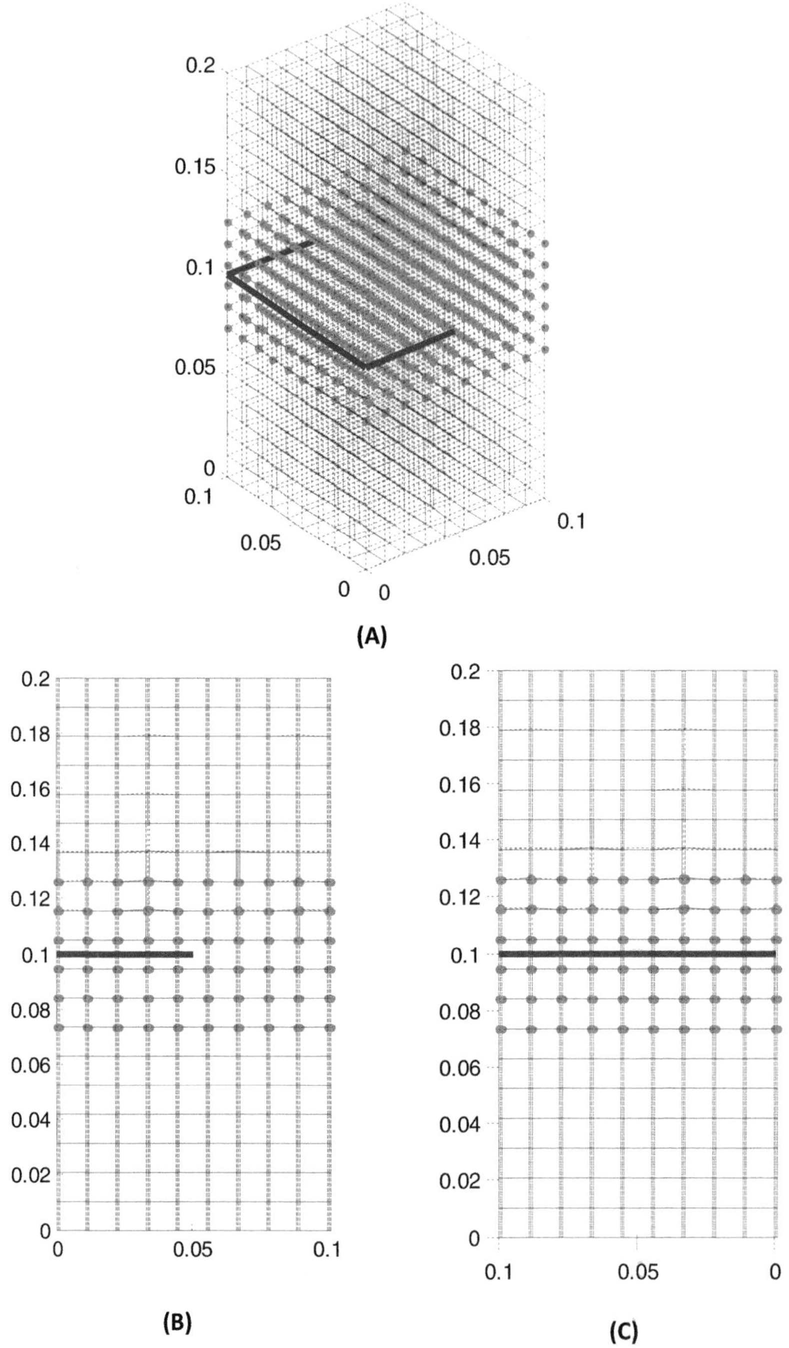

FIGURE 3.3 XFEM mesh for a plane edge crack: (A) 3D view; (B) side view; (C) front view. *XFEM*, Extended finite element method.

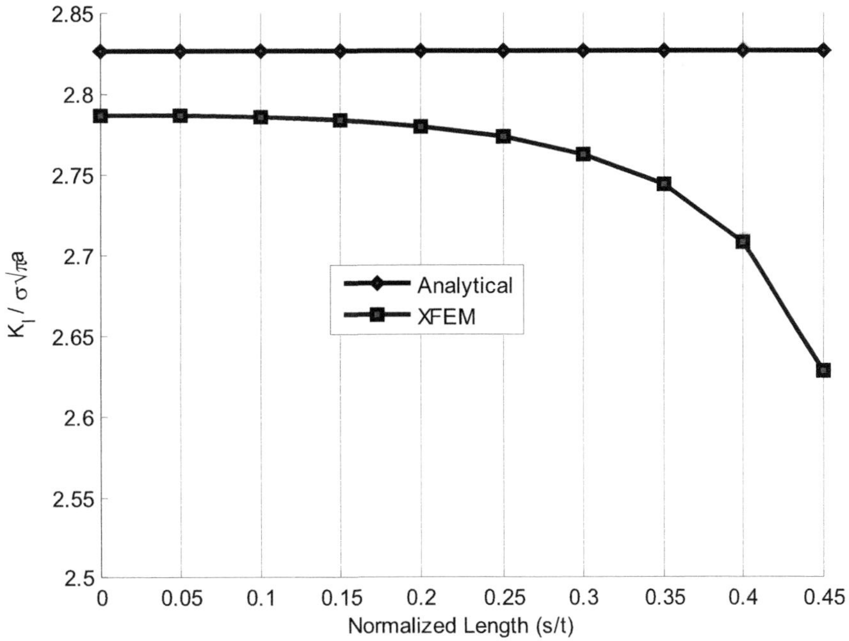

FIGURE 3.4 Variation of normalized mode-I stress intensity factors.

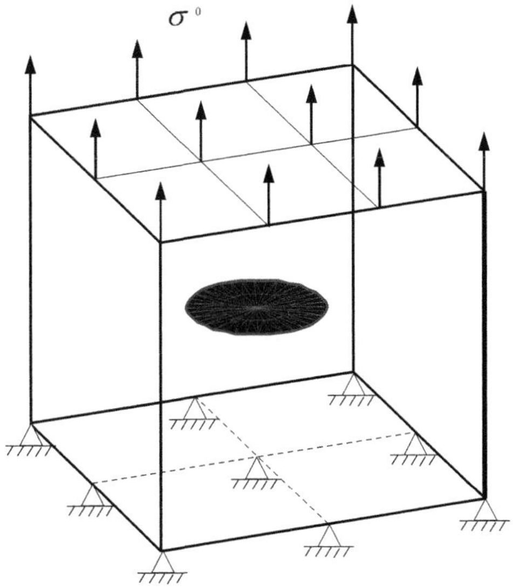

FIGURE 3.5 Cubical specimen with a horizontal penny crack.

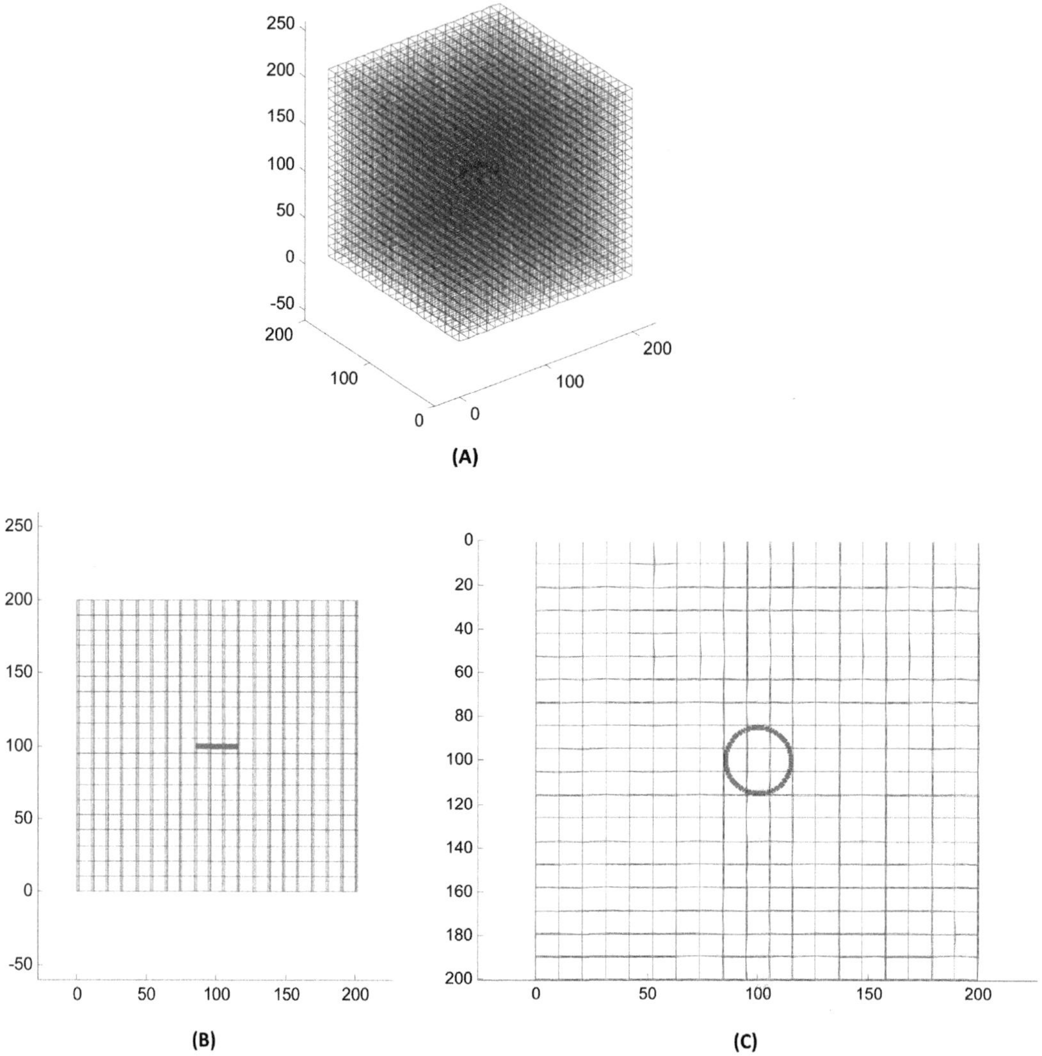

FIGURE 3.6 XFEM mesh for horizontal penny crack: (A) 3D view; (B) side view; (C) top view. *XFEM*, Extended finite element method.

load application. The XFEM mesh of inclined penny-cracked specimen is shown in Fig. 3.9. The variations of tearing mode SIFs with crack angle are shown in Fig. 3.10. The numerical simulations carried out in the current study show a close agreement with the analytical or benchmark solutions published in literature [40]. The numerical simulations also show the accuracy and potential of the proposed approach in modeling three dimensional crack growth in engineering materials.

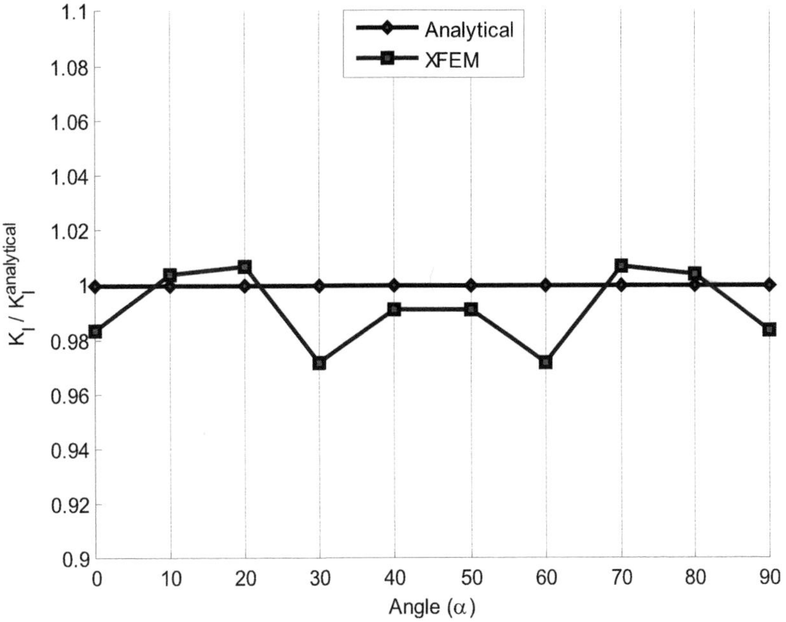

FIGURE 3.7 Normalized mode-I SIFs along the crack surface. *SIFs*, Stress intensity factors.

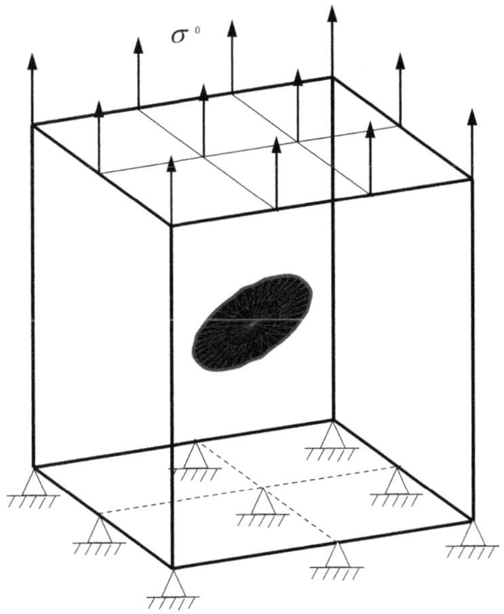

FIGURE 3.8 Cubical specimen with an inclined penny crack.

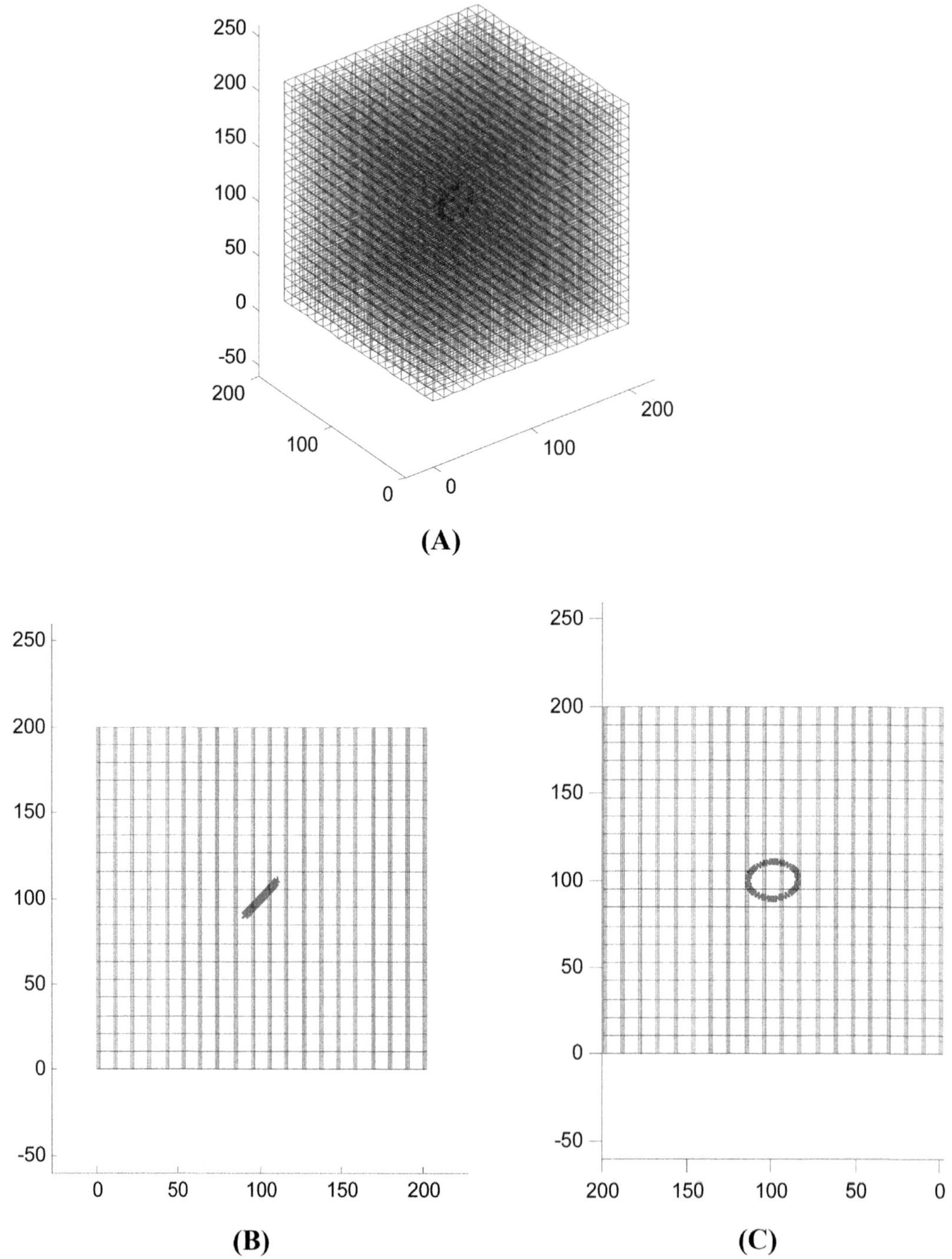

FIGURE 3.9 XFEM mesh for a 3D domain with an inclined penny crack: (A) 3D view; (B) side view; (C) front view. *XFEM*, Extended finite element method.

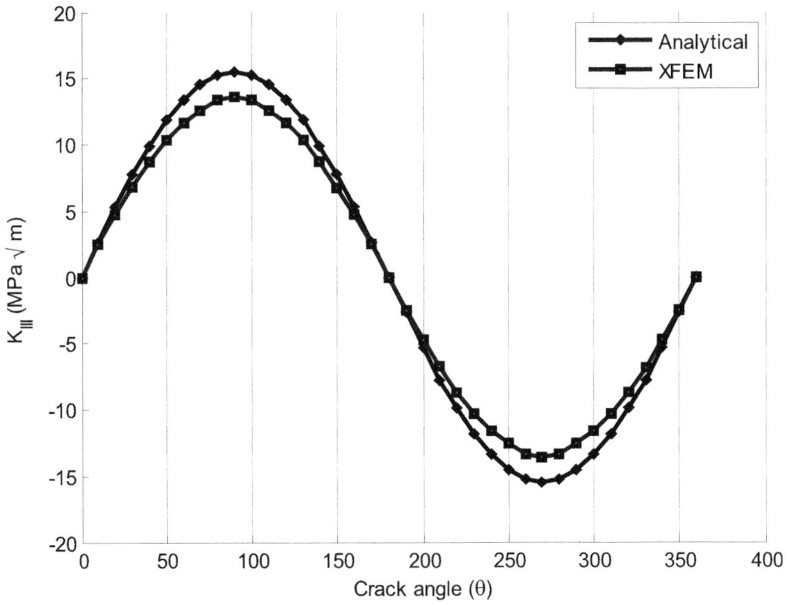

FIGURE 3.10 Mode-III stress intensity factors with the angle of crack.

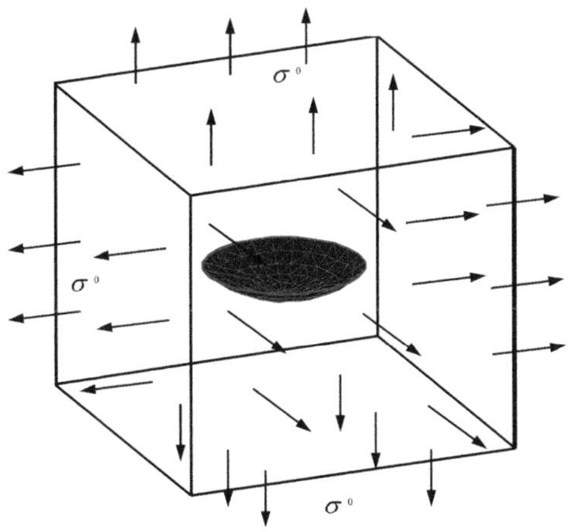

FIGURE 3.11 Problem description for a lens-shaped crack.

3.4.4 Cuboidal domain with a lens crack

A cubical domain with each side equal to 200 mm and containing a lens crack has been considered for analysis as shown in Fig. 3.11. In this study, a sphere having a radius of 20 mm has been used to define the lens crack, which has been cut through an angle of $\pi/4$ radians inside the cubical structure. A hydrostatic load equal to 100 MPa is applied to the cracked domain as shown in Fig. 3.11. Young's modulus and Poisson's ratio of the three-dimensional cracked specimen are taken as 68.2 GPa and 0.22, respectively. The domain representation for the cuboidal domain with a lens crack is shown in Fig. 3.12. The

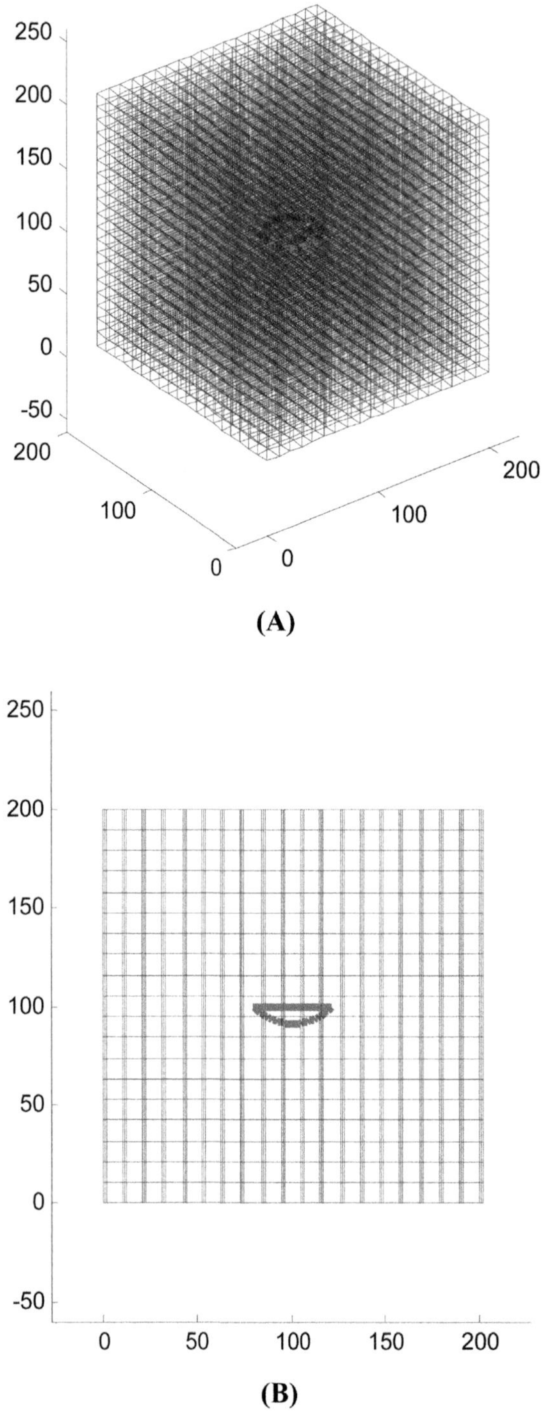

FIGURE 3.12 XFEM meshes the lens crack: (A) 3D view; (B) front view. *XFEM*, Extended finite element method.

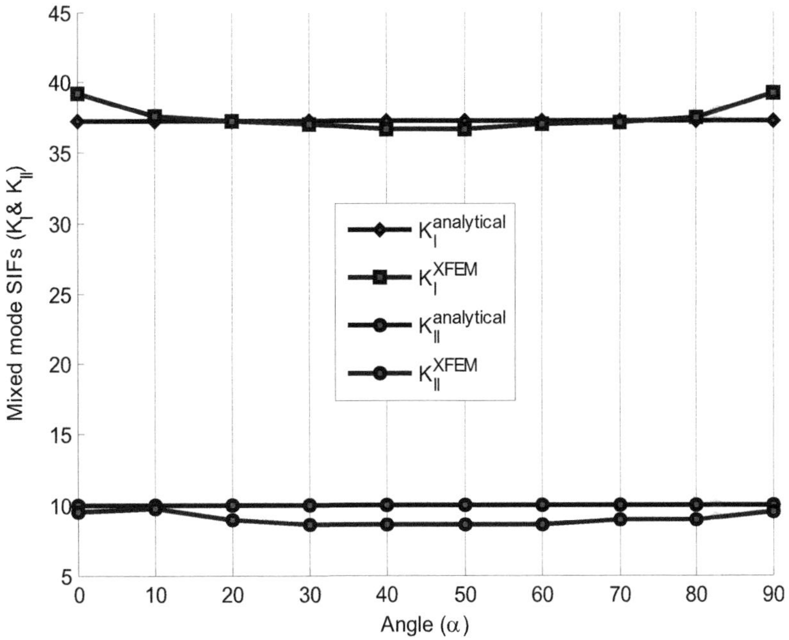

FIGURE 3.13 Variation of mixed mode SIFs for the lens crack. *SIFs*, Stress intensity factors.

variations of mixed mode SIFs along the boundary of the crack can be seen in Fig. 3.13. The numerical simulations carried out in the current study show a close agreement with the analytical or benchmark solutions published in literature [40]. The numerical simulations also show the accuracy and potential of the proposed approach in modeling three-dimensional crack growth in engineering materials.

3.5 Conclusion

The current study invokes the use of XFEM to simulate the effect of presence of cracks in three-dimensional structural components subjected to monotonic tensile loads. Unlike FEM, the issues of remeshing, mesh adaption, and other mesh-related drawbacks do not arise in XFEM. The mathematical models on XFEM and other related algorithms have been coded in MATLAB to perform different numerical tests on three-dimensional cracked specimens. The analytical or exact solutions published in literature serve as the benchmark or reference solutions for validating the MATLAB codes developed in the current work. Several numerical problems on cracked specimens have been considered in the current work to demonstrate the capabilities of the XFEM in modeling the behavior of cracks in 3D structural components. Numerical simulations carried out in the current study show a remarkable agreement with the analytical solutions published in literature. The results also show the accuracy, applicability, and potential of the proposed approach in modeling three-dimensional crack growth in engineering materials.

References

[1] Aliabadi MH, Brebbia CA. Boundary element formulations in fracture mechanics: a review. Transactions on Engineering Sciences 1998;17:589–98.

[2] Wen PH, Aliabadi MH, Young A. Dual boundary element methods for three-dimensional dynamic crack problems. Journal of Strain Analysis for Engineering Design 1999;34:373–94. Available from: https://doi.org/10.1177/030932479903400601.

[3] Leonel ED, Venturini WS. Non-linear boundary element formulation applied to contact analysis using tangent operator. Engineering Analysis with Boundary Elements 2011;35:1237–47. Available from: https://doi.org/10.1016/j.enganabound.2011.06.005.

[4] Spangenberger AG, Lados DA. Extended finite element modeling of fatigue crack growth microstructural mechanisms in alloys with secondary/reinforcing phases: model development and validation. Computational Mechanics 2020;. Available from: https://doi.org/10.1007/s00466-020-01921-2.

[5] Lone AS, Kanth SA, Jameel A, Harmain GA. A state of art review on the modeling of contact type nonlinearities by extended finite element method. Materials Today: Proceedings 2019;18:3462–71. Available from: https://doi.org/10.1016/j.matpr.2019.07.274.

[6] Sukumar N, Chopp DL, Moës N, Belytschko T. Modeling holes and inclusions by level sets in the extended finite-element method. Computer Methods in Applied Mechanics and Engineering 2001;190:6183–200. Available from: https://doi.org/10.1016/S0045-7825(01)00215-8.

[7] Jameel A, Harmain GA. Fatigue crack growth in presence of material discontinuities by EFGM. International Journal of Fatigue 2015;81:105–16. Available from: https://doi.org/10.1016/j.ijfatigue.2015.07.021.

[8] Lone AS, Jameel A, Harmain GA. A coupled finite element-element free Galerkin approach for modeling frictional contact in engineering components. Materials Today: Proceedings 2018;5:18745–54. Available from: https://doi.org/10.1016/j.matpr.2018.06.221.

[9] Kanth SA, Harmain GA, Jameel A. Modeling of nonlinear crack growth in steel and aluminum alloys by the element free Galerkin method. Materials Today: Proceedings 2018;5:18805–14. Available from: https://doi.org/10.1016/j.matpr.2018.06.227.

[10] Jameel A, Harmain GA. Fatigue crack growth analysis of cracked specimens by the coupled finite element-element free Galerkin method. Mechanics of Advanced Materials and Structures 2019;26:1343–56. Available from: https://doi.org/10.1080/15376494.2018.1432800.

[11] Lone AS, Jameel A, Harmain GA. Enriched element free Galerkin method for solving frictional contact between solid bodies. Mechanics of Advanced Materials and Structures 2022;. Available from: https://doi.org/10.1080/15376494.2022.2092791.

[12] Lone AS, Jameel A, Harmain GA. Modelling of contact interfaces by penalty based enriched finite element method. Mechanics of Advanced Materials and Structures 2022;. Available from: https://doi.org/10.1080/15376494.2022.2034075.

[13] Jameel A, Harmain GA. Large deformation in bi-material components by XIGA and coupled FE-IGA techniques. Mechanics of Advanced Materials and Structures 2022;29:850–72. Available from: https://doi.org/10.1080/15376494.2020.1799120.

[14] Kanth SA, Jameel A, Harmain GA. Investigation of fatigue crack growth in engineering components containing different types of material irregularities by XFEM. Mechanics of Advanced Materials and Structures 2022;29(24):3570–87. Available from: https://doi.org/10.1080/15376494.2021.1907003.

[15] Gupta V, Jameel A, Anand S, Anand Y. Analysis of composite plates using isogeometric analysis: a discussion. Materials Today: Proceedings 2021;44:1190–4. Available from: https://doi.org/10.1016/j.matpr.2020.11.238.

[16] Jameel A, Harmain GA. Effect of material irregularities on fatigue crack growth by enriched techniques. International Journal for Computational Methods in Engineering Science and Mechanics 2020;21:109–33. Available from: https://doi.org/10.1080/15502287.2020.1772902.

[17] Lone AS, Kanth SA, Jameel A, Harmain GA. XFEM modelling of frictional contact between elliptical inclusions and solid bodies. Materials Today: Proceedings 2020;26:819–24. Available from: https://doi.org/10.1016/j.matpr.2019.12.424.

[18] Sheikh UA, Jameel A. Elasto-plastic large deformation analysis of bi-material components by FEM. Materials Today: Proceedings 2020;26:1795–802. Available from: https://doi.org/10.1016/j.matpr.2020.02.377.

[19] Simpson R, Trevelyan J. A partition of unity enriched dual boundary element method for accurate computations in fracture mechanics. Computer Methods in Applied Mechanics and Engineering 2011;200:1–10. Available from: https://doi.org/10.1016/j.cma.2010.06.015.
[20] Noda NA, Oda K. Numerical solutions of the singular integral equations in the crack analysis using the body force method. International Journal of Fracture 1992;58:285–304. Available from: https://doi.org/10.1007/BF00048950.
[21] Rao BN, Rahman S. A coupled meshless-finite element method for fracture analysis of cracks. International Journal of Pressure Vessels and Piping 2001;78:647–57. Available from: https://doi.org/10.1016/S0308-0161(01)00076-X.
[22] Kanth SA, Lone AS, Harmain GA, Jameel A. Modelling of embedded and edge cracks in steel alloys by XFEM. Materials Today: Proceedings 2020;26:814–18. Available from: https://doi.org/10.1016/j.matpr.2019.12.423.
[23] Kanth SA, Lone AS, Harmain GA, Jameel A. Elasto plastic crack growth by XFEM: a review. Materials Today: Proceedings 2019;18:3472–81. Available from: https://doi.org/10.1016/j.matpr.2019.07.275.
[24] Jameel A, Harmain GA. Modeling and numerical simulation of fatigue crack growth in cracked specimens containing material discontinuities. Strength of Materials 2016;48(2):294–307. Available from: https://doi.org/10.1007/s11223-016-9765-0.
[25] Eberhard P, Gaugele T. Simulation of cutting processes using mesh-free Lagrangian particle methods. Computational Mechanics 2013;51:261–78. Available from: https://doi.org/10.1007/s00466-012-0720-z.
[26] De Lorenzis L, Wriggers P, Zavarise G. A mortar formulation for 3D large deformation contact using NURBS-based isogeometric analysis and the augmented Lagrangian method. Computational Mechanics 2012;49:1–20. Available from: https://doi.org/10.1007/s00466-011-0623-4.
[27] Bhardwaj G, Singh IV, Mishra BK, Bui TQ. Numerical simulation of functionally graded cracked plates using NURBS based XIGA under different loads and boundary conditions. Composite Structures 2015;126:347–59. Available from: https://doi.org/10.1016/j.compstruct.2015.02.066.
[28] Gu YT, Wang QX, Lam KY. A meshless local Kriging method for large deformation analyses. Computer Methods in Applied Mechanics and Engineering 2007;196:1673–84. Available from: https://doi.org/10.1016/j.cma.2006.09.017.
[29] Belytschko T, Krongauz Y, Organ D, Fleming M, Krysl P. Meshless methods: an overview and recent developments. Computer Methods in Applied Mechanics and Engineering 1996;139:3–47. Available from: https://doi.org/10.1016/S0045-7825(96)01078-X.
[30] Rao BN, Rahman S. An enriched meshless method for non-linear fracture mechanics. International Journal for Numerical Methods in Engineering 2004;59:197–223. Available from: https://doi.org/10.1002/nme.868.
[31] Duflot M, Nguyen-Dang H. A meshless method with enriched weight functions for fatigue crack growth. International Journal for Numerical Methods in Engineering 2004;59:1945–61. Available from: https://doi.org/10.1002/nme.948.
[32] Jameel A, Harmain GA. Extended iso-geometric analysis for modeling three-dimensional cracks. Mechanics of Advanced Materials and Structures 2019;26:915–23. Available from: https://doi.org/10.1080/15376494.2018.1430275.
[33] Singh IV, Bhardwaj G, Mishra BK. A new criterion for modeling multiple discontinuities passing through an element using XIGA. Journal of Mechanical Science and Technology 2015;29:1131–43. Available from: https://doi.org/10.1007/s12206-015-0225-8.
[34] Bhardwaj G, Singh IV, Mishra BK. Numerical simulation of plane crack problems using extended isogeometric analysis. Procedia Engineering 2013;64:661–70. Available from: https://doi.org/10.1016/j.proeng.2013.09.141.
[35] Jameel A, Harmain GA. A coupled FE-IGA technique for modeling fatigue crack growth in engineering materials. Mechanics of Advanced Materials and Structures 2019;26:1764–75. Available from: https://doi.org/10.1080/15376494.2018.1446571.
[36] Gupta V, Jameel A, Verma SK, Anand S, Anand Y. Transient isogeometric heat conduction analysis of stationary fluid in a container. Part E: Journal of Process Mechanical Engineering 2022;. Available from: https://doi.org/10.1177/09544089221112.
[37] Gupta JV, Jameel A, Verma SK, Anand S, Anand Y. An insight on NURBS based isogeometric analysis, its current status and involvement in mechanical applications. Archives of Computational Methods in Engineering 2022;. Available from: https://doi.org/10.1007/s11831-022-09838-0.

[38] Singh AK, Jameel A, Harmain GA. Investigations on crack tip plastic zones by the extended iso-geometric analysis. Materials Today: Proceedings 2018;5:19284–93. Available from: https://doi.org/10.1016/j.matpr.2018.06.287.

[39] Melenk JM, Babuška I. The partition of unity finite element method: basic theory and applications. Computer Methods in Applied Mechanics and Engineering 1996;139:289–314. Available from: https://doi.org/10.1016/S0045-7825(96)01087-0.

[40] Jameel A, Harmain GA. Extended iso-geometric analysis for modeling three dimensional cracks. Mechanics of Advanced Materials and Structures 2019;26:915–23. Available from: https://doi.org/10.1080/15376494.2018.1430275.

CHAPTER

4

Extended finite element method for free vibration analyses of cracked plate based on higher order shear deformation theory

Ahmed Raza, Himanshu Pathak and Mohammad Talha

Design Against Failure and Fracture Group, School of Mechanical and Materials Engineering, Indian Institute of Technology Mandi, Mandi, India

4.1 Introduction

This chapter introduces the review of various enrichment techniques, their need for different kinds of discontinuity problems, and the implementation of the extended finite element method (FEM) (XFEM) for the analysis of the cracked plates. This chapter also includes a brief review of the implementation of enrichment functions in different kinds of discontinuity problems. Over the last decades, enrichment techniques have been used in modeling the problem of discontinuity in the material domain. Initially, researchers implemented an analytical approach to study the cracked structures [1–4]. After the introduction of the FEM [5–7], there was some improvement in computation. Modeling of the crack in the structure and solving other discontinuity problems using the FEM is a tedious and time-consuming process. Stress singularities appear near the crack tip, which requires fine mesh for solving the crack problem using FEM. In case the crack front is moving, continuous modification of mesh near the crack is required, increasing the computation process. There are various advanced numerical techniques available to address the crack in the material domain, such as the mesh-free method [8–11], finite element–element-free Galerkin method (FE-EFGM) [12], extended isogeometric analysis [13–16], phase-field method, and fragile point method, XFEM [17–27].

Over the last several decades, researchers have explored and implemented the meshless method. The simulation of extremely large deformations such as extrusion, molding, and failure processes with moving crack is very complex, and conventional computational method is not suited to solve this. To deal with such a problem, the meshless method is best option. In this, the discretization error, which eventually leads to inaccuracy in solution, can

Enriched Numerical Techniques
DOI: https://doi.org/10.1016/B978-0-443-15362-4.00003-6

be eliminated. In the meshless method, mess is not required for the simulation of domain. Interconnection between the nodes is absent, and simulation is done by the interaction of neighboring nodes [9]. Although the computational cost of the meshless method is high as compared to the conventional method, it is used because of its potential advantages. Since mesh is absent and nodal data is sufficient for interpolation, hence remeshing is not required. It has high convergence rate and is good in terms of solution accuracy. It does not face the problem of element degeneration during simulation of large deformation.

Simulation of crack initiation, multiple crack nucleation, and complex crack trajectories is challenging for the conventional computational method. Phase-field method is a novel technique to address such types of fracture issues. The technique of minimization of total system energy is implemented in the phase-field method [28] to determine the feasible crack nucleation and propagation. Phase-field methods are frequently used to solve a variety of problems in physics [29]. Amiri et al. [30] implemented a phase-field model in fracture mechanics problem to simulate the crack propagation in brittle material. Researchers implemented a phase-field model to simulate the crack propagation in homogenous materials for various loading conditions [31–34].

This chapter explains the XFEM and its implementation in solving the plate's problem having crack in vibration environment. The XFEM is very efficient in solving the progressive crack problem. In progressive crack [17], discretization of the domain remains the same; only the local enrichment is done in the elements near the crack. A technique called the partition of unity (PU) method [35] enables the enrichment process in conventional FEM. In XFEM, the discretization of the domain remains the same. The cracks/discontinuities in the material domain are traced using the level set method [35]; consequently, the nodes near the crack are selected for enrichment.

The advantages of XFEM over FEM are given as follows:

- Progressive crack can be easily handled;
- Fine mesh near the crack and crack tip is not needed to capture stress singularity;
- Even coarse mesh can have higher convergence rate with good accuracy;
- Remeshing process is not needed;
- XFEM is efficient and robust in addressing the fracture mechanics problem.

Researchers have extensively used XFEM for the fracture mechanics problem to investigate the stress intensity factor, J-integral, energy release rate, dynamic stress intensity factor, and crack growth [36–39]. The propagating cracks in the material domain can be easily investigated using XFEM. The detailed literature review of XFEM and its implementation is discussed here.

Daux et al. [40] developed extended finite element formulation for arbitrary geometries, such as complex branched cracks and voids without the requirement of special mesh for the discontinuities. Sukumar et al. [41] implemented PU for enrichment to model three-dimensional cracks using XFEM. Dolbow et al. [42] developed the XFEM formulation to simulate the crack growth in plates using Mindlin-Reissner plates theory. Dolbow et al. [43] developed a new technique to simulate the crack growth with frictional contact on crack surface using XFEM. Areias and Belytschko [44] presented new formulation by coupling linear tetrahedral elements with XFEM for the simulation and quasistatic analysis of crack growth in brittle materials. Nagashima et al. [45] implemented XFEM proposed by Belytschko et al. [17] to simulate and investigate the stress intensity factor of interface cracks between two dissimilar materials. Liu et al. [46] developed XFEM to compute the stress intensity factor for homogenous materials and bi-materials. Sukumar and Prevost [47] modeled the crack growth (quasistatic) in isotropic and bimaterial media by implementing XFEM. Zi and Belytschko [48] developed XFEM formulation by introducing new crack tip elements to simulate the cohesive crack. Sukumar et al. [49] introduced an enrichment technique for modeling the interfacial crack within the extended finite element framework. Sukumar et al. [50] introduced a new enrichment technique called level set enrichment to model holes and inclusion in the XFEM framework. Pathak et al. [51] presented the modeling and simulation of three-dimensional interfacial cracks using XFEM. Shedbale et al. [52] implemented updated Lagrangian formulation with extended finite formulation to model large deformation to simulate the metal-forming process. Singh et al. [53] implemented XFEM in a thermo-elastic medium for the simulation of crack. Singh et al. [54] implemented XFEM to simulate the cracked functionally graded materials (FGMs) having holes and inclusions. Sharma et al. [55] reported the interaction of edge-cracked two-dimensional domain with holes and hard and soft inclusions. Sharma et al. [56] implemented an XFEM in fracture mechanics problem to investigate the stress intensity factor. Kumar et al. [57] solved nonlinear stable crack growth in ductile materials using updated Lagrangian approach in XFEM framework. Bhattacharya et al. [58] computed the mixed-mode fatigue life of FGMs under plane strain condition in XFEM framework. Bhattacharya et al. [59] investigated the interaction of holes, inclusions, and minor crack in FGMs under the influence of cyclic thermal load to compute fatigue life using the XFEM enrichment technique. Kumar et al. [60] implemented J-integral decomposition approach in XFEM to simulate fatigue crack growth in plastically graded materials. Shedbale et al. [61] performed nonlinear analysis to compute fatigue life by implementing the XFEM under plane stress condition. Singh et al. [62] implemented XFEM to compute the fatigue life of structures having crack, holes, and inclusions. Bhattacharya et al. [63] computed the fatigue life of bi-material FGMs with interfacial cracks. Pathak et al. [38] computed fatigue life for 3-D crack growth in XFEM framework. Kumar et al. [23] implemented virtual node XFEM to simulate kinked crack growth. Pathak et al. [64] implemented coupled EFGM and XFEM approach to simulate interfacial cracked bi-material domain. Patil et al. [65] developed a new multiscale XFEM to compute mechanical properties. They incorporated a representative volume element-based homogenization technique to compute material properties, and the proposed methodology was found to be more efficient as compared to standard XFEM. Kumar et al. [39] developed a new enrichment technique to solve dynamic crack and for blast loading. They evaluated the dynamic stress intensity factor using the proposed enrichment scheme in XFEM. Bansal et al. [66] proposed multisplit XFEM for multiple cracks in an element to simulate 3-D heterogeneous materials. Singh et al. [54] simulated cracked FGMs having cracks, holes, and inclusions and computed the stress intensity factor. They also reported the influence of major cracks and minor cracks on SIF. Pandey et al. [67] proposed new methodology based on damage evaluation to simulate high cycle fatigue crack growth in XFEM framework. Kumar et al. [68] implemented the XFEM and performed elasto-plastic simulation to evaluate fatigue life. Kumar et al. [69] proposed a homogenized multigrid XFEM to simulate the crack growth in ductile material with microstructural defects. They also computed material properties using a strain energy homogenization approach. Kumar et al. [70] XFEM to simulate creep crack growth. Dwivedi et al. [71] estimated fatigue life for fiber-reinforced polymer composite material at different volume fractions of fiber in XFEM framework considering multiple discontinuities like cracks and holes. Suman et al. [72] implemented artificial neural network approach to predict the fatigue performance in XFEM framework. Dwivedi et al. [73] investigated the influence of size and shape of patch repair of cracked aluminum aircraft panel under the thermo-mechanical loading environment. Kumar et al. [74] simulated stable crack growth of two-dimensional domain under plane stress condition and performed nonlinear analysis using updated Lagrangian approach for large deformation. They incorporated J−R criterion, von Mises, elastic-predictor and plastic-corrector algorithms, and the Newton−Raphson iterative method in XFEM framework. Pathak et al. [75] implemented XFEM for the analysis of crack growth of three-dimensional domain under fatigue-thermal loading. Patil et al. [76] proposed a multiscale phase-field method and XFEM to analyze crack growth in heterogeneous materials.

In the present study, the authors have incorporated XFEM to model cracked plates using high-order shear deformation plate kinematics. Free flexural vibration study has been performed for the cracked functionally graded plates in a hygrothermal environment. FGMs are basically composites developed for the advanced applications. In the last few decades, composites have been used in the aviation industry and other applications because they provide high specific strength compared to metallic alloys. In composites, fibers are reinforced into the matrix to enhance mechanical properties. In fiber-reinforced composite, discontinuity in the material domain is present. To overcome the discontinuity in the material domain and the demand for enhanced mechanical and thermo-mechanical properties for the desired application, a different material called FGMs emerged.

Manufacturing of functionally graded structures or engineering components is quite complex. Cracks may arise in the domain during its manufacturing or may arise during operational stage. Sometimes cracks may be invisible. The cracks developed in the structural components may reduce the structural stiffness and the residual life of structures. Therefore the study of cracks in engineering components has always been an area of interest for engineers. There are various fracture mechanics parameters that provide the detailed information regarding the behavior of cracks present in the domain, such as the stress intensity factors, J-integrals, and crack opening displacements; however, structural analysis such as free vibration, bending, and buckling in the presence of crack, voids, and inclusions is essential in design and analysis part. Stress intensity factors are crucial parameters that quantify the intensity of stress singularity at crack tips of a cracked material. This quantity is used in linear elastic fracture mechanics to determine crack propagation and stability. Cracks may affect the structural integrity and structural response during operation. In the present work, vibrational analysis is reported as structural analysis. Knowledge of vibration is essential to avoid resonance disasters. Bachene et al. [26] implemented XFEM to investigate free vibration of cracked plate using FSDT plate kinematics. Natarajan et al. [19] implemented XFEM to perform free vibration analysis of cracked FGM plate. Raza et al. [77] performed in-plane free vibration analysis of the cracked FG plates using XFEM. Raza et al. [78] studied the in-plane free vibration of a bi-material FGM plate in deterministic and stochastic environments by implementing XFEM. Raza et al. [79] implemented XFEM to simulate the free vibration of cracked porous FG plates. Raza et al. [80] incorporated higher order shear deformation theory (HSDT) in XFEM formulation to solve the cracked porous plate in thermal environment. Raza et al. [18] implemented stochastic XFEM approach to quantify the uncertainty in free vibration with material randomness. Natarajan et al. [81] studied the free vibration of the cracked FGM plate in thermal environment. FGMs are widely used in aviation because of their high mechanical strength for large temperature differences. Including temperature and moisture in the free vibration analysis of the cracked plate is essential. Huang and Shen [82] studied the free vibration of FGM plates by implementing an analytical approach considering temperature effect. Talha and Singh [83] reported the FE analysis of the free vibration of the FGM plate considering temperature effect.

The literature survey shows that the enrichment technique is widely used to model the engineering components having crack in the domain. Enrichment technique reduces the complexity in the modeling and simulating the crack problem. By implementing the extended finite element approach, the issue of crack and material interface can be solved efficiently and accurately. In this chapter, XFEM has been explained and implemented for free vibration of cracked FGM plate.

This chapter is organized as follows: abstract, introduction, theoretical formulation, numerical results, and summary.

4.2 Theoretical formulation

This section includes the concept, implementation of enrichment function, and formulation of the extended finite element technique for the plate. Kinematics of plate implemented in XFEM formulation is based on higher-order shear deformation theory. The PU technique facilitates enrichment in the primary variable, whereas the level set method is used to trace the discontinuity in the domain; consequently, these two techniques are integral parts of the XFEM. PU technique facilitates the enrichment procedure by adding required local enrichment function in the primary variable. The augmented primary variable field is defined as an extended finite element approximation, which is explained later on.

4.2.1 Partition of unity

PU [35] is a strong mathematical technique that gives the background of the XFEM. The PU is a set of functions. The PU helps to augment the displacement approximation. PU states that:

$$\sum_{i \in I} f_i(x) = 1 \tag{4.1}$$

It is evident that by selecting any arbitrary function $g(x)$, the following properties will be satisfied:

$$\sum_{i \in I} f_i(x) g(x) = g(x) \tag{4.2}$$

4.2.2 Level set method

The level set method [35] is an essential component of XFEM, which is required to track the discontinuities in the domain. The level set method gives the information of the crack surface and cracks front in the domain. Level set method defines the split element, tip element, split nodes, and tip nodes. Once the split nodes and tip nodes tracked down in the domain, the local enrichment is done to that corresponding nodes. Consider a domain Ω having an interface dividing the domain into Ω_A and Ω_B. The level set function can be defined as (Fig. 4.1)

$$\begin{aligned} \psi(x) > 0 \quad & \text{if } x \in \Omega_A \\ \psi(x) = 0 \quad & \text{if } x \in \Gamma \\ \psi(x) < 0 \quad & \text{if } x \in \Omega_B \end{aligned}$$

4.2.3 Enrichment

The basic purpose of enrichment is to increase the order of completeness that can be achieved. Enrichment procedure is done by augmenting the primary variable with Heaviside and tip enrichment (asymptomatic) functions locally. These enrichment functions are stated as follows:

Heaviside function: Enrichment of the nodes corresponding to split elements (elements completely separated into two halves by crack) is done using the Heaviside enrichment function [40,42,43] and is defined as

$$\text{Heaviside function}, H(x) = \begin{cases} 1 & \mathit{if}(x - x^*) \cdot n \geq 0 \\ -1 & (x - x^*) \quad \mathit{otherwise} \end{cases} \tag{4.3}$$

The value of Heaviside is 1 at one side and -1 at other side; otherwise, the value is 0 at separating boundary.

Asymptomatic functions: Asymptomatic functions [42,49,50,78] or tip enrichment functions are used in enriching the nodes corresponding to tip elements to take care of the stress singularities near the crack front. There are various tip enrichments for isotropic materials, orthotropic materials, plane stress condition, interfacial crack tip, and for plate kinematics.

There are four asymptotic enrichment functions required to model crack tip in isotropic two-dimensional domain and are obtained from displacement field. To increase the solution accuracy, the isotropic crack tip enrichment function is augmented [71]. The isotropic crack tip enrichment functions obtained from displacement fields are defined as

$$\gamma(\boldsymbol{x}) = \left[r^{1/2}\sin\frac{\phi}{2}, r^{1/2}\sin\frac{\phi}{2}\sin\phi, r^{1/2}\cos\frac{\phi}{2}, r^{1/2}\cos\frac{\phi}{2}\sin\phi \right] \tag{4.4}$$

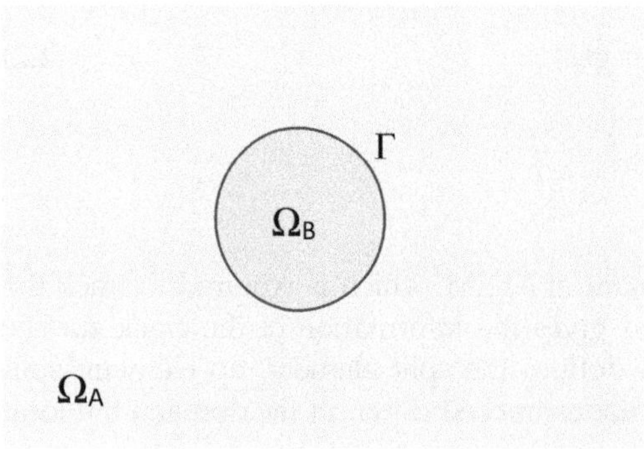

FIGURE 4.1 Level set function.

Different enrichment functions are used to enrich the displacement field in the case of orthotropic materials. The enrichment functions for the crack tip elements in the orthotropic material medium are defined as [84]

$$\gamma(x) = \left[r^{1/2}\cos\frac{\phi_1}{2}\left(g_1(\phi)\right)^{1/2}, r^{1/2}\cos\frac{\phi_2}{2}\left(g_2(\phi)\right)^{1/2}, r^{1/2}\sin\frac{\phi_1}{2}\left(g_1(\phi)\right)^{1/2}, r^{1/2}\sin\frac{\phi_2}{2}\left(g_2(\phi)\right)^{1/2}\right] \quad (4.5)$$

There are 12 asymptotic enrichment functions required to model interfacial crack tip in bi-material two-dimensional domain [49].

Crack tip enrichment function is different for the plate kinematics as the plate experiences both moment and shear field. A through crack in the plate is considered for the present study. Tip enrichment function for plate: enrichment function for displacement term is defined as G_α and enrichment function for rotation term is defined as F_α. Following are the enrichment functions (crack tip) to model the cracked plate having transverse shear deformation:

$$G_\alpha(r,\theta) = \left\{r^{1/2}\sin\left(\frac{\phi}{2}\right), r^{3/2}\sin\left(\frac{\phi}{2}\right), r^{3/2}\cos\left(\frac{\phi}{2}\right), r^{3/2}\sin\left(\frac{3\phi}{2}\right), r^{3/2}\cos\left(\frac{3\phi}{2}\right)\right\} \quad (4.6)$$

$$F_\alpha(r,\theta) = \left\{r^{1/2}\sin\left(\frac{\phi}{2}\right), r^{1/2}\cos\left(\frac{\phi}{2}\right), r^{1/2}\sin\left(\frac{\phi}{2}\right)\sin(\theta), r^{1/2}\cos\left(\frac{\phi}{2}\right)\sin(\theta)\right\} \quad (4.7)$$

where (r, ϕ) are polar coordinates (origin is at the crack tip) in the local coordinate system.

For defining the tip enrichment function for plate theory, only the term propertional to $\sqrt{r}$ is considered for the rotation, and the terms propertional to $\sqrt{r}$ and $r^{3/2}$ are considered for transverse displacement [42].

4.2.4 Selection of enriched nodes

The node required to enrich is selected with the help of the level set functions ϕ and ψ. Consider Γ_c be interior of crack; x_c any point on crack (Γ_c); x^* is closet to point x. If the level set $\phi(x) > 0$, then the point x is above the crack, and if $\phi(x) < 0$, then the point x is below the crack (Fig. 4.2). The orthogonal nature of zero-level sets ϕ and ψ defines the

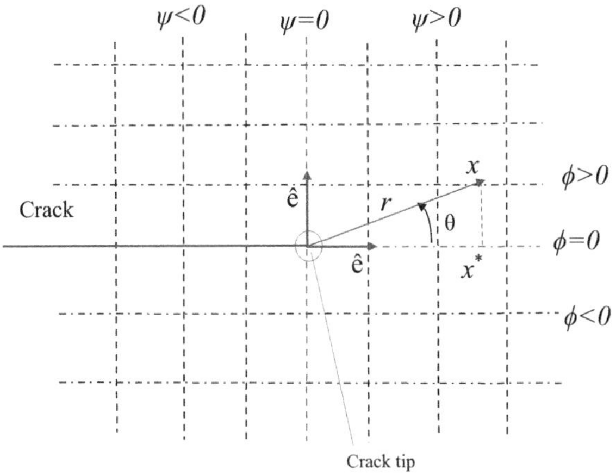

FIGURE 4.2 Level set functions for the selection of enriching nodes in the 2D domain.

crack tip. Let the maximum and minimum level set functions corresponding to ϕ and ψ be ϕ_{max}, ϕ_{min}, ψ_{max}, and ψ_{min}. Necessary conditions for the selection of split nodes are $\phi_{max} \times \phi_{min} \leq 0$ and $\psi < 0$. Whereas for the selection of tip nodes, necessary conditions are $\phi_{max} \times \phi_{min} \leq 0$ and $\psi_{max} \times \psi_{min} \leq 0$. A simple MATLAB code for the selection of enriched nodes is given.

MATLAB code for the selection of enriched nodes (enrichment) *due to crack discontinuity is given as follows:*

```
% -----------------------------------------------------------------------------------------------
% Level set initialization for all nodes
x0 = Cr_cord(1,1); y0 = Cr_cord(1,2);  % Cr_cord- crack coordinate in the domain
x1 = Cr_cord(2,1); y1 = Cr_cord(2,2);
t  = 1/norm(Cr_seg) * Cr_seg;
for i = 1 : num_nodes                        % num_nodes- number of nodes
 x = node_cord(i,1);                      % extract x-cordinate for all nodes
 y = node_cord(i,2);                      % extract y-cordinate for all nodes
 l  = sqrt((x1-x0)^2+(y1-y0)^2);       % Find the crack length
 phi = (y0-y1)*x + (x1-x0)*y + (x0*y1-x1*y0);   % Level set function
 ls(i,1) = phi/l ;                        % normal Level set
 ls(i,2) = ([x y]-Cr_Tip)*t';             % tangent Level set
end

% Find enriched nodes
enriched_nodes = zeros(num_nodes,1);
incre1 = 0;
incre2 = 0;
for iel = 1 : num_elems                   % num_elems is number of elements
  sctr = element_connectivity(iel,:); % Extract Elemental nodes
  phi  = ls(sctr,1);         % perpendicular dist from crack tip to mesh point for y coordinate
  psi  = ls(sctr,2);         % perpendicular dist from crack tip to mesh point for x coordinate
  if ( max(phi)*min(phi) < 0 )       % find the element along the crack
    if max(psi) < 0                  % find the element completely cut by crack
      incre1 = incre1 + 1 ;
      split_elem(incre1) = iel;      % locate the split element
      enrich_nodes(sctr)  = 1;       % mask for split element
    elseif max(psi)*min(psi) < 0     % find the element partially cut by crack
      incre2 = incre2 + 1 ;
      tip_elem(incre2) = iel;        % locate the tip element
      enrich_nodes(sctr)  = 2;       % mask for tip element
    end
  end
end
split_nodes = find(enrich_nodes == 1);  % enriched nodes corresponding to split element
tip_nodes   = find(enrich_nodes == 2);  % enriched nodes corresponding to tip element
%-----------------------------------------------------------------------------------------------
```

4.2.5 Extended finite element approximation

In the extended finite element approach, the discretization of the domain remains the same, and the crack is modeled separately by enriching the nodes near the crack. Enrichment is done by augmenting the displacement field by adding some additional enrichment functions to encounter the problem of stress singularities. The level set method [40] administers the enrichment by tracking the crack location. Fig. 4.3 shows the discretized domain with various crack terms.

The displacement field in Eq. (4.9) is augmented by adding the Heaviside and tip enrichment functions. The XFEM displacement field for the cracked domain is given as [80]

$$\Theta = u_{FEM} + u_{Heaviside\ enrichment} + u_{Tip\ enrichment} \tag{4.8}$$

where $\Theta = \{u, v, w\}^T$

$$\Theta(x) = \sum_{i=1}^{n} N_i(x)\left(u_i + \underbrace{(H(x) - H(x_i))u_i^a}_{i \varepsilon n_s} + \underbrace{\sum(\varphi_\alpha(x) - \varphi_\alpha(x_i))u_i^b}_{i \varepsilon n_t} \right) \tag{4.9}$$

where N is shape function, u_i is nodal displacement associated with finite element (nonenriched nodes), u_i^a is the unknown associated with nodes corresponding to split elements (the Heaviside enrichment, $H(x)$) [41], and u_i^b is the unknown associated with nodes corresponding to the tip element (tip enrichment function/asymptotic function). The asymptotic function, φ_α, is defined in Eqs. (4.6 and 4.7). In Eq. (4.9) n, n_s, and n_t represent nodes for finite elements, enriched nodes for split element, and enriched nodes for tip element, respectively.

4.2.6 Plate kinematics

4.2.6.1 *Displacement field*

Reddy [85] proposed a shear deformation theory for the flexural deformation of plate called the HSDT. In his theory, the displacement field is considered up to third order. Shear locking phenomena do not appear in this theory as compared to other theories. The displacement field based on HSDT is expressed as

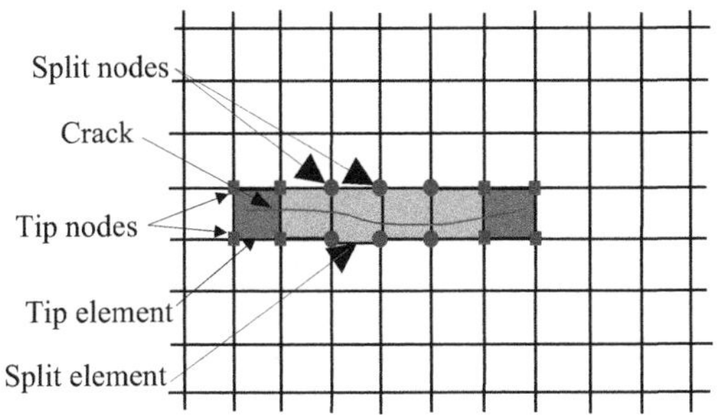

FIGURE 4.3 Schematic diagram of discretized cracked domain showing tip elements, split elements.

$$\begin{aligned} u(x,y,z,t) &= u_0(x,y,t) + \left(z - \frac{4z^3}{3h^2}\right)\Psi_x(x,y,t) - \frac{4z^3}{3h^2}\Phi_x(x,y,t) \\ v(x,y,z,t) &= v_0(x,y,t) + \left(z - \frac{4z^3}{3h^2}\right)\Psi_y(x,y,t) - \frac{4z^3}{3h^2}\Phi_y(x,y,t) \\ w(x,y,z,t) &= w_0(x,y,t) \end{aligned} \tag{4.10}$$

In matrix form, Eq. (4.10) can be written as

$$\begin{Bmatrix} u \\ v \\ w \end{Bmatrix} = \underbrace{\begin{bmatrix} 1 & 0 & 0 & z - c_1 z^3 & 0 & -c_1 z^3 & 0 \\ 0 & 1 & 0 & 0 & z - c_1 z^3 & 0 & -c_1 z^3 \\ 0 & 0 & 1 & 0 & 0 & 0 & 0 \end{bmatrix}}_{T_N} \begin{Bmatrix} u_0 \\ v_0 \\ w_0 \\ \Psi_x \\ \Psi_y \\ \Phi_x \\ \Phi_y \end{Bmatrix}$$

$$\{\Theta\} = [\boldsymbol{T_N}]\{\Xi\} \tag{4.11}$$

However, $c_1 = \frac{4}{3h^2}; \Psi_x = \frac{\partial w_0}{\partial x}; \Psi_y = \frac{\partial w_0}{\partial y}$

4.2.6.2 *Strain–displacement relation*

The strain–displacement relationship obtained from the abovementioned equations can be expressed as

$$\{\bar{\varepsilon}\} = \begin{Bmatrix} \varepsilon_{xx} \\ \varepsilon_{yy} \\ \gamma_{yz} \\ \gamma_{xz} \\ \gamma_{xy} \end{Bmatrix} = \begin{Bmatrix} \frac{\partial u}{\partial x} \\ \frac{\partial v}{\partial y} \\ \frac{\partial v}{\partial z} + \frac{\partial w}{\partial y} \\ \frac{\partial u}{\partial z} + \frac{\partial w}{\partial x} \\ \frac{\partial u}{\partial y} + \frac{\partial v}{\partial x} \end{Bmatrix} = \begin{Bmatrix} \underbrace{\frac{\partial u_0}{\partial x}}_{\varepsilon_1^0} + z \underbrace{\frac{\partial \Psi_x}{\partial x}}_{\varepsilon_1^1} - z^3 c_1 \underbrace{\left(\frac{\partial \Psi_x}{\partial x} + \frac{\partial \Phi_x}{\partial x}\right)}_{\varepsilon_1^3} \\ \underbrace{\frac{\partial v_0}{\partial y}}_{\varepsilon_2^0} + z \underbrace{\frac{\partial \Psi_y}{\partial y}}_{\varepsilon_2^1} - z^3 c_1 \underbrace{\left(\frac{\partial \Psi_y}{\partial y} + \frac{\partial \Phi_y}{\partial y}\right)}_{\varepsilon_2^3} \\ \underbrace{\Psi_y + \frac{\partial w_0}{\partial y}}_{\varepsilon_4^0} - z^2 \underbrace{3c_1(\Psi_y + \Phi_y)}_{\varepsilon_4^2} \\ \underbrace{\Psi_x + \frac{\partial w_0}{\partial x}}_{\varepsilon_5^0} - z^2 \underbrace{3c_1(\Psi_x + \Phi_x)}_{\varepsilon_5^2} \\ \underbrace{\frac{\partial u_0}{\partial y} + \frac{\partial v_0}{\partial x}}_{\varepsilon_6^0} + z \underbrace{\left(\frac{\partial \Psi_x}{\partial y} + \frac{\partial \Psi_y}{\partial x}\right)}_{\varepsilon_6^1} - z^3 c_1 \underbrace{\left(\frac{\partial \Psi_x}{\partial y} + \frac{\partial \Psi_y}{\partial x} + \frac{\partial \Phi_x}{\partial y} + \frac{\partial \Phi_y}{\partial x}\right)}_{\varepsilon_6^3} \end{Bmatrix} \tag{4.12}$$

$$\{\bar{\varepsilon}\} = \begin{Bmatrix} \varepsilon_{xx} \\ \varepsilon_{yy} \\ \gamma_{yz} \\ \gamma_{xz} \\ \gamma_{xy} \end{Bmatrix} = \begin{bmatrix} \varepsilon_1^0 + z\varepsilon_1^1 + z^3\varepsilon_1^3 \\ \varepsilon_2^0 + z\varepsilon_2^1 + z^3\varepsilon_2^3 \\ \varepsilon_4^0 + z^2\varepsilon_4^2 \\ \varepsilon_5^0 + z^2\varepsilon_5^2 \\ \varepsilon_6^0 + z\varepsilon_6^1 + z^3\varepsilon_6^3 \end{bmatrix} \tag{4.13}$$

Further strain can be written as

$$\begin{Bmatrix} \varepsilon_{xx} \\ \varepsilon_{yy} \\ \gamma_{yz} \\ \gamma_{xz} \\ \gamma_{xy} \end{Bmatrix} = \begin{Bmatrix} \varepsilon_1^0 \\ \varepsilon_2^0 \\ \varepsilon_4^0 \\ \varepsilon_5^0 \\ \varepsilon_6^0 \end{Bmatrix} + z\begin{Bmatrix} \varepsilon_1^1 \\ \varepsilon_2^1 \\ 0 \\ 0 \\ \varepsilon_6^1 \end{Bmatrix} + z^2\begin{Bmatrix} 0 \\ 0 \\ \varepsilon_4^2 \\ \varepsilon_5^2 \\ 0 \end{Bmatrix} + z^3\begin{Bmatrix} \varepsilon_1^3 \\ \varepsilon_2^3 \\ 0 \\ 0 \\ \varepsilon_6^3 \end{Bmatrix} \tag{4.14}$$

Eq. (4.19) can be written as

$$\{\bar{\varepsilon}\} = [T_B]\{\varepsilon\} \tag{4.15}$$

4.2.6.3 *Material constitutive relations*

FGMs are microscopically inhomogeneous. Effective material properties are calculated at each point in the domain using a power law and the rule of mixture. Hence, Hooke's law for homogenous and isotropic materials can be implemented. The stress displacement relation for FGM is given as

$$\begin{Bmatrix} \sigma_{xx} \\ \sigma_{yy} \\ \tau_{yz} \\ \tau_{xz} \\ \tau_{xy} \end{Bmatrix} = \begin{bmatrix} \bar{q}_{11} & \bar{q}_{12} & 0 & 0 & 0 \\ \bar{q}_{12} & \bar{q}_{22} & 0 & 0 & 0 \\ 0 & 0 & \bar{q}_{44} & 0 & 0 \\ 0 & 0 & 0 & \bar{q}_{55} & 0 \\ 0 & 0 & 0 & 0 & \bar{q}_{66} \end{bmatrix} \begin{Bmatrix} \varepsilon_{xx} \\ \varepsilon_{yy} \\ \gamma_{yz} \\ \gamma_{xz} \\ \gamma_{xy} \end{Bmatrix} \tag{4.16}$$

$$\bar{q}_{11} = \bar{q}_{22} = \frac{E(z,T)}{1-\nu^2}; \bar{q}_{12} = \frac{\nu E(z,T)}{1-\nu^2}; \bar{q}_{44} = \bar{q}_{55} = \bar{q}_{66} = \frac{E(z,T)}{2(1+\nu)}$$

4.2.7 Modeling of functionally graded plate

FGM plates are made of two or more materials, and their volume varies with some function. Generally, FGMs are made of ceramic and metal for heat-resistant applications. Single porous material can also be called FGM, where the size of pores and density of pores vary along the particular direction. There are various investigations available in literature considering porous material as FGM. In the present chapter, the study is focused on FGM made of metal and ceramic. The mathematical modeling of the functionally graded plate is done using the power law. The schematic diagram of the FGM plate is shown in Fig. 4.4. The mid-plane represented by the dotted line is assumed to be the origin. The effective Young

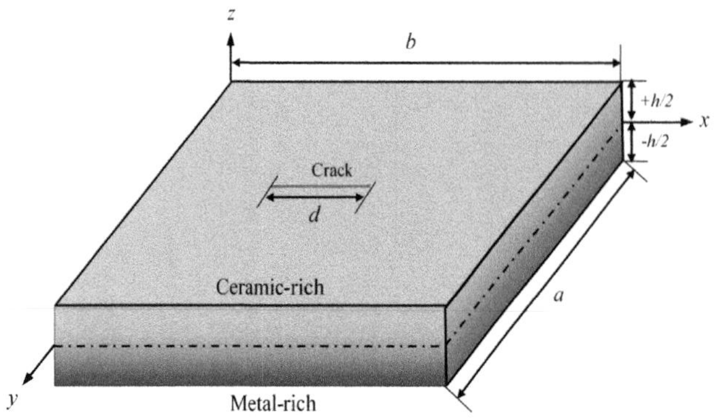

FIGURE 4.4 The schematic diagram of the FGM plate. *FGM*, Functionally graded material.

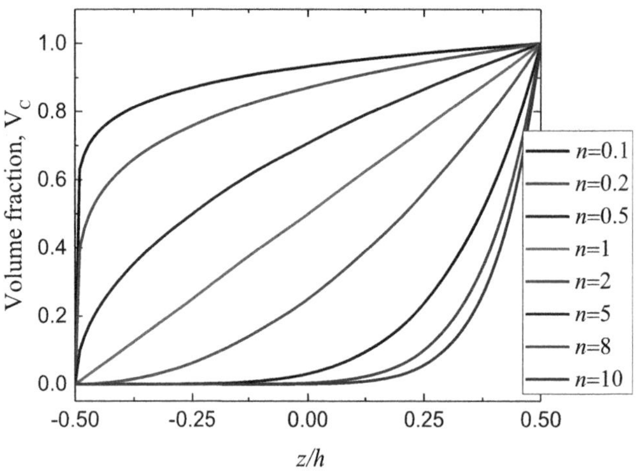

FIGURE 4.5 Gradation of volume fraction from bottom to top surface.

modulus E, thermal expansion coefficient α, *and* moisture expansion coefficient β are assumed to be temperature-dependent properties, whereas the mass density ρ and thermal conductivity κ are assumed as temperature-independent properties. The plot shown in Fig. 4.5 defines the gradation of volume fraction using power law.

The effective material properties of Young's modulus E, thermal expansion coefficient α, and moisture expansion coefficient can be computed using the rule of mixture, which is given as

$$P(z,T) = [P_t(T) - P_b(T)]\left(\frac{h+2z}{2h}\right)^n + P_b(T) \tag{4.17}$$

where P is the generalized term for Young's modulus, thermal expansion coefficient, and moisture expansion coefficient. However, n is the volume fraction index, t and b denote the bottom surface of the plate, respectively. The given empirical relation can evaluate the temperature-dependent properties [43]:

$$P(T) = P_0\left(P_{-1}T^{-1} + P_1T + P_2T^2 + P_3T^3 + 1\right) \tag{4.18}$$

The temperature and moisture distribution across the thickness can be expressed as

$$T(z) = T_b + (T_t - T_b)\left(\frac{2z + h}{2h}\right) \tag{4.19}$$

$$\overline{M}(z) = \overline{M}_b + \left(\overline{M}_t - \overline{M}_b\right)\left(\frac{2z + h}{2h}\right) \tag{4.20}$$

where $\overline{M}_b = 0$; $T_b = 300$ K. Temperature and moisture rise is considered on the ceramic side.

The material properties implemented in the present work is given in Table 4.4 and Table 4.5. The temperature-dependent material coefficient given in Table 4.5 is used in Eq. (4.17). Eqs. (4.17)–(4.22) can be used to find out the effective material properties.

4.2.8 Governing equation

4.2.8.1 *Strain energy*

The total strain energy of the initially stressed plate is given as $U = \Pi + \Pi_{th}$
However,

$$\Pi = \frac{1}{2}\int_v [\sigma\varepsilon] dV \tag{4.21}$$

4.2.8.2 *Energy due to temperature rise and moisture rise*

Energy due to temperature and moisture change is written as

$$\Pi_{th} = \frac{1}{2}\int_A \left[F_{xx}(w_{0,x})^2 + F_{yy}(w_{0,y})^2 + 2F_{xy}(w_{0,x})(w_{0,y})\right] dA \tag{4.22}$$

Where, $\begin{Bmatrix} F_{xx} \\ F_{yy} \\ F_{xy} \end{Bmatrix} = \{F^T\} = \int_{-h/2}^{h/2} \{\forall\} dz$; $\{S^T\}$ is load due to temperature and moisture change

$$\{\forall\} = -\begin{bmatrix} \overline{q}_{11} & \overline{q}_{12} & 0 \\ \overline{q}_{12} & \overline{q}_{22} & 0 \\ 0 & 0 & \overline{q}_{66} \end{bmatrix} \begin{bmatrix} 1 & 0 \\ 0 & 1 \\ 0 & 0 \end{bmatrix} \begin{bmatrix} \alpha(z,T)\Delta T + \beta(z,C)\Delta C \\ \alpha(z,T)\Delta T + \beta(z,C)\Delta C \\ 0 \end{bmatrix} \tag{4.23}$$

4.2.8.3 *Kinetic energy*

The kinetic energy of the plate is given as

$$T = \frac{1}{2}\int_v \rho(z)\left[\dot{u}^2 + \dot{v}^2 + \dot{w}^2\right] dv \tag{4.24}$$

Using all energy equation mentioned earlier, the final equation can be derived as

$$[M]\{\ddot{\Xi}\} + [K - K_{th}]\{\Xi\} = 0 \tag{4.25}$$

4.2.9 Numerical integration

In the XFEM, discretization of the domain is done, and the discrete governing equation governs the physical behavior of the discretized domain. A four-noded quadrilateral element is implemented to discretize the domain. The physical behavior of the whole domain can be obtained by implementing the numerical integration technique. The Gauss quadrature rule can be implemented to evaluate the integral of the mass matrix and stiffness matrix. With the introduction of the crack in the domain and the presence of split and tip elements, integration is interrupted; hence, gauss quadrature rule cannot be implemented. To overcome this problem, a necessary modification is required in the numerical integration for the split and tip element. Gauss integration points are increased, and the integration is performed by doing subtriangulation in the split elements and partially cut elements with cracks. The present study uses four-noded quadrilateral element to discretize the domain. Gauss integration points with subtriangulation are shown in Fig. 4.6.

4.2.10 Boundary conditions

The various boundary condition used in the present work is expressed as follows (Fig. 4.7):

Simply supported (SSSS)

v_0, w_0, Φ_y, and Ψ_y are zero at the left and right edges of the plate, whereas u_0, w_0, Φ_x, and Ψ_x are zero at the bottom and top edges of the plate, as shown in Fig. 4.7.

Clamped support (CCCC)

u_0, v_0, w_0, Φ_x, Ψ_x, Φ_y, and Ψ_y are zero at the left, right, bottom, and top edges.

Cantilever support (FFFC)

u_0, v_0, w_0, Φ_x, Ψ_x, Φ_y, and Ψ_y are zero at the right edge.

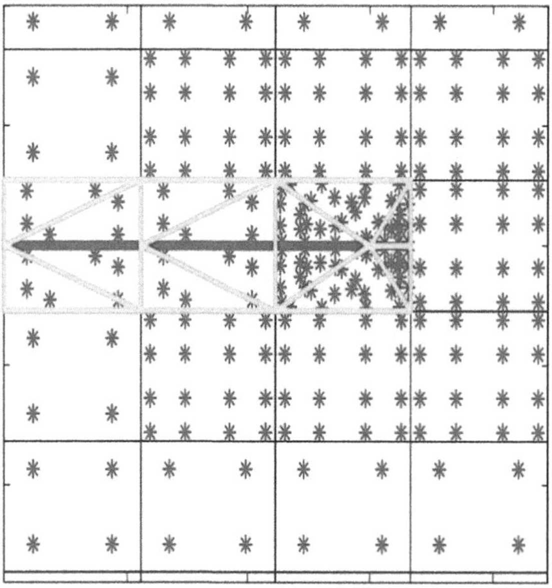

FIGURE 4.6 Subtriangulation of tip element and split elements with Gauss integration points.

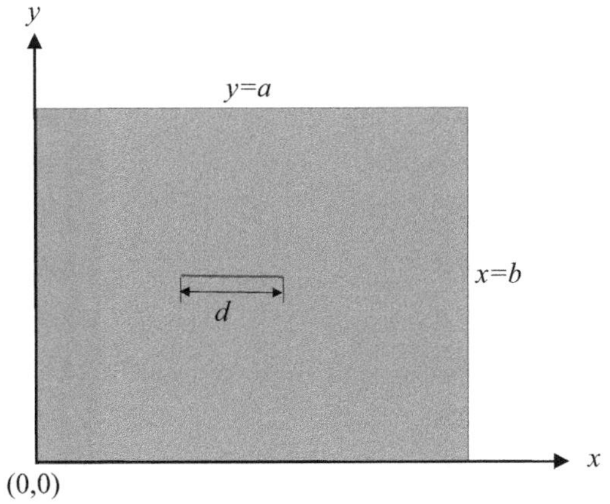

FIGURE 4.7 Cracked plate with dimension.

4.3 Numerical results

The extended finite element formulation is developed for the given physical problem and implemented to perform free vibration analysis of the cracked FG plates in a hygrothermal environment. The developed XFEM formulation is implemented in in-house MATLAB code to compute the numerical problem. The in-house MATLAB code based on the developed formulation is tested by solving some numeral problems. The convergence and comparison studies are done to check the accuracy of the formulation. Further, parametric study is reported for various variable parameters such as volume fraction, crack, temperature, moisture, and various boundary conditions in detail.

4.3.1 Convergence and validation study

Example 1: A simply supported square FGM (SUS304/Si_3N_4) plate with center crack is solved, the geometry and material properties of which are given as $b/a=1$; $a/h=10$; $E_c=322$ GPa; $\rho_c=2370\ \text{kg/m}^3$; $\nu_c=0.24$; $E_m=207$ GPa; $\rho_m=8166\ \text{kg/m}^3$; $\nu_m=0.3177$. The problem is solved for free flexural vibration response. The results obtained for various gradient indices and crack sizes are presented in Table 4.1 and compared with the reference results. The obtained results are presented in nondimensional frequencies $\omega^*=\omega a^2/h\sqrt{\rho_c/E_c}$. The results exhibited are very close to the reference results.

Example 2: This problem is introduced for the validation of free vibration of noncracked FGM plate. The problem is solved for various temperature differences across the thickness and simply supported boundary condition. The problem is defined here as $a/b=1$; $a/h=8$; material SUS304/Si_3N_4. The nondimensional frequency $\omega^*=\omega(a^2/h)\sqrt{\rho_m(1-\nu^2)/E_m}$ is evaluated. The convergence study is performed and presented in Fig. 4.8. The comparisons of results with the analytical solutions [44] (reference results) are presented in Table 4.2. It is observed that the present results are near the analytical results, which shows the accuracy of the present formulation.

TABLE 4.1 Comparison of linear frequency of the plate with center crack.

Gradient Index, n	Crack length, d/a	Natarajan et al. [19]	Present
2	0	3.0016	3.0893
	0.4	2.7383	2.7984
	0.6	2.5769	2.6738
	0.8	2.4747	2.5987
5	0	2.7221	2.7894
	0.4	2.4833	2.5267
	0.6	2.3371	2.4128
	0.8	2.2445	2.3442

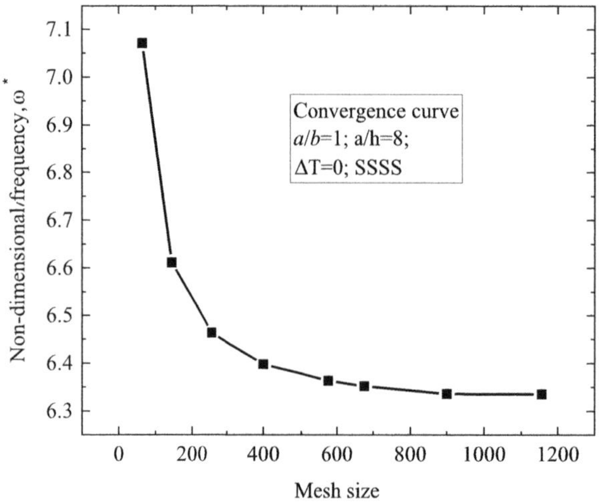

FIGURE 4.8 Convergence analysis of vibration of cracked FGM plate ($n = 2$); $T_t = 300$K; $T_b = 300$K. *FGM*, Functionally graded material.

TABLE 4.2 Comparison study of vibration of functionally graded material plate e, $a/b = 1$; $a/h = 8$; SSSS.

	ΔT K	$n = 0.5$	$n = 1$	$n = 2$
Reference [82]	0	7.139	6.657	6.286
Present		7.168	6.709	6.325
Reference [82]	100	6.876	6.437	6.101
Present		6.939	6.534	6.197
Reference [82]	300	6.123	5.819	5.612
Present		6.211	5.912	5.695

Example 3: This problem is introduced for validating the free vibration of FGM plates having crack in it. Geometry of the plate is considered $a/b = 1$ and $a/h = 10$. The problem is solved for various gradient indices and crack sizes in temperature environment. The material $SUS304/Si_3N_4$ is used for the analysis. The nondimensional frequency of the plate is evaluated as $\omega^* = \omega(a^2/h)\sqrt{12(1-\nu^2)\rho_c/E_c}$. The obtained results are presented in Table 4.3 and compared with the reference results, which shows the accuracy of the present formulation.

4.3.2 Parametric study

The computational investigation of the temperature and moisture effect on the free vibration of the cracked FG plates is presented in detail. The XFEM formulation based on HSDT plate kinematics is employed in MATLAB code development. The MATLAB code is developed to solve the free vibration problem in the hygrothermal environment. The material properties given in Tables 4.4 and 4.5 are implemented in the present study. To check the accuracy of the developed formulation, two numerical values are introduced, and the results are compared with the reference results. The influence of crack, volume fraction, temperature, moisture, and various boundary conditions has been studied in detail. The geometry of the cracked FGM ($a/b = 1$; $a/h = 10$) plate shown in Fig. 4.4 is used. Edge-cracked FGM plate is solved for free vibration, and results are reported in Table 4.6

TABLE 4.3 Comparison study of vibration of functionally graded material plate, $a/b = 1$; $a/h = 10$; $SUS304/Si_3N_4$; $T_t = 300K$; $T_b = 300K$; SSSS.

		Crack size d/a			
	n	**0**	**0.2**	**0.4**	**0.6**
Reference [18]		18.357	17.815	16.805	15.829
Present	0	18.601	18.045	17.143	16.181
Reference [18]		12.584	12.214	11.522	10.853
Present	0.5	12.812	12.631	12.011	11.486
Reference [18]		11.069	10.742	10.132	9.544
Present	1	11.260	11.068	10.524	10.063
Reference [18]		9.954	9.659	9.111	8.582
Present	2	10.113	9.933	9.441	9.024

TABLE 4.4 Material properties of the FG plate [86].

	Young's modulus, E (GPa)	**Poisson's ratio, ν**	**Density, ρ (kg/m^3)**
Si_3N_4	322.27	0.28	2370
SUS304	207.78	0.28	8166

TABLE 4.5 Temperature-dependent material coefficient [86].

	Properties	P_0	P_{-1}	P_1	P_2	P_3
Si_3N_4	E (Pa)	348.43×10^9	0	-3.070×10^{-4}	2.160×10^{-7}	-8.946×10^{-11}
	α $(1/\kappa)$	5.8723×10^{-6}	0	9.095×10^{-4}	0	0
	β	0	0	0	0	0
SUS304	E (Pa)	201.04×10^9	0	3.079×10^{-4}	-6.534×10^{-7}	0
	α $(1/\kappa)$	12.330×10^{-6}	0	8.086×10^{-4}	0	0
	β	0.0005	0	0	0	0

TABLE 4.6 Nondimensional natural frequencies for clamped (CCCC) square functionally graded material ($SUS304/Si_3N_4$); $b/a = 1$; $a/h = 10$; edge crack.

		Nondimensional frequencies		
n	d/a	CCCC	SSSS	FFFC
0	0	9.9050	5.8036	1.0356
	0.1	9.8584	5.7960	1.0355
	0.2	9.8353	5.7746	1.0354
	0.3	9.7749	5.7173	1.0353
	0.4	9.6336	5.6111	1.0352
	0.5	9.4053	5.4622	1.0351
0.5	0	6.8336	3.9962	0.7137
	0.1	6.8070	3.9944	0.7138
	0.2	6.7918	3.9799	0.7137
	0.3	6.7501	3.9403	0.7136
	0.4	6.6473	3.8641	0.7135
	0.5	6.4953	3.7657	0.7134
5	0	4.8490	2.8555	0.5096
	0.1	4.8282	2.8531	0.5096
	0.2	4.8171	2.8426	0.5095
	0.3	4.7872	2.8139	0.5095
	0.4	4.7142	2.7587	0.5094
	0.5	4.6063	2.6871	0.5093
10	0	4.6222	2.7208	0.4854
	0.1	4.6010	2.7177	0.4854
	0.2	4.5904	2.7076	0.4853
	0.3	4.5619	2.6803	0.4853
	0.4	4.4922	2.6277	0.4852
	0.5	4.3892	2.5595	0.4851

(nondimensional linear frequency $\omega^* = \omega(a^2/h)\sqrt{\rho_c/E_c}$). Thickness ratio is assumed to be $a/h = 10$. There is considerable change in the nondimensional frequency with the temperature variation. Also, various crack sizes alter the vibration response. With the increase in gradient index, the volume fraction of ceramic increases, consequently decreasing the stiffness; hence, the linear frequency decreases.

Further free flexural vibration analysis of FGM plate with center crack is performed in thermal environment. The results obtained are reported in Table 4.7. Boundary condition on all four sides of plate is considered fully clamped (CCCC). The analysis is performed for various temperature environments. Here, the temperature on metallic side is fixed at 300K, and the temperature on ceramic side is increasing. The material of FGM is considered SUS304/Si_3N_4. The numerical value of linear frequency is decreasing with the rise in temperature on ceramic side. Increase in the length of the crack in plate greatly influences the linear frequency. The numerical value of linear frequency decreases with the increase in the length of the crack.

Further, the analysis is done to investigate the influence of hygrothermal on the linear frequency of FG plates. A crack is assumed at the center of the plate at, and investigation is done for various temperature and moisture conditions. The results obtained are reported in Table 4.8. Boundary condition on all four sides of plate is considered simply supported (SSSS). Here, it clear that the temperature variation has a significant influence on free vibration. Similar studies have been done for mixed boundary conditions (SSCC) and all sides clamped boundary conditions. Results obtained for mixed boundary condition are given in Table 4.9. The results obtained for the fully clamped support boundary condition are given in Table 4.10.

TABLE 4.7 Nondimensional frequency ($\omega^* = \omega(a^2/h)\sqrt{\rho_m/E_m}$) of the plate; $h/a = 0.1$; $a/b = 1$; SUS304/Si_3N_4; CCCC; $T_b = 300K$.

		d/a			
	n	0	0.2	0.4	0.6
$T_t = 300K$	0.5	15.9681	15.0593	14.7811	14.3670
	1	13.9798	13.1845	12.9405	12.5773
	2	12.5084	11.7974	11.5782	11.2513
	10	10.7945	10.1808	9.9904	9.7062
$T_t = 400K$	0.5	15.8850	14.9814	14.7047	14.2928
	1	13.9098	13.1188	12.8762	12.5149
	2	12.4487	11.7414	11.5234	11.1982
	10	10.7480	10.1373	9.9479	9.6651
$T_t = 600K$	0.5	15.7283	14.8344	14.5607	14.1530
	1	13.7714	12.9890	12.7490	12.3915
	2	12.3215	11.6222	11.4066	11.0848
	10	10.6180	10.0153	9.8285	9.5493

TABLE 4.8 Nondimensional frequency ($\omega^* = \omega(a^2/h)\sqrt{\rho_m/E_m}$) of the plate; $a/b = 1$; $a/h = 10$; SUS304/Si_3N_4; SSSS.

		Nondimensional frequency		
Gradient index, *n*	**Crack size, *d/a***	**(ΔT K, ΔC %) (0,2)**	**(ΔT K, ΔC %) (20,2)**	**(ΔT K, ΔC %) (50,2)**
0.5	0.2	9.1019	9.0922	9.0779
	0.3	8.8927	8.8832	8.8693
	0.4	8.6810	8.6718	8.6582
	0.6	8.3232	8.3144	8.3014
2	0.2	7.0033	6.9966	6.9866
	0.3	6.8980	6.8914	6.8815
	0.4	6.7792	6.7728	6.7631
	0.6	6.5344	6.5283	6.5190
10	0.2	6.0581	6.0532	6.0455
	0.3	5.9661	5.9614	5.9538
	0.4	5.8622	5.8576	5.8501
	0.6	5.6478	5.6433	5.6363

TABLE 4.9 Nondimensional frequency ($\omega^* = \omega(a^2/h)\sqrt{\rho_m/E_m}$) of the plate; a/b = 1; $a/h = 10$; SUS304/Si_3N_4; SSCC.

		Nondimensional frequency		
Gradient index, *n*	**Crack size, *d/a***	**(ΔT K, ΔC %) (0,2)**	**(ΔT K, ΔC %) (20,2)**	**(ΔT K, ΔC %) (50,2)**
0.5	0.2	13.0265	13.0124	12.9916
	0.3	12.8215	12.8076	12.7871
	0.4	12.6186	12.6050	12.5849
	0.6	12.2826	12.2694	12.2498
2	0.2	10.0634	10.0536	10.0389
	0.3	9.9588	9.9492	9.9346
	0.4	9.8483	9.8388	9.8244
	0.6	9.6215	9.6122	9.5982
10	0.2	8.6876	8.6806	8.6695
	0.3	8.5964	8.5895	8.5785
	0.4	8.4998	8.4930	8.4821
	0.6	8.3010	8.2944	8.2839

TABLE 4.10 Nondimensional frequency $\omega^* = \omega(a^2/h)\sqrt{\rho_m/E_m}$ of the functionally graded material plate; $a/h = 10$; $a/b = 1$; SUS304/Si_3N_4; CCCC

		Nondimensional frequency		
Gradient index, *n*	**Crack size, *d*/*a***	**(ΔT K, ΔC %) (0,2)**	**(ΔT K, ΔC %) (20,2)**	**(ΔT K, ΔC %) (50,2)**
0.5	0.2	15.7107	15.6940	15.6694
	0.3	15.3738	15.3575	15.3335
	0.4	15.0500	15.0342	15.0107
	0.6	14.5461	14.5308	14.5081
2	0.2	12.0745	12.0632	12.0461
	0.3	11.9042	11.8931	11.8762
	0.4	11.7257	11.7147	11.6982
	0.6	11.3858	11.3752	11.3592
10	0.2	10.4155	10.4075	10.3947
	0.3	10.2682	10.2604	10.2478
	0.4	10.1133	10.1056	10.0933
	0.6	9.8173	9.8098	9.7979

4.4 Summary

This chapter provides a detailed extended finite element formulation for solving the cracked plate. An extensive literature review of XFEM and its implementation is reported. The potential of the enrichment technique is presented by explaining the advantages of XFEM over FEM. XFEM formulation is developed by incorporating higher-order shear deformation to solve the cracked FGM plate problem. The developed extended finite element formulation is used to develop MATLAB code to solve the considered problem. To assess the efficiency and accuracy of the developed formulation, some numerical problems are introduced. The convergence and validation of the work are presented. The study of the free vibration of the cracked plate is done to investigate the influence of crack size, gradient index, and temperature hygrothermal on linear frequency. The temperature and moisture variations have a significant impact on the linear frequency. Following are the major conclusions from the numerical analysis.

- The numerical value of linear frequency is decreasing with the rise in temperature and moisture
- With the increase in the length of crack in plate, the natural frequency of the plate is decreasing.
- With the increase in gradient index, the volume fraction of ceramic increases, consequently decreasing the stiffness; hence, the linear frequency decreases.

References

[1] Lynn PP, Kumbasar N. Free vibrations of thin rectangular plates having narrow cracks with simply supported edges. In: Proc. of Tenth Midwestern Mechanics Conf., Colorado: Developments in Mechanics, 1967;4:911–28.

[2] Stahl B, Keer L. Vibration and stability of cracked rectangular plates. International Journal of Solids and Structures 1972;8:69–91. Available from: https://doi.org/10.1016/0020-7683(72)90052-2.

[3] Hirano Y, Okazaki K. Bulletin of JSME 1980;23:732–40.

[4] Joshi PV, Jain NK, Ramtekkar GD. Analytical modeling for vibration analysis of thin rectangular orthotropic/functionally graded plates with an internal crack. Journal of Sound and Vibration 2015;344:377–98. Available from: https://doi.org/10.1016/j.jsv.2015.01.026.

[5] Jameel A, Harmain GA. Fatigue crack growth in presence of material discontinuities by EFGM. International Journal of Fatigue 2015;81:105–16. Available from: https://doi.org/10.1016/j.ijfatigue.2015.07.021.

[6] Qian GL, Gu SN, Jiang JS. The dynamic behavior of beam with crack and crack detection. Journal of Sound and Vibration 1990;138:233–43. Available from: https://doi.org/10.1016/0022-460X(90)90540-G.

[7] Qian GL, Gu SN, Jiang JS. A finite element model of cracked plates and application to vibration problems. Computers & Structures 1991;39:483–7. Available from: https://doi.org/10.1016/0045-7949(91)90056-R.

[8] Belytschko T, Lu YY, Gu L. Crack propagation by element-free Galerkin methods. Engineering Fracture Mechanics 1995;51:295–315. Available from: https://doi.org/10.1016/0013-7944(94)00153-9.

[9] Belytschko T, Gu L, Lu YY. Fracture and crack growth by element-free Galerkin methods. Modelling and Simulation in Materials Science and Engineering 1994;2:519–34. Available from: https://doi.org/10.1088/0965-0393/2/3A/007.

[10] Kanth SA, Harmain GA, Jameel A. Modeling of nonlinear crack growth in steel and aluminum alloys by the element free Galerkin method. Materials Today: Proceedings 2018;5(9):18805–14. Available from: https://doi.org/10.1016/j.matpr.2018.06.227.

[11] Zhang H, Wu J, Wang D. Free vibration analysis of cracked thin plates by quasi-convex coupled isogeometric-meshfree method. Frontiers of Structural and Civil Engineering 2015;9:405–19. Available from: https://doi.org/10.1007/s11709-015-0310-1.

[12] Pathak H, Singh A, Singh IV, Brahmankar M. Three-dimensional stochastic quasi-static fatigue crack growth simulations using coupled FE-EFG approach. Computers & Structures 2015;160:1–19. Available from: https://doi.org/10.1016/j.compstruc.2015.08.002.

[13] Singh SK, Singh IV, Mishra BK, Bhardwaj G, Bui TQ. A simple, efficient and accurate Bézier extraction based T-spline XIGA for crack simulations. Theoretical and Applied Fracture Mechanics 2017;88:74–96. Available from: https://doi.org/10.1016/j.tafmec.2016.12.002.

[14] Jameel A, Harmain GA. Extended iso-geometric analysis for modeling three dimensional cracks. Mechanics of Advanced Materials and Structures 2019;26:915–23. Available from: https://doi.org/10.1080/15376494.2018.1430275.

[15] Jameel A, Harmain GA. A coupled FE-IGA technique for modeling fatigue crack growth in engineering materials. Mechanics of Advanced Materials and Structures 2019;26:1764–75. Available from: https://doi.org/10.1080/15376494.2018.1446571.

[16] Tran LV, Ly HA, Lee J, Wahab MA, Nguyen-Xuan H. Vibration analysis of cracked FGM plates using higher-order shear deformation theory and extended isogeometric approach. International Journal of Mechanical Sciences 2015;96–97:65–78. Available from: https://doi.org/10.1016/j.ijmecsci.2015.03.003.

[17] Belytschko T, Black T. Elastic crack growth in finite elements with minimal remeshing. International Journal for Numerical Methods in Engineering 1999;45(5):601–20. Available from: https://doi.org/10.1002/(SICI)1097-0207(19990620)45:5 < 601::AID-NME598 > 3.0.CO;2-S.

[18] Raza A, Pathak H, Talha M. Influence of material uncertainty on vibration characteristics of higher-order cracked functionally gradient plates using XFEM. International Journal of Applied Mechanics 2021;13(5) 2150062. Available from: https://doi.org/10.1142/S1758825121500629.

[19] Natarajan S, Baiz PM, Bordas S, Rabczuk T, Kerfriden P. Natural frequencies of cracked functionally graded material plates by the extended finite element method. Composite Structures 2011;93:3082–92. Available from: https://doi.org/10.1016/j.compstruct.2011.04.007.

[20] Nasirmanesh A, Mohammadi S. An extended finite element framework for vibration analysis of cracked FGM shells. Composite Structures 2017;180:298–315. Available from: https://doi.org/10.1016/j.compstruct.2017.08.019.

[21] Pathak H, Singh A, Singh IV, Yadav SK. Fatigue crack growth simulations of 3-D linear elastic cracks under thermal load by XFEM. Frontiers of Structural and Civil Engineering 2015;9:359–82. Available from: https://doi.org/10.1007/s11709-015-0304-z.

[22] Kumar S, Shedbale AS, Singh IV, Mishra BK. Elasto-plastic fatigue crack growth analysis of plane problems in the presence of flaws using XFEM. Frontiers of Structural and Civil Engineering 2015;9:420–40. Available from: https://doi.org/10.1007/s11709-015-0305-y.

[23] Kumar S, Singh IV, Mishra BK, Rabczuk T. Modeling and simulation of kinked cracks by virtual node XFEM. Computer Methods in Applied Mechanics and Engineering 2015;283:1425–66. Available from: https://doi.org/10.1016/j.cma.2014.10.019.

[24] Kanth SA, Jameel A, Harmain GA. Investigation of fatigue crack growth in engineering components containing different types of material irregularities by XFEM. Mechanics of Advanced Materials and Structures 2022;29(24):3570–87. Available from: https://doi.org/10.1080/15376494.2021.1907003.

[25] Nguyena VP, Anitescu C, Bordasa SPA, Rabczuk T. Isogeometric analysis: An overview and computer implementation aspects. Mathematics and Computers in Simulation 2015;117:89–116. Available from: https://doi.org/10.1016/j.matcom.2015.05.008.

[26] Bachene M, Tiberkak R, Rechak S. Vibration analysis of cracked plates using the extended finite element method. Archive of Applied Mechanics 2009;79:249–62. Available from: https://doi.org/10.1007/s00419-008-0224-7.

[27] Moës N, Dolbow J, Belytschko T. A finite element method for crack growth without remeshing. International Journal for Numerical Methods in Engineering 1999;46:131–50. Available from: https://doi.org/10.1002/(SICI)1097-0207(19990910)46:1 < 131::AID-NME726 > 3.0.CO;2-J.

[28] Schneider D, Schoof E, Huang Y, Selzer M, Nestler B. Phase-field modeling of crack propagation in multiphase system. Computer Methods in Applied Mechanics and Engineering 2016;312(1):186–95. Available from: https://doi.org/10.1016/j.cma.2016.04.009.

[29] Gomez H, Calo VM, Bazilevs Y, Hughes TJR. Isogeometric analysis of the Cahn–Hilliard phase-field model. Computer Methods in Applied Mechanics and Engineering 2008;197:4333–52. Available from: https://doi.org/10.1016/j.cma.2008.05.003.

[30] Amiri F, Millán D, Shen Y, Rabczuk T, Arroyo M. Phase-field modeling of fracture in linear thin shells. Theoretical and Applied Fracture Mechanics 2014;69:102–9. Available from: https://doi.org/10.1016/j.tafmec.2013.12.002.

[31] Karma A, Kessler D, Levine H. Phase-field model of mode III dynamic fracture. Physical Review Letters 2001;87(4):3–6. Available from: https://doi.org/10.1103/PhysRevLett.87.045501.

[32] Bourdin B, Francfort G, Marigo J. The variational approach to fracture, Vol. 91. Netherlands, Dordrecht: Springer; 2008. Available from: http://doi.org/10.1007/978-1-4020-6395-4.

[33] Kuhn C, Muller R. A continuum phase field model for fracture. Engineering Fracture Mechanics 2010;77(18):3625–34. Available from: https://doi.org/10.1016/j.engfracmech.2010.08.009.

[34] Abdollahi A, Arias I. Phase-field modeling of the coupled microstructure and fracture evolution in ferroelectric single crystals. Acta Materialia 2011;59(12):4733–46. Available from: https://doi.org/10.1016/j.actamat.2011.03.030.

[35] Melenk JM, Babuska I. The partition of unity finite element method. basic theory and applications. Computer Methods in Applied Mechanics and Engineering 1996;39:289–314. Available from: https://doi.org/10.1016/S0045-7825(96)01087-0.

[36] Elmeguenni I, Mazari M. Numerical investigation on Stress Intensity Factor and J Integral in Friction Stir Welded Joint through XFEM method. Frattura ed Integrità Strutturale 2019;47:54–64. Available from: https://doi.org/10.3221/IGF-ESIS.47.05.

[37] Pathak H, Singh A, Singh IV. A simple and efficient XFEM approach for 3-D cracks simulations. International Journal of Fracture 2013;181:189–208. Available from: https://doi.org/10.1007/s10704-013-9835-2.

[38] Pathak H, Singh A, Singh IV. Fatigue crack growth simulations of 3-D problems using XFEM. International Journal of Mechanical Sciences 2013;76:112–31. Available from: https://doi.org/10.1016/j.ijmecsci.2013.09.001.

[39] Kumar S, Singh IV, Mishra BK, Singh A. New enrichments in XFEM to model dynamic crack response of 2-D elastic solids. International Journal of Impact Engineering 2016;87:198–211. Available from: https://doi.org/10.1016/j.ijimpeng.2015.03.005.

[40] Daux C, Moes N, Dolbow J, Sukumar N, Belytschko T. Arbitrary branched and intersecting cracks with the extended finite element method. International Journal for Numerical Methods in Engineering 2000;48:1741–60. Available from: https://doi.org/10.1002/1097-0207(20000830)48:12 < 1741::AID-NME956 > 3.0.CO;2-L.

[41] Sukumar N, Moes N, Moran B, Belytschko T. Extended finite element method for three-dimensional crack modelling. International Journal for Numerical Methods in Engineering 2000;48:1549–70. Available from: https://doi.org/10.1002/1097-0207(20000820)48:11 < 1549::AID-NME955 > 3.0.CO;2-A.

[42] Dolbow J, Moes N, Belytschko T. Modeling fracture in Mindlin–Reissner plates with the extended finite element method. International Journal of Solids and Structures 2000;37(48–50):7161–83. Available from: https://doi.org/10.1016/S0020-7683(00)00194-3.

[43] Dolbow J, Moes N, Belytschko T. An extended finite element method for modeling crack growth with frictional contact. Computer Methods in Applied Mechanics and Engineering 2001;190(51–52):6825–46. Available from: https://doi.org/10.1016/S0045-7825(01)00260-2.

[44] Areias PMA, Belytschko T. Analysis of three-dimensional crack initiation and propagation using the extended finite element method. International Journal for Numerical Methods in Engineering 2005;63 (5):760–88. Available from: https://doi.org/10.1002/nme.1305.

[45] Nagashima T, Omoto Y, Tani S. Stress intensity factor analysis of interface cracks using X-FEM. International Journal for Numerical Methods in Engineering 2003;56(8):1151–73. Available from: https://doi.org/10.1002/nme.604.

[46] Liu XY, Xiao QZ, Karihaloo BL. XFEM for direct evaluation of mixed mode SIFs in homogeneous and bi-materials. International Journal for Numerical Methods in Engineering 2004;56(8):1103–18. Available from: https://doi.org/10.1002/nme.906.

[47] Sukumar N, Prevost H. Modeling quasi-static crack growth with the extended finite element method Part I: Computer implementation. International Journal of Solids and Structures 2003;40(26):7513–37. Available from: https://doi.org/10.1016/j.ijsolstr.2003.08.002.

[48] Zi G, Belytschko T. New crack-tip elements for XFEM and applications to cohesive cracks. International Journal for Numerical Methods in Engineering 2003;57(15):2221–40. Available from: https://doi.org/10.1002/nme.849.

[49] Sukumar N, Huang ZY, Prevost JH, Suo Z. Partition of unity enrichment for bimaterial interface cracks. International Journal for Numerical Methods in Engineering 2004;59(8):1075–102. Available from: https://doi.org/10.1002/nme.902.

[50] Sukumar N, Chopp DL, Moës N, Belytschko T. Modeling holes and inclusions by level sets in the extended finite-element method. Computer Methods in Applied Mechanics and Engineering 2001;190 (46–47):6183–200. Available from: https://doi.org/10.1016/S0045-7825(01)00215-8.

[51] Pathak H, Singh A, Singh IV, Zafar S. Modeling and simulation of 3-D interfacial cracks by XFEM. Proceedings; Int. Mechanical Engineering Congress and Exposition, IMECE2017–70275, V009T12A036. Available from: https://doi.org/10.1115/IMECE2017-70275.

[52] Shedbale AS, Sharma AK, Singh IV, Mishra BK. Modeling and simulation of metal forming processes by XFEM. Applied Mechanics and Materials 2016;829:41–5. Available from: https://doi.org/10.4028/http://www.scientific.net/AMM.829.41 ISSN: 1662–7482.

[53] Singh A, Pathak H, Singh IV. Crack interaction study under thermo-mechanical loading by XFEM. Multiphysics Modeling of Solids an Int. Colloquium (MPMS 2014) ENSTA ParisTech, 6–8 October 2014.

[54] Singh IV, Mishra BK, Bhattacharya S. XFEM simulation of cracks, holes and inclusions in functionally graded materials. International Journal of Mechanics and Materials in Design 2011;7:199–218. Available from: https://doi.org/10.1007/s10999-011-9159-1.

[55] Sharma K, Bhasin V, Singh IV, Mishra BK, Vaze KK. X-Fem simulation of 2-D fracture mechanics problem. Transactions, SMiRT 21, 6–11 November, 2011, New Delhi, India.

[56] Sharma K, Singh IV, Mishra BK, Maurya SK. Numerical simulation of semi-elliptical axial crack in pipe bend using XFEM. Journal of the Mechanics and Physics of Solids 2014;6(2):208–28.

[57] Kumar S, Singh IV, Mishra BK. Numerical investigation of stable crack growth in ductile materials using XFEM. Procedia Engineering 2013;64:652–60.

[58] Bhattacharya S, Singh IV, Mishra BK. Mixed-mode fatigue crack growth analysis of functionally graded materials by XFEM. International Journal of Fracture 2013;183:81–97. Available from: https://doi.org/10.1007/s10704-013-9877-5.

[59] Bhattacharya S, Singh IV, Mishra BK. Fatigue life simulation of functionally graded materials under cyclic thermal load using XFEM. International Journal of Mechanical Sciences 2014;82:41–59. Available from: https://doi.org/10.1016/j.ijmecsci.2014.03.005.

[60] Kumar M, Singh IV, Mishra BK. Fatigue crack growth simulations of plastically graded materials using XFEM and J-integral decomposition approach. Engineering Fracture Mechanics 2019;216106470. Available from: https://doi.org/10.1016/j.engfracmech.2019.05.002.

[61] Shedbale AS, Singh IV, Mishra BK. Nonlinear simulation of an embedded crack in the presence of holes and inclusions by XFEM. Procedia Engineering 2013;64:642–51.

[62] Singh IV, Mishra BK, Bhattacharya S, Patil RU. The numerical simulation of fatigue crack growth using extended finite element method. International Journal of Fatigue 2012;36(1):109–19. Available from: https://doi.org/10.1016/j.ijfatigue.2011.08.010.

[63] Bhattacharya S, Singh IV, Mishra BK, Bui TQ. Fatigue crack growth simulations of interfacial cracks in bi-layered FGMs using XFEM. Computational Mechanics 2013;52:799–814. Available from: https://doi.org/10.1007/s00466-013-0845-8.

[64] Pathak H, Singh A, Singh IV. Numerical simulation of bi-material interfacial cracks using EFGM and XFEM. International Journal of Mechanics and Materials in Design 2012;8:9–36. Available from: https://doi.org/10.1007/s10999-011-9173-3.

[65] Patil RU, Mishra BK, Singh IV. A new multiscale XFEM for the elastic properties evaluation of heterogeneous materials. International Journal of Mechanical Sciences 2017;122:277–87. Available from: https://doi.org/10.1016/j.ijmecsci.2017.01.028.

[66] Bansal M, Singh IV, Mishra BK, Bordas SPA. A parallel and efficient multi-split XFEM for 3-D analysis of heterogeneous materials. Computer Methods in Applied Mechanics and Engineering 2019;347:365–401. Available from: https://doi.org/10.1016/j.cma.2018.12.023.

[67] Pandey VB, Singh IV, Mishra BK, Ahmad S, Rao AV, Kumar V. A new framework based on continuum damage mechanics and XFEM for high cycle fatigue crack growth simulations. Engineering Fracture Mechanics 2019;206:172–200. Available from: https://doi.org/10.1016/j.engfracmech.2018.11.021.

[68] Kumar S, Shedbale AS, Singh IV, Mishra BK. Elasto-plastic fatigue crack growth analysis of plane problems in the presence of flaws using XFEM. Frontiers of Structural and Civil Engineering 2015;9:420–40. Available from: https://doi.org/10.1007/s11709-015-0305-y.

[69] Kumar S, Singh IV, Mishra BK, Sharma K, Khan IA. A homogenized multigrid XFEM to predict the crack growth behavior of ductile material in the presence of microstructural defects. Engineering Fracture Mechanics 2019;205:577–602. Available from: https://doi.org/10.1016/j.engfracmech.2016.03.051.

[70] Kumar M, Singh IV, Mishra BK, Ahmad S, Rao AV, Kumar V. Mixed mode crack growth in elasto-plastic-creeping solids using XFEM. Engineering Fracture Mechanics 2018;199:489–517. Available from: https://doi.org/10.1016/j.engfracmech.2018.05.014.

[71] Dwivedi K, Arora G, Pathak H. Fatigue crack growth in CNT-reinforced polymer composite. Journal of Micromechanics and Molecular Physics 2022;7:173–4. Available from: https://doi.org/10.1142/S242491302241003X.

[72] Suman S, Dwivedi K, Anand S, Pathak H. XFEM-ANN approach to predict the fatigue performance of a composite patch repaired aluminum panel. Composite Part C: Open Access 2022;9100326. Available from: https://doi.org/10.1016/j.jcomc.2022.100326.

[73] Dwivedi K, Pathak H. Patch shape and size effect on performance of in-situ scarf patch repair in aircraft panel. IOP Conference Series Materials Science and Engineering 2020;1004:012002. Available from: https://doi.org/10.1088/1757-899X/1004/1/012002.

[74] Kumar S, Singh IV, Mishra BK. XFEM simulation of stable crack growth using J–R curve under finite strain plasticity. International Journal of Mechanics and Materials in Design 2014;10:165–77. Available from: https://doi.org/10.1007/s10999-014-9238-1.

[75] Pathak H, Singh A, Singh IV, Yadav SK. Fatigue crack growth simulations of 3-D linear elastic cracks under thermal load by XFEM. Frontiers of Structural and Civil Engineering 2015;9:359–82. Available from: https://doi.org/10.1007/s11709-015-0304-z.

[76] Patil RU, Mishra BK, Singh IV. A multiscale framework based on phase field method and XFEM to simulate fracture in highly heterogeneous materials. Theoretical and Applied Fracture Mechanics 2019;100:390–415. Available from: https://doi.org/10.1016/j.tafmec.2019.02.002.

[77] Raza A, Pathak H, Talha M. Vibration characteristics of cracked functionally graded structures using XFEM. Journal of Physics: Conference Series 2019;1240:012028. Available from: https://doi.org/10.1088/1742-6596/1240/1/012028.

[78] Raza A, Pathak H, Talha M. Stochastic extended finite element implementation for natural frequency of cracked functionally gradient and bi-material structure. International Journal of Structural Stability and Dynamics 2021;21(03):2150044. Available from: https://doi.org/10.1142/S0219455421500449.

[79] Raza A, Pathak H, Talha M. Computational investigation of porosity effect on free vibration of cracked functionally graded plates using XFEM. Materials Today: Proceedings 2022;61(1):96–102. Available from: https://doi.org/10.1016/j.matpr.2022.03.654.

[80] Raza A, Pathak H, Talha M. Influence of microstructural defects on free flexural vibration of cracked functionally graded plates in thermal medium using XFEM. Mechanics Based Design of Structures and Machines 2022;. Available from: https://doi.org/10.1080/15397734.2022.2066544.

[81] Natarajan S, Baiz PM, Ganapathi M, Kerfriden P, Bordas S. Linear free flexural vibration of cracked functionally graded plates in thermal environment. Computers and Structures 2011;89:1535–46. Available from: https://doi.org/10.1016/j.compstruc.2011.04.002.

[82] Huang X-L, Shen H-S. Nonlinear vibration and dynamic response of functionally graded plates in thermal environments. International Journal of Solids and Structures 2004;41:2403–27. Available from: https://doi.org/10.1016/j.ijsolstr.2003.11.012.

[83] Talha M, Singh BN. Thermo-mechanical induced vibration characteristics of shear deformable functionally graded ceramic-metal plates using the finite element method. Proceedings of the Institution of Mechanical Engineers, Part C: Journal of Mechanical Engineering Science 2011;225(1):50–65. Available from: https://doi.org/10.1243/09544062JMES2115.

[84] Ghorashi SS, Mohammadi S, Sabbagh-Yazdi SR. Orthotropic enriched element free Galerkin method for fracture analysis of composites. Engineering Fracture Mechanics 2011;78:1906–27. Available from: https://doi.org/10.1016/j.engfracmech.2011.03.011.

[85] Reddy JN. A simple higher-order theory for laminated composite plates. Journal of Applied Mechanics 1984;51(4):745–52. Available from: https://doi.org/10.1115/1.3167719.

[86] Nguyen T-K, Nguyen B-D, Vo TP, Thai H-T. Hygro-thermal effects on vibration and thermal buckling behaviours of functionally graded beams. Composite Structures 2017;176:1050–60. Available from: https://doi.org/10.1016/j.compstruct.2017.06.036.

CHAPTER

5

Extended finite element method for stability analysis of stiffened trapezoidal composite panels

Mehnaz Rasool[1], *Showkat Ahmad Kanth*[2], *Mohd Junaid Mir*[1] *and Ovais Gulzar*[1]

[1]Department of Mechanical Engineering, Islamic University of Science and Technology, Awantipora, Jammu and Kashmir, India [2]Department of Mechanical Engineering, National Institute of Technology Srinagar, Hazratbal, Srinagar, Jammu and Kashmir, India

5.1 Introduction

Trapezoidal panels serve as load-bearing parts in airplanes, spacecraft, submarines, naval ships, vehicles, civil engineering buildings, and other applications. When thickness is reduced in these structures to achieve a light weight and cost-effective structure for various applications, the flexural rigidity of these structures becomes low as compared to axial/membrane rigidity. As a result, under in-plane compressive/shear loads, such structures may become unstable. Hence, stability issue due to various loading conditions in such structures is an important design criterion. One of the primary challenges in numerous sectors of engineering is the stability behavior of structural panels, which has been comprehensively examined in [1,2]. To explore the stability of the panels, various numerical techniques, such as the Rayleigh–Ritz method [3], the extended Kantorovich method [4,5], the finite strip method [6–9], and the finite element method, have been used. Reddy and Phan (1985) evaluated the buckling loads of isotropic, orthotropic, and laminated rectangular plates with simply supported edge conditions by using higher order shear deformation theory [10]. Using the finite element method, Singh et al. (1996) performed a stability analysis on rectangular laminated composite plates that had been subjected to uniaxial and biaxial compression, positive and negative shear, and biaxial compression–tension and tension–compression loads [11]. Babu and Kant (1999) carried out the

Enriched Numerical Techniques
DOI: https://doi.org/10.1016/B978-0-443-15362-4.00013-9

buckling analysis of skew fiber-reinforced composite and sandwich plates by using shear deformable finite element models [12]. Huyton and York (2001) investigated the buckling analysis of thin isotropic skew plates under uniform uniaxial compression [13].

However, the literature on buckling analysis of trapezoidal panels is limited. Although few investigations have been reported in the literature, Azhari et al. (2004) carried out the analysis of postlocal buckling of simply supported skew and trapezoidal plates after the development of numerical procedure by the use of virtual work method in conjunction with the natural coordinates [14]. Radoslaw Mania (2005) used the Galerkin Method to perform buckling analysis on trapezoidal plates subjected to in-plane compression and shear [15]. Recently, Kumar et al. (2019) carried out the analysis of stability behavior of shear deformable trapezoidal composite plates by using the shear deformable smooth finite element method [16].

Stiffened plates of various shapes and forms, such as rectangular, skew, or curved, are widely employed in modern engineering. Bridges, airplane structures, and so on are examples of rib-reinforced plating structures that are designed to endure both transverse and in-plane bending. Buckling study of such structures aids in understanding their behavior when subjected to in-plane forces. The investigation of the stability of stiffened plates has a long history. Cox and Riddel (1949) discussed the case of multiple stiffeners in depth [17]. The analytical findings for the buckling loads of rectangular plates stiffened by longitudinal and transverse ribs are reported in [18]. The application of the finite element approach to the evaluation of stiffened plate problems is presented [19]. The initial developments in this field were also discussed in [20]. Shastry et al. (1976) presented the solutions for the plates with arbitrarily oriented stiffeners [21]. Mizusawa et al. (1980) were the first to study the stability of skew-stiffened plates with various bending and torsional stiffnesses of the stiffeners [22]. Mukhopadhyay and Mukherjee (1989) presented a finite element algorithm for the buckling analysis of stiffened plates and the iso-parametric stiffened plate bending element used in this formulation can accommodate the stiffener anywhere within the plate element [23]. Patel et al. (2006) performed the buckling analysis for different types of stiffened shell panels, such as flat plate, cylindrical shell panel, and spherical shell panel, which are subjected to uniform in-plane harmonic edge loading; the effects of various parameters on buckling, like shell geometry, stiffening scheme, static and dynamic load factors, stiffener size and position, are investigated [24].

The stiffened trapezoidal structural panels have found the application in vast engineering fields varying from aerospace to mechanical to civil engineering. The stability of these panels is an important design criterion. According to a comprehensive literature survey, no studies on the stability of stiffened trapezoidal composite plates have been documented. A large number of computational tools have been developed over time to model various engineering problems, such as the boundary element method [25–27], extended finite element method (XFEM) [3,28–33], mesh-free techniques [34–38], peridynamic methods [39–41], phase field methods [42–44], and conventional finite element method [45–48]. Generation of a conformal mesh for different shapes of columns is computationally more demanding and costly, and hence it establishes a limit on the application of the conventional finite element method for modeling different types of material irregularities in columns. Conformal meshing, mesh refinement, and other mesh-related challenges do not emerge in the extended finite element. Appropriate enrichment functions are added to the standard finite element approximations to model various material irregularities occurring in the domain. The problems of re-meshing, mesh adaption, and conformal meshing do not arise in the XFEM, which makes it computationally more efficient as compared to the conventional finite element method [49–55]. The additional enrichment functions are incorporated in the mathematical models by employing by the partition of unity approaches [56–62].

The present book chapter investigates the stability of stiffened and unstiffened trapezoidal plates under different types of in-plane loading conditions by using the XFEM. The enriched numerical techniques such as extended finite element are effective in modeling and analyzing the intricate shapes in comparison to the conventional finite element technique. Unlike FEM, XFEM does not have re-meshing, mesh adaptation, or conformal meshing difficulties. Trapezoidal panels are key load-bearing elements in aircraft, ships, and numerous civil engineering constructions. Thus effective and safe design of such panels is extremely important to ensure the safety and reliability of engineering structures. The trapezoidal panels are investigated by using the first-order shear deformation theory (FSDT) in conjunction with the XFEM for various loading and boundary conditions. The current study additionally explores the influence of ply orientation and stiffener arrangement on the buckling stress of a panel. To explore multiple trapezoidal panels under varied loads and boundary conditions, accurate and efficient codes based on the proposed methodology have been built in MATLAB®.

5.2 Theory and formulation

The governing finite element equilibrium equations are obtained by using the FSDT. The displacement field of a rectangular panel can be written in standard form as

$$\begin{aligned} u &= u_i + z\theta_x \\ v &= v_i + z\theta_y \\ w &= w_i \end{aligned} \tag{5.1}$$

where u_i, v_i, and w_i are mid surface displacements, θ_x and θ_y are the nodal rotations. The strain components can be written as

$$\begin{Bmatrix} \varepsilon_{xx} \\ \varepsilon_{yy} \\ \gamma_{xy} \end{Bmatrix} = \begin{Bmatrix} \dfrac{\partial u_i}{\partial x} \\ \dfrac{\partial v_i}{\partial y} \\ \dfrac{\partial v_i}{\partial x} + \dfrac{\partial u_i}{\partial y} \end{Bmatrix} + z \begin{Bmatrix} \dfrac{\partial \theta_x}{\partial x} \\ \dfrac{\partial \theta_y}{\partial y} \\ \dfrac{\partial \theta_y}{\partial x} + \dfrac{\partial \theta_x}{\partial y} \end{Bmatrix} = \{\varepsilon_m\} + z\{k\} \tag{5.2}$$

$$\begin{Bmatrix} \gamma_{xz} \\ \gamma_{yz} \end{Bmatrix} = \begin{Bmatrix} \theta_x + \dfrac{\partial w}{\partial x} \\ \theta_y + \dfrac{\partial w}{\partial y} \end{Bmatrix} \tag{5.3}$$

where ε_{xx} and ε_{yy} are normal strains; γ_{xy}, γ_{xz}, and γ_{yz} are the shear strains; ε_m is membrane component of strain; and $k$is the curvature. The strains can be evaluated using the following relations obtained from the finite element displacement assumptions:

$$\{\varepsilon_m\} = [B_M]\{\delta\} \quad \{\kappa\} = [B_B]\{\delta\} \quad \{\gamma\} = [B_S]\{\delta\} \tag{5.4}$$

The stiffness matrices corresponding to each strain component are given by

$$[K_M] = \int [B_M]^T A[B_M]dA \quad [K_B] = \int [B_B]^T D[B_B]dA \tag{5.5}$$

where

$$A = \frac{Eh}{(1-\mu^2)}\begin{bmatrix} 1 & \mu & 0 \\ \mu & 1 & 0 \\ 0 & 0 & \dfrac{1-\mu}{2} \end{bmatrix} \text{ and } D = \frac{Eh^3}{12(1-\mu^2)}\begin{bmatrix} 1 & \mu & 0 \\ \mu & 1 & 0 \\ 0 & 0 & \dfrac{1-\mu}{2} \end{bmatrix}$$

The problem of shear locking is observed in the panel; the elements lock due to spurious zero energy modes in in-plane shear giving faulty results. To overcome the problem of shear locking, mixed interpolated shear strains are obtained following the procedure given in [25] for MITC9 element. Here, assumed shear strains $\{\gamma\}^{AS}$ are obtained by interpolating the shear strains $\{\gamma\}^{DI}$ obtained by direct interpolation of displacement variables as

$$\gamma_{ij}^{AS}(r,s) = \sum_{k=1}^{n_{ij}} N_k^{ij}(r,s)\,\gamma_{ij}^{DI}(r,s) \tag{5.6}$$

where $N_k^{ij}(r,s)$ are interpolation functions.

The shear stiffness matrix is calculated as

$$[K_S] = \int [B_S]^T D_S[B_S]dA \tag{5.7}$$

where $D_S = \frac{Eh\alpha}{2(1+\mu)}\begin{bmatrix} 1 & 0 \\ 0 & 1 \end{bmatrix}$ $\alpha =$ shear correction factor

The nonlinear bending analysis is carried out using the Newton–Raphson technique. A tangent stiffness matrix K_T is introduced, which is given by the following relation:

$$K_T = \begin{bmatrix} K_M & 0 \\ 0 & K_B + K_S \end{bmatrix} + \begin{bmatrix} 0 & 1.5\, B_M^T A\, B_{NL} \\ 1.5\, B_{NL}^T A\, B_M & 0 \end{bmatrix} + \begin{bmatrix} 0 & 0 \\ 0 & 1.5\, B_{NL}^T A\, B_{NL} \end{bmatrix} \tag{5.8}$$

The stability analysis is carried out for static and dynamic loads under static load. The governing equation is represented as

$$[K]\{\delta\} - p[K_G]\{\delta\} = \{F(t)\} \tag{5.9}$$

where $[K_G]$ is prebuckling stiffness matrix, p is axial load, and $\{F(t)\}$ is the transverse load. Prebuckling stiffness matrix may be calculated as

$$K_G = \int \frac{\partial w^T}{\partial x}\sigma\frac{\partial w}{\partial x}.dx \text{ or } K_G = \int G^T \sigma G dx \tag{5.10}$$

The above problem is reduced to an eigenvalue problem that can be solved for critical buckling load as

$$[K - \sigma K_G]\{\delta\} = 0 \tag{5.11}$$

5.3 The extended finite element method

The extended finite element approach was created to eliminate the reliance of numerical computation on the finite element mesh used for analysis. XFEM has consistently proven to be a reliable and versatile method for analyzing material science problems with complicated geometries, localized deformations, and discontinuities, such as cracks, inclusions, contact surfaces, and holes. In the conventional finite element method, a considerable amount of time and effort is consumed for the re-meshing of domains while modeling evolving discontinuities as the geometry of the interfaces continuously changes. Such problems do not arise in XFEM as a uniform mesh is considered for modeling any type of discontinuity present in the engineering component. XFEM does not have conformal meshing, mesh adaptation, or mesh refinement problems because it models all different kinds of discontinuities regardless of mesh. Any deformable body's governing static equilibrium equation under a certain combination of loading and boundary conditions can be expressed in strong form as

$$\nabla.\boldsymbol{\sigma} + \boldsymbol{b} = 0 \tag{5.12}$$

where ∇ is gradient operator, $\boldsymbol{b}$ is body force vector, $\boldsymbol{\sigma}$ is stress tensor. The boundary conditions can be expressed as

$$\boldsymbol{u} = \overline{\boldsymbol{u}} \quad on \quad \Gamma_u \tag{5.13}$$

$$\boldsymbol{\sigma}.\boldsymbol{n}_\Gamma = \overline{\boldsymbol{t}} \quad on \quad \Gamma_t \tag{5.14}$$

where $\boldsymbol{n}_\Gamma$ is normal vector to domain boundary. To investigate any engineering phenomenon in the XFEM, the partition of unity method is used to enrich standard finite element approximation with appropriate enrichment functions for modeling of discontinuities such as cracks, holes, or inclusions. The enriched displacement-based approximation in the XFEM can be written as

$$\boldsymbol{u}^h(\boldsymbol{x}) = \sum_{i=1}^{n} N_i(\boldsymbol{x})\boldsymbol{u}_i + \sum_{j=1}^{n_s} N_j(\boldsymbol{x})\{f(\boldsymbol{x}) - f(\boldsymbol{x}_j)\}\boldsymbol{a}_j \tag{5.15}$$

where $\boldsymbol{u}_i$ represents nodal degree of freedom and $\boldsymbol{a}_j$ represents enriched degree of freedom. $\boldsymbol{n}$ denotes the standard nodes and $\boldsymbol{n}_s$ denotes enriched nodes having additional degrees of freedom. Effect of bi-material interface is incorporated into formulation by using enrichment function $f(\boldsymbol{x})$. The type of enrichment function included into a finite element approximation depends on the nature of the discontinuity. N represents standard finite element shape function. In order to develop the numerical models in the XFEM, the equilibrium equation can be expressed in variational form as

$$\int_{\Omega} \delta \varepsilon^T \sigma d\Omega - \int_{\Omega} \delta u^T b d\Omega - \int_{\Gamma_t} \delta u^T t d\Gamma_t = 0 \tag{5.16}$$

The equilibrium equation can also be written as

$$\int_{\Omega} \delta \left[\varepsilon_x \varepsilon_y \gamma_{xy} \right] \left\{ \begin{array}{c} \sigma_x \\ \sigma_y \\ \sigma_{xy} \end{array} \right\} d\Omega - \int_{\Omega} \delta[uv] \left\{ \begin{array}{c} b_x \\ b_y \end{array} \right\} d\Omega - \int_{\Gamma_t} \delta[uv] \left\{ \begin{array}{c} t_x \\ t_y \end{array} \right\} d\Gamma_t = 0 \tag{5.17}$$

On substituting XFEM approximation in the equilibrium equation, we get

$$\left(\int_{\Omega} \mathbf{B}_{std}^T \boldsymbol{\sigma} d\Omega - \int_{\Omega} \mathbf{N}_{std}^T \mathbf{b} d\Omega - \int_{\Gamma_t} \mathbf{N}_{std}^T \mathbf{t} d\Gamma_t \right) \delta \mathbf{u}^T + \left(\int_{\Omega} \mathbf{B}_{enr}^T \boldsymbol{\sigma} d\Omega - \int_{\Omega} \mathbf{N}_{enr}^T \mathbf{b} d\Omega - \int_{\Gamma_t} \mathbf{N}_{enr}^T \mathbf{t} d\Gamma_t \right) \delta \mathbf{a}^T = 0 \tag{5.18}$$

where,

$$\boldsymbol{B}_{std} = \begin{bmatrix} \dfrac{\partial N_i}{\partial x} & 0 \\ 0 & \dfrac{\partial N_i}{\partial y} \\ \dfrac{\partial N_i}{\partial y} & \dfrac{\partial N_i}{\partial x} \end{bmatrix} \tag{5.19}$$

$$\boldsymbol{B}_{enr} = \begin{bmatrix} \dfrac{\partial \overline{N}_j}{\partial x} & 0 \\ 0 & \dfrac{\partial \overline{N}_j}{\partial y} \\ \dfrac{\partial \overline{N}_j}{\partial y} & \dfrac{\partial \overline{N}_j}{\partial x} \end{bmatrix} \tag{5.20}$$

The equilibrium equation can also be expressed as follows:

$$\left(\int_{\Omega} \mathbf{B}_{std}^T \boldsymbol{\sigma} d\Omega - f_{std} \right) \delta \boldsymbol{u}^T + \left(\int_{\Omega} \mathbf{B}_{enr}^T \boldsymbol{\sigma} d\Omega - f_{enr} \right) \delta \boldsymbol{a}^T = 0 \tag{5.21}$$

where

$$f_{std} = \int_{\Omega} \mathbf{N}_{std}^T \mathbf{b} d\Omega + \int_{\Gamma_t} \mathbf{N}_{std}^T \mathbf{t} d\Gamma_t; \quad f_{enr} = \int_{\Omega} \mathbf{N}_{enr}^T \mathbf{b} d\Omega + \int_{\Gamma_t} \mathbf{N}_{enr}^T \mathbf{t} d\Gamma_t \tag{5.22}$$

The final discrete model in XFEM for any two-dimensional problem can be obtained by substituting $\boldsymbol{\sigma} = \boldsymbol{C}_e\varepsilon$ and $\varepsilon = \boldsymbol{Bd}$ as

$$\begin{bmatrix} \int_\Omega \boldsymbol{B}_{std}^T \boldsymbol{C}_e \boldsymbol{B}_{std} \mathrm{d}\Omega & \int_\Omega \boldsymbol{B}_{std}^T \boldsymbol{C}_e \boldsymbol{B}_{enr} \mathrm{d}\Omega \\ \int_\Omega \boldsymbol{B}_{enr}^T \boldsymbol{C}_e \boldsymbol{B}_{std} \mathrm{d}\Omega & \int_\Omega \boldsymbol{B}_{enr}^T \boldsymbol{C}_e \boldsymbol{B}_{enr} \mathrm{d}\Omega \end{bmatrix} \begin{Bmatrix} \boldsymbol{u} \\ \boldsymbol{a} \end{Bmatrix} = \begin{Bmatrix} \boldsymbol{f}_{std} \\ \boldsymbol{f}_{enr} \end{Bmatrix} \tag{5.23}$$

The Hookean matrix $\boldsymbol{C}_e$ is defined as

$$\boldsymbol{C}_e = \frac{E}{(1+\nu)(1-2\nu)} \begin{bmatrix} 1-\nu & \nu & 0 \\ \nu & 1-\nu & 0 \\ 0 & 0 & \frac{1-2\nu}{2} \end{bmatrix} \text{(plane stress)} \tag{5.24}$$

$$\boldsymbol{C}_e = \frac{\mathrm{E}}{(1-\nu^2)} \begin{bmatrix} 1 & \nu & 0 \\ \nu & 1 & 0 \\ 0 & 0 & \frac{1-\nu}{2} \end{bmatrix} \text{(plane strain)} \tag{5.25}$$

The final XFEM model can be written in compact form as

$$\begin{bmatrix} \boldsymbol{K}^{uu} & \boldsymbol{K}^{ua} \\ \boldsymbol{K}^{au} & \boldsymbol{K}^{aa} \end{bmatrix} \begin{Bmatrix} \boldsymbol{u} \\ \boldsymbol{a} \end{Bmatrix} = \begin{Bmatrix} \boldsymbol{f}^u \\ \boldsymbol{f}^a \end{Bmatrix} \tag{5.26}$$

$$\boldsymbol{Kd} = \boldsymbol{f} \tag{5.27}$$

where $\boldsymbol{K}$ represents XFEM stiffness matrix and $\boldsymbol{f}$ represents global force vector, such that

$$\boldsymbol{K}^{\alpha\beta} = \int_\Omega (\boldsymbol{B}^\alpha)^T \boldsymbol{C}_e \quad \boldsymbol{B}^\beta d\Omega; (\alpha, \beta = \mathrm{u}, \mathrm{a}) \tag{5.28}$$

$$\boldsymbol{f}^\alpha = \int_\Gamma \boldsymbol{N}^\alpha \boldsymbol{t} \quad d\Gamma + \int_\Omega \boldsymbol{N}^\alpha \boldsymbol{b} \quad d\Omega; (\alpha = \mathrm{u}, \mathrm{a}) \tag{5.29}$$

Here $\boldsymbol{N}^u = \boldsymbol{N}_{std}, \boldsymbol{N}^a = \boldsymbol{N}_{enr}$, $\quad \mathbf{B}^u = \boldsymbol{B}_{std}$, $\quad \mathbf{B}^a = \boldsymbol{B}_{enr}, \boldsymbol{f}^u = \boldsymbol{f}_{std}$ and $\boldsymbol{f}^a = \boldsymbol{f}_{enr}$

5.4 Results and discussions

As schematically depicted in Fig. 5.1, the stability behavior of stiffened symmetric isotropic and composite trapezoidal panels that undergo in-plane compression and shear loads is studied using the FSDT and the extended finite element technique.

Unless otherwise stated, the panel's and the stiffener's dimensions and material composition are as follows. For isotropic, we have $E = 200$ GPa and $\mu = 0.3$. For composite material, we have $E_x/E_y = 10$, $G_{xy} = G_{xz} = 0.33\ E_y$, $G_{yz} = 0.2E_y$, $\upsilon_{xy} = \upsilon_{xz} = \upsilon_{yz} = 0.22$,

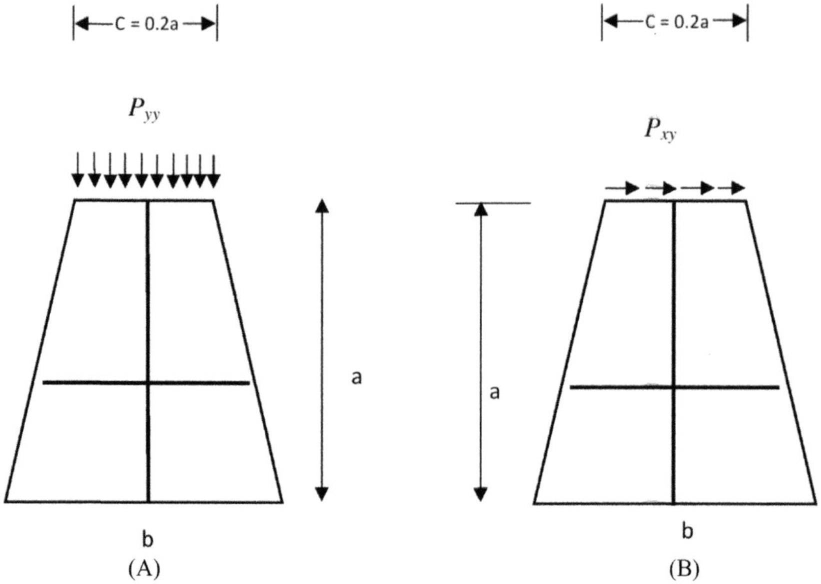

FIGURE 5.1 Schematic representation of stiffened trapezoidal composite plate with two stiffeners at center under in-plane compression and shear force. The dimensions of the plate are $a/b = 1$, $a/h = 100$, where h is the thickness of plate. In this figure, part (A) represents the in-plane shear case and part (B) denotes in-plane shear.

TABLE 5.1 Nondimensional critical buckling load for isotopic $(\lambda_{cr} = \lambda a/\pi^2 D)$ and composite $(\lambda_{cr} = \lambda a/E_T h^3)$ trapezoidal plate.

	Compression		Shear	
	SSSS	CCCC	SSSS	CCCC
Isotropic plate				
Present Study	5.962	12.769	5.597	11.073
Kumar et al. [17]	5.993	12.863	5.653	11.242
Composite plate $[0^0/90^0/90^0/0^0]$				
Present study	9.939	21.964	9.646	19.859
Kumar et al. [17]	10.017	22.354	9.788	20.511

$E_y = 20$ GPa, where E, G, and $\upsilon$are Young's Modulus, Shear modulus, and Poisson's ratio. The current analysis takes into account two out of plane boundary conditions. We have a simply supported case (SSSS), for which $u_z = 0$ along the edges. Another case is the clamped (CCCC) condition, for which $u_z = \theta_x = \theta_y = 0$ along the edges. As shown in Table 5.1, the results of the buckling analysis of isotropic and composite trapezoidal panels for in-plane compression and shear under two different boundary conditions are compared first to determine the effectiveness of the technique. Thereafter, the technique is validated for the stability analysis of rectangular stiffened plate (aspect ratio = 1; As/bt = 0.05; $EI_s/Db = 10$) with Mukhopadhyay and Mukherjee (1990) (Table 5.2).

The buckling mode shapes for trapezoidal panels stiffened horizontally and vertically are shown in Fig. 5.2. It is observed that the stress concentration is toward the bottom corner of the trapezoidal panel in the case of shear-loaded panel. However, the stress is uniformly distributed with maximum at the center for a compressed trapezoidal panel. The nondimensional critical buckling loads ($\lambda_{cr} = \lambda a/\pi^2 D$) of thin ($a/b = 1, a/h = 100$) stiffened isotropic plate with three different orientations of stiffener under pure compression and shear force are studied in Table 5.3. Under pure compression, the maximum buckling capacity has been found for the trapezoidal plate consisting of stiffeners along horizontal and vertical directions. The increased capacity is because of the stiffening action of the panel, hence resistance to the load. However, when the stiffener is positioned vertically, the influence of the stiffener on the buckling resistance of the panel is observed under shear force, whereas the horizontal stiffener has little impact. Next, the buckling capacity ($\lambda_{cr} = \lambda a/E_T \quad h^3$) of thin ($a/b = 1, a/h = 100$) stiffened cross-ply laminated composite trapezoidal plate with three different orientations of stiffeners under pure compression and

TABLE 5.2 Nondimensional critical buckling load ($\lambda_{cr} = \lambda a/\pi^2 D$) of rectangular stiffened isotropic plate under uniform compression:

		EI_S/Db **10**
A_s/bt	Present study	Mukhopadhyay and Mukherjee [24]
0.05	15.83	16

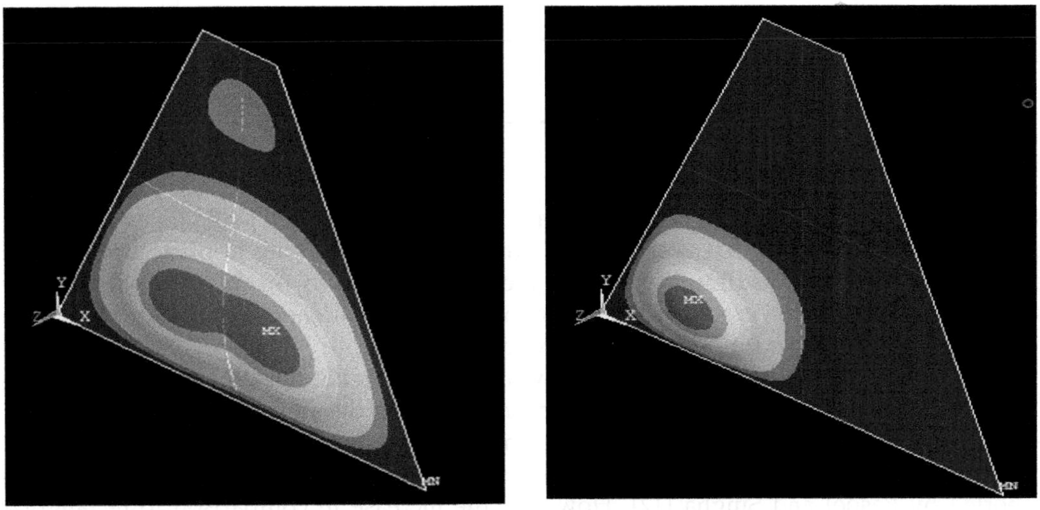

FIGURE 5.2 Compressive and shear buckling mode shapes of stiffened trapezoidal panels.

TABLE 5.3 Nondimensional critical buckling load ($\lambda_{cr} = = \lambda a/ \pi^2 D$) of symmetric isotropic stiffened trapezoidal plate ($a/b = 1, a/h = 100$) under pure compression and shear ($d_s = b_s = 4$ mm, $A_s = 16$ mm^2, $d_s/b_s = 1$).

	Under compression		Under shear	
Position of stiffener	SSSS	CCCC	SSSS	CCCC
Horizontal	7.346	14.788	5.698	11.684
Vertical	11.938	34.419	7.276	13.051
Horizontal and vertical	19.569	43.084	7.356	13.488

TABLE 5.4 Nondimensional critical buckling load ($\lambda_{cr} = \lambda a/E_T h^3$) of stiffened cross-ply $[0°/90°]_{2S}$ laminated composite trapezoidal plate ($a/b = 1, a/h = 100$) under compression and shear ($d_s = b_s = 4$ mm, $A_s = 16$ mm^2, $d_s/b_s = 1$).

	Compression		Shear	
Position of stiffener	SSSS	CCCC	SSSS	CCCC
Horizontal	24.188	51.095	19.159	43.408
Vertical	51.255	140.975	19.464	37.179
Horizontal and vertical	93.175	176.515	21.619	45.679

TABLE 5.5 Nondimensional critical buckling load ($\lambda_{cr} = \lambda a/E_T \quad h^3$) of stiffened angle-ply $[45° / -45° / -45° /45°]$ laminated composite trapezoidal plate under pure compression and shear ($d_s = b_s = 4$ mm, $A_s = 16$ mm^2, $d_s/b_s = 1$).

	Compression		Shear	
Position of stiffener	SSSS	CCCC	SSSS	CCCC
Horizontal	18.254	28.334	13.266	23.125
Vertical	40.912	116.955	17.707	29.38
Horizontal and vertical	94.13	149.905	18.084	26.984

shear force is studied in Table 5.4. The behavior of critical buckling loads for various stiffener positions is same as that observed for isotropic panels for both in-plane compressive and shear forces. The maximum value for the buckling load is for in-plane compressive stiffened panel is for the cross-ply configuration with two stiffeners, one horizontally, and other vertically at center for both the edge conditions.

Table 5.5 now examines the impact of ply orientation on the buckling behavior of composite trapezoidal composite panels. It has been noted that the angle-ply composite panels' shear buckling capacity has increased. The increase is possibly due to the formation of diagonal compression and tension strips that enhance the buckling load of the panels, as reported by Rasool and Singha [12]. However, the increase in comparison to rectangular panels is less possibly due to the varying geometry of the panels.

TABLE 5.6 Nondimensional critical buckling load ($\lambda_{cr} = \lambda a/E_T \quad h^3$) of stiffened angle-ply $[45^\circ/-45^\circ]_{2S}$ laminated composite plate under compression and shear ($d_s = b_s = 4$ mm, $A_s = 16$ mm^2, $d_s/b_s = 1$).

	Compression		Shear	
Position of stiffener	SSSS	CCCC	SSSS	CCCC
Horizontal	20.773	32.580	16.132	27.716
Vertical	43.901	122.3	20.781	33.467
Horizontal and vertical	98.805	164.305	20.632	31.744

Finally, the effect of the number of layers on the buckling capacity of the trapezoidal composite panels is presented in Table 5.6. It is seen that the number of layers plays a vital role in the design of the panels. As is obvious that the numbers of layers increases the load-bearing capacity. However, the effect is marginal when the stiffeners along horizontal and vertical directions are placed. Thus for composite stiffened panels, there is no need to increase the number of layers. Thus for an effective and optimized design of trapezoidal panels, it is better to use stiffeners rather than increasing the number of layers for the panel.

5.5 Conclusion

Using the extended finite element approach, the stability behavior of stiffened isotropic and composite trapezoidal plates is discussed here, and the results of the investigations are summarized as follows:

- The geometry of the panel plays a significant role in understanding and evaluating the stability characteristics of the panels.
- The stiffened panels offer more resistance to the bending loads rather than the in-plane loads.
- In-plane compression stiffeners are of great importance and effect on buckling is prominent. However, for in-plane positive shear stiffeners, they do not play a significant role, but for negative shear, the buckling capacity increases.
- For stiffened isotropic trapezoidal plate under pure compression and shear force, the maximum critical buckling load has been found for the plate consisting of horizontal and vertical stiffener at center and under in-plane compression and clamped at the edges.
- On comparing stiffened cross-ply $[0^\circ/90^\circ/90^\circ/0^\circ]$ and angle-ply $[45^\circ/-45^\circ/-45/45]$ composite trapezoidal plate, the maximum critical buckling load has been found for the angle-ply $[45^\circ/-45^\circ/-45/45]$ consisting of a combination of stiffeners at center and under in-plane compression and clamped at the edges.
- The increase in the number of layers for a composite trapezoidal stiffened panel has a minimal effect on the buckling capacity of the panel.

References

[1] Leissa AW. A review of laminated composite plate buckling. Applied Mechanics Reviews 1987;40:579–91.
[2] Kapania RK, Raciti S. Recent advances in analysis of laminated beams and plates. Part I-Shear effects and buckling. AIAA Journal 1989;27(7):923–35.
[3] Wang CM, Xiang Y, Kitipornchai S. Buckling solutions of rectangular Mindlin plates under uniform shear. Journal of Engineering Mechanics 1994;120(11):2462–70.
[4] Smith ST, Bradford MA, Oehlers DJ. Elastic buckling of unilaterally constrained rectangular plates in pure shear. Engineering Structures 1999;21(5):443–53.
[5] Yuan S, Jin Y. Computation of elastic buckling loads of rectangular thin plates using the extended Kantorovich method. Computers & Structures 1998;66(6):861–7.
[6] Shufrin I, Rabinovitch O, Eisenberger M. Buckling of symmetrically laminated rectangular plates with general boundary conditions—a semi analytical approach. Composite Structures 2008;82(4):521–31.
[7] Smith TG, Sridharan S. A finite strip method for the buckling of plate structures under arbitrary loading. International Journal of Mechanical Sciences 1978;20(10):685–93.
[8] Mahendran M, Murray NW. Elastic buckling analysis of ideal thin-walled structures under combined loading using a finite strip method. Thin-Walled Structures 1986;4(5):329–62.
[9] Dawe DJ, Wang S. Buckling of composite plates and plate structures using the spline finite strip method. Composites Engineering 1994;4(11):1099–117.
[10] Loughlan J. The influence of bend–twist coupling on the shear buckling response of thin laminated composite plates. Thin-Walled Structures 1999;34(2):97–114.
[11] Reddy JN, Phan N. Stability and vibration of isotropic, orthotropic and laminated plates according to a higher-order shear deformation theory. Journal of Sound and Vibration 1985;98(2):157–70.
[12] Rasool M, Singha MK. A finite element study on the nonlinear behavior of rectangular shear panels. Thin-Walled Structures 2016;104:248–58.
[13] Babu CS, Kant T. Two shear deformable finite element models for buckling analysis of skew fibre-reinforced composite and sandwich panels. Composite Structures 1999;46(2):115–24.
[14] Huyton P, York CB. Buckling of skew plates with continuity or rotational edge restraint. Journal of Aerospace Engineering 2001;14(3):92–101.
[15] Azhari M, Shahidi AR, Saadatpour MM. Post local buckling of skew and trapezoidal plates. Advances in Structural Engineering 2004;7(1):61–70.
[16] Mania R. Buckling analysis of trapezoidal composite sandwich plate subjected to in-plane compression. Composite Structures 2005;69(4):482–90.
[17] Kumar A, Singha MK, Tiwari V. Stability analysis of shear deformable trapezoidal composite plates. International Journal of Structural Stability and Dynamics 2019;19(08):1971004.
[18] Cox HL, Riddell JR. Buckling of a longitudinally stiffened flat panel. Aeronautical Q 1949;1(3):225–44.
[19] Timošenko SP, Gere JM. Theory of elastic stability. McGraw-Hill; 1961.
[20] Dawe DJ. Application of the discrete element method to the buckling analysis of rectangular plates under arbitrary membrane loading. Aeronautical Q 1969;20(2):114–28.
[21] Troitsky MS. Stiffened plates: bending, stability, and vibrations. Elsevier Scientific Publishing Company; 1976.
[22] Shastry BP, Rao GV, Reddy MN. Stability of stiffened plates using high precision finite elements. Nuclear Engineering and Design 1976;36(1):91–5.
[23] Mizusawa T, Kajita T, Naruoka M. Buckling of skew plate structures using B-spline functions. International Journal for Numerical Methods in Engineering 1980;15(1):87–96.
[24] Mukhopadhyay M, Mukherjee A. Finite element buckling analysis of stiffened plates. Computers & Structures 1990;34(6):795–803.
[25] Patel SN, Datta PK, Sheikh AH. Buckling and dynamic instability analysis of stiffened shell panels. Thin-Walled Structures 2006;44(3):321–33.
[26] Aliabadi MH, Brebbia CA. Boundary element formulations in fracture mechanics: a review. Transactions on Engineering Sciences 1998;17:589–98.
[27] Wen PH, Aliabadi MH, Young A. Dual boundary element methods for three-dimensional dynamic crack problems. Journal of Strain Analysis for Engineering Design 1999;34:373–94. Available from: https://doi.org/10.1177/030932479903400601.

[28] Leonel ED, Venturini WS. Non-linear boundary element formulation applied to contact analysis using tangent operator. Engineering Analysis with Boundary Elements 2011;35:1237–47. Available from: https://doi.org/10.1016/j.enganabound.2011.06.005.
[29] Spangenberger AG, Lados DA. Extended finite element modeling of fatigue crack growth microstructural mechanisms in alloys with secondary/reinforcing phases: model development and validation. Computational Mechanics 2020;. Available from: https://doi.org/10.1007/s00466-020-01921-2.
[30] Lone AS, Kanth SA, Jameel A, Harmain GA. A state of art review on the modeling of contact type nonlinearities by extended finite element method. Materials Today: Proceedings 2019;18:3462–71. Available from: https://doi.org/10.1016/j.matpr.2019.07.274.
[31] Sukumar N, Chopp DL, Moës N, Belytschko T. Modeling holes and inclusions by level sets in the extended finite-element method. Computer Methods in Applied Mechanics and Engineering 2001;190:6183–200. Available from: https://doi.org/10.1016/S0045-7825(01)00215-8.
[32] Jameel A, Harmain GA. "Fatigue crack growth in presence of material discontinuities by EFGM,". International Journal of Fatigue 2015;81:105–16. Available from: https://doi.org/10.1016/j.ijfatigue.2015.07.021.
[33] Lone AS, Jameel A, Harmain GA. "A coupled finite element-element free Galerkin approach for modeling frictional contact in engineering components,". Materials Today: Proceedings 2018;5:18745–54. Available from: https://doi.org/10.1016/j.matpr.2018.06.221.
[34] Kanth SA, Harmain GA, Jameel A. "Modeling of nonlinear crack growth in steel and aluminum alloys by the element free Galerkin method,". Materials Today: Proceedings 2018;5:18805–14. Available from: https://doi.org/10.1016/j.matpr.2018.06.227.
[35] Jameel A, Harmain GA. "Fatigue crack growth analysis of cracked specimens by the coupled finite element-element free Galerkin method,". Mechanics of Advanced Materials and Structures 2019;26:1343–56. Available from: https://doi.org/10.1080/15376494.2018.1432800.
[36] Lone AS, Jameel A, Harmain GA. "Enriched element free Galerkin method for solving frictional contact between solid bodies,". Mechanics of Advanced Materials and Structures 2022;. Available from: https://doi.org/10.1080/15376494.2022.2092791.
[37] Lone AS, Jameel A, Harmain GA. "Modelling of contact interfaces by penalty based enriched finite element method,". Mechanics of Advanced Materials and Structures 2022;. Available from: https://doi.org/10.1080/15376494.2022.2034075.
[38] Jameel A, Harmain GA. "Large deformation in bi-material components by XIGA and coupled FE-IGA techniques,". Mechanics of Advanced Materials and Structures 2022;29:850–72. Available from: https://doi.org/10.1080/15376494.2020.1799120.
[39] Kanth SA, Jameel A, Harmain GA. "Investigation of fatigue crack growth in engineering components containing different types of material irregularities by XFEM,". Mechanics of Advanced Materials and Structures 2022;29(24):3570–87. Available from: https://doi.org/10.1080/15376494.2021.1907003.
[40] Gupta V, Jameel A, Anand S, Anand Y. "Analysis of composite plates using isogeometric analysis: a discussion,". Materials Today: Proceedings 2021;44:1190–4. Available from: https://doi.org/10.1016/j.matpr.2020.11.238.
[41] Jameel A, Harmain GA. "Effect of material irregularities on fatigue crack growth by enriched techniques,". International Journal for Computational Methods in Engineering Science and Mechanics 2020;21:109–33. Available from: https://doi.org/10.1080/15502287.2020.1772902.
[42] Lone AS, Kanth SA, Jameel A, Harmain GA. "XFEM Modelling of frictional contact between elliptical inclusions and solid bodies,". Materials Today: Proceedings 2020;26:819–24. Available from: https://doi.org/10.1016/j.matpr.2019.12.424.
[43] Sheikh UA, Jameel A. "Elasto-plastic large deformation analysis of bi-material components by FEM,". Materials Today: Proceedings 2020;26:1795–802. Available from: https://doi.org/10.1016/j.matpr.2020.02.377.
[44] Simpson R, Trevelyan J. A partition of unity enriched dual boundary element method for accurate computations in fracture mechanics. Computer Methods in Applied Mechanics and Engineering 2011;200:1–10. Available from: https://doi.org/10.1016/j.cma.2010.06.015.
[45] Noda NA, Oda K. Numerical solutions of the singular integral equations in the crack analysis using the body force method. International Journal of Fracture 1992;58:285–304. Available from: https://doi.org/10.1007/BF00048950.
[46] Rao BN, Rahman S. A coupled meshless-finite element method for fracture analysis of cracks. International Journal of Pressure Vessels and Piping 2001;78:647–57. Available from: https://doi.org/10.1016/S0308-0161(01)00076-X.

[47] Kanth SA, Lone AS, Harmain GA, Jameel A. "Modelling of embedded and edge cracks in steel alloys by XFEM,". Materials Today: Proceedings 2020;26:814–18. Available from: https://doi.org/10.1016/j.matpr.2019.12.423.
[48] Kanth SA, Lone AS, Harmain GA, Jameel A. "Elasto plastic crack growth by XFEM: a review,". Materials Today: Proceedings 2019;18:3472–81. Available from: https://doi.org/10.1016/j.matpr.2019.07.275.
[49] Jameel A, Harmain GA. "Modeling and numerical simulation of fatigue crack growth in cracked specimens containing material discontinuities,". Strength of Materials 2016;48(2):294–307. Available from: https://doi.org/10.1007/s11223-016-9765-0.
[50] Eberhard P, Gaugele T. Simulation of cutting processes using mesh-free Lagrangian particle methods. Computational Mechanics 2013;51:261–78. Available from: https://doi.org/10.1007/s00466-012-0720-z.
[51] De Lorenzis L, Wriggers P, Zavarise G. A mortar formulation for 3D large deformation contact using NURBS-based isogeometric analysis and the augmented Lagrangian method. Computational Mechanics 2012;49:1–20. Available from: https://doi.org/10.1007/s00466-011-0623-4.
[52] Bhardwaj G, Singh IV, Mishra BK, Bui TQ. Numerical simulation of functionally graded cracked plates using NURBS based XIGA under different loads and boundary conditions. Composite Structures 2015;126:347–59. Available from: https://doi.org/10.1016/j.compstruct.2015.02.066.
[53] Gu YT, Wang QX, Lam KY. A meshless local Kriging method for large deformation analyses. Computer Methods in Applied Mechanics and Engineering 2007;196:1673–84. Available from: https://doi.org/10.1016/j.cma.2006.09.017.
[54] Belytschko T, Krongauz Y, Organ D, Fleming M, Krysl P. Meshless methods: an overview and recent developments. Computer Methods in Applied Mechanics and Engineering 1996;139:3–47. Available from: https://doi.org/10.1016/S0045-7825(96)01078-X.
[55] Rao BN, Rahman S. An enriched meshless method for non-linear fracture mechanics. International Journal for Numerical Methods in Engineering 2004;59:197–223. Available from: https://doi.org/10.1002/nme.868.
[56] Duflot M, Nguyen-Dang H. A meshless method with enriched weight functions for fatigue crack growth. International Journal for Numerical Methods in Engineering 2004;59:1945–61. Available from: https://doi.org/10.1002/nme.948.
[57] Jameel A, Harmain GA. Extended iso-geometric analysis for modeling three-dimensional cracks. Mechanics of Advanced Materials and Structures 2019;26:915–23. Available from: https://doi.org/10.1080/15376494.2018.1430275.
[58] Singh IV, Bhardwaj G, Mishra BK. A new criterion for modeling multiple discontinuities passing through an element using XIGA. Journal of Mechanical Science and Technology 2015;29:1131–43. Available from: https://doi.org/10.1007/s12206-015-0225-8.
[59] Bhardwaj G, Singh IV, Mishra BK. Numerical simulation of plane crack problems using extended isogeometric analysis. Procedia Engineering 2013;64:661–70. Available from: https://doi.org/10.1016/j.proeng.2013.09.141.
[60] Jameel A, Harmain GA. "A coupled FE-IGA technique for modeling fatigue crack growth in engineering materials,". Mechanics of Advanced Materials and Structures 2019;26:1764–75. Available from: https://doi.org/10.1080/15376494.2018.1446571.
[61] Gupta V, Jameel A, Verma SK, Anand S, Anand Y. "Transient isogeometric heat conduction analysis of stationary fluid in a container,". Part E: Journal of Process Mechanical Engineering 2022;. Available from: https://doi.org/10.1177/0954408922112.
[62] Gupta V, Jameel A, Verma SK, Anand S, Anand Y. "An insight on NURBS based isogeometric analysis, its current status and involvement in mechanical applications,". Archives of Computational Methods in Engineering 2022;. Available from: https://doi.org/10.1007/s11831-022-09838-0.

CHAPTER

6

Implementation issues in element-free Galerkin method

Azher Jameel[1], Aazim Shafi Lone[1], Qazi Junaid Ashraf[2] and G.A. Harmain[1]

[1]Department of Mechanical Engineering, National Institute of Technology Srinagar, Hazratbal, Srinagar, Jammu and Kashmir, India [2]Department of Mechanical Engineering, University of Kashmir, Hazratbal, Srinagar, Jammu and Kashmir, India

6.1 Introduction

A huge number of numerical techniques have been proposed to investigate different classes of nonlinearities arising because of large geometrical changes, material property variations, or frictional contact among solid bodies. Out of these available tools, FEM is the most widely applied technique in computational solid mechanics. Although FEM is a very strong and potential tool for solving solid mechanics problems, it faces several problems while modeling various engineering problems. FEM requires conformal meshing, that is, the finite element (FE) mesh should conform to the geometry of discontinuity. If interface changes with time, as in the case of sliding contact between two bodies, crack propagation, and so on, the application of FEM becomes very difficult because remeshing of domain is required after each stage of simulation. FEM also faces several problems while modeling large deformation in structural components, where mesh distortion occurs due to large deformation in the domain, and hence, domain demands remeshing after each step of simulation. From a computational point of view, it is very cumbersome, costly, and results in a decrease of accuracy because, after remeshing, the data needs to be transformed from old mesh to the updated one. In order to eliminate or reduce such issues, several enriched computational tools, such as extended finite element method (XFEM) and element-free Galerkin method (EFGM), were developed in recent times to investigate different discontinuities present in engineering components. The motivation for enriched methods is that the discontinuities are not considered during mesh or node generation. Instead, the classical variable approximation is updated by adding

Enriched Numerical Techniques
DOI: https://doi.org/10.1016/B978-0-443-15362-4.00005-X

additional enrichment functions to introduce the effect of such interfaces in mathematical formulations. XFEM offers many advantages over classical FEM in modeling solid mechanics problems containing different types of discontinuities. XFEM does not consider any discontinuity present in the domain during mesh generation. Instead, we add some additional functions to classical variable approximations, and it is these enrichment functions that take different types of discontinuities into consideration [1–5]. For a contact problem, the displacement field is discontinuous at contact interface [6,7]. Therefore the Heaviside jump function is appropriate to model contact surfaces.

Smooth particle hydrodynamics (SPH), developed by Lucy [8] and Gingold and Monaghan [9], was the first meshless technique. Such Lagrangian formulation-based meshless particle method defines a continuum by the set of moving particles. SPH was first invoked for solid mechanics problems by Libersky et al. [10]. SPH has been applied to simulate metal formation [11], underwater explosion [12], astrophysical problems [8,13], shock problems [14], impact problem [15], and computational fluid hydrodynamics [16]. Classical SPH method found limited application in solid mechanics due to instabilities and inconsistencies [12,17,18], due to which many modified versions were proposed to increase its suitability in computational solid mechanics [11,19–23]. Other problems suffered by SPH include difficulty in choosing artificial viscosity and imposing essential boundary conditions, inconsistency of approximation, and tensile instability problems. In recent years, various authors [12,15,24] have modified the SPH method to overcome these drawbacks. moving least square (MLS) approximation was invoked by Lancaster and Salkauskas [25] to develop smooth surface or curve fitting for scattered nodal data. Later on, Nayroles et al. [26] proposed a diffuse element method in which MLS approximation were used to develop field approximation. For solving solid mechanics problems, Belytschko et al. [27] proposed EFGM in which MLS interpolants are introduced into weak Galerkin forms. EFGM found tremendous applicability in solid mechanics problems. MLS approximation was invoked to develop shape functions that are directly introduced into governing equations in a meshless technique called the finite point method [28]. An improvement over EFGM and SPH, Liu et al. [29] developed the reproducing kernel particle method with the use of wavelet functions. Sukumar introduced concepts of the natural neighbor method [30], natural element method [31], and the meshless finite element method [32] based on Sibsonion interpolations. Meshless approximation and finite element interpolation are used to construct approximation in the partition of unity (PUM) concept developed by Melenk and Babuška [33]. Direct implementation of essential boundary condition is achieved in the point interpolation method [34–37], which is based on point local interpolation. Other important families of meshless techniques called local Petrov–Galerkin methods [38–42] based on MLS approximation and local Petrov–Galerkin weak forms have been proposed in literature. Computational cost in meshless methods is generally more compared to finite element method due to computational time involved in obtaining the shape functions.

EFGM is different from FEM as no element and element connectivity data is needed, but only a grid of nodes along with the boundary description is required to develop the approximate function. The engineering domain is discretized into a grid of nodes. There is no requirement for generating a FE mesh in this method [43–45]. This method offers many advantages while solving large deformation problems, crack propagation problems, and other discontinuous problems [46–49]. EFGM techniques were developed to eliminate the limitation associated

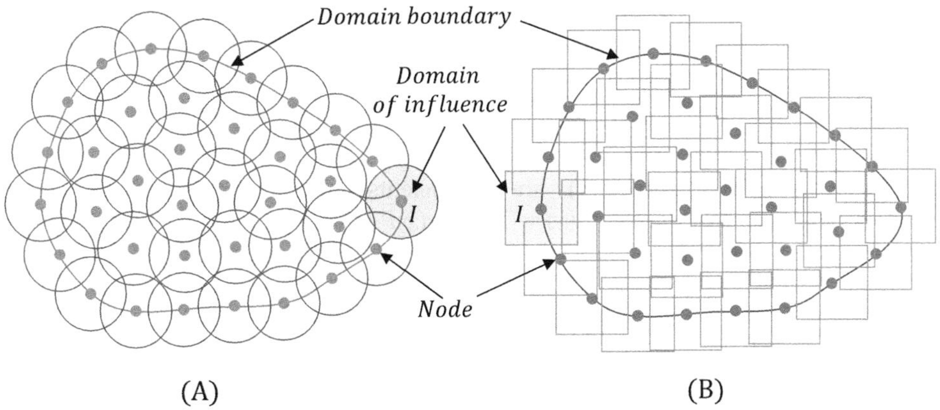

FIGURE 6.1 Domain of influence in element free Galerkin method: (A) circular and (B) rectangular.

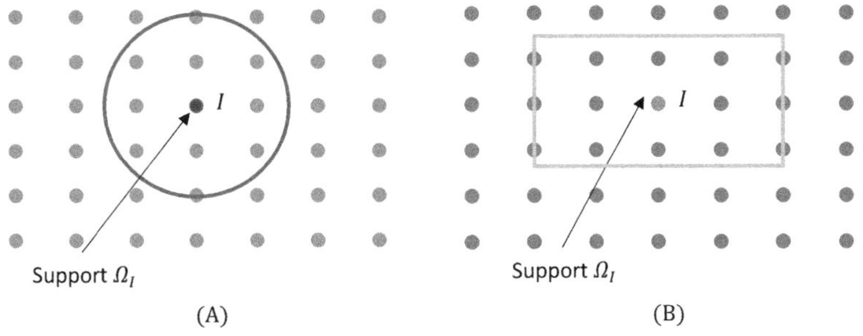

FIGURE 6.2 Compact supports in EFGM: (A) circular support and (B) rectangular support. *EFGM*, Element-free Galerkin method.

with the dependence of elemental mesh to construct approximation solution. EFGM technique was first proposed by Belytschko [50], where the engineering domain was represented by a set of nodes as shown in Fig. 6.1. This set of nodes along with boundary conditions are required to construct the approximation function. Mesh distortion issues, as observed in large deformation problems, are also eliminated by the meshless techniques [43,44,51–53].

Each weight function used in EFGM has a compact support, which defines the domain of influence of the nodes. Such weight functions remain zero outside and nonzero inside the compact support. Circular and rectangular domains of support have found extensive applications in EFGM, as shown in Fig. 6.2.

6.2 Approximation function in element-free Galerkin method

Moving least squares approximation was introduced by Shepard [54] and is mostly used in EFGM to approximate the primary variables in EFGM. Such approximations include polynomial basis function, weight function with a compact support, and coefficients varying with node position. The approximation $u^h(x)$ of primary variable function $\boldsymbol{u}$

is obtained by the product of the polynomial of order m with nonconstant coefficients $a(x)$ and is expressed as

$$u^h(\boldsymbol{x}) = \sum_{j=1}^{m} p_j(\boldsymbol{x})a_j(\boldsymbol{x}) = \boldsymbol{P}^T(\boldsymbol{x})\boldsymbol{a}(\boldsymbol{x}) \tag{6.1}$$

where $p_j(\boldsymbol{x})$ polynomials and $a_j(\boldsymbol{x})$ are coefficients and n represents the nodes in the compact support. We have $P^T(x) = [P_1(x)P_2(x)P_3(x)\ldots\ldots\ldots P_m(x)]$ and $\boldsymbol{a}^T(\boldsymbol{x}) = [a_1(\boldsymbol{x})$ $a_2(\boldsymbol{x})a_3(\boldsymbol{x})\ldots\ldots\ldots a_m(\boldsymbol{x})]$. Polynomial basis functions commonly used in EFGM are as follows:

$$P^T(x) = \left\{ \begin{array}{c} \text{linear-one dimensional } [1, x] \\ \text{linear-two dimensional } [1, x, y] \\ \text{quadratic-two dimensional } [1, x, y, x^2, y^2, xy] \\ \text{cubic-two dimensional } [1, x, y, x^2, y^2, xy, x^3, y^3, x^2y, xy^2] \end{array} \right\} \tag{6.2}$$

For obtaining coefficients "$\mathbf{a}(\boldsymbol{x})$," a weighted error norm is obtained to minimize the weighted least square sum "$L(\boldsymbol{x})$" as

$$L(\boldsymbol{x}) = \sum_{i=1}^{n} w(\boldsymbol{x} - \boldsymbol{x}_i)[u^h(\boldsymbol{x}_i) - u_i]^2 = \sum_{i=1}^{n} w_i(\boldsymbol{x})[\boldsymbol{P}^T(\boldsymbol{x}_i)\boldsymbol{a}(\boldsymbol{x}) - u_i]^2 \tag{6.3}$$

where $\boldsymbol{x}_i$ is the coordinate of the ith node, x is the coordinate of a generic sampling point, and n is the number of nodes in the domain of influence. The extremum of L can be obtained by minimizing error norm with respect to $\boldsymbol{a}(\boldsymbol{x})$, that is, $\frac{\partial L(x)}{\partial a(x)} = 0$, which results in following equations:

$$\begin{aligned} \frac{\partial L(\boldsymbol{x})}{\partial a_1(\boldsymbol{x})} &= \sum_{i=1}^{n} w_i(\boldsymbol{x}) 2P_1(\boldsymbol{x}_i)\left[\boldsymbol{P}^T(\boldsymbol{x}_i)\boldsymbol{a}(\boldsymbol{x}) - u_i\right] = 0 \\ \frac{\partial L(\boldsymbol{x})}{\partial a_2(\boldsymbol{x})} &= \sum_{i=1}^{n} w_i(\boldsymbol{x}) 2P_2(\boldsymbol{x}_i)\left[\boldsymbol{P}^T(\boldsymbol{x}_i)\boldsymbol{a}(\boldsymbol{x}) - u_i\right] = 0 \\ &\vdots \\ \frac{\partial L(\boldsymbol{x})}{\partial a_m(\boldsymbol{x})} &= \sum_{i=1}^{n} w_i(\boldsymbol{x}) 2P_m(\boldsymbol{x}_i)\left[\boldsymbol{P}^T(\boldsymbol{x}_i)\boldsymbol{a}(\boldsymbol{x}) - u_i\right] = 0 \end{aligned} \tag{6.4}$$

After rearrangement, we get

$$\sum_{i=1}^{n} w_i(\boldsymbol{x})\boldsymbol{P}(\boldsymbol{x}_i)\boldsymbol{P}^T(\boldsymbol{x}_i)\boldsymbol{a}(\boldsymbol{x}) = \sum_{i=1}^{n} w_i(\boldsymbol{x})\boldsymbol{P}(\boldsymbol{x}_i)u_i \tag{6.5}$$

More compactly linear set of equation for $\boldsymbol{a}(\boldsymbol{x})$ becomes

$$\boldsymbol{A}(\boldsymbol{x})\boldsymbol{a}(\boldsymbol{x}) = \boldsymbol{B}(\boldsymbol{x})\boldsymbol{u} \tag{6.6}$$

where

$$\boldsymbol{A}(\boldsymbol{x}) = \sum_{j=1}^{n} w_j(\boldsymbol{x})\mathbf{P}(\boldsymbol{x}_j)\mathbf{P}^T(\boldsymbol{x}_j); \ \boldsymbol{B}(\boldsymbol{x}) = [w_1(\boldsymbol{x})\boldsymbol{P}(\boldsymbol{x}_1), w_2(\boldsymbol{x})\boldsymbol{P}(\boldsymbol{x}_2), \ldots, w_n(\boldsymbol{x})\boldsymbol{P}(\boldsymbol{x}_n)] \tag{6.7}$$

and

$$\boldsymbol{u} = \{u_1, u_2, \ldots, u_n\} \tag{6.8}$$

The coefficients, $a(x)$, can be solved as

$$a(x) = [A(x)]^{-1}B(x)u \tag{6.9}$$

MLS approximation can be further expressed as

$$u^h(x) = P^T(x)a(x) = P^T(x)[A(x)]^{-1}B(x)u = \sum_{i=1}^{n} \Psi_i(x)u_i = \Psi^T(x)u \tag{6.10}$$

where $\Psi(x)$ is MLS shape function written as

$$\Psi(x) = P^T(x)[A(x)]^{-1}B(x) \tag{6.11}$$

Derivatives of shape functions are written as

$$\Psi_{,x} = \left(P^T(x)[A(x)]^{-1}B(x)\right)_{,x} = \left(P^T\right)_{,x}A^{-1}B + P^T\left(A^{-1}\right)_{,x}B + P^TA^{-1}B_{,x} \tag{6.12}$$

where

$$B_{,x} = w_{i,x}P(x_i);\ \left(A^{-1}\right)_{,x} = -A^{-1}A_{,x}A^{-1};\ A_{,x} = \sum_{j=1}^{n} w_{i,x}P(x_j)P^T(x_j) \tag{6.13}$$

If the weight function and its first kth derivatives are continuous, then the MLS shape functions and their first kth derivatives are also continuous.

6.3 Choice of weight (Kernel) function

Proper selection of weight functions has a great impact on solution accuracy, performance, and stability of EFGM. Weight functions are positive, differentiable, continuous in their support, and are monotonically decreasing functions, as we move away from the node. Commonly used weight functions are given as follows [55,56]:

6.3.1 Quartic spline weight function

$$w(\overline{r}_i) = \begin{cases} 1 - 6\overline{r}_i^2 + 8\overline{r}_i^3 - 3\overline{r}_i^4, \overline{r}_i \leq 1, \\ 0, \overline{r}_i > 1 \end{cases} \tag{6.14}$$

6.3.2 Cubic spline weight function

$$w(\overline{r}_i) = \begin{cases} \frac{2}{3} - 4\overline{r}_i^2 + 4\overline{r}_i^3, \overline{r}_i \leq 0.5, \\ \frac{4}{3} - 4\overline{r}_i + 4\overline{r}_i^2 - \frac{4}{3}\overline{r}_i^3, 0.5 < \overline{r}_i \leq 1, \\ 0\overline{r}_i > 1 \end{cases} \tag{6.15}$$

6.3.3 Exponential spline weight function

$$w(\overline{r_i}) = \begin{cases} e^{-(\overline{r_i}/\alpha)^2}, \overline{r_i} \le 1, \\ 0, \overline{r_i} > 1 \end{cases} \tag{6.16}$$

where $\overline{r_i} = \frac{\|x_I - x\|}{d_I}$, x_I is a given node, and d_I denotes size of compact support. Since there is no proper justification to assess the choice of weight function, the selection of weight function is somewhat arbitrary. Studies on the performance of different types of weight functions have been compared on error convergence rate in [27,55] with results indicating that one type of weight function is better than the other type of weight function for some particular problem, although further research is needed. The weight functions and their corresponding EFGM shape functions are shown for a 1D domain where $0 \le x \le 8$, with nine nodes that are equally spaced. In this case, the domain of influence is 2. The quartic spline and cubic spline weight functions are shown in Fig. 6.3, whereas their corresponding derivatives are presented in Fig. 6.4.

Quartic and cubic spline weight functions for 2D domains are shown in Fig. 6.5. The weight functions and the derivatives are shown for a central node present in a rectangular domain, and the size of compact support is assumed to be 1.4. The MLS shape functions for two-dimensional problems are shown in Fig. 6.6. The derivatives of quartic spline and the cubic spline weight functions are presented in Fig. 6.7, whereas derivatives of MLS shape functions are presented in Fig. 6.8. The size of the compact support of a node plays a very important role in EFGM. Dolbow and Belytschko obtained the dilatation parameter that is used to define the size of compact support as $d_I = d_{\max} c_I$, where c_I is large enough so that a good number of nodes are present in the influence domain [57] and $d_{\max}$ is an empirical factor lying between 2.0 and 4.0.

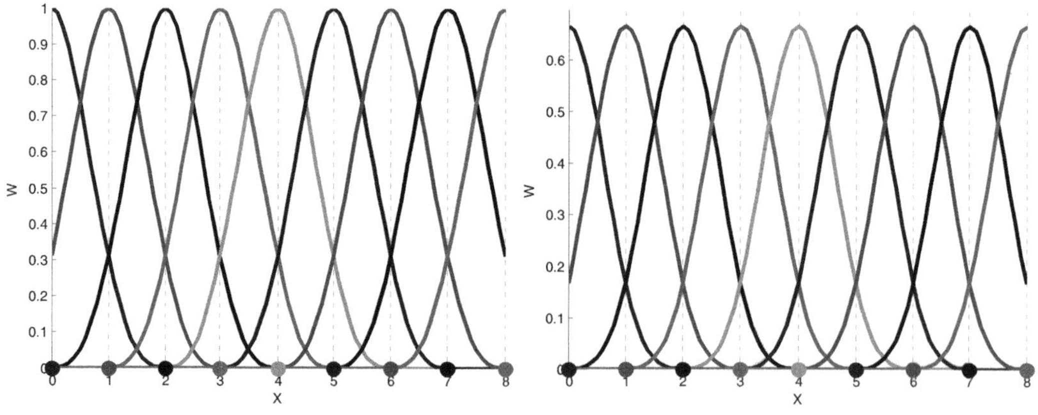

(A) (B)

FIGURE 6.3 Weight function for 1D: (A) quartic spline weight function and (B) cubic spline weight function.

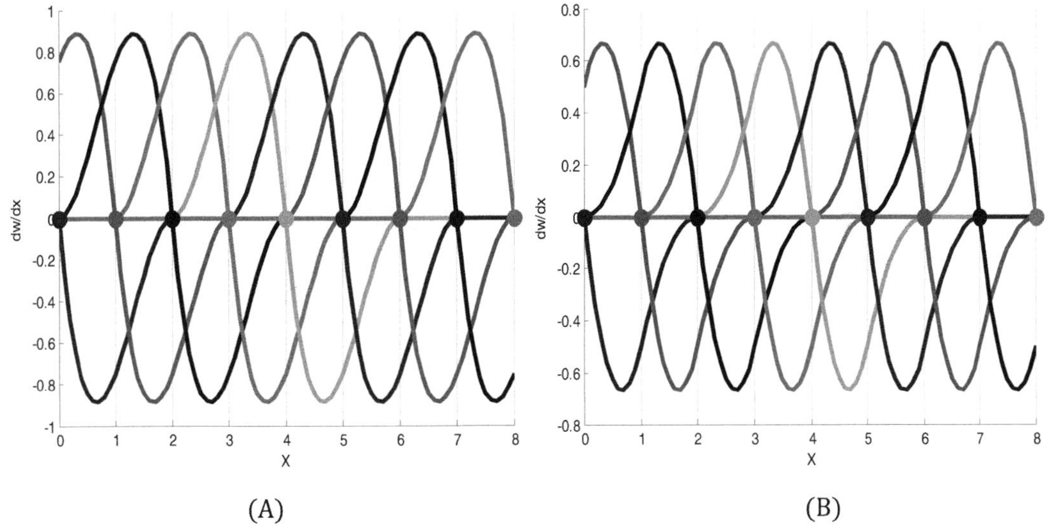

FIGURE 6.4 Derivatives of weight functions for 1D: (A) quartic spline and (B) cubic spline.

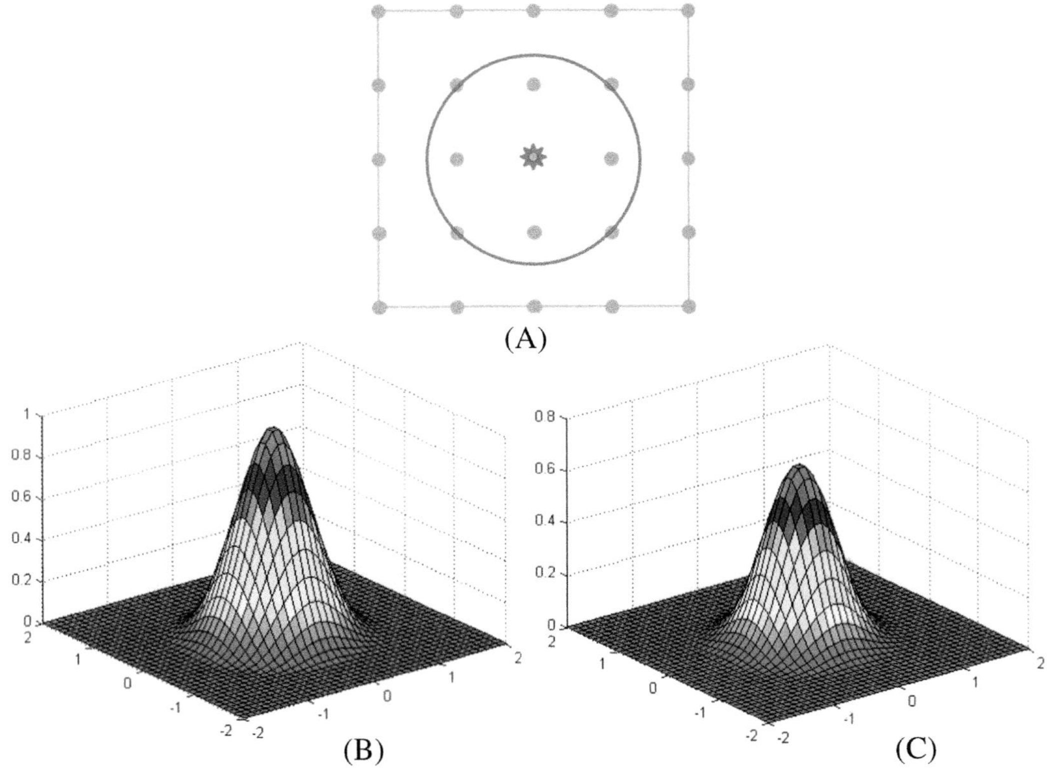

FIGURE 6.5 Weight functions for 2D domain: (A) domain representation, (B) quartic spline weight function, and (C) cubic spline weight function.

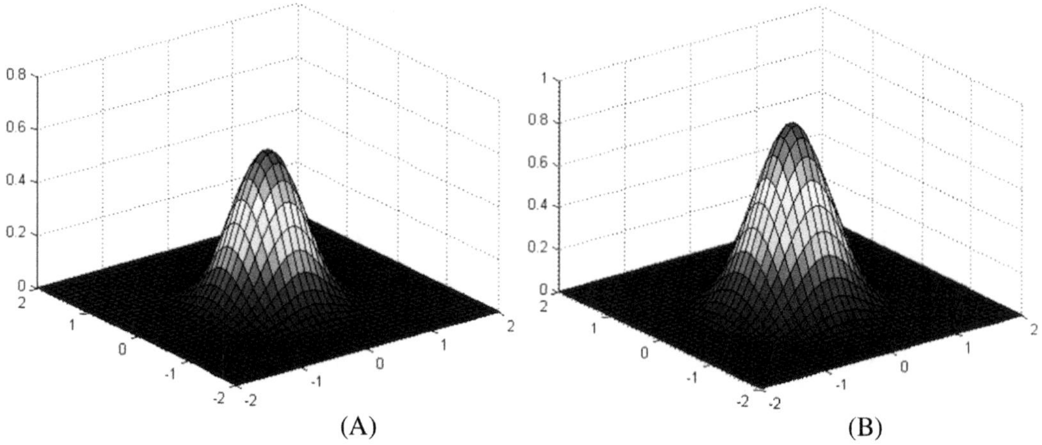

FIGURE 6.6 Moving least square shape functions for 2D using: (A) quartic spline and (B) cubic spline.

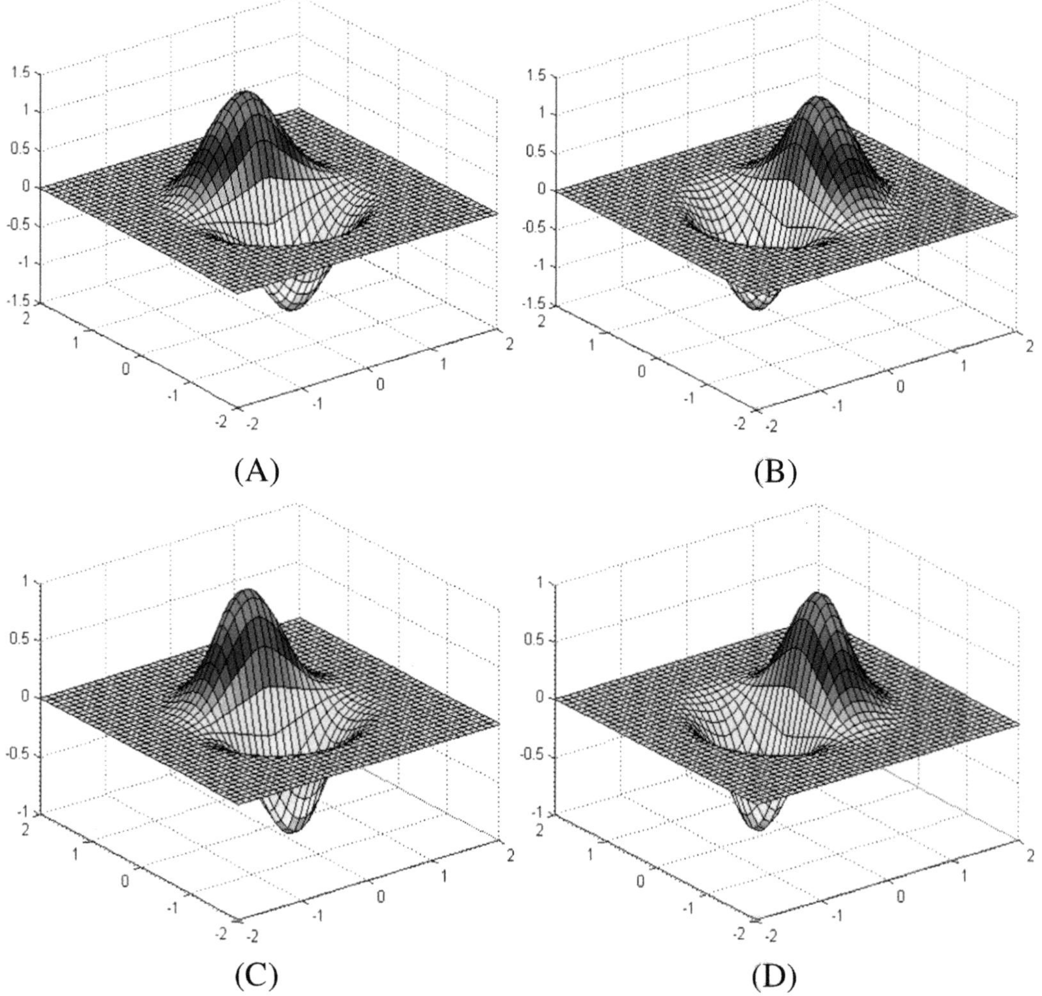

FIGURE 6.7 Derivatives of weight functions for 2D domain: (A) quartic spline $(w_{,x})$, (B) quartic spline $(w_{,y})$, (C) cubic spline $(w_{,x})$, and (D) cubic spline $(w_{,y})$.

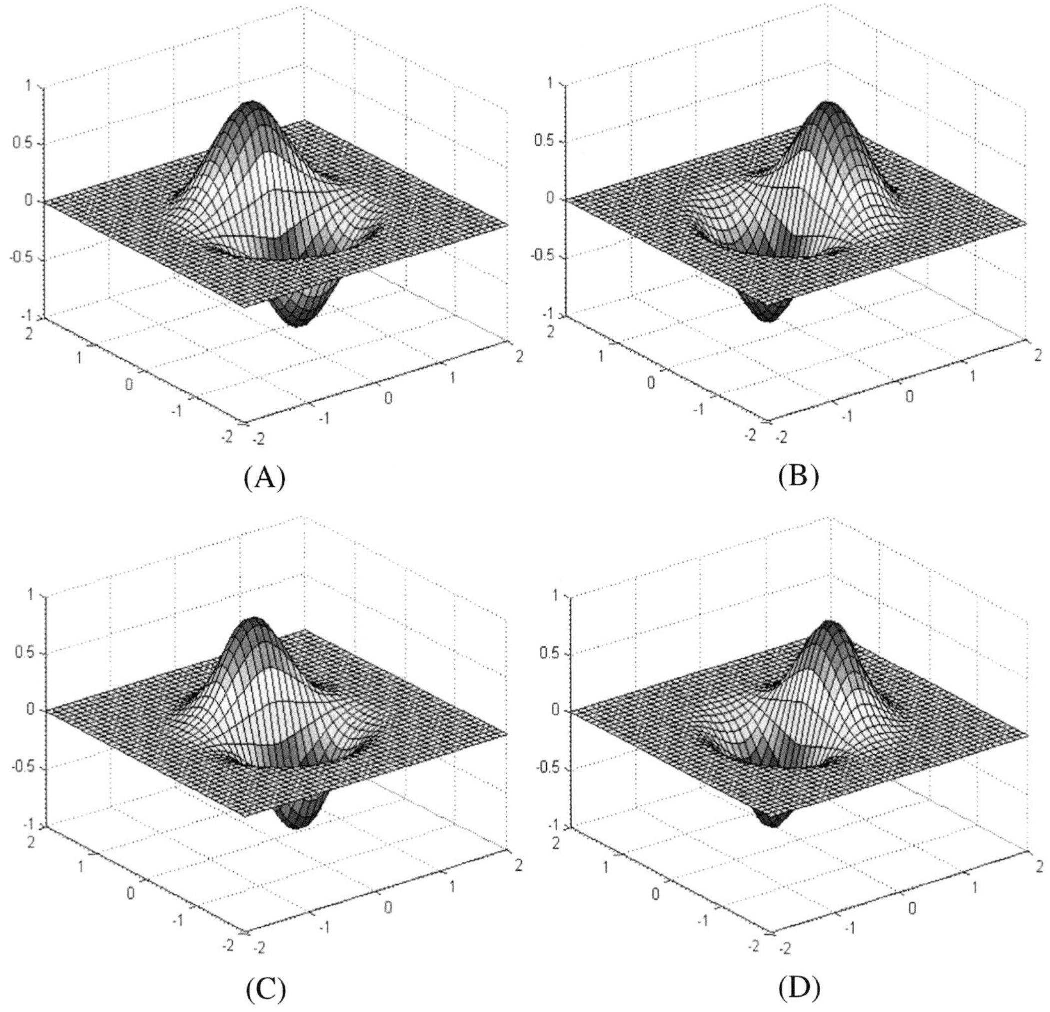

FIGURE 6.8 Derivatives of moving least square shape functions for 2D domain using: (A) quartic spline $(\Psi_{,x})$, (B) quartic spline $(\Psi_{,y})$, (C) cubic spline $(\Psi_{,x})$, and (D) cubic spline $(\Psi_{,y})$.

6.4 Imposition of boundary conditions in element-free Galerkin method

Direct application of essential boundary conditions is not possible in EFGM, which demands an additional and special treatment for the imposition of essential boundary conditions. Various techniques were proposed from time to time to apply boundary conditions in EFGM. Some of these methods are based on weak form modifications, and some techniques are based on the modification of shape functions to impose the boundary conditions [58]. Out of all proposed methods, the Lagrangian multiplier and penalty approaches have found extensive use in EFGM. A variational principle can be used to specify a scalar quantity, which can be defined in an integral form as

$$\Pi = \int_{\Omega} F(\boldsymbol{u})d\Omega + \int_{\Gamma} E(\boldsymbol{u})d\Gamma \tag{6.17}$$

where Π is the functional, $\boldsymbol{u}$ is the unknown variable, and F and E are differential operators. The solution will be a function $\mathbf{u}$, which keeps the functional Π stationary as

$$\delta\Pi = 0 \ \textit{for any} \ \delta\boldsymbol{u} \tag{6.18}$$

6.4.1 Lagrangian multiplier method

Consider a particular case of making some functional Π stationary with following constraints:

$$C(u) = o \text{ on } \Gamma \tag{6.19}$$

Following functional can be built to satisfy the above constraints:

$$\overline{\Pi}(\mathbf{u},\boldsymbol{\lambda}) = \Pi(\mathbf{u}) + \int_{\Gamma} \boldsymbol{\lambda}^T \mathbf{C}(\mathbf{u})d\Gamma \tag{6.20}$$

Thus we obtain a new functional, the variation of which is given by

$$\delta\overline{\Pi} = \delta\Pi + \int_{\Gamma} \delta\boldsymbol{\lambda}^T \mathbf{C}(\mathbf{u})d\Gamma + \int_{\Gamma} \boldsymbol{\lambda}^T \delta\mathbf{C}(\mathbf{u})d\Gamma \tag{6.21}$$

where $\boldsymbol{\lambda}$ represents the Lagrangian multipliers, which should conform to governing differential equations of the problem under consideration. Lagrangian multipliers can be obtained as

$$\boldsymbol{\lambda}(x) = \sum_{i=1}^{p} N_i(x)\boldsymbol{\lambda}_i \tag{6.22}$$

where $\boldsymbol{p}$ represents the nodes where displacement boundary conditions are imposed and N denotes shape functions. The shape functions used for approximating Lagrangian multipliers can be either the Lagrangian functions or MLS-based shape functions on the boundary Γ. The variations of the Lagrangian multipliers can be given by

$$\delta\boldsymbol{\lambda}(x) = \sum_{i=1}^{p} N_i(x)\delta\boldsymbol{\lambda}_i \tag{6.23}$$

The Lagrangian multiplier method proves to be very efficient in imposing the essential boundary conditions, but this technique has one major drawback of introducing additional unknowns ($\boldsymbol{\lambda}$) to the problem.

6.4.2 Penalty method

For the penalty approach, we develop the functional as follows:

$$\overline{\Pi}(\mathbf{u},\alpha) = \Pi(\mathbf{u}) + \frac{\alpha}{2}\int_{\Gamma} \mathbf{C}(\mathbf{u})^T\mathbf{C}(\mathbf{u})d\Gamma \tag{6.24}$$

where α represents penalty factors. Variation of the above functional can be given by

$$\delta\overline{\Pi} = \delta\Pi + \frac{\alpha}{2}\int_{\Gamma} \delta\mathbf{C}(\mathbf{u})^T\mathbf{C}(\mathbf{u})d\Gamma + \frac{\alpha}{2}\int_{\Gamma} \mathbf{C}(\mathbf{u})^T\delta\mathbf{C}(\mathbf{u})d\Gamma \tag{6.25}$$

The penalty method does not add any additional unknown to the problem, when compared with the Lagrangian method. But the choice of penalty factor strongly affects the conditioning of stiffness matrix.

6.5 Governing equations for element-free Galerkin method

Consider an engineering domain Ω, surrounded by the boundary Γ, as can be seen in Fig. 6.9. Essential boundary conditions are applied on Γ_u and the tractions at Γ_t. Equilibrium equation for a simple elastic problem can be expressed in standard form as

$$C(u) = o \text{ on } \Gamma \tag{6.26}$$

where $\boldsymbol{\sigma}$ denotes the stresses, and $\mathbf{b}$ represents body force vector. Boundary conditions for the given problem can be expressed as

$$\sigma \times n = t \text{ on } \Gamma_{t'} \text{ (Natural B.C)} \tag{6.27}$$

$$u = \overline{\text{u}} \text{ on } \Gamma_{\text{u}}, \text{(Essential B.C)} \tag{6.28}$$

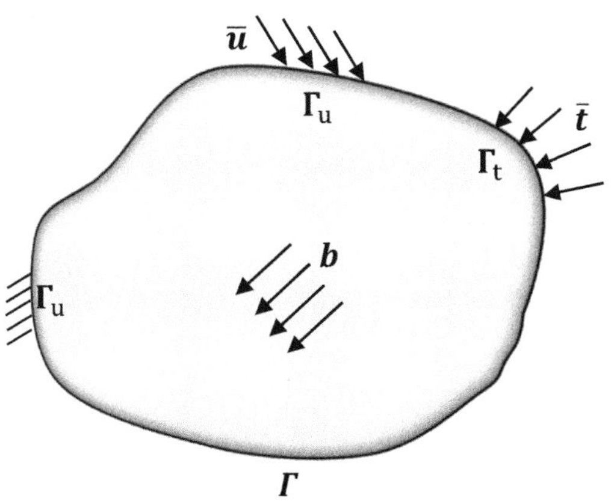

FIGURE 6.9 Engineering domain subjected to displacements and tractions.

Using the Lagrangian multiplier method, the equilibrium equation can be written as [59]

$$\int_{\Omega} \boldsymbol{\varepsilon}^T \boldsymbol{\sigma} d\Omega - \int_{\Omega} \mathbf{u}^T \mathbf{b} d\Omega - \int_{\Gamma_t} \mathbf{u}^T \mathbf{t} d\Gamma_t + \int_{\Gamma_u} \boldsymbol{\lambda}^T (\mathbf{u} - \overline{\mathbf{u}}) d\Gamma_u = 0 \tag{6.29}$$

where ε denotes the strains, and the fourth term is added to weak formulation for the imposition of boundary conditions, as required by the Lagrangian method. The approximate solution in EFGM eliminates all mesh-related drawbacks that occur in FEM. The EFGM approximation of the field variable and its variations can be obtained as

$$\mathbf{u}(\mathbf{x}) = \sum_{i=1}^{n} \Psi_i(\mathbf{x}) \mathbf{u}_i = \Psi \boldsymbol{u} \tag{6.30}$$

$$\delta \mathbf{u}(\mathbf{x}) = \sum_{i=1}^{n} \Psi_i(\mathbf{x}) \delta \mathbf{u}_i = \Psi \delta \boldsymbol{u} \tag{6.31}$$

where Ψ denotes MLS shape functions and n denotes the nodes in the compact support. We can further arrive at

$$\int_{\Omega} \boldsymbol{u}^T \mathbf{B}^T \boldsymbol{\sigma} d\Omega - \int_{\Omega} \boldsymbol{u}^T \Psi^T \mathbf{b} d\Omega - \int_{\Gamma_t} \boldsymbol{u}^T \Psi^T \mathbf{t} d\Gamma_t + \int_{\Gamma_u} (\Psi \boldsymbol{u} - \overline{\mathbf{u}})^T \mathbf{N} \boldsymbol{\lambda} \mathbf{d} \Gamma_u = 0 \tag{6.32}$$

The variational form of equilibrium equation is expressed as

$$\delta \boldsymbol{u}^T \left(\int_{\Omega} \mathbf{B}^T \mathbf{D} \mathbf{B} \mathbf{u} d\Omega - \int_{\Omega} \Psi^T \mathbf{b} d\Omega - \int_{\Gamma_t} \Psi^T \mathbf{t} d\Gamma_t \right) + \int_{\Gamma_u} \Psi^T \delta \boldsymbol{u}^T \mathbf{N} \boldsymbol{\lambda} \mathbf{d} \Gamma_u + \int_{\Gamma_u} (\Psi \boldsymbol{u} - \overline{\mathbf{u}})^T \mathbf{N} \delta \boldsymbol{\lambda} \mathbf{d} \Gamma_u = \boldsymbol{0} \tag{6.33}$$

which can be further written as

$$\delta \boldsymbol{u}^T \left[\boldsymbol{K} \boldsymbol{u} - \boldsymbol{f} + \boldsymbol{G} \boldsymbol{\lambda} \right] + \delta \boldsymbol{\lambda}^T \left[\boldsymbol{G}^T \boldsymbol{u} - \boldsymbol{q} \right] = \mathbf{0} \tag{6.34}$$

Discrete system of equations become

$$\boldsymbol{K} \boldsymbol{u} + \boldsymbol{G} \boldsymbol{\lambda} = \boldsymbol{f} \tag{6.35}$$

$$\boldsymbol{G}^T \boldsymbol{u} = \boldsymbol{q} \tag{6.36}$$

which can be expressed as

$$\begin{bmatrix} K & G \\ G^T & 0 \end{bmatrix} \begin{Bmatrix} u \\ \lambda \end{Bmatrix} = \begin{Bmatrix} f \\ q \end{Bmatrix} \tag{6.37}$$

where

$$\boldsymbol{K} = \int_{\Omega} \mathbf{B}^T \mathbf{D} \mathbf{B} \mathbf{u} d\Omega \tag{6.38}$$

$$\boldsymbol{f} = \int_{\Omega} \Psi^T \mathbf{b} d\Omega + \int_{\Gamma_t} \Psi^T \mathbf{t} d\Gamma_t \tag{6.39}$$

$$G = \int_{\Gamma_u} \Psi^T \mathbf{N} d\Gamma_u \tag{6.40}$$

$$q = \int_{\Gamma_u} N^T \overline{\mathbf{u}} \mathbf{d}\Gamma_u \tag{6.41}$$

$$\mathbf{B} = \begin{bmatrix} \Psi_{i,x} & 0 \\ 0 & \Psi_{i,y} \\ \Psi_{i,y} & \Psi_{i,x} \end{bmatrix} \tag{6.42}$$

$$\mathbf{D} = \frac{\mathrm{E}}{(1+\mu)(1-2\mu)} \begin{bmatrix} 1-\mu & \mu & 0 \\ \mu & 1-\mu & 0 \\ 0 & 0 & \frac{1-2\mu}{2} \end{bmatrix} (\textit{plane strain}) \tag{6.43}$$

$$\mathbf{D} = \frac{\mathrm{E}}{(1-\mu^2)} \begin{bmatrix} 1 & \mu & 0 \\ \mu & 1 & 0 \\ 0 & 0 & \frac{1-\mu}{2} \end{bmatrix} (\textit{plane stress}) \tag{6.44}$$

6.6 Modeling of discontinuities in element-free Galerkin method

Several methods were proposed to model different discontinuities occurring in engineering domains. Some of these methods are based on the modification of weight functions, such as the visibility method [60], diffraction method [61], and transparency method [62], whereas some methods are based on the modification of intrinsic basis [63] to incorporate enrichment functions [64–69].

6.6.1 Methods based on modification of weight function

Visibility method has been used to model strong discontinuities in meshless methods. In this method, crack interface is assumed to be opaque, and the nodes that lie on the opposite side of crack are excluded and not considered in the development of variable approximations. The nodal support is truncated and restricted to that portion of domain only, which is visible from the node. However, this technique of modeling strong discontinuities suffered various drawbacks since the shape function is intersected abruptly and undesired discontinuities occur for the nodes that lie close to the crack tip [60]. The visibility criterion cannot be used to model the nonconvex boundaries.

Diffraction method is the modification of the visibility method where undesired interior discontinuities are removed and no-convex boundaries can be easily accommodated in this technique [62]. The diffraction method treats crack interface as an opaque entity and evaluates the length of the ray of light by a path that passes around the corner of interface [62]. Sudden truncation of the nodal support does not occur in this method because the

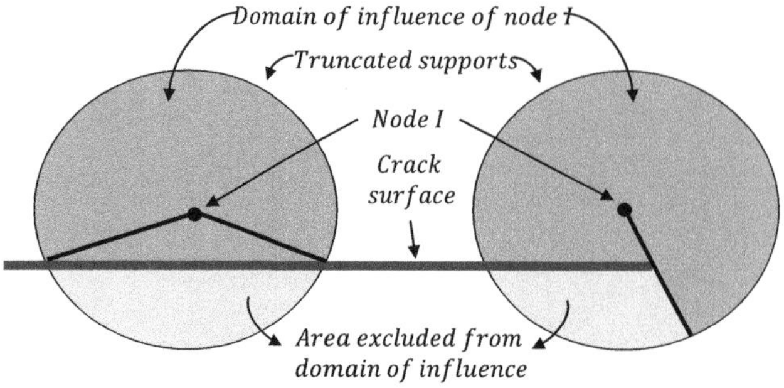

FIGURE 6.10 Scheme of visibility method.

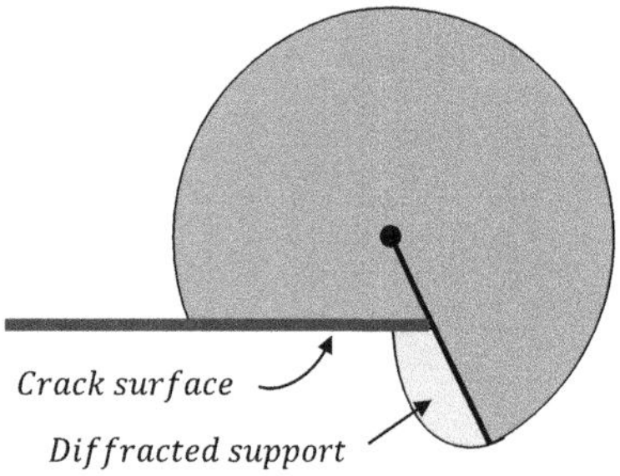

FIGURE 6.11 Scheme of diffraction and transparency method.

support is free to bend or diffract around the crack tip, as can be seen in Fig. 6.10. The shape functions obtained in the diffraction method are very complex, and this method experiences severe complexities while modeling three-dimensional problems.

The transparency method was first developed by Organ et al. [62]. This method is very much suitable for three-dimensional problems. Here, crack is considered to be transparent near crack tip, and an additional condition is required for the nodes located close to the crack, as shown in Fig. 6.11. Usually, a sharp gradient in weight function is introduced in the region ahead of the crack because the angle between crack and the ray of light from the node to crack tip is very small. All three methods discussed above are not entirely satisfactory because neither of them corresponds to the physical behavior of the cracks.

6.6.2 Methods based on modification of intrinsic basis

Techniques based on the modification of intrinsic basis use modification schemes based on the crack kinematics. In LEFM, modified intrinsic basis can be written for two-dimensional crack problems as

$$\mathbf{P}^T(x) = \left[1, x, y, \sqrt{r}\cos\frac{\theta}{2}, \sqrt{r}\sin\frac{\theta}{2}, \sqrt{r}\cos\frac{\theta}{2}\sin\theta, \sqrt{r}\sin\frac{\theta}{2}\sin\theta\right] \tag{6.45}$$

The first three terms of the above enriched basis represent the linear polynomial basis, and the other four terms denote the enrichment functions. If we use quadratic basis, then the modified intrinsic basis can be written as

$$\mathbf{P}^T(x) = \left[1, x, y, x^2, y^2, xy, \sqrt{r}\cos\frac{\theta}{2}, \sqrt{r}\sin\frac{\theta}{2}, \sqrt{r}\cos\frac{\theta}{2}\sin\theta, \sqrt{r}\sin\frac{\theta}{2}\sin\theta\right] \tag{6.46}$$

One of the main drawbacks of the intrinsic enrichment is that it is applied in the entire domain in order to eliminate the undesired discontinuities. Since there are additional terms in the modified basis functions, additional computational cost is introduced.

6.6.3 Methods based on extrinsic MLS enrichment

The methods based on extrinsic MLS enrichment model discontinuities by introducing analytical solutions extrinsically [63]. For modeling the discontinuities created by cracks, modified variable approximation can be written as

$$u^h(x) = \sum_{i=1}^{n} P(x_i)a(x) + \sum_{j=1}^{n_c} k_I^j Q_I^j(x_j) + k_{II}^j Q_{II}^j(x_j) \tag{6.47}$$

where n_c represents cracks, P is polynomial basis, and k_I and k_{II} are the additional enriched degrees of freedom associated with mode-I and mode-II fractures, respectively. Q_I^i and Q_{II}^i define the crack tip displacement field and can be obtained as [70]

$$Q_I^1(x) = \frac{1}{2G}\sqrt{\frac{r}{2\pi}}\cos\frac{\theta}{2}\left(\kappa - 1 + 2\sin^2\frac{\theta}{2}\right) \tag{6.48}$$

$$Q_I^2(x) = \frac{1}{2G}\sqrt{\frac{r}{2\pi}}\sin\frac{\theta}{2}\left(\kappa + 1 - 2\cos^2\frac{\theta}{2}\right) \tag{6.49}$$

$$Q_{II}^1(x) = \frac{1}{2G}\sqrt{\frac{r}{2\pi}}\sin\frac{\theta}{2}\left(\kappa + 1 + 2\cos^2\frac{\theta}{2}\right) \tag{6.50}$$

$$Q_{II}^2(x) = -\frac{1}{2G}\sqrt{\frac{r}{2\pi}}\cos\frac{\theta}{2}\left(\kappa - 1 - 2\sin^2\frac{\theta}{2}\right) \tag{6.51}$$

where G is shear modulus, and κ is the Kolosov constant. The application of extrinsic enrichment to the standard displacement-based approximation enables us to obtain the SIFs directly, without evaluating the J-integral. However, this type of enrichment is introduced throughout the domain, which makes it computationally more demanding.

6.6.4 Methods based on extrinsic PUM enrichment

The methods based on extrinsic PU method enrichment enable us to enrich the approximations for the nodes that lie close to the discontinuity only. This proves to be highly advantageous from the computational viewpoint because enrichments are introduced in the vicinity of the discontinuity only. Based on extrinsic PUM enrichment, the approximation employed to model bi-material interfaces is expressed as

$$\mathbf{u}^h(x) = \sum_{i=1}^{n} \Psi_i(x)\mathbf{u}_i + \sum_{j=1}^{n_s} \Psi_j(x)\{f(x) - f(x_j)\}\mathbf{a}_j \tag{6.52}$$

where $\mathbf{u}_i$ and $\mathbf{a}_j$ are nodal and enriched degrees of freedom, respectively. $\boldsymbol{n}$ represents standard nodes, and $\boldsymbol{n_s}$ denotes enriched nodes. $f(x)$ is the enrichment function. Based on above approximations, the final EFGM model for modeling bi-material problems can be derived as $\mathbf{Kd} = f$, where $\mathbf{d} = [\mathbf{u} \quad \mathbf{a}]^T$. We have

$$\boldsymbol{K} = \begin{bmatrix} \mathrm{K}^{\mathrm{uu}} & \mathrm{K}^{\mathrm{ua}} \\ \mathrm{K}^{\mathrm{au}} & \mathrm{K}^{\mathrm{aa}} \end{bmatrix} \text{ and } \boldsymbol{f} = \begin{Bmatrix} f^u \\ f^a \end{Bmatrix} \tag{6.53}$$

such that

$$\mathrm{K}^{\alpha\beta} = \int_{\Omega^e} (\boldsymbol{B}^\alpha)^{\mathrm{T}} \boldsymbol{D} \; and \; \mathrm{B}^\beta d\Omega(\alpha, \beta = u, \alpha) \tag{6.54}$$

$$f^\alpha = \int_{\Gamma^e} \Psi^\alpha \mathbf{t} d\Gamma + \int_{\Omega^e} \Psi^\alpha \mathbf{b} d\Omega(\alpha = u, a) \tag{6.55}$$

where $\mathbf{t}$ and $\mathbf{b}$ are the applied tractions and the body force vector, respectively. We have

$$\Psi^u = \begin{bmatrix} \Psi_i & 0 \\ 0 & \Psi_i \end{bmatrix} \text{ and } \Psi^a = \begin{bmatrix} \overline{\Psi}_j & 0 \\ 0 & \overline{\Psi}_j \end{bmatrix}; \overline{\Psi}_j = \Psi_j(x)\{f(x) - f(x_j)\} \tag{6.56}$$

$$\mathbf{B}^u = \begin{bmatrix} \dfrac{\partial \Psi_i}{\partial x} & 0 \\ 0 & \dfrac{\partial \Psi_i}{\partial y} \\ \dfrac{\partial \Psi_i}{\partial y} & \dfrac{\partial \Psi_i}{\partial x} \end{bmatrix} \; and \; \mathbf{B}^u = \begin{bmatrix} \dfrac{\partial \overline{\Psi}_j}{\partial x} & 0 \\ 0 & \dfrac{\partial \overline{\Psi}_j}{\partial y} \\ \dfrac{\partial \overline{\Psi}_j}{\partial y} & \dfrac{\partial \overline{\Psi}_j}{\partial x} \end{bmatrix} \tag{6.57}$$

Now, based on extrinsic PUM enrichment, the enriched EFGM approximation for modeling the discontinuities produced by cracks can be written as

$$\mathbf{u}^h(x) = \sum_{i=1}^{n} \Psi_i(x)\mathbf{u}_i + \sum_{j=1}^{n_s} \Psi_j(x)\{H(x) - H(x_j)\}\mathbf{a}_j + \sum_{k=1}^{n_T} \Psi_k(x) \sum_{\alpha=1}^{4} [\beta_\alpha(x) - \beta_\alpha(x_k)] b_k^\alpha \tag{6.58}$$

where $\mathbf{u}_i$ are standard degrees of freedom and $\mathbf{a}_j$ the enriched degrees of freedom of split nodes, b_k^α denotes the enriched degrees of freedom corresponding to tip nodes. $\boldsymbol{n}$ denotes standard nodes, $\boldsymbol{n_s}$ the split nodes, and $\boldsymbol{n_T}$ denotes the tip nodes. $H(x)$ is the Heaviside jump function. $\beta_\alpha(x)$ represents the enrichment function applied at crack tip. The crack tip enrichment functions have been discussed in earlier sections. Based on the above approximations, the final EFGM model can be obtained as $\mathbf{Kd} = f$, where $\mathbf{d} = [\mathbf{u} \quad \mathbf{a}\mathbf{b}^\alpha]^T$. We have

$$[\mathrm{K}] = \begin{bmatrix} \mathbf{K}^{uu} & \mathbf{K}^{ua} & \mathbf{K}^{ub} \\ \mathbf{K}^{au} & \mathbf{K}^{aa} & \mathbf{K}^{ab} \\ \mathbf{K}^{bu} & \mathbf{K}^{ba} & \mathbf{K}^{bb} \end{bmatrix} \tag{6.59}$$

$$\{f\} = \left\{f_{std}\, f_{jump}\, f_{tip}\right\}^T \tag{6.60}$$

$$\mathbf{K}^{rs} = \int_\Omega (\mathbf{B}^r)^T \mathbf{D}\mathbf{B}^s d\Omega \tag{6.61}$$

where

$$r, s = u, a, b \tag{6.62}$$

$$f_{std} = \int_\Omega \Psi_{std}^T \mathbf{b} d\Omega + \int_{\Gamma_t} \Psi_{std}^T \mathbf{t} d\Gamma_t \tag{6.63}$$

$$f_{jump} = \int_\Omega \Psi_{jump}^T \mathbf{b} d\Omega + \int_{\Gamma_t} \Psi_{jump}^T \mathbf{t} d\Gamma_t \tag{6.64}$$

$$f_{tip} = \int_\Omega \Psi_{tip}^T \mathbf{b} d\Omega + \int_{\Gamma_t} \Psi_{tip}^T \mathbf{t} d\Gamma_t \tag{6.65}$$

$$\Psi_{std} = \begin{bmatrix} \Psi_i & 0 \\ 0 & \Psi_i \end{bmatrix}, \Psi_{JUMP} = \begin{bmatrix} \overline{\Psi}_j & 0 \\ 0 & \overline{\Psi}_j \end{bmatrix} \; and \; \Psi_{tip} = \begin{bmatrix} \overline{\overline{\Psi}}_{k\alpha} & 0 \\ 0 & \overline{\overline{\Psi}}_{k\alpha} \end{bmatrix} \tag{6.66}$$

$$\mathbf{B}^u = \begin{bmatrix} \frac{\partial \Psi_i}{\partial x} & 0 \\ 0 & \frac{\partial \Psi_i}{\partial y} \\ \frac{\partial \Psi_i}{\partial y} & \frac{\partial \Psi_i}{\partial x} \end{bmatrix}, \mathbf{B}^a = \begin{bmatrix} \frac{\partial \overline{\Psi}_j}{\partial x} & 0 \\ 0 & \frac{\partial \overline{\Psi}_j}{\partial y} \\ \frac{\partial \overline{\Psi}_j}{\partial y} & \frac{\partial \overline{\Psi}_j}{\partial x} \end{bmatrix}, \mathbf{B}^b = \begin{bmatrix} \frac{\partial \overline{\overline{\Psi}}_{k\alpha}}{\partial x} & 0 \\ 0 & \frac{\partial \overline{\overline{\Psi}}_{k\alpha}}{\partial y} \\ \frac{\partial \overline{\overline{\Psi}}_{k\alpha}}{\partial y} & \frac{\partial \overline{\overline{\Psi}}_{k\alpha}}{\partial x} \end{bmatrix} \tag{6.67}$$

such that $\overline{\Psi}_J = \Psi_J(x)\{H(x) - H\}$ *and* $\overline{\overline{\Psi}}_{k\alpha} = \Psi_k(x)\{\beta_\alpha(x) - \beta_\alpha(x_k)\}$.

6.7 Level set method in element-free Galerkin method

The level set methodology employed in EFGM is the same as that in XFEM, but there are no split elements or tip elements in EFGM as no mesh is generated here. In XFEM, there are two types of nodes along with standard nodes. The nodes of elements completely cut by the interface area are called split nodes and nodes belonging to elements that contain crack tip are called the tip nodes. In EFGM, the domain of influence of a particular node is equivalent to an element. Then we define split nodes as the nodes, the domain of influence or support of which is completely cut by the discontinuity, and the tip nodes are the nodes, the supports of which contain the crack tip, as shown in Fig. 6.12. In the given figure, there are two split nodes whose domain of influence is completely cut by the crack, one tip node whose domain of influence contains the crack tip, and one standard node that lies away from the crack surface.

Evaluation of split and tip nodes in EFGM is different from that of XFEM. However, the level set methodology is same in EFGM as that of XFEM. Closed discontinuities

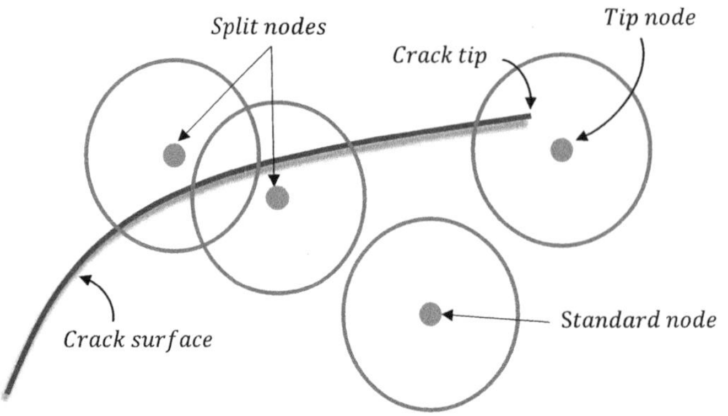

FIGURE 6.12 Standard nodes, split nodes, and tip nodes in EFGM. *EFGM*, Element-free Galerkin method.

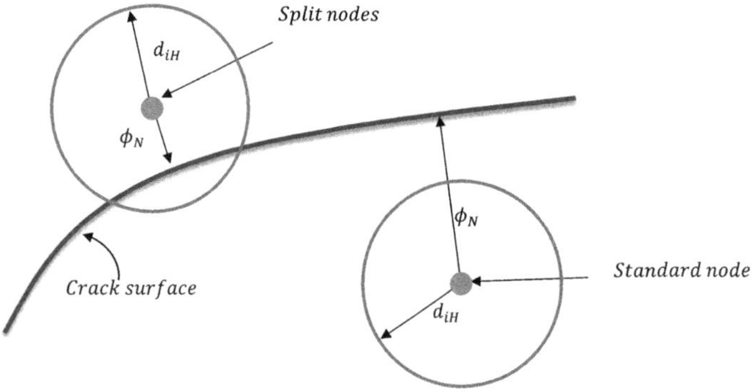

FIGURE 6.13 Evaluation of split nodes in EFGM. *EFGM*, Element-free Galerkin method.

are completely defined by the normal level sets, whereas the cracks are completely defined by the normal and tangential level sets. After calculating the normal and tangential level sets for all nodes of the domain, appropriate algorithms are used to find out the split nodes and the tip nodes. The evaluation of split nodes in EFGM is shown in Fig. 6.13, whereas the evaluation of tip nodes is presented in Fig. 6.14. If $\varnothing_N$ is normal level set and d_{iH} is the size of compact support, then this node is a split node if

$$\varnothing_N - d_{iH} < 0 \tag{6.68}$$

In order to find out the tip nodes, we calculate the distance of each node from crack tip. If d_{tip} is the distance of a node from crack tip and d_{iT} is the size of the compact support, then this node will be considered as a tip node if

$$d_{tip} - d_{iT} < 0 \tag{6.69}$$

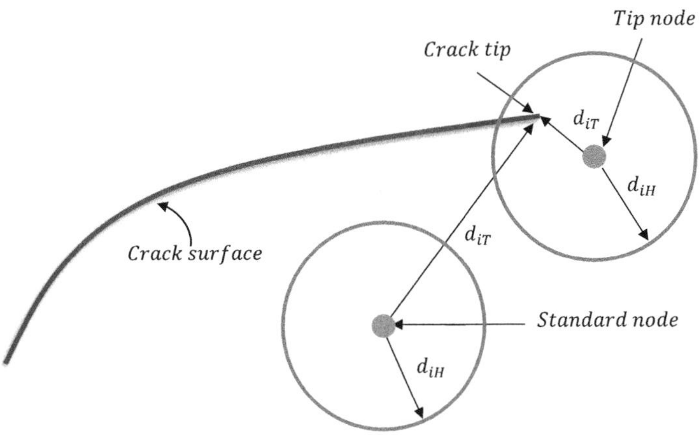

FIGURE 6.14 Evaluation of tip nodes in EFGM. *EFGM*, Element-free Galerkin method.

6.8 h- and p-refinement

Compared to FEM, meshless methods do not need highly structured mesh. For critical regions in FEM such as crack tip zone, we enhance the region with the help of p-refinement and/or h-refinement. The critical regions are finely meshed by using h-refinement approach. Higher order polynomials are used for the critical region in p-refinement approach, that is, elements of region are upgraded to higher order elements such as eight-node element is used in place of four-node element in a plane problem. Tremendous cost is used to re-mesh the finite element mesh in both p-refinement and h-refinement. In meshless methods, h-refinement for critical region is achieved by simply placing an arbitrary number of nodes. Thus for approximation in meshless methods, we require the location of extra-added nodes. In the case of p-refinement in meshless methods, we simply need to modify the order of polynomial. Four refinement approaches were summarized by Chen et al. [71] as

I. h-refinement achieved by adding nodes: In this approach, new nodes are added, but the size of the domain of the influence of newly added nodes as well as of existing nodes is kept unchanged. Accuracy of meshless method is improved by this approach, but the domain of influence becomes saturated after adding certain quantity of nodes.

II. p-refinement achieved by increasing the polynomial order: There is no addition of nodes, but in order to ensure invertability of matrix A, the size of the domain of the influence of certain nodes may need to be increased.

III. h-refinement achieved by adding nodes with the smaller domain of influence: In this refinement approach, the domain of the influence of newly added nodes is comparatively smaller than existing nodes. This approach improves the accuracy in the enriched region and is an effective way of improving accuracy by h-refinement with minimum cost.

IV. h-refinement achieved by adding nodes and reducing the size of the domain of the influence of all nodes: Although this refinement approach improves the accuracy of enriched region, it is computationally costlier. In this approach, decrease in the size of support is directly proportional to increase in number of nodes.

6.9 Numerical integration

Numerical integration is one of the main drawback of EFGM due to the rational form of MLS shape function. As a result, exact integration is a challenging task in most of the mesh-free techniques. Most commonly used technique includes the following:

6.9.1 Direct nodal integration

In this technique, integrals of weak form are evaluated to obtain partial integration at the nodes [72]. In the case of EFGM, nodes are used for evaluating integrals, and these nodes serve as integration points. The integral is evaluated as

$$\int_{\Omega} f(\mathbf{X})d\Omega = \sum_{J=1}^{n} f(\mathbf{X}_J)V_J \tag{6.70}$$

where V_J denotes quadrature weights, which describes the volume associated with the nodes. This approach does not need any background mesh and is more efficient compared to full integration. However, nodal integration would result in spatial instability, which occurs due to rank deficiency, and stabilization procedure is required. Lagrangian kernels with stress points have proven to be best method to stabilize particle discretization in fluids and solids [17].

6.9.2 Stress point integration

To avoid spatial instability, this technique adds extra stress points in addition to nodes for evaluating integrals. In stress point integration technique, the integral is evaluated as

$$\int_{\Omega} f(\mathbf{X})d\Omega = \sum_{J=1}^{n} f(\mathbf{X}_J)V_J + \sum_{I=1}^{s} f(\mathbf{X}_I)V_I \tag{6.71}$$

where n represents nodes, and s represents stress points. These stress points are used to calculate the value of stresses only, whereas nodes are used to calculate all kinematic values. This integration technique was first used for one-dimensional problem using smoothed particle hydrodynamics [73] and was extended to higher order problems [74].

6.9.3 Cell structure or background mesh

The engineering domain is divided into regular integration cells in the background, which are used to perform Gaussian quadrature, as can be seen in Fig. 6.15. This technique is also known as the Octree quadrature method, and integral is evaluated as

$$\int_{\Omega} f(\mathbf{X})d\Omega = \sum_{J=1}^{n} f(\boldsymbol{\xi}_J)w_J \det \boldsymbol{J}^{\xi}(\boldsymbol{\xi}) \tag{6.72}$$

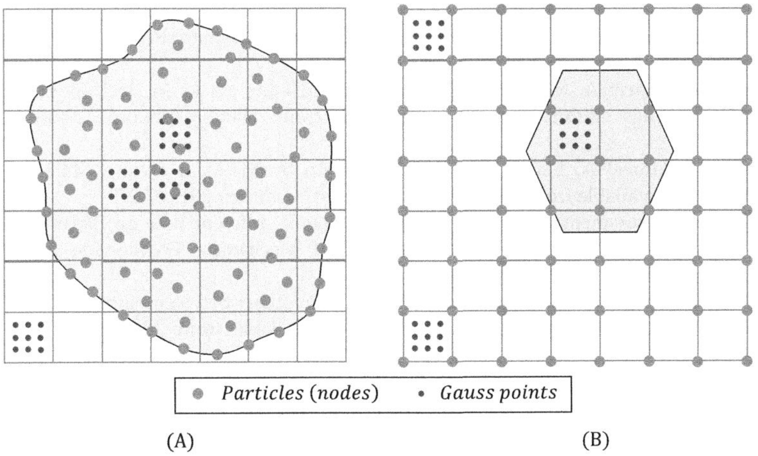

FIGURE 6.15 Integration scheme: (A) background cell structure and (B) finite element mesh.

where det $J^{\xi}(\xi)$ and ξ represent the determinant of the Jacobian and local coordinates, respectively. In case cell structures are used for integration, domain is divided into a regular array of small domains, irrespective of particle position, as shown in Fig. 6.15A. Another way is to use conventional FE mesh as the background mesh is depicted shown in Fig. 6.15B. In this case, integral is evaluated by using Gauss quadrature and is more accurate. In our study, we have employed Gauss quadrature on background mesh.

6.10 Conclusion

Implementation and formulation of the EFGM have been reported in the chapter. EFGM employs the moving least squares method for approximating the displacement field across the engineering domain. Unlike FEM, the implementation of EFGM is not straightforward, as the EFGM approximations do not satisfy the interpolation properties. Imposition of boundary conditions requires a special treatment in EFGM. A number of numerical procedures have been developed from time to time for imposing the boundary conditions in EFGM, such as the Lagrange multiplier approach and penalty approach. The postprocessing also requires special treatment in EFGM. Compared to FEM, computational cost involved in obtained EFGM shape functions is more due to matrix inversion. The present work discusses the implementation issues associated with the EFGM, such as the development of MLS approximations, the imposition of boundary conditions, and postprocessing methodologies.

References

[1] Jameel A, Harmain GA. Modeling and numerical simulation of fatigue crack growth in cracked specimens containing material discontinuities. Strength of Materials 2016;48:294–307. Available from: https://doi.org/10.1007/s11223-016-9765-0.

[2] Kanth SA, Harmain GA, Jameel A. Investigation of fatigue crack growth in engineering components containing different types of material irregularities by XFEM. Mechanics of Advanced Materials and Structures 2021. Available from: https://doi.org/10.1080/15376494.2021.1907003.
[3] Kanth SA, Lone AS, Harmain GA, Jameel A. Modeling of embedded and edge cracks in steel alloys by XFEM. Materials Today: Proceedings 2020;26:814–18. Available from: https://doi.org/10.1016/j.matpr.2019.12.423.
[4] Kanth SA, Lone AS, Harmain GA, Jameel A. Elasto plastic crack growth by XFEM: a review. Materials Today: Proceedings 2019;18:3472–81. Available from: https://doi.org/10.1016/j.matpr.2019.07.275.
[5] Kanth SA, Harmain GA, Jameel A. Assessment of fatigue life in presence of different hole geometries by X-FEM. Iranian Journal of Science and Technology – Transactions of Mechanical Engineering 2022;. Available from: https://doi.org/10.1007/s40997-022-00569-y.
[6] Lone AS, Kanth SA, Harmain GA, Jameel A. XFEM modeling of frictional contact between elliptical inclusions and solid bodies. Materials Today: Proceedings 2020;26:819–24. Available from: https://doi.org/10.1016/j.matpr.2019.12.424.
[7] Lone AS, Harmain GA, Jameel A. Modeling of contact interfaces by penalty based enriched finite element method. Mechanics of Advanced Materials and Structures 2022;1–19. Available from: https://doi.org/10.1080/15376494.2022.2034075.
[8] Lucy LB. A numerical approach to the testing of the fission hypothesis. The Astronomical Journal 1977;82:1013. Available from: https://doi.org/10.1086/112164.
[9] Gingold RA, Monaghan JJ. Smoothed particle hydrodynamics: theory and application to non-spherical stars. Monthly Notices of the Royal Astronomical Society 1977;181:375–89. Available from: https://doi.org/10.1093/mnras/181.3.375.
[10] Libersky LD, Petschek AG, Carney TC, Hipp JR, Allahdadi FA. High strain Lagrangian hydrodynamics: A three-dimensional SPH code for dynamic material response. Journal of Computational Physics 1993;109(1):67–75. Available from: https://doi.org/10.1006/jcph.1993.1199.
[11] Bonet J, Kulasegaram S. Correction and stabilization of smooth particle hydrodynamics methods with applications in metal forming simulations. International Journal for Numerical Methods in Engineering 2000;47:1189–214. https://doi.org/10.1002/(SICI)1097-0207(20000228)47:6 < 1189::AID-NME830 > 3.0.CO;2-I.
[12] Swegle JW, Hicks DL, Attaway SW. Smoothed particle hydrodynamics stability analysis. Journal of Computational Physics 1995;116:123–34. Available from: https://doi.org/10.1006/jcph.1995.1010.
[13] Federrath C, Banerjee R, Clark PC, Klessen RS. Modeling collapse and accretion in turbulent gas clouds: Implementation and comparison of sink particles in AMR and SPH. The Astrophysical Journal 2010;713. Available from: https://doi.org/10.1088/0004-637X/713/1/269.
[14] Monaghan JJ, Gingold RA. Shock simulation by the particle method SPH. Journal of Computational Physics 1983;52. Available from: https://doi.org/10.1016/0021-9991(83)90036-0.
[15] Johnson GR, Beissel SR. Normalized smoothing functions for SPH impact computations. International Journal for Numerical Methods in Engineering 1996;39. https://doi.org/10.1002/(SICI)1097-0207(19960830)39:16 < 2725::AID-NME973 > 3.0.CO;2–9.
[16] Gingold RA, Monaghan JJ. Kernel estimates as a basis for general particle methods in hydrodynamics. Journal of Computational Physics 1982;46:429–53. Available from: https://doi.org/10.1016/0021-9991(82)90025-0.
[17] Belytschko T, Guo Y, Liu WK, Xiao SP. A unified stability analysis of meshless particle methods. International Journal for Numerical Methods in Engineering 2000;48. https://doi.org/10.1002/1097-0207(20000730)48:9 < 1359::AID-NME829 > 3.0.CO;2-U.
[18] Xiao SP, Belytschko T. Material stability analysis of particle methods. Advances in Computational Mathematics 2005;23:171–90. Available from: https://doi.org/10.1007/s10444-004-1817-5.
[19] Gupta V, Jameel A, Anand S, Anand Y. Analysis of composite plates using isogeometric analysis: a discussion. Materials Today: Proceedings 2021;44:1190–4. Available from: https://doi.org/10.1016/j.matpr.2020.11.238.
[20] Bonet J, Lok TSL. Variational and momentum preservation aspects of Smooth Particle Hydrodynamic formulations. Computer Methods in Applied Mechanics and Engineering 1999;180:97–115. Available from: https://doi.org/10.1016/S0045-7825(99)00051-1.

[21] Libersky LD, Petschek AG. Smooth particle hydrodynamics with strength of materials. Advances in the free-Lagrange method including contributions on adaptive gridding and the smooth particle hydrodynamics method, Lecture Notes in Physics, 2008, p. 248–257. Available from: https://doi.org/10.1007/3-540-54960-9_58.

[22] Vila JP. On particle weighted methods and smooth particle hydrodynamics. Mathematical Models and Methods in Applied Sciences 1999;9:161–209. Available from: https://doi.org/10.1142/S0218202599000117.

[23] Libersky LD, Randles PW, Carney TC, Dickinson DL. Recent improvements in SPH modeling of hypervelocity impact. International Journal of Impact Engineering 1997;20:525–32. Available from: https://doi.org/10.1016/s0734-743x(97)87441-6.

[24] Dyka CT, Ingel RP. Addressing tension instability in SPH Methods. Naval Research Laboratory; 1994.

[25] Lancaster P, Salkauskas K. Surfaces generated by moving least squares methods. Mathematics of Computation 1981;37:141. Available from: https://doi.org/10.1090/s0025-5718-1981-0616367-1.

[26] Nayroles B, Touzot G, Villon P. Generalizing the finite element method: Diffuse approximation and diffuse elements. Computational Mechanics 1992;10:307–18. Available from: https://doi.org/10.1007/BF00364252.

[27] Belytschko T, Lu YY, Gu L. Element-free Galerkin methods. International Journal for Numerical Methods in Engineering 1994;37:229–56. Available from: https://doi.org/10.1002/nme.1620370205.

[28] Oñate E, Idelsohn S, Zienkiewicz OC, Taylor RL. A finite point method in computational mechanics. Applications to convective transport and fluid flow. International Journal for Numerical Methods in Engineering 1996;39. https://doi.org/10.1002/(SICI)1097-0207(19961130)39:22 < 3839::AID-NME27 > 3.0.CO;2-R.

[29] Carpinteri A, Chiaia B, Cornetti P. A scale-invariant cohesive crack model for quasi-brittle materials. Engineering Fracture Mechanics 2001;69. Available from: https://doi.org/10.1016/S0013-7944(01)00085-6.

[30] Sukumar N, Moran B, Semenov AY, Belikov VV. Natural neighbour Galerkin methods. International Journal for Numerical Methods in Engineering 2000;50. https://doi.org/10.1002/1097-0207(20010110)50:1 < 1::AID-NME14 > 3.0.CO;2-P.

[31] Braun J, Sambridge M. A numerical method for solving partial differential equations on highly irregular evolving grids. Nature 1995;376. Available from: https://doi.org/10.1038/376655a0.

[32] Idelsohn SR, Onate E, Calvo N, Del Pin F. The meshless finite element method. International Journal for Numerical Methods in Engineering 2003;58. Available from: https://doi.org/10.1002/nme.798.

[33] Melenk JM, Babuška I. The partition of unity finite element method: basic theory and applications. Computer Methods in Applied Mechanics and Engineering 1996;139:289–314. Available from: https://doi.org/10.1016/S0045-7825(96)01087-0.

[34] Liu GR, Gu YT. A point interpolation method for two-dimensional solids. International Journal for Numerical Methods in Engineering 2001;50:937–51. https://doi.org/10.1002/1097-0207(20010210)50:4 < 937::AID-NME62 > 3.0.CO;2-X.

[35] Gupta V, Jameel A, Verma SK, Anand S, Anand Y. Transient isogeometric heat conduction analysis of stationary fluid in a container. Part E: Journal of Process Mechanical Engineering 2022. Available from: https://doi.org/10.1177/0954408922112.

[36] Wang JG, Liu GR. A point interpolation meshless method based on radial basis functions. International Journal for Numerical Methods in Engineering 2002;54. Available from: https://doi.org/10.1002/nme.489.

[37] Liu GR, Gu YT, Dai KY. Assessment and applications of point interpolation methods for computational mechanics. International Journal for Numerical Methods in Engineering 2004;59:1373–97. Available from: https://doi.org/10.1002/nme.925.

[38] Gupta V, Jameel A, Verma SK, Anand S, Anand Y. An insight on NURBS based Isogeometric Analysis, its current status and involvement in Mechanical Applications. Archives of Computational Methods in Engineering 2022;. Available from: https://doi.org/10.1007/s11831-022-09838-0.

[39] Zhu T, Zhang J, Atluri SN. Meshless numerical method based on the local boundary integral equation (LBIE) to solve linear and non-linear boundary value problems. Engineering Analysis with Boundary Elements 1999;23:375–89. Available from: https://doi.org/10.1016/S0955-7997(98)00096-4.

[40] Atluri SN, Zhu T. A new meshless local Petrov-Galerkin (MLPG) approach in computational mechanics. Computational Mechanics 1998;22:117–27. Available from: https://doi.org/10.1007/s004660050346.

[41] Atluri SN, Kim HG, Cho JY. Critical assessment of the truly Meshless Local Petrov-Galerkin (MLPG), and Local Boundary Integral Equation (LBIE) methods. Computational Mechanics 1999;24. Available from: https://doi.org/10.1007/s004660050457.

[42] Atluri SN, Zhu T. New concepts in meshless methods. International Journal for Numerical Methods in Engineering 2000;47. https://doi.org/10.1002/(SICI)1097-0207(20000110/30)47:1/3 < 537::AID-NME783 > 3.0.CO;2-E.
[43] Jameel A, Harmain GA. A coupled FE-IGA technique for modeling fatigue crack growth in engineering materials. Mechanics of Advanced Materials and Structures 2019;26:1764–75. Available from: https://doi.org/10.1080/15376494.2018.1446571.
[44] Jameel A, Harmain GA. Extended iso-geometric analysis for modeling three-dimensional cracks. Mechanics of Advanced Materials and Structures 2019;26:915–23. Available from: https://doi.org/10.1080/15376494.2018.1430275.
[45] Jameel A, Harmain GA. Effect of material irregularities on fatigue crack growth by enriched techniques. International Journal for Computational Methods in Engineering Science and Mechanics 2020;21:109–33. Available from: https://doi.org/10.1080/15502287.2020.1772902.
[46] Lone AS, Jameel A, Harmain GA. A coupled finite element-element free Galerkin approach for modeling frictional contact in engineering components. Materials Today: Proceedings 2018;5:18745–54. Available from: https://doi.org/10.1016/j.matpr.2018.06.221.
[47] Lone AS, Kanth SA, Jameel A, Harmain GA. A state of art review on the modeling of Contact type Nonlinearities by Extended Finite Element method. Materials Today: Proceedings 2019;18:3462–71. Available from: https://doi.org/10.1016/j.matpr.2019.07.274.
[48] Lone AS, Harmain GA, Jameel A. Enriched element free Galerkin method for solving frictional contact between solid bodies. Mechanics of Advanced Materials and Structures 2022;. Available from: https://doi.org/10.1080/15376494.2022.2092791.
[49] Kanth SA, Harmain GA, Jameel A. Modeling of nonlinear crack growth in steel and aluminum alloys by the element free Galerkin method. Materials Today: Proceedings 2018;5:18805–14. Available from: https://doi.org/10.1016/j.matpr.2018.06.227.
[50] Belytschko T, Gu L, Lu YY. Fracture and crack growth by element free Galerkin methods. Modelling and Simulation in Materials Science and Engineering 1994;2:519–34. Available from: https://doi.org/10.1088/0965-0393/2/3A/007.
[51] Jameel A, Harmain GA. Fatigue crack growth analysis of cracked specimens by the coupled finite element-element free Galerkin method. Mechanics of Advanced Materials and Structures 2019;26:1343–56. Available from: https://doi.org/10.1080/15376494.2018.1432800.
[52] Jameel A, Harmain GA. Fatigue crack growth in presence of material discontinuities by EFGM. International Journal of Fatigue 2015;81:105–16. Available from: https://doi.org/10.1016/j.ijfatigue.2015.07.021.
[53] Jameel A, Harmain GA. Large deformation in bi-material components by XIGA and coupled FE-IGA techniques. Mechanics of Advanced Materials and Structures 2020. Available from: https://doi.org/10.1080/15376494.2020.1799120.
[54] Shepard D. Two-dimensional interpolation function for irregularly- spaced data. Proceedings of the 23rd National Conference, 1968, p. 517–24.
[55] Liu GR. Mesh free methods: moving beyond the finite element method. Taylor and Francis; 2002. Available from: https://doi.org/10.1299/jsmecmd.2003.16.937.
[56] Liu GR, Gu YT. An introduction to meshfree methods and their programming. Springer; 2005. Available from: https://doi.org/10.1007/1-4020-3468-7.
[57] Dolbow J, Belytschko T. An introduction to programming the meshless element free Galerkin method. Archives of Computational Methods in Engineering 1998;5:207–41. Available from: https://doi.org/10.1007/bf02897874.
[58] Cai YC, Zhu HH. Direct imposition of essential boundary conditions and treatment of material discontinuities in the EFG method. Computational Mechanics 2004;34:330–8. Available from: https://doi.org/10.1007/s00466-004-0577-x.
[59] Chen JS, Wang HP. New boundary condition treatments in meshfree computation of contact problems. Computer Methods in Applied Mechanics and Engineering 2000;187:441–68. Available from: https://doi.org/10.1016/S0045-7825(00)80004-3.
[60] Belytschko T, Krongauz Y, Fleming M, Organ D, Liu WKS. Smoothing and accelerated computations in the element free Galerkin method. Journal of Computational and Applied Mathematics 1996;74. Available from: https://doi.org/10.1016/0377-0427(96)00020-9.

[61] Krongauz Y, Belytschko T. EFG approximation with discontinuous derivatives. International Journal for Numerical Methods in Engineering 1998;41. https://doi.org/10.1002/(sici)1097-0207(19980415)41:7 < 1215::aid-nme330 > 3.0.co;2-%23.

[62] Organ D, Fleming M, Terry T, Belytschko T. Continuous meshless approximations for nonconvex bodies by diffraction and transparency. Computational Mechanics 1996;18. Available from: https://doi.org/10.1007/BF00369940.

[63] Fleming M, Chu YA, Moran B, Belytschko T. Enriched element-free Galerkin methods for crack tip fields. International Journal for Numerical Methods in Engineering 1997;40:1483–504. https://doi.org/10.1002/(SICI)1097-0207(19970430)40:8 < 1483::AID-NME123 > 3.0.CO;2–6.

[64] Singh AK, Jameel A, Harmain GA. Investigations on crack tip plastic zones by the extended iso-geometric analysis. Materials Today: Proceedings 2018;5:19284–93. Available from: https://doi.org/10.1016/j.matpr.2018.06.287.

[65] Rabczuk T, Belytschko T. Cracking particles: a simplified meshfree method for arbitrary evolving cracks. International Journal for Numerical Methods in Engineering 2004;61:2316–43. Available from: https://doi.org/10.1002/nme.1151.

[66] Rabczuk T, Areias P. A meshfree thin shell for arbitrary evolving cracks based on an extrinsic basis. CMES-Computer Modeling in Engineering & Sciences 2006;16.

[67] Rabczuk T, Zi G. A meshfree method based on the local partition of unity for cohesive cracks. Computational Mechanics 2007;39:743–60. Available from: https://doi.org/10.1007/s00466-006-0067-4.

[68] Rabczuk T, Belytschko T. A three-dimensional large deformation meshfree method for arbitrary evolving cracks. Computer Methods in Applied Mechanics and Engineering 2007;196:2777–99. Available from: https://doi.org/10.1016/j.cma.2006.06.020.

[69] Zi G, Rabczuk T, Wall W. Extended meshfree methods without branch enrichment for cohesive cracks. Computational Mechanics 2007;40:367–82. Available from: https://doi.org/10.1007/s00466-006-0115-0.

[70] Belytschko T, Krongauz Y, Dolbow J, Gerlach C. On the completeness of meshfree particle methods. International Journal for Numerical Methods in Engineering 1998;43. https://doi.org/10.1002/(SICI)1097-0207(19981115)43:5 < 785::AID-NME420 > 3.0.CO;2–9.

[71] Chen Y, Lee JD, Eskandarian A. Meshless methods in solid mechanics. Springer; 2006. Available from: https://doi.org/10.1007/0-387-33368-1.

[72] Beissel S, Belytschko T. Nodal integration of the element-free Galerkin method. Computer Methods in Applied Mechanics and Engineering 1996;139. Available from: https://doi.org/10.1016/S0045-7825(96)01079-1.

[73] Dyka CT, Ingel RP. An approach for tension instability in smoothed particle hydrodynamics (SPH). Computers & Structures 1995;57. Available from: https://doi.org/10.1016/0045-7949(95)00059-P.

[74] Randles PW, Libersky LD. Normalized SPH with stress points. International Journal for Numerical Methods in Engineering 2000;48. https://doi.org/10.1002/1097-0207(20000810)48:10 < 1445::AID-NME831 > 3.0.CO;2–9.

CHAPTER

7

Enriched element-free Galerkin method for three-dimensional cracks

Azher Jameel[1], *Mohd Junaid Mir*[2], *Shuhaib Mushtaq*[2] *and Mehnaz Rasool*[2]

[1]Department of Mechanical Engineering, National Institute of Technology Srinagar, Hazratbal, Srinagar, Jammu and Kashmir, India [2]Department of Mechanical Engineering, Islamic University of Science and Technology, Awantipora, Jammu and Kashmir, India

7.1 Introduction

Investigation of cracks in three-dimensional structural components has always posed a challenge in conventional finite element method (FEM) as it requires mesh conformation for simulating the presence of cracks in engineering structures. It is understood that cracks are inherently present in all structural components irrespective of the quality of manufacturing processes employed to fabricate the components. Crack may also develop during the operational stage of the engineering components. The propagation of cracks in engineering components reduces their service life due to which the study of three-dimensional cracks becomes extremely important. Different computational techniques have been developed in the past few decades to model different engineering phenomena, such as crack growth, large elasto-plastic deformations, frictional contact problems, and many other important engineering problems. Some of the extensively used computational models include the boundary element technique [1–3], enriched FEM [4–6], meshless approaches [7–11], models based on peridynamics [12–14], phase field techniques [15–17], and standard finite element method [18–21]. Although the conventional FEM has always remained a dominant computational tool in engineering, it faces various issues while modeling material irregularities in structural components, such as advancing cracks, inclusions, and holes. Finite element techniques also suffer extreme element distortion issues while modeling geometric nonlinearities caused due to large deformations occurring in the domain. Geometrically nonlinear problems demand re-meshing of the domain after each stage of simulation.

Enriched Numerical Techniques
DOI: https://doi.org/10.1016/B978-0-443-15362-4.00006-1

Meshless methods, including enriched element-free Galerkin method (EFGM), eliminate all mesh-related issues and model all types of material interfaces occurring in structural components independent of nodal grid selected for investigation [22–24]. Meshless methods provide a strong computational framework for modeling geometric nonlinearities occurring in the domain because of their meshless nature and model. In enriched EFGM, the conventional moving least squares approximations are updated by adding extra enrichment functions to include the influence created by the presence of discontinuities in structural components. The discontinuities are represented by a strong numerical tool known as the level set method, which remains zero along the interface and changes its sign across the discontinuity [25–34]. The addition of enrichment functions to classical displacement approximations is possible by employing the partition of unity approaches [35–40].

The current work employs enriched EFGM to investigate the behavior of cracks in three-dimensional domains subjected to monotonic tensile loads. As discussed earlier, the conventional moving least square (MLS) approximations are modified by adding extra enrichment functions to include the influence created by the presence of discontinuities in structural components. Due to the use of such enrichments, drawbacks of re-meshing, mesh adaption, and mesh conformation do not occur in enriched EFGM, which makes it an efficient computational tool for modeling static and advancing cracks in structural components. Mixed mode stress intensity factors (SIFs) corresponding to different fracture modes are estimated by employing the domain form of interaction integral. Mathematical models derived on enriched EFGM have been coded in MATLAB® to simulate different engineering problems involving the investigation of static cracks in 3D engineering components. Simulation results derived in the current study are validated with respect to benchmark solutions already reported in literature. Different types of cracks in 3D structural components, such as plane edge cracks, horizontal and inclined penny cracks, and lens-type cracks, have been considered for simulation by enriched EFGM.

7.2 The element-free Galerkin method

EFGM is one of the mesh-free techniques that represents the structural domain by an array of nodes and displacement approximations are defined from this nodal array only. Generation of finite element mesh is not required in this method. Owing to its meshless nature, EFGM eliminates all mesh-related issues that we see in conventional FE-based techniques. Moreover, the standard finite element method suffers extreme element distortion while modeling large deformations occurring in structural components. During large deformation, the geometry of elements changes significantly, which introduces geometric nonlinearities in the formulations. In past few decades, EFGM has evolved as a strong computational framework for modeling materials and geometric nonlinearities in engineering materials. Since no mesh is generated in EFGM, mesh distortion does not occur, which makes it a potential tool for investigating such problems. The introduction of various enrichment strategies in classical EFGM makes it suitable for modeling internal material interfaces present in structural components, independent of the nodal distribution selected for investigation. Enriched EFGM provides a strong and efficient computational framework for modeling the discontinuities created by cracks, holes, and other material irregularities. The proposed approach has found tremendous applications in the areas of computational fracture mechanics, heat transfer, large deformations, elasto-plastic analysis, and other nonlinear structural problems.

As discussed earlier, EFGM constructs the variable approximations from the set of nodes only without creating a need for the generation of the finite element mesh, which further eliminates all mesh-related problems. The displacement approximations are constructed from the moving least squares approximation functions that consist of three basic components, including a polynomial basis function, weight function with a compact support, and a set of coefficients that are dependent of the nodal position. MLS approximations used in EFGM to approximate the displacement field can be written as

$$u^h(x) = \sum_{j=1}^{n} \mathrm{p}_j(x)\mathrm{a}_j(x) = \mathbf{P}^T(x)\mathbf{a}(x) \tag{7.1}$$

where $\mathbf{P}(x)$ is a matrix of polynomial basis functions defined as

$$\mathbf{P}^T(x) = [1, x, y, z, xy, yz, zx \ldots, x^k, y^k, z^k] \tag{7.2}$$

k is the degree of polynomial, and $\mathbf{a}(x)$ represents the matrix coefficients described as

$$\mathbf{a}(x) = [\mathrm{a}_1(x), \mathrm{a}_2(x), \mathrm{a}_3(x) \ldots, \mathrm{a}_m(x)] \tag{7.3}$$

The weighted least squares sum in MLS approximations can be defined as

$$\mathrm{L}(x) = \sum_{i=1}^{n} w(x - x_i)[\mathbf{P}^T(x)\mathbf{a}(x) - u_i]^2 \tag{7.4}$$

where u_i is the nodal parameter. u_i does not represent the nodal values because $u^h(x)$ is not an interpolant. It represents an approximant. To find coefficient $\mathbf{a}(x)$, $\mathbf{L}(x)$ is to be minimized with respect to $\mathbf{a}(x)$, which can be defined as

$$\frac{\partial \mathrm{L}}{\partial \mathbf{a}} = 0 \tag{7.5}$$

Using the above conditions, the system of linear equations is written as

$$\mathbf{A}(x)\mathbf{a}(x) = \mathbf{B}(x)\mathbf{u} \tag{7.6}$$

where

$$\mathbf{u}^T = [u_1, u_2, u_3 \ldots, u_n] \tag{7.7}$$

$$\mathbf{A} = \sum_{i=1}^{n} w_i(\mathrm{x})\mathbf{P}(\mathrm{x}_i)\mathbf{P}^T(\mathrm{x}_i) \tag{7.8}$$

$$\mathbf{B} = [w_1(x)\mathbf{P}(x_1), w_2(x)\mathbf{P}(x_2) \ldots, w_n(x)\mathbf{P}(x_n)] \tag{7.9}$$

The matrix of coefficient $\mathbf{a}(x)$ can be obtained as $\mathbf{a}(x) = \mathbf{A}^{-1}(x)\mathbf{B}(x)\mathbf{u}$, which is defined only if the matrix $\mathbf{A}$ is nonsingular for all values of x. Further, the approximation of displacement variable $u(x)$ can be described in standard form as

$$u^h(x) = \mathbf{P}^T(x)\mathbf{A}^{-1}(x)\mathbf{B}(x)\mathbf{u} \tag{7.10}$$

where $\mathbf{A}(x)$ and $\mathbf{B}(x)$ are defined as

$$\mathbf{A}(x) = \sum_{i=1}^{n} w(x - x_i)\mathbf{P}(x_i)\mathbf{P}^T(x_i) \tag{7.11}$$

$$\mathbf{B}(x) = [w(x - x_1)\mathbf{P}(x_1), w(x - x_2)\mathbf{P}(x_2)\ldots, w(x - x_n)\mathbf{P}(x_n)] \tag{7.12}$$

In the above equations, weight function has an important role to play in EFGM, and it is represented by w. The value of weight function remains zero outside the compact support, due to which the MLS shape functions vanish outside the domain of influence. Weight function provides proper weightings to the residuals at the nodes in their domains of influence. Weight functions used in EFGM also ensure that nodes enter or leave the support domain in a smooth and gradual manner when the sampling point x moves. In terms of mesh-free shape functions, MLS-based variable approximation can be written as

$$u^h(\mathrm{x}) = \sum_{i=1}^{n} N(x)u_i = \mathbf{N}^T(x)\mathbf{u} \tag{7.13}$$

where

$$\mathbf{N}^T(x) = [N_1(x), N_2(x), N_3(x)\ldots, N_n(x)] \tag{7.14}$$

$$\mathbf{u}^T = [u_1, u_2, u_3\ldots, u_n] \tag{7.15}$$

The mesh-free shape functions $N_i(\mathbf{x})$ are obtained as

$$\mathrm{N}_i(x) = \mathbf{P}^T(x)\mathbf{A}^{-1}(x)\mathbf{B}(x) \tag{7.16}$$

Mesh-free shape function derivatives are derived as

$$\mathrm{N}_{i,x}(x) = \left(\mathbf{P}^T(x)\mathbf{A}^{-1}(x)\mathbf{B}(x)\right)_{,x} = \mathbf{p}_{,x}^T\mathbf{A}^{-1}\mathbf{B} + \mathbf{P}^T(\mathbf{A}^{-1})_{,x}\mathbf{B} + \mathbf{P}^T\mathbf{A}^{-1}\mathbf{B}_{,x} \tag{7.17}$$

where

$$\mathbf{B}_{,x}(x) = \frac{\partial \mathrm{w}}{\partial \mathrm{x}}(x - x_i)\mathbf{P}(x_i) \tag{7.18}$$

$$(\mathbf{A}^{-1})_{,x} = -\mathbf{A}^{-1}\mathbf{A}_{,x}\mathbf{A}^{-1} \tag{7.19}$$

where

$$\mathbf{A}_{,x} = \sum_{i=1}^{n} \frac{\partial \mathrm{w}}{\partial \mathrm{x}}(\mathrm{x} - \mathrm{x_i})\mathbf{P}(x_i)\mathbf{P}^T(x_i) \tag{7.20}$$

7.2.1 Basic element-free Galerkin method models

To develop the mathematical models for modeling different engineering problems in EFGM, the equilibrium equation for a simple elastic problem can be described in standard form as

$$\nabla \cdot \boldsymbol{\sigma} + \mathbf{b} = \mathbf{0} \text{ in } \Omega \tag{7.21}$$

where $\boldsymbol{\sigma}$ denotes stresses, and $\mathbf{b}$ represents body forces. Natural and essential boundary conditions (BCs) can be written as

$$\text{(Natural B.C)}\ \boldsymbol{\sigma}\cdot\mathbf{n}=\mathbf{t}\quad on\ \Gamma_t \tag{7.22}$$

$$\text{(Essential B.C)}\ \mathbf{u}=\overline{\mathbf{u}}\ on\ \Gamma_u \tag{7.23}$$

Equilibrium equation can be expressed in variational form as

$$\int_\Omega \delta\boldsymbol{\varepsilon}{:}\boldsymbol{\sigma}\mathbf{d}\Omega - \int_\Omega \delta\mathbf{u}{:}\mathbf{b}\mathbf{d}\Omega - \int_{\Gamma_t} \delta\mathbf{u}{:}\mathbf{t}\mathbf{d}\Gamma - \delta\mathbf{W}_u(\mathbf{u},\nabla_s\boldsymbol{\lambda}) = \mathbf{0} \tag{7.24}$$

This equilibrium equation has been slightly modified in EFGM as the MLS approximations do not possess the Kronecker delta property. Direct imposition of essential BCs is not possible in EFGM. To tackle this issue, several numerical approaches have been developed in the past few decades for imposing such BCs in EFGM. The Lagrangian approach has found tremendous application in EFGM for imposing essential BCs. In the equilibrium equation expressed above, ε denotes the strain field, and $\delta\mathbf{W}_u$ is the additional term that is required for the imposition of BCs in the given problem. The additional term used in the Lagrangian approach for imposing BCs can be written as

$$\mathbf{W}_u(\mathbf{u},\boldsymbol{\lambda}) = \int_\Gamma \boldsymbol{\lambda}(\mathbf{u}-\overline{\mathbf{u}})\mathbf{d}\Gamma \tag{7.25}$$

$$\delta\mathbf{W}_u(\mathbf{u},\boldsymbol{\lambda}) = \int_\Gamma \delta\boldsymbol{\lambda}(\mathbf{u}-\overline{\mathbf{u}})\mathbf{d}\Gamma + \int_\Gamma \delta\mathbf{u}.\boldsymbol{\lambda}\mathbf{d}\Gamma \tag{7.26}$$

In linear elastic materials, the stress and strain fields can be written as

$$\varepsilon = \nabla_s\mathbf{u} \tag{7.27}$$

$$\boldsymbol{\sigma} = \mathbf{D}\varepsilon \tag{7.28}$$

where ε denotes the strain, $\boldsymbol{\sigma}$ denotes stresses, and $\mathbf{D}$ is constitutive matrix used in generalized Hooke's law. The Lagrange multiplier used in EFGM can be defined as

$$\boldsymbol{\lambda}(x) = N_{fi}(\mathrm{s})\lambda_i \tag{7.29}$$

$$\delta\boldsymbol{\lambda}(x) = N_{fi}(\mathrm{s})\delta\lambda_i \tag{7.30}$$

where $N_{fi}(\mathrm{s})$ represents the Lagrange shape function, and s is the arc length. Displacement approximation in EFGM is written in terms of MLS shape functions as

$$u^h(\mathbf{x}) = \sum_{i=1}^{n} N_i(\mathbf{x})u_i \tag{7.31}$$

$$\delta u^h(\mathbf{x}) = \sum_{i=1}^{n} N_i(\mathbf{x})\delta u_i \tag{7.32}$$

The final EFGM model for simulating different engineering problems can be written as

$$\begin{bmatrix} \mathbf{K} & \mathbf{G} \\ \mathbf{G}^T & 0 \end{bmatrix} \begin{Bmatrix} \mathbf{u} \\ \boldsymbol{\lambda} \end{Bmatrix} = \begin{Bmatrix} \mathbf{f} \\ \mathbf{q} \end{Bmatrix} \tag{7.33}$$

where

$$\mathbf{K}_{ij} = \int_{\Omega} \mathbf{B}^T \mathbf{D} \mathbf{B} d\Omega \tag{7.34}$$

$$\mathbf{G}_{ik=} \int_{\Omega} \mathbf{N}_i \mathbf{N}_k \mathbf{d}\Gamma \tag{7.35}$$

$$\mathbf{f}_{i=} \int_{\Gamma_t} \mathbf{N}_i \mathbf{t} \mathbf{d}\Gamma_t + \int_{\Omega} \mathbf{N}_i \mathbf{b} \mathrm{d}\Omega \tag{7.36}$$

$$\mathbf{q}_{k=} \int_{\Gamma_t} \mathbf{N}_k \bar{\mathbf{u}} \mathrm{d}\Gamma_u \tag{7.37}$$

$$\mathbf{B}_i = \begin{bmatrix} N_{i,x} & 0 \\ 0 & N_{i,y} \\ N_{i,y} & N_{i,x} \end{bmatrix} \tag{7.38}$$

$$\mathbf{N}_k = \begin{bmatrix} N_k & 0 \\ 0 & N_k \end{bmatrix} \tag{7.39}$$

$$\mathrm{D} = \frac{\mathrm{E}}{(1+\upsilon)(1-2\upsilon)} \begin{bmatrix} 1-\upsilon & \upsilon & 0 \\ \upsilon & 1-\upsilon & 0 \\ 0 & 0 & \dfrac{1-2\upsilon}{2} \end{bmatrix} \textit{for plane strain} \tag{7.40}$$

$$\mathrm{D} = \frac{\mathrm{E}}{(1-\upsilon^2)} \begin{bmatrix} 1 & \upsilon & 0 \\ \upsilon & 1 & 0 \\ 0 & 0 & \dfrac{1-\upsilon}{2} \end{bmatrix} \textit{for plane stress} \tag{7.41}$$

7.2.2 Enriched element-free Galerkin method for cracks

As discussed earlier, enriched EFGM will model all geometries and types of cracks found in structural components, irrespective of the nodal distribution chosen for investigation. Consider a loaded elastic body with domain "Ω" and surrounded by boundary "Γ" as depicted in Fig. 7.1, where body forces are shown as "*b*," and the applied tractions are represented by "*t*." Traction-free crack surfaces are shown as Γ_c, and traction boundaries are represented by Γ_t. The prescribed displacements are imposed at the displacement boundary given by Γ_u.

For a given loaded deformable body, the equilibrium equations are

$$\nabla.\boldsymbol{\sigma} + \boldsymbol{b} = 0 \text{ in } \Omega \tag{7.42}$$

$$\boldsymbol{\sigma}.\boldsymbol{n} = \boldsymbol{t} \text{ on } \Gamma_t \tag{7.43}$$

$$\boldsymbol{\sigma}.\boldsymbol{n} = 0 \text{ on } \Gamma_c^+ \tag{7.44}$$

$$\boldsymbol{\sigma}.\boldsymbol{n} = 0 \text{ on } \Gamma_c^- \tag{7.45}$$

$$\boldsymbol{u} = \boldsymbol{U} \text{ on } \Gamma_u \tag{7.46}$$

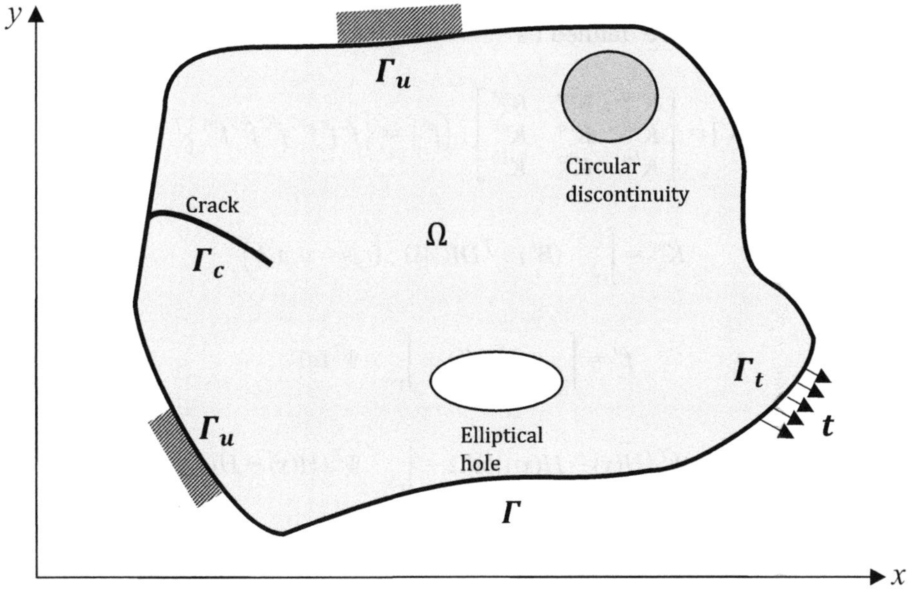

FIGURE 7.1 Domain with multiple material irregularities.

where ε is the strain tensor defined as $\varepsilon = \boldsymbol{Bu}$. Here "$\boldsymbol{B}$" represents the strain–displacement matrix and '$\boldsymbol{u}$' represents the displacements. Equilibrium equation is described in variational form as

$$\int_{\Omega} \delta\varepsilon^T \sigma d\Omega - \int_{\Omega} \delta \boldsymbol{u}^T \boldsymbol{b} d\Omega - \int_{\Gamma_t} \delta \boldsymbol{u}^T \boldsymbol{t} d\Gamma_t = 0 \tag{7.47}$$

The enriched MLS approximation for modeling the behavior of cracks in EFGM can be written as

$$\boldsymbol{u}^h(\boldsymbol{x}) = \sum_{i=1}^{n} N_i(\boldsymbol{x})\boldsymbol{u}_i + \sum_{j=1}^{n_s} N_j(\boldsymbol{x})[H(\boldsymbol{x}) - H(\boldsymbol{x}_j)]\boldsymbol{a}_j + \sum_{k=1}^{n_T} N_k(\boldsymbol{x}) \sum_{\alpha=1}^{4} [\beta_\alpha(\boldsymbol{x}) - \beta_\alpha(\boldsymbol{x}_k)]\boldsymbol{b}_k^\alpha \tag{7.48}$$

where N_i are the MLS shape functions, and "$\boldsymbol{u}_i$" are standard nodal displacements corresponding to nodes "n". The split and tip nodes are given by "n_s" and "n_T", respectively. Additional degrees of freedom for split and tip nodes are given by "$\boldsymbol{a}_j$" and "$\boldsymbol{b}_k^{\alpha}$", respectively. The Heaviside jump function has been employed to enrich the variable approximations of split nodes, and its value remains positive on one side and negative on other side of crack. In local polar coordinates (r, θ), crack tip enrichment functions for linear elastic cracked bodies are described as

$$\beta_\alpha(\boldsymbol{x}) = \left[\sqrt{r}\cos\frac{\theta}{2}, \sqrt{r}\sin\frac{\theta}{2}, \sqrt{r}\cos\frac{\theta}{2}\sin\theta, \sqrt{r}\sin\frac{\theta}{2}\cos\theta\right] \tag{7.49}$$

Substitution of enriched MLS approximation in variational form of equilibrium equation gives the final EFGM model as $[\boldsymbol{K}^e]\{\boldsymbol{d}^e\} = \{\boldsymbol{f}^e\}$. Here $\{\boldsymbol{d}^e\} = \{\boldsymbol{u}\ \boldsymbol{a}\ \boldsymbol{b}_1\ \boldsymbol{b}_2\ \boldsymbol{b}_3\ \boldsymbol{b}_4\}^T$ includes the nodal displacements for standard and enriched nodes. Different matrices in the final enriched EFGM model can be defined as

$$[K^e] = \begin{bmatrix} K^{uu} & K^{ua} & K^{ub} \\ K^{au} & K^{aa} & K^{ab} \\ K^{bu} & K^{ba} & K^{bb} \end{bmatrix}; \{f^e\} = \left\{f^u f^a f^{b1} f^{b2} f^{b3} f^{b4}\right\}^T \tag{7.50}$$

$$K^{rs} = \int_{\Omega^e} (\mathbf{B}^r)^T D B^s d\Omega \; ; (r, s = u, a, b) \tag{7.51}$$

$$f^u = \int_{\Omega^e} \Psi^T b d\Omega + \int_{\Gamma^e} \Psi^T t d\Gamma \tag{7.52}$$

$$f^a = \int_{\Omega^e} \Psi^T (H(x) - H(x_j)) b d\Omega + \int_{\Gamma^e} \Psi^T (H(x) - H(x_j)) t d\Gamma \tag{7.53}$$

$$f^{b\alpha} = \int_{\Omega^e} \Psi^T(\beta_\alpha(x) - \beta_\alpha(x_k)) b d\Omega + \int_{\Gamma^e} \Psi^T(\beta_\alpha(x) - \beta_\alpha(x_k)) t d\Gamma; (\alpha = 1,2,3,4) \tag{7.54}$$

$$B^u = \begin{bmatrix} \Psi_{i,x} & 0 & 0 \\ 0 & \Psi_{i,y} & 0 \\ 0 & 0 & \Psi_{i,z} \\ \Psi_{i,y} & \Psi_{i,x} & 0 \\ 0 & \Psi_{i,z} & \Psi_{i,y} \\ \Psi_{i,z} & 0 & \Psi_{i,x} \end{bmatrix} \tag{7.55}$$

$$B^a = \begin{bmatrix} (\Psi_j(H(x) - H(x_j)))_x & 0 & 0 \\ 0 & (\Psi_j(H(x) - H(x_j)))_y & 0 \\ 0 & 0 & (\Psi_j(H(x) - H(x_j)))_z \\ (\Psi_j(H(x) - H(x_j)))_y & (\Psi_j(H(x) - H(x_j)))_x & 0 \\ 0 & (\Psi_j(H(x) - H(x_j)))_z & (\Psi_j(H(x) - H(x_j)))_y \\ (\Psi_j(H(x) - H(x_j)))_z & 0 & (\Psi_j(H(x) - H(x_j)))_x \end{bmatrix} \tag{7.56}$$

$$B^{b\alpha} = \begin{bmatrix} (\Psi_k(\beta_\alpha(x) - \beta_\alpha(x_k)))_x & 0 & 0 \\ 0 & (\Psi_k(\beta_\alpha(x) - \beta_\alpha(x_k)))_y & 0 \\ 0 & 0 & (\Psi_k(\beta_\alpha(x) - \beta_\alpha(x_k)))_z \\ (\Psi_k(\beta_\alpha(x) - \beta_\alpha(x_k)))_y & (\Psi_k(\beta_\alpha(x) - \beta_\alpha(x_k)))_x & 0 \\ 0 & (\Psi_k(\beta_\alpha(x) - \beta_\alpha(x_k)))_z & (\Psi_k(\beta_\alpha(x) - \beta_\alpha(x_k)))_y \\ (\Psi_k(\beta_\alpha(x) - \beta_\alpha(x_k)))_z & 0 & (\Psi_k(\beta_\alpha(x) - \beta_\alpha(x_k)))_x \end{bmatrix} \tag{7.57}$$

7.2.3 Enrichment functions for cracks

Enriched EFGM models static and advancing cracks irrespective of the nodal distribution selected for analysis. Classical moving least squares approximation for displacements is enriched by adding compatible enrichment functions, which add the effect of these internal interfaces into the formulations. Choice of enrichment functions is very crucial in the investigation of crack growth problems. Several enrichment strategies were proposed from time to time for investigating the presence of cracks in structural components. There are two types of nodes in the nodal distribution of the EFGM. The nodes whose influence domain is completely cut by crack interface are known as split nodes, whereas the nodes whose compact support carries the crack tip are known as tip nodes. Separate enrichment strategies are adopted for split and tip nodes as severe stress singularities exist near the crack tip compared to the regions lying around the crack interface. For applying appropriate enrichments, one additional enriched node with extra three degrees of freedom is added at the split node. At each tip node, four additional nodes with three degrees of freedom are added. As discussed earlier, split nodes are enriched with the Heaviside jump function, which has been defined below. Let us take a point **x** in the engineering domain, and let x_Γ be the point nearest to **x** on the crack surface, and **n** be the unit normal to crack surface. The Heaviside jump function H(**x**) can be defined as

$$\mathrm{H}(\mathbf{x}) = \left\{ \begin{array}{cc} +1 & \text{if } (\mathbf{x} - x_\Gamma)\cdot\mathbf{n} \geq 0 \\ -1 & \text{otherwise} \end{array} \right\} \tag{7.58}$$

Heaviside enrichment function provides a strong numerical framework for modeling the discontinuities developed by crack surfaces, and this type of enrichment finds a wide range of applications in computational fracture mechanics. Heaviside jump function cannot be used for the enrichment of displacement fields of split nodes as their domain of influence is not completely cut by the crack interface. It is also known that extreme stress singularities exist at the crack tip, due to which extreme care should be taken before selecting the appropriate enrichment functions for the tip nodes so that the effects of these stress singularities are captured more accurately and efficiently. Linear elastic fracture mechanics is based on the assumption that material behavior remains linear elastic during simulation. The elasto-plastic effects near the crack tip are neglected as long as the size of crack tip plastic zone is very small in comparison to the dimensions of cracked component. The expressions for displacement fields around the crack tip using LEFM principles are well known in literature. From the expressions of the displacement fields, it can be easily shown that the displacement field lies within the span of following four trigonometric functions, as given in the following equation:

$$\beta_\alpha(r,\theta) = \left[\sqrt{r}\cos\frac{\theta}{2}, \sqrt{r}\sin\frac{\theta}{2}, \sqrt{r}\cos\frac{\theta}{2}\sin\theta, \sqrt{r}\sin\frac{\theta}{2}\sin\theta\right] \tag{7.59}$$

The four enrichment functions as given above form the basis for the enrichment of the displacement fields around crack tip. For LEFM problems, these four functions are used for enriching the displacement fields near the crack tip. Since four enrichment functions have to be added to the conventional displacement-based approximation, 4 additional enriched nodes have to be placed on the standard nodes, giving rise to 12 extra enriched degrees of freedom at each node in three-dimensional engineering components. Second enrichment function $\sqrt{r}\sin\frac{\theta}{2}$ produces the discontinuity across the crack surface. Remaining three enrichment functions are introduced to improve the quality of approximation in the domain around the crack tip.

The enrichment functions discussed above cannot be used if the elasto-plastic region around the crack tip is large in size compared to other dimensions of the cracked specimen. Therefore, the application of these enrichment functions remains limited to linear elastic fracture mechanics only. The displacement fields are not known analytically in elasto-plastic fracture mechanics, and hence, it has always been a challenge to obtain the appropriate enrichment functions for the elasto-plastic crack growth problems. One of the most commonly used solutions describing the nature of the singularities near crack tip in the case of power-hardening materials is the Hutchinson–Rice–Rosengren (HRR) solution. The enrichment functions for elasto-plastic fracture mechanics problems are developed from these HRR solutions, by carrying out the Fourier analysis of the HRR fields. Thus HRR solutions provide a strong basis for developing the enrichment functions for crack growth problems using elasto-plastic fracture mechanics. In the region near the crack tip, the elastic strain rates are negligible in comparison to the plastic strain rates. The HRR solutions for the asymptotic fields in the vicinity of the crack tip can be written as

$$\sigma_{ij} = \sigma_0 \left(\frac{J}{\alpha \sigma_0 \varepsilon_0 I_n r} \right)^{\frac{1}{n+1}} \tilde{\sigma}_{ij}(\theta, n) \tag{7.60}$$

$$\varepsilon_{ij} = \alpha \varepsilon_0 \left(\frac{J}{\alpha \sigma_0 \varepsilon_0 I_n r} \right)^{\frac{n}{n+1}} \tilde{\varepsilon}_{ij}(\theta, n) \tag{7.61}$$

$$u_i = \alpha \varepsilon_0 r \left(\frac{J}{\alpha \sigma_0 \varepsilon_0 I_n r} \right)^{\frac{n}{n+1}} \tilde{u}_i(\theta, n) \tag{7.62}$$

In the above equations, I_n is a dimensionless constant, α and n are the Ramberg–Osgood coefficients, and $\tilde{\sigma}_{ij}$, $\tilde{\varepsilon}_{ij}$, and $\tilde{u}_i$ are dimensionless angular functions of θ and n. J represents the J-integral. σ_{ij} and ε_{ij} are the stresses and strains, and u_i denotes the displacement field. The Ramberg–Osgood model for representing the power-law hardening in the material for uni-axial stress–strain relationship can be written as

$$\frac{\varepsilon}{\varepsilon_0} = \frac{\sigma}{\sigma_0} + \alpha \left(\frac{\sigma}{\sigma_0} \right)^n \tag{7.63}$$

where σ_0 is the initial yield stress, $\varepsilon_0 = \sigma_0 / E$ is the elastic strain at yield, α is the material constant, and n is the hardening exponent. The linear elastic materials have the hardening exponent equal to 1. On the other hand, the rigid perfectly plastic materials tend to have the hardening exponent equal to infinity. The elasto-plastic fields around the crack tip are obtained by carrying out the Fourier decomposition of the functions $\tilde{u}_i(\theta, n)$. It can be shown that the displacement field under pure mode-I and pure mode-II loadings can be represented by expanding the HRR functions on the following basis:

$$r^{\frac{n}{n+1}} \left\{ \left(\cos\frac{k\theta}{2}, \sin\frac{k\theta}{2} \right); k \epsilon [1, 3, 5, 7] \right\} \tag{7.64}$$

The Fourier decomposition of the displacement fields near the crack tip led to the development of the enrichment functions for elasto-plastic fracture mechanics problems. Three different types of enrichment schemes have been proposed for modeling crack tip singularities in elasto-plastic fracture analysis. The three types of enrichment functions used for modeling crack growth in elasto-plastic analysis can be written in standard form as

$$\beta_1(r,\theta) = r^{\frac{n}{n+1}}\left\{\sin\frac{\theta}{2}, \cos\frac{\theta}{2}, \sin\frac{\theta}{2}\sin\theta, \cos\frac{\theta}{2}\sin\theta, \sin\frac{\theta}{2}\sin2\theta, \cos\frac{\theta}{2}\sin2\theta\right\} \tag{7.65}$$

$$\beta_2(r,\theta) = r^{\frac{n}{n+1}}\left\{\sin\frac{\theta}{2}, \cos\frac{\theta}{2}, \sin\frac{\theta}{2}\sin\theta, \cos\frac{\theta}{2}\sin\theta, \sin\frac{\theta}{2}\sin3\theta, \cos\frac{\theta}{2}\sin3\theta\right\} \tag{7.66}$$

$$\beta_3(r,\theta) = r^{\frac{n}{n+1}}\left\{\sin\frac{\theta}{2}, \cos\frac{\theta}{2}, \sin\frac{\theta}{2}\sin\theta, \cos\frac{\theta}{2}\sin\theta, \sin\frac{\theta}{2}\sin2\theta, \cos\frac{\theta}{2}\sin2\theta, \sin\frac{\theta}{2}\sin3\theta, \cos\frac{\theta}{2}\sin3\theta\right\} \tag{7.67}$$

It was later on observed that the addition of the enrichment functions $\sin\frac{\theta}{2}\sin2\theta$ and $\cos\frac{\theta}{2}\sin2\theta$ improves the quality of the displacement-based approximations, but simultaneously they result in an ill-conditioned stiffness matrix, which creates problems in inverting the stiffness matrix and thereby complicates the solution process. However, it has also been observed that the addition of the two extra enrichment functions $\sin\frac{\theta}{2}\sin3\theta$ and $\cos\frac{\theta}{2}\sin3\theta$ increases the quality of the displacement-based approximations, and, at the same time, they results in a well-conditioned stiffness matrix and facilitates the solution process. However, both enrichment strategies result in the same amount of accuracy in the displacement-based approximations. In view of above discussions, two types of enrichment functions were proposed for modeling elasto-plastic crack growth problems:

$$\beta_1(r,\theta) = r^{\frac{n}{n+1}}\left\{\sin\frac{\theta}{2}, \cos\frac{\theta}{2}, \sin\frac{\theta}{2}\sin\theta, \cos\frac{\theta}{2}\sin\theta\right\} \tag{7.68}$$

$$\beta_2(r,\theta) = r^{\frac{n}{n+1}}\left\{\sin\frac{\theta}{2}, \cos\frac{\theta}{2}, \sin\frac{\theta}{2}\sin\theta, \cos\frac{\theta}{2}\sin\theta, \sin\frac{\theta}{2}\sin3\theta, \cos\frac{\theta}{2}\sin3\theta\right\} \tag{7.69}$$

7.3 Numerical results and discussions

Several numerical problems on the investigation of nonpropagating cracks in 3D structural components have been solved by invoking the use of enriched EFGM, in which additional enrichment functions are added to classical moving least squares approximations to model different irregularities present in structural components. SIFs corresponding to different fracture modes are evaluated by using the interaction integral. Mathematical models derived on enriched EFGM have been coded in MATLAB to simulate different engineering problems involving the investigation of static cracks in 3D engineering components. Simulation results derived in the current study are validated with respect to benchmark solutions already reported in literature. Different types of cracks in 3D structural components, such as plane edge cracks, horizontal and inclined penny cracks, and lens-type cracks, have been considered for simulation by enriched EFGM.

7.3.1 Analysis of plane edge crack

The first numerical example presents the investigation of a plane edge crack having a length of 50mm present in a structural component of cuboidal geometry with dimensions $100 \times 175 \times 150\text{mm}^3$, as shown in Fig. 7.2. The analysis is performed using enriched EFGM, in which the domain is represented by an array of nodes only. The specimen has Young's modulus of 200 GPa and Poisson's ratio equal to 0.3. The bottom surface remains fixed during simulation, whereas the top face of cracked specimen is subjected to a tensile load of 100 MPa. Discretization of the three-dimensional cracked domain in EFGM is shown in Fig. 7.3. Enriched EFGM generates a uniform array of nodes for engineering components, which is independent of the nature of crack present in structural components. Enriched EFGM eliminates the problems of re-meshing, element refinements, and conformation of FE meshes, which makes it a potential computational tool as compared to conventional nom-enriched method such as classical FEM. The distribution of SIFs with normalized length (s/t) can be seen in Fig. 7.4, where "s" denotes the arc length from crack center measured along the crack front. Comparison of enriched EFGM results with the benchmark or reference solutions has been carried out, and a very close agreement was observed between the two solutions [40]. Numerical simulations performed in this study clearly demonstrate that the enriched EFGM has a strong potential in modeling the influence caused by the presence of the cracks in three-dimensional structural components.

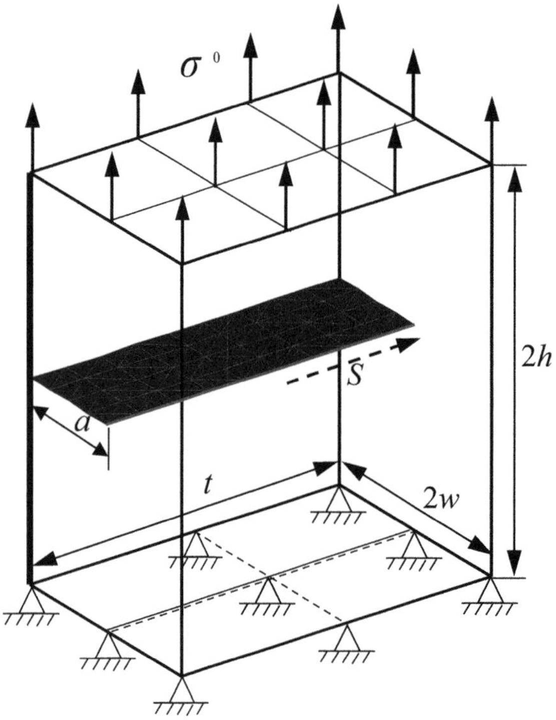

FIGURE 7.2 Plane edge crack in cuboidal domain.

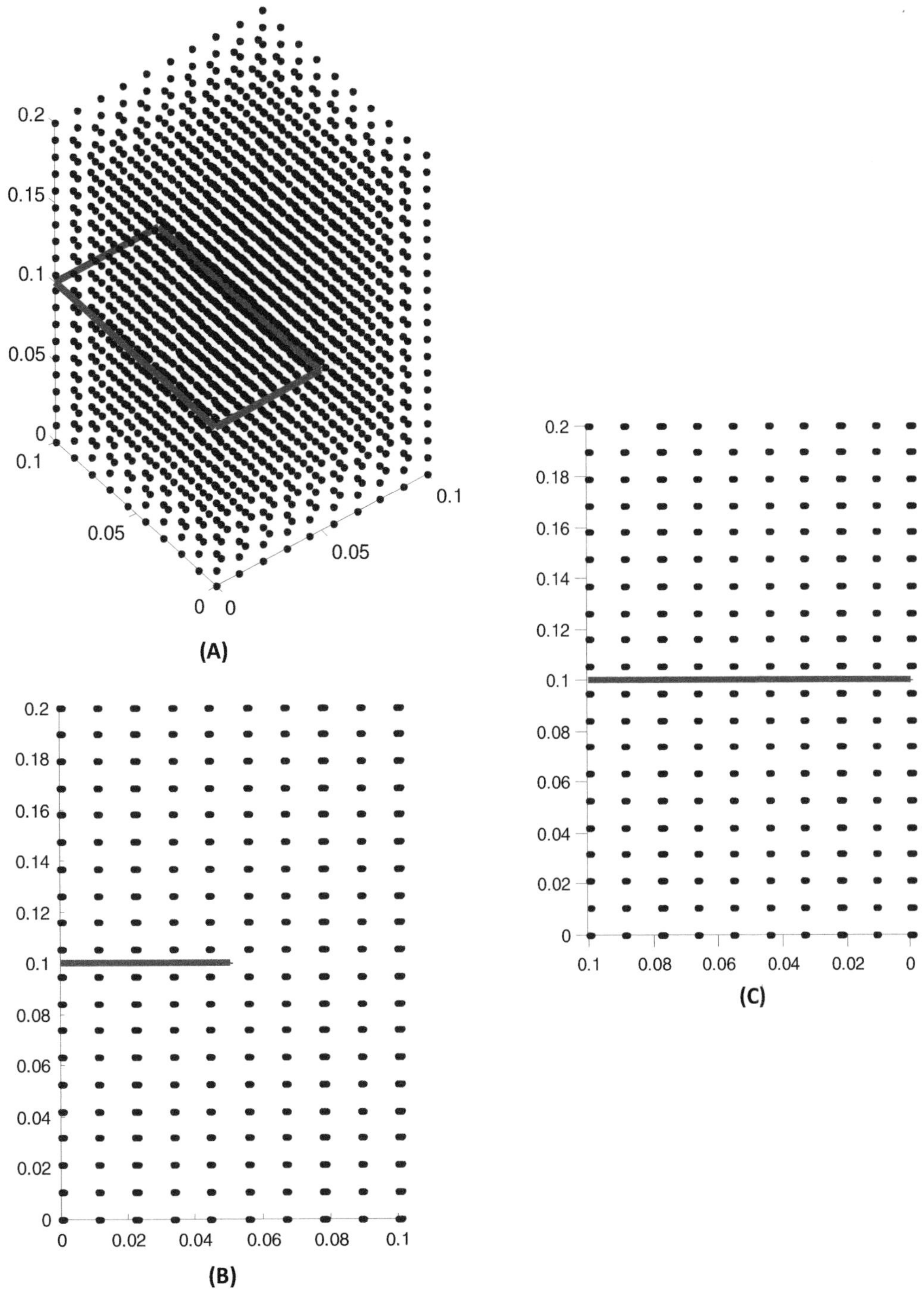

FIGURE 7.3 Nodal distribution for a plane edge crack: (A) 3D view, (B) side view, and (C) front view.

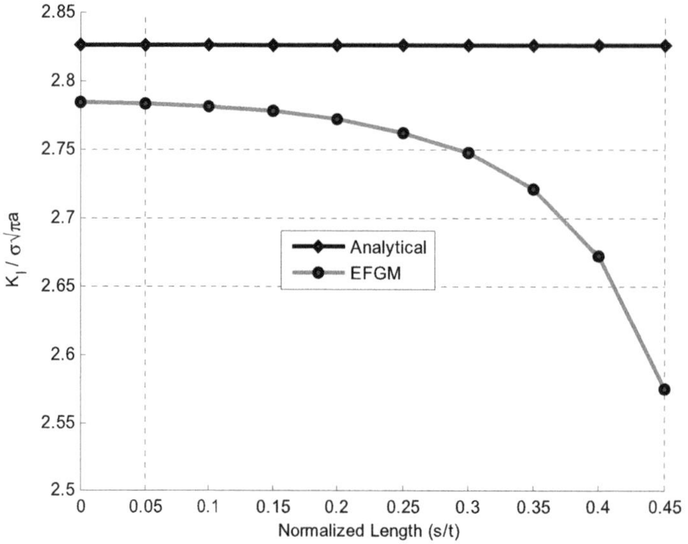

FIGURE 7.4 Variation of normalized mode-I stress intensity factors by EFGM. *EFGM*, Element-free Galerkin method.

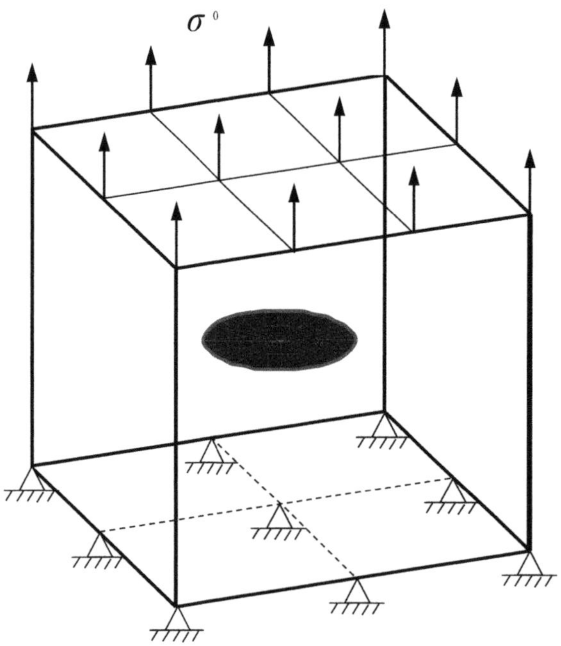

FIGURE 7.5 Cubical specimen containing a horizontal penny crack at is center.

7.3.2 Horizontal penny crack in a cubical domain

A cubical domain, with each side equal 200 mm and containing a horizontal penny crack at the center, which is having a radius of 10 mm, is considered for analysis as presented in Fig. 7.5. Same material properties, imposed BCs, and loads have been assumed in this numerical example as chosen in previous example. EFGM domain representation of the given penny crack in a three-dimensional component can be seen in Fig. 7.6. It can be easily observed that enriched EFGM creates a uniform array of nodes for the cracked engineering domain, which eliminates the drawbacks of mesh conformation, refinements, and element distortions. Distribution of normalized mode-I SIFs along the crack surface can be seen in Fig. 7.7. Enriched EFGM results obtained from the simulations carried out in the current study are compared with benchmark solutions reported in literature [40]. Very close agreement can be observed between current simulations and reference solutions. It is concluded that the proposed enriched approach provides an efficient computational tool for modeling the behavior of cracks in 3D structural components.

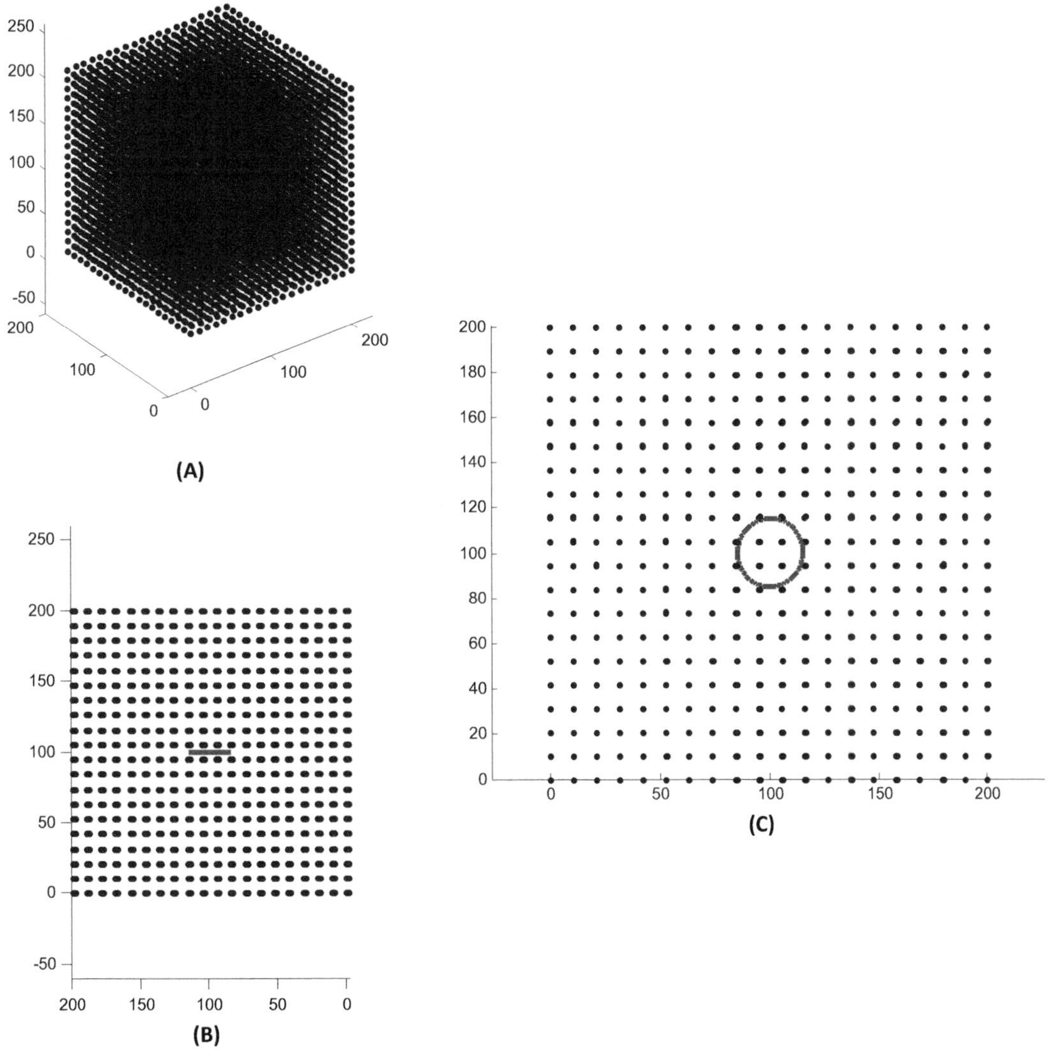

FIGURE 7.6 Nodal distribution for a horizontal penny crack: (A) 3D view, (B) side view, and (C) top view.

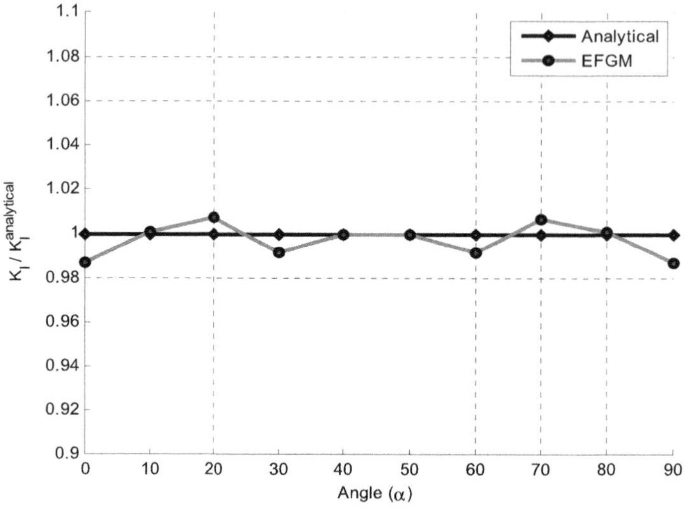

FIGURE 7.7 Normalized mode-I stress intensity factors along crack surface by Element-free Galerkin method (EFGM)

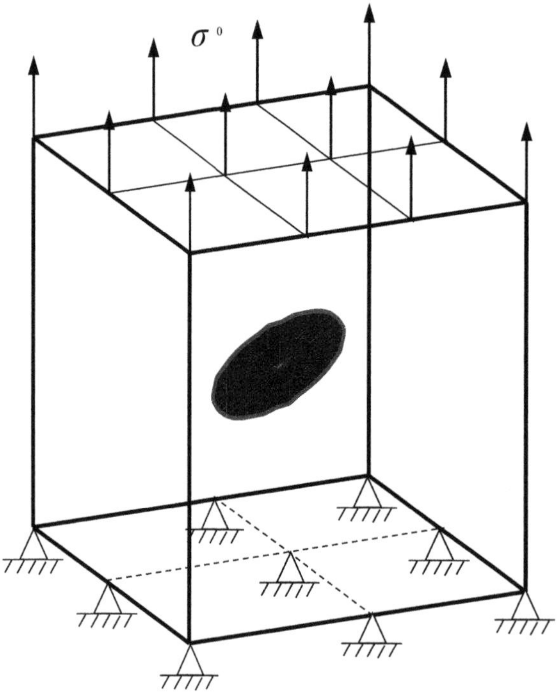

FIGURE 7.8 Cubical specimen containing an embedded inclined penny crack.

7.3.3 Inclined penny crack in a cubical domain

A cubical domain containing an inclined penny crack (Fig. 7.8) at the center is investigated here. The geometry of the cracked specimen, crack length, material parameters,

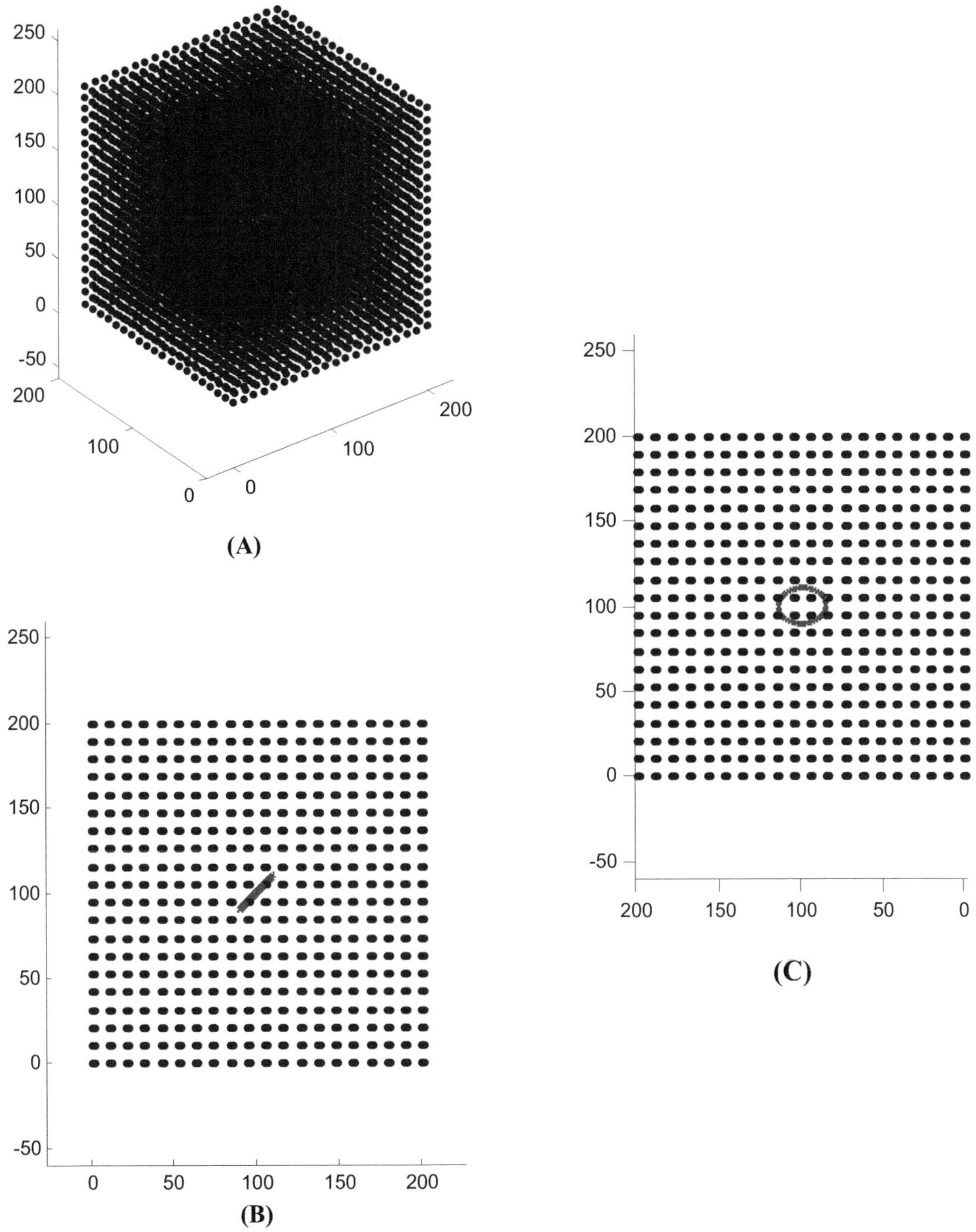

FIGURE 7.9 Nodal distribution for an inclined penny crack: (a) 3D view; (b) side view; (c) front view.

imposed boundary, and loading conditions are chosen to be the same as considered earlier. EFGM domain representation of inclined penny crack in a three-dimensional structural component can be seen in Fig. 7.9. The plots of mode-III SIFs with crack angles can be seen in Fig. 7.10. Enriched EFGM results obtained from the simulations carried out in the current work are validated with benchmark solutions reported in literature [40]. Very close agreement was observed between the current simulations and reference solutions. It

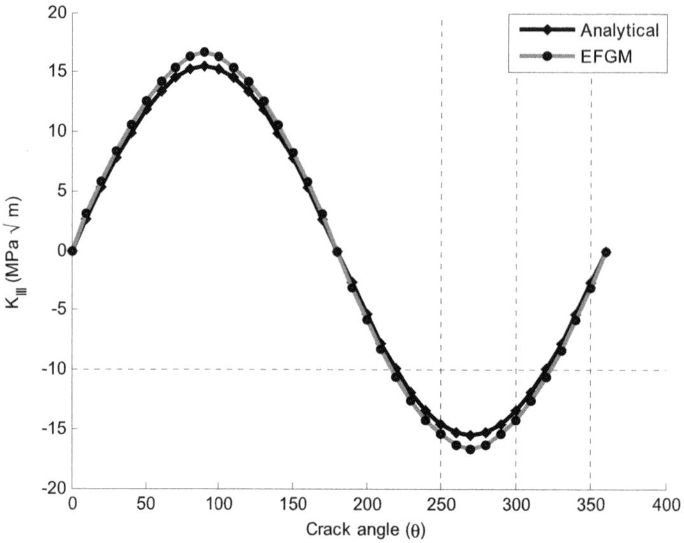

FIGURE 7.10 Variations of mode-III stress intensity factors with crack angle by Element-free Galerkin method.

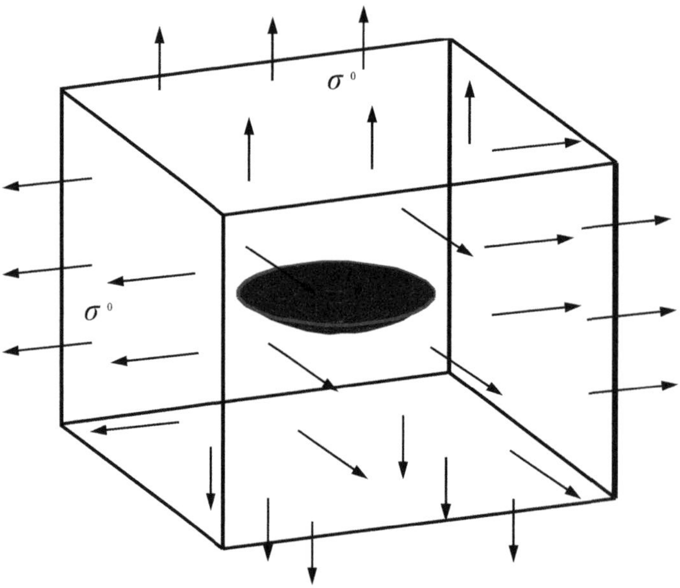

FIGURE 7.11 Domain for a lens-shaped crack.

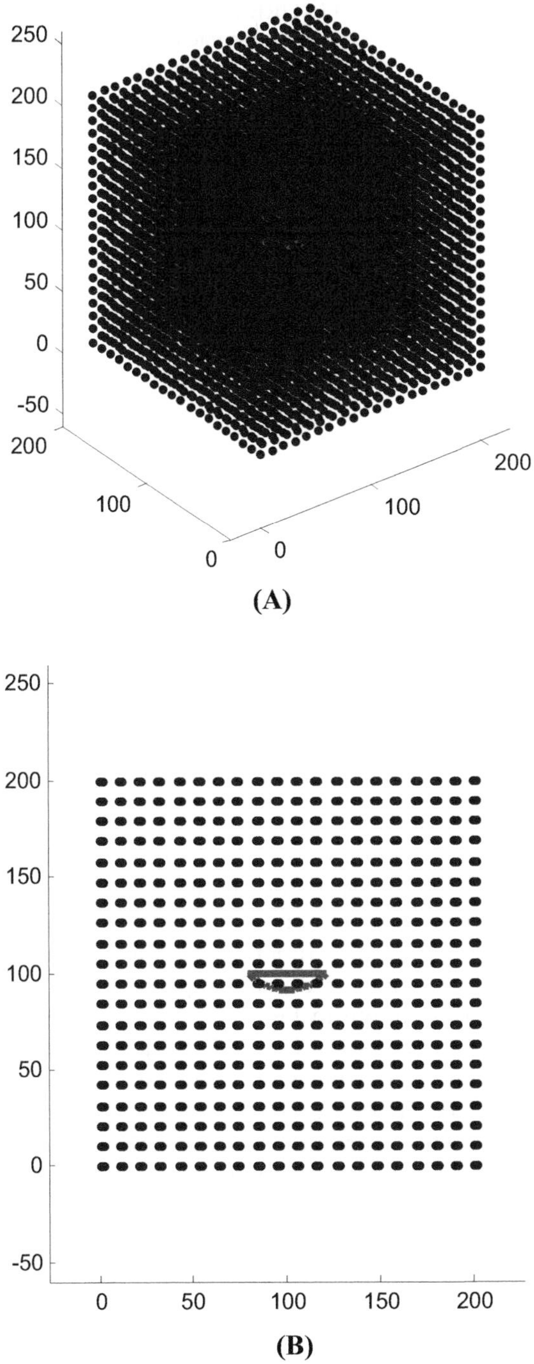

FIGURE 7.12 Nodal distribution for the lens crack: (A) 3D view and (B) front view.

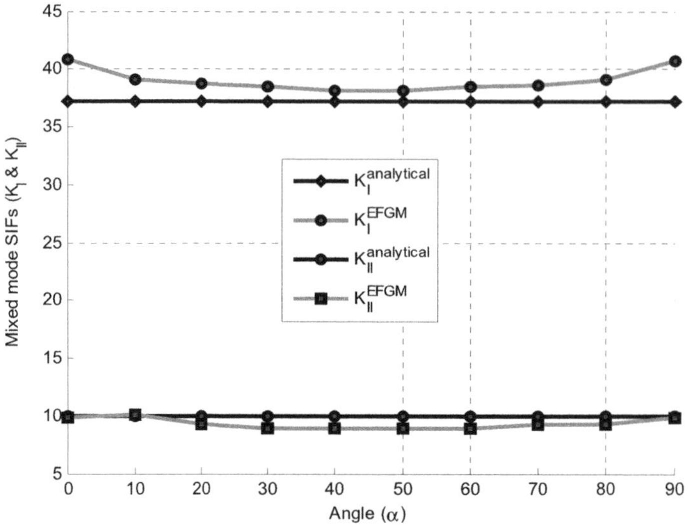

FIGURE 7.13 Variation of mixed mode stress intensity factors for the lens crack by Element-free Galerkin method.

is concluded that the enriched EFGM provides an efficient computational tool for modeling the behavior of cracks in 3D structural components. Enriched EFGM eliminates problems of re-meshing, mesh conformation, and mesh adaption by making simulations independent of the grid chosen for analysis.

7.3.4 Lens crack in a cubical domain

A cubical domain with each side equal to 200 mm and having a lens crack at the center is selected for investigation, as shown in Fig. 7.11. A sphere of radius 20 mm is used to define the lens crack by cutting it at an angle of 45degrees at the center of the cubical structural component. The cracked domain is acted upon by a monotonic load of 100 MPa at all sides. Young's modulus and Poisson's ratio of the three-dimensional cracked specimen are chosen to be 68.2 GPa and 0.22, respectively. EFGM domain representation for the three-dimensional component with a lens crack is shown in Fig. 7.12. Distributions of Mode-I and mode-II SIFs in the cracked specimen can be seen in Fig. 7.13. Enriched EFGM results obtained from the simulations carried out in the current work are compared with benchmark solutions reported in literature [40]. Very close agreement can be observed between the current simulations and reference solutions. Current simulations also demonstrate the efficiency and potential of enriched EFGM in investigating three-dimensional crack growth in engineering materials.

7.4 Conclusion

Enriched EFGM is employed in the current work to investigate the behavior of different types of cracks in 3D structural components subjected to monotonic tensile loads.

Enriched EFGM models different cracks present in the cracked specimen irrespective of the set of nodes chosen for investigation. Enriched EFGM eliminates problems of re-meshing, mesh adaption, and conformal meshing, which makes it a strong and efficient computational framework as compared to conventional nom-enriched methods such as the classical FEM. Different types of cracks in 3D structural components were investigated in the current work, such as the plane edge cracks, penny cracks, and lens cracks. Very close agreement was observed between the current simulations and reference solutions. Current simulations also demonstrate and establish the efficiency, applicability, and potential of enriched EFGM in modeling three-dimensional crack growth in engineering materials.

References

[1] Aliabadi MH, Brebbia CA. Boundary element formulations in fracture mechanics: a review. Transactions on Engineering Sciences 1998;17:589–98.

[2] Wen PH, Aliabadi MH, Young A. Dual boundary element methods for three-dimensional dynamic crack problems. Journal of Strain Analysis for Engineering Design 1999;34:373–94. Available from: https://doi.org/10.1177/030932479903400601.

[3] Leonel ED, Venturini WS. Non-linear boundary element formulation applied to contact analysis using tangent operator. Engineering Analysis with Boundary Elements 2011;35:1237–47. Available from: https://doi.org/10.1016/j.enganabound.2011.06.005.

[4] Spangenberger AG, Lados DA. Extended finite element modeling of fatigue crack growth microstructural mechanisms in alloys with secondary/reinforcing phases: model development and validation. Computational Mechanics 2020;. Available from: https://doi.org/10.1007/s00466-020-01921-2.

[5] Lone AS, Kanth SA, Jameel A, Harmain GA. A state of art review on the modeling of contact type nonlinearities by extended finite element method. Materials Today: Proceedings 2019;18:3462–71. Available from: https://doi.org/10.1016/j.matpr.2019.07.274.

[6] Sukumar N, Chopp DL, Moës N, Belytschko T. Modeling holes and inclusions by level sets in the extended finite-element method. Computer Methods in Applied Mechanics and Engineering 2001;190:6183–200. Available from: https://doi.org/10.1016/S0045-7825(01)00215-8.

[7] Jameel A, Harmain GA. Fatigue crack growth in presence of material discontinuities by EFGM. International Journal of Fatigue 2015;81:105–16. Available from: https://doi.org/10.1016/j.ijfatigue.2015.07.021.

[8] Lone AS, Jameel A, Harmain GA. A coupled finite element-element free Galerkin approach for modeling frictional contact in engineering components. Materials Today: Proceedings 2018;5:18745–54. Available from: https://doi.org/10.1016/j.matpr.2018.06.221.

[9] Kanth SA, Harmain GA, Jameel A. Modeling of nonlinear crack growth in steel and aluminum alloys by the element free Galerkin method. Materials Today: Proceedings 2018;5:18805–14. Available from: https://doi.org/10.1016/j.matpr.2018.06.227.

[10] Jameel A, Harmain GA. Fatigue crack growth analysis of cracked specimens by the coupled finite element-element free Galerkin method. Mechanics of Advanced Materials and Structures 2019;26:1343–56. Available from: https://doi.org/10.1080/15376494.2018.1432800.

[11] Lone AS, Jameel A, Harmain GA. Enriched element free Galerkin method for solving frictional contact between solid bodies. Mechanics of Advanced Materials and Structures 2022. Available from: https://doi.org/10.1080/15376494.2022.2092791.

[12] Lone AS, Jameel A, Harmain GA. "Modelling of contact interfaces by penalty based enriched finite element method. Mechanics of Advanced Materials and Structures 2022. Available from: https://doi.org/10.1080/15376494.2022.2034075.

[13] Jameel A, Harmain GA. Large deformation in bi-material components by XIGA and coupled FE-IGA techniques. Mechanics of Advanced Materials and Structures 2022;29:850–72. Available from: https://doi.org/10.1080/15376494.2020.1799120.

[14] Kanth SA, Jameel A, Harmain GA. Investigation of fatigue crack growth in engineering components containing different types of material irregularities by XFEM. Mechanics of Advanced Materials and Structures 2022;29(24):3570–87. Available from: https://doi.org/10.1080/15376494.2021.1907003.

[15] Gupta V, Jameel A, Anand S, Anand Y. Analysis of composite plates using isogeometric analysis: a discussion. Materials Today: Proceedings 2021;44:1190–4. Available from: https://doi.org/10.1016/j.matpr.2020.11.238.

[16] Jameel A, Harmain GA. Effect of material irregularities on fatigue crack growth by enriched techniques. International Journal for Computational Methods in Engineering Science and Mechanics 2020;21:109–33. Available from: https://doi.org/10.1080/15502287.2020.1772902.

[17] Lone AS, Kanth SA, Jameel A, Harmain GA. "XFEM modelling of frictional contact between elliptical inclusions and solid bodies. Materials Today: Proceedings 2020;26:819–24. Available from: https://doi.org/10.1016/j.matpr.2019.12.424.

[18] Sheikh UA, Jameel A. Elasto-plastic large deformation analysis of bi-material components by FEM. Materials Today: Proceedings 2020;26:1795–802. Available from: https://doi.org/10.1016/j.matpr.2020.02.377.

[19] Simpson R, Trevelyan J. A partition of unity enriched dual boundary element method for accurate computations in fracture mechanics. Computer Methods in Applied Mechanics and Engineering 2011;200:1–10. Available from: https://doi.org/10.1016/j.cma.2010.06.015.

[20] Noda NA, Oda K. Numerical solutions of the singular integral equations in the crack analysis using the body force method. International Journal of Fracture 1992;58:285–304. Available from: https://doi.org/10.1007/BF00048950.

[21] Rao BN, Rahman S. A coupled meshless-finite element method for fracture analysis of cracks. International Journal of Pressure Vessels and Piping 2001;78:647–57. Available from: https://doi.org/10.1016/S0308-0161(01)00076-X.

[22] Kanth SA, Lone AS, Harmain GA, Jameel A. Modelling of embedded and edge cracks in steel alloys by XFEM,". Materials Today: Proceedings 2020;26:814–18. Available from: https://doi.org/10.1016/j.matpr.2019.12.423.

[23] Kanth SA, Lone AS, Harmain GA, Jameel A. Elasto plastic crack growth by XFEM: a review. Materials Today: Proceedings 2019;18:3472–81. Available from: https://doi.org/10.1016/j.matpr.2019.07.275.

[24] Jameel A, Harmain GA. Modeling and numerical simulation of fatigue crack growth in cracked specimens containing material discontinuities. Strength of Materials 2016;48(2):294–307. Available from: https://doi.org/10.1007/s11223-016-9765-0.

[25] Eberhard P, Gaugele T. Simulation of cutting processes using mesh-free Lagrangian particle methods. Computational Mechanics 2013;51:261–78. Available from: https://doi.org/10.1007/s00466-012-0720-z.

[26] De Lorenzis L, Wriggers P, Zavarise G. A mortar formulation for 3D large deformation contact using NURBS-based isogeometric analysis and the augmented Lagrangian method. Computational Mechanics 2012;49:1–20. Available from: https://doi.org/10.1007/s00466-011-0623-4.

[27] Bhardwaj G, Singh IV, Mishra BK, Bui TQ. Numerical simulation of functionally graded cracked plates using NURBS based XIGA under different loads and boundary conditions. Composite Structures 2015;126:347–59. Available from: https://doi.org/10.1016/j.compstruct.2015.02.066.

[28] Gu YT, Wang QX, Lam KY. A meshless local Kriging method for large deformation analyses. Computer Methods in Applied Mechanics and Engineering 2007;196:1673–84. Available from: https://doi.org/10.1016/j.cma.2006.09.017.

[29] Belytschko T, Krongauz Y, Organ D, Fleming M, Krysl P. Meshless methods: an overview and recent developments. Computer Methods in Applied Mechanics and Engineering 1996;139:3–47. Available from: https://doi.org/10.1016/S0045-7825(96)01078-X.

[30] Rao BN, Rahman S. An enriched meshless method for non-linear fracture mechanics. International Journal for Numerical Methods in Engineering 2004;59:197–223. Available from: https://doi.org/10.1002/nme.868.

[31] Duflot M, Nguyen-Dang H. A meshless method with enriched weight functions for fatigue crack growth. International Journal for Numerical Methods in Engineering 2004;59:1945–61. Available from: https://doi.org/10.1002/nme.948.

[32] Jameel A, Harmain GA. Extended iso-geometric analysis for modeling three-dimensional cracks. Mechanics of Advanced Materials and Structures 2019;26:915–23. Available from: https://doi.org/10.1080/15376494.2018.1430275.

[33] Singh IV, Bhardwaj G, Mishra BK. A new criterion for modeling multiple discontinuities passing through an element using XIGA. Journal of Mechanical Science and Technology 2015;29:1131–43. Available from: https://doi.org/10.1007/s12206-015-0225-8.

[34] Bhardwaj G, Singh IV, Mishra BK. Numerical simulation of plane crack problems using extended isogeometric analysis. Procedia Engineering 2013;64:661–70. Available from: https://doi.org/10.1016/j.proeng.2013.09.141.

[35] Jameel A, Harmain GA. A coupled FE-IGA technique for modeling fatigue crack growth in engineering materials. Mechanics of Advanced Materials and Structures 2019;26:1764–75. Available from: https://doi.org/10.1080/15376494.2018.1446571.

[36] Gupta V, Jameel A, Verma SK, Anand S, Anand Y. "Transient isogeometric heat conduction analysis of stationary fluid in a container,". Part E: Journal of Process Mechanical Engineering 2022. Available from: https://doi.org/10.1177/0954408922112.

[37] Gupta V, Jameel A, Verma SK, Anand S, Anand Y. An insight on NURBS based isogeometric analysis, its current status and involvement in mechanical applications. Archives of Computational Methods in Engineering 2022. Available from: https://doi.org/10.1007/s11831-022-09838-0.

[38] Singh AK, Jameel A, Harmain GA. Investigations on crack tip plastic zones by the extended iso-geometric analysis. Materials Today: Proceedings 2018;5:19284–93. Available from: https://doi.org/10.1016/j.matpr.2018.06.287.

[39] Melenk JM, Babuška I. The partition of unity finite element method: basic theory and applications. Computer Methods in Applied Mechanics and Engineering 1996;139:289–314. Available from: https://doi.org/10.1016/S0045-7825(96)01087-0.

[40] Jameel A, Harmain GA. Extended iso-geometric analysis for modeling three dimensional cracks. Mechanics of Advanced Materials and Structures 2019;26:915–23. Available from: https://doi.org/10.1080/15376494.2018.1430275.

CHAPTER

8

Enriched element-free Galerkin method for elastoplastic crack growth in steel alloys

Showkat Ahmad Kanth[1], *G.A. Harmain*[1], *Mumtaz Ahmad*[1] *and Shuhaib Mushtaq*[2]

[1]Department of Mechanical Engineering, National Institute of Technology Srinagar, Hazratbal, Srinagar, Jammu and Kashmir, India [2]Department of Mechanical Engineering, Islamic University of Science and Technology, Awantipora, Jammu and Kashmir, India

8.1 Introduction

The static and crack growth simulation have always been a problem of attention for researchers. Since the ductile and bi-materials are widely employed for various engineering purposes, therefore it is necessary to monitor the structures and components, such as turbine blades, aircrafts, and shafts experiencing different loading conditions. Fatigue loading has been one of the prime causes of failures in various engineering structures subjected to cyclic loading. To predict the fatigue life of these components, it is essential to perform the fracture analysis accurately. The modeling of the stable crack growth under mode-I loading conditions has been investigated by various fracture parameters like crack opening displacement [1,2], crack tip opening angle [3,4], J-integral [5,6], and energy release rate [7,8]. Among the various fracture parameters, J-integral is the most preferred fracture parameter for the crack growth simulation as it can characterize the stress—strain fields in the vicinity of crack tip in the presence of plasticity. Based on the material response, two approaches such as linear elastic fracture mechanics (LEFM) and elastoplastic fracture mechanics (EPFM) can be utilized for the analysis of static and propagating cracks. LEFM approach has been widely applied to investigate the prediction of service life and reliability of the engineering components. This approach assumes the

Enriched Numerical Techniques
DOI: https://doi.org/10.1016/B978-0-443-15362-4.00002-4

linear or elastic material response. In LEFM, the yielding near the crack tip is neglected due to the assumption of small-scale yielding [9]. Therefore the utility of LEFM is confined to elastic problems and is not applicable to the materials exhibiting elastoplastic behavior or exposed to large deformation under loading conditions. EPFM is a better approach to understand the behavior of the ductile materials containing discontinuities subjected to service loads. EPFM considers the yield effect and also characterizes plasticity effect ahead of the crack tip [10]. Hence, EPFM is an accurate approach to analyze the reliability and life of ductile materials. To investigate the crack propagation in the materials exhibiting yielding, J-integral has been widely employed as it characterizes the crack tip plasticity. The structures subjected to fatigue loading fail as a result of multiple crack initiation [11,12]. Therefore to check the structural integrity of the components, it is inevitable to investigate the analysis of crack interaction [7]. Among the various computational techniques, the finite element method (FEM) has been widely employed for the modeling and simulation of engineering problems. FEM has been accurately used to solve different static, elastic crack growth and contact problems in engineering components subjected to different loading conditions [13–16]. However, FEM faces several issues while modeling large deformation and elastoplastic crack growth problems. The problem of mesh distortion is encountered during large deformation and the issue of remeshing while solving crack propagating problems. Another major limitation of FEM is conformal meshing, that is, the mesh topology needs to conform the edges of the discontinuity present in the domain. To overcome the issues of mesh refinement, mesh distortion, and mesh dependency, that is, conformal meshing faced in FEM, several computational techniques, such as boundary element method [17–19], cracking particle method [20], strain smoothing FEM [21], element-free Galerkin method (EFGM) [22–25], coupled FE-EFGM [26], extended iso-geometric analysis [27,28], and extended FEM (XFEM) [29–33] have been proposed from time to time. The EFGM is an improved numerical technique introduced to solve the fracture-related engineering problems. In EFGM, to capture the effect of the discontinuities like cracks, holes, inclusions, and bi-materials in the physical domain, the primary variable is approximated with extrinsic enrichment [34–39]. The results obtained by employing EFGM showed significant computational efficiency with enhanced accuracy. Several researchers have investigated elastoplastic crack propagation problems with large deformations using the FEM [40,41]. However, a limited amount of research has been conducted on elastoplastic crack growth problems with large deformation using EFGM.

This work aims at modeling and the simulation of elastoplastic crack growth using EFGM. The Newton–Raphson method has been utilized for the approximation of nonlinear equations. Several algorithms like elastic predictor and plastic corrector [42], the radial return method [43,44], and plane stress plasticity algorithm [45] have been employed to determine the stress computation in elastoplastic crack growth. For the modeling of plastic behavior, von Mises yield criteria have been used in conjunction with the isotropic hardening phenomenon, which states that there is no translation in yield surface; however, the yield surface expands uniformly. Finally, several elastoplastic crack growths have been solved to check the fidelity and the accuracy of the EFGM codes developed in MATLAB®.

8.2 Governing formulations

The weak form of the equilibrium equation can be written as

$$\int_{\Omega}\sigma:\varepsilon(u)d\Omega=\int_{\Omega}b:ud\Omega+\int_{\Omega}t:ud\Gamma \tag{8.1}$$

By further utilizing the arbitrariness of nodal variations and substituting the trial and test functions, we obtain the discrete system of equations as

$$[K^e]\{d^e\}=\{f^e\} \tag{8.2}$$

where K^e denotes the global stiffness matrix, f^e represents the force vector, and d^e is the displacement vector. The enriched displacement approximation for 2-D body in presence of crack can be written in generalized form as

$$u^h(x)=\sum_{i=1}^{n}\Psi_i(x)u_i+\sum_{j=1}^{n_s}\Psi_j(x)[H(x)-H(x_j)]a_j+\sum_{k=1}^{n_T}\Psi_k(x)\sum_{\alpha=1}^{4}[\beta_\alpha(x)-\beta_\alpha(x_k)]b_k^\alpha \tag{8.3}$$

where Ψ_i denotes the standard finite element shape function, and "u_i" is the standard nodal displacement vector. Standard nodes in the mesh are represented by "n", the set of nodes for the elements that are completely split by the crack are denoted by "n_s" and "n_T" symbolizes the set of nodes for tip elements. The enriched degrees of freedom for elements completely cut by crack are given by nodes "a_j". $H(x)$ denotes the Heaviside jump function defined for split elements, which is discontinuous across the crack surface. Its value is 0 for the crack surface. But it has a value of $+1$ and -1 on different sides of the crack. The enriched degrees of freedom associated with crack tip enrichment are given by b_k^α. Also, $\beta_\alpha(x)$ are the enrichment functions in polar coordinates (r,θ) near the vicinity of crack tip. In the case of elastoplastic behavior, the crack tip enrichment functions $\beta_\alpha(x)$ can be expressed as

$$\beta_1(r,\theta)=r^{\frac{1}{n+1}}\left\{\sin\frac{\theta}{2},\cos\frac{\theta}{2},\sin\frac{\theta}{2}\sin\theta,\cos\frac{\theta}{2}\sin\theta\right\} \tag{8.4}$$

where n is the strain-hardening exponent, which is a material property. Using the approximation function defined in Eq. (8.18), the elemental stiffness matrix and force vector can be obtained as

$$[K^e]=\begin{bmatrix}K^{uu} & K^{ua} & K^{ub}\\ K^{au} & K^{aa} & K^{ab}\\ K^{bu} & K^{ba} & K^{bb}\end{bmatrix} \tag{8.5}$$

$$\{f^e\}=\left\{f^u\ f^a\ f^{b2}\ f^{b3}\ f^{b4}\right\}^T \tag{8.6}$$

The submatrices are given by

$$K^{rs}=\int_{\Omega^e}(\mathbf{B}^r)^T C\mathbf{B}^s d\Omega;(r,s=u,a,b) \tag{8.7}$$

$$f^u = \int_{\Omega^e} \Psi^T \boldsymbol{b} d\Omega + \int_{\Gamma^e} \Psi^T \boldsymbol{t} d\Gamma \tag{8.8}$$

$$f^a = \int_{\Omega^e} \Psi^T \left(H(\boldsymbol{x}) - H(\boldsymbol{x}_j)\right) \boldsymbol{b} d\Omega + \int_{\Gamma^e} \Psi^T \left(H(\boldsymbol{x}) - H(\boldsymbol{x}_j)\right) \boldsymbol{t} d\Gamma \tag{8.9}$$

$$\int \mathrm{ba} = \int_{\Omega^e} \Psi^{\mathrm{T}}(\beta_a(x) - \beta_a(x_{\mathrm{k}}))bd\Omega + \int_{re} \Psi^{\mathrm{T}}(\beta_a(x) - \beta_a(x_{\mathrm{k}})td\Gamma; (a = 1,2,3,4) \tag{8.10}$$

where Ψ_i are the standard finite element shape functions. ($\mathbf{B}^u$), ($\mathbf{B}^a$) and ($\mathbf{B}^{b\alpha}$) are the matrices containing derivatives of shape function for normal and enriched conditions:

$$\boldsymbol{B}^u = \begin{bmatrix} \Psi_{i,x} & 0 \\ 0 & \Psi_{i,y} \\ \Psi_{i,y} & \Psi_{i,x} \end{bmatrix} \tag{8.11}$$

$$\boldsymbol{B}^a = \begin{bmatrix} (\Psi_j(H(\boldsymbol{x}) - H(\boldsymbol{x}_j)))_x & 0 \\ 0 & (\Psi_j(H(\boldsymbol{x}) - H(\boldsymbol{x}_j)))_y \\ (\Psi_j(H(\boldsymbol{x}) - H(\boldsymbol{x}_j)))_y & (\Psi_j(H(\boldsymbol{x}) - H(\boldsymbol{x}_j)))_x \end{bmatrix} \tag{8.12}$$

$$\boldsymbol{B}^{b\alpha} = \begin{bmatrix} (\Psi_k(\beta_\alpha(\boldsymbol{x}) - \beta_\alpha(\boldsymbol{x}_k)))_x & 0 \\ 0 & (\Psi_k(\beta_\alpha(\boldsymbol{x}) - \beta_\alpha(\boldsymbol{x}_k)))_y \\ (\Psi_k(\beta_\alpha(\boldsymbol{x}) - \beta_\alpha(\boldsymbol{x}_k)))_y & (\Psi_k(\beta_\alpha(\boldsymbol{x}) - \beta_\alpha(\boldsymbol{x}_k)))_x \end{bmatrix} \tag{8.13}$$

8.3 Numerical formulation in elastoplastic analysis

8.3.1 Plasticity modeling

The type of nonlinearities that occur due to nonlinear material behavior leads to material nonlinearities. The nonlinear stress–strain relationship gives rise to material nonlinearities. Material nonlinearity generally follows postyielding, leading to elastoplastic material behavior. The stress–strain relationship for the elastoplastic response of the material under plane stress or plane strain conditions is provided by the mathematical theory of plasticity. Before the initiation of yielding, that is, during the elastic conditions, the stress–strain relationship is determined by standard Hooke's law as $\sigma_{ij} = C_{ijkl}\varepsilon_{kl}$, where σ_{ij} and ε_{kl} are the stress and stress components and C_{ijkl} is the constitutive matrix or elastic constant tensor relating stresses to strains.

The stress level at which yielding begins has been determined by the von Mises yield criterion, which states that a material under a given set of loading begins to yield when the second invariant of the deviatoric stress tensor reaches a critical value. After the onset of yielding, appropriate relationship between stress and strain must be developed for modeling the elastoplastic behavior of the body. The yielding criterion can be written as $f(\sigma_{ij}) = K(k)$ where f is some general function, k is hardening parameter, and K is a material parameter. When $f(\sigma_{ij}) < K(k)$, the state of stress is said to elastic. However, the stress state is undefined for $f(\sigma_{ij}) > K(k)$.

As the yield criterion is independent of the coordinate system, it should be the function of three stress invariants, that is, (J_1, J_2, J_3) written as

$$J_1 = \sigma_{ii};\ J_2 = \frac{1}{2}\sigma_{ij}\sigma_{ij};\ J_3 = \frac{1}{3}\sigma_{ij}\sigma_{jk}\sigma_{ki} \tag{8.14}$$

Through experimental evidences, it has been reported that the plastic deformation is independent of the hydrostatic stress tensor. Therefore the yield function can be expressed in terms of deviatoric stress components, that is, $f(J_2', J_3') = K(k)$, where J_2' and J_3' are the second and third invariants of the deviatoric stress tensor $\left(\sigma_{ij}'\right)$, stated as $\sigma_{ij}' = \sigma_{ij} - \frac{1}{3}\delta_{ij}\sigma_{kk}$. According to von Mises yield criterion, the material subjected to a given set of loading conditions will start to yield when the second invariant J_2' of the deviatoric stress tensor reaches a critical value, that is, $\sqrt{J_2'} = K(k)$. In 2-D, the second invariant of deviatoric stress tensor can be expressed as

$$J_2' = \frac{1}{2}\sigma_{ij}'\sigma_{ij}' = \frac{1}{2}\left[\sigma_x'^2 + \sigma_y'^2 + \sigma_z'^2\right] + \tau_{xy}'^2 + \tau_{yz}'^2 + \tau_{xz}'^2 \tag{8.15}$$

Therefore the von Mises criterion can be further represented as $\overline{\sigma} = \sqrt{3J_2'} = K(k)$, where $\overline{\sigma}$ denotes the effective or von Mises stress.

8.3.2 Analysis of elastoplastic stress−strain relations

Since during plastic deformation, the yield surface changes continuously after every stage of simulation. This phenomenon is known as strain hardening or work hardening. The hardening behavior after initial yielding can be categorized into perfectly plastic behavior, isotropic hardening, and kinematic hardening. The isotropic strain hardening has been assumed to determine the hardening behavior during elastoplastic analysis. During isotropic hardening, the initial yield surface expands uniformly without translation. The relationship between the material parameter K and the hardening parameter k is helpful in determining the evolving yield surface during isotropic hardening. There are two approaches to obtain the relationship. The hardening parameter k can be described as a function of the total plastic work (W_p), that is, $k = W_p = \int_\Omega \sigma_{ij}\left(d\varepsilon_{ij}\right)_p$, where $\left(d\varepsilon_{ij}\right)_p$ denotes the plastic strain component for any strain increment. The hardening parameter "k" can also be explained with respect to effective plastic strain $\left(\overline{\varepsilon_p}\right)$ • as • $k = \overline{\varepsilon} = \int d\overline{\varepsilon}_p$, where

$$d\overline{\varepsilon}_p = \sqrt{\frac{2}{3}\left(d\varepsilon_{ij}\right)_p\left(d\varepsilon_{ij}\right)_p} = \sqrt{\frac{2}{3}\left(d\varepsilon_{ij}'\right)_p\left(d\varepsilon_{ij}'\right)_p} \tag{8.16}$$

The incremental change in yield function $f\left(\sigma_{ij}\right)$ for any change in stress increment can be written as $df = \frac{\partial F}{\partial \sigma_{ij}} d\sigma_{ij}$. If the function $df < 0$, the material is unloading elastically, that is, the stress point reaches inside the yield surface. If $df = 0$, the incremental stress change has no impact on the yield surface and the stress point lies on the same surface. If $df > 0$, there is a uniform or gradual increase in the yield surface during the current change in stress increment and the new stress point lies on the new evolved yield surface. This case characterizes the strain-hardening behavior for an elastoplastic material.

After a certain amount of initial yielding, the deformable body exhibits both elastic and plastic behavior [46]. Therefore the total increment in strain $d\varepsilon_{ij}$ can be composed as the sum of elastic $(d\varepsilon_{ij})_e$ and plastic $(d\varepsilon_{ij})_p$ strain components, that is, $d\varepsilon_{ij} = (d\varepsilon_{ij})_e + (d\varepsilon_{ij})_p$

The elastic strain component $(d\varepsilon_{ij})_e$ can be determined by the constitutive law as, $(d\varepsilon_{ij})_e = [C]^{-1} d\sigma_{ij}$, where $\boldsymbol{C}$ denotes the elastic constitutive matrix.

The plastic incremental strain and stress relationship can be obtained after further assumptions on the material behavior. Therefore a quantity termed plastic potential (Q) is introduced such that the plastic strain increment component is proportional to the stress gradient of the quantity, that is, plastic potential, and is expressed as

$$(d\varepsilon_{ij})_p = d\lambda \frac{\partial Q}{d\sigma_{ij}} \tag{8.17}$$

where $d\lambda$ is the constant of proportionality and is called the plastic multiplier. Post yielding, the plastic flow is determined by the above equation, known as flow rule. It is quite difficult to determine the general form of plastic potential. However, in the mathematical plastic theory, the plastic potential in terms of the yield function has been found to be significant, giving rise to the associated theory of plasticity, which defines the plastic potential as $Q = f$. Thus above equation can be composed as

$$(d\varepsilon_{ij})_p = d\lambda \frac{\partial f}{d\sigma_{ij}} \tag{8.18}$$

where the quantity $\frac{\partial f}{d\sigma_{ij}}$ denotes a vector normal to the yield surface at the stress point.

8.3.3 Matrix representation of elastoplastic stress–strain relations

The governing equilibrium equation for any deformable body in elasto-statics defined as [29,35] is written as

$$\nabla.\boldsymbol{\sigma} + \boldsymbol{b} = 0 \text{in} \Omega \tag{8.19}$$

where $\boldsymbol{\sigma}$ represents the Cauchy stress tensor and "$\boldsymbol{b}$" is the uniform body force per unit volume.

We can rearrange the von Mises yield function as

$$F(\boldsymbol{\sigma}, k) = f(\boldsymbol{\sigma}) - K(k) \tag{8.20}$$

For the yield function to lie on the yield surface, the following consistency condition must be satisfied:

$$dF = \frac{\partial F}{\partial \boldsymbol{\sigma}} d\boldsymbol{\sigma} + \frac{\partial F}{\partial k} dk = \boldsymbol{a}^T d\boldsymbol{\sigma} - A d\lambda \tag{8.21}$$

where $\boldsymbol{a}$ is called the flow vector and is written as

$$a = \frac{\partial F}{\partial \sigma} = \left[\frac{\partial F}{\partial \sigma_{xx}} \frac{\partial F}{\partial \sigma_{YY}} \frac{\partial F}{\partial \sigma_{zz}} \frac{\partial F}{\partial \sigma_{yx}} \frac{\partial F}{\partial \sigma_{zx}} \frac{\partial F}{\partial \sigma_{xy}} \right]^T \text{ and } A = -\frac{1}{d\lambda} \frac{\partial F}{\partial k} dk \tag{8.22}$$

Elastoplastic stress–strain relationship can be represented in matrix form as

$$d\varepsilon = (d\varepsilon_{ij})_e + (d\varepsilon_{ij})_p = \boldsymbol{C}^{-1}d\boldsymbol{\sigma} + d\lambda\frac{\partial F}{\partial \boldsymbol{\sigma}} \tag{8.23}$$

Above equation can be further solved to obtain the plastic multiplier $d\lambda$ as

$$d\lambda = \frac{1}{[A + \boldsymbol{a}^T \boldsymbol{C}_e \boldsymbol{a}]}\boldsymbol{a}^T \boldsymbol{C}_e d\varepsilon \tag{8.24}$$

Thus the complete stress–strain relationship for elastoplastic behavior can be obtained as

$$d\boldsymbol{\sigma} = \left(\boldsymbol{C}_e - \frac{\boldsymbol{C}_e \boldsymbol{a} \boldsymbol{C}_e \boldsymbol{a}^T}{A + \boldsymbol{C}_e \boldsymbol{a}^T \boldsymbol{a}}\right) d\varepsilon \tag{8.25}$$

$$d\boldsymbol{\sigma} = \boldsymbol{C}_{ep} d\varepsilon \tag{8.26}$$

where $\boldsymbol{C}_{ep} = \boldsymbol{C}_e - \frac{d_C d_C^T}{A + d_C^T \boldsymbol{a}}$; $\boldsymbol{d}_C = \boldsymbol{C}_e \boldsymbol{a}$

As stated earlier, $Q = F$ for the associated theory of plasticity, which implies that the elastoplastic constitutive matrix $\boldsymbol{C}_{ep}$ becomes symmetric, which is otherwise unsymmetric for nonassociated theory.

8.4 Stress calculation algorithms in elastoplasticity

During the application of load increment, an element or any part of it may yield. The stresses and strains are calculated at all Gauss integration points, and hence, it can be found whether yielding has occurred at any Gauss point or not. For the Gauss points that have yielded during the application of a particular load increment, it is essential to adjust the stresses and strains so that the yield criterion and the constitutive laws are satisfied. Various algorithms can be utilized to achieve the values of stress levels. In this work, we have employed three algorithms for stress calculations, as discussed in the following section.

8.4.1 Elastic predictor and plastic corrector algorithm

This algorithm predicts the values of stresses for any particular load step by assuming elastic behavior, and then plastic corrections are applied to reduce the stresses to the yield surface [47]. Let us consider the rth iteration of any particular load step. The elastic predictor plastic corrector algorithm can be applied as follows:

The applied loads for the rth iteration of any particular load step are the residual forces f_{res}^{r-1}, which give rise to displacement increments $\Delta \boldsymbol{u}^r$, and the corresponding strain increments $\Delta\varepsilon^r$. Assuming the linear elastic behavior, compute the incremental stresses using generalized Hooke's law. If the material did not yield during the present iteration, then the elastically obtained stresses are the true actual stresses. However, if yielding occurs during the present iteration, then the stresses obtained here have to be corrected. The stress increments are obtained by $\Delta\sigma_e^r = \boldsymbol{C}_e \Delta\varepsilon^r$. Accumulate the total stress for the rth iteration, assuming the elastic behavior as $\boldsymbol{\sigma}_e^r = \boldsymbol{\sigma}^{r-1} + \Delta\boldsymbol{\sigma}_e^r$, where $\boldsymbol{\sigma}^{r-1}$ is the final corrected stress of $(r-1)^{th}$ iteration. Now, we have to check whether the element had yielded in the $(r-1)^{th}$ iteration or not. The yield stress for the $(r-1)^{th}$ iteration is given by $\sigma_Y^{r-1} = \sigma_Y + H\bar{\varepsilon}_p{}^{r-1}$, where $\bar{\varepsilon}_p$ denotes the accumulated effective plastic strain, H is the strain hardening parameter, and σ_Y is the initial yield stress. Thus we have to check if $\bar{\sigma}^{r-1} > \sigma_Y^{r-1}$, where $\bar{\sigma}^{r-1}$ is the effective stress obtained at the end of the $(r-1)^{th}$ iteration.

1. If $\bar{\sigma}^{r-1} > \sigma_Y^{r-1}$, the element had yielded in the previous iteration. Now, we have to check whether $\bar{\sigma}_e^r > \bar{\sigma}^{r-1}$, where $\bar{\sigma}_e^r$ is the effective stress for the rth iteration, obtained elastically.
   - If $\bar{\sigma}_e^r > \bar{\sigma}^{r-1}$, the element had yielded previously and is still yielding since the stress in increasing further. In this case, all the excess stress $(\boldsymbol{\sigma}_e^r - \boldsymbol{\sigma}^{r-1})$ must be reduced to the yield surface, as shown in Fig. 8.1. The factor "R" which defines the portion of the stress that is to be reduced to the actual value is equal to one.
   - If $\bar{\sigma}_e^r < \bar{\sigma}^{r-1}$, elastic unloading has taken place during this iteration. According to the plastic theory, the unloading is always elastic. In this case, $\boldsymbol{\sigma}_e^r$ represents the actual stress, and no modification is required, that is, $\boldsymbol{\sigma}^r = \boldsymbol{\sigma}_e^r$.
2. If $\bar{\sigma}^{r-1} < \sigma_Y^{r-1}$, the Gauss point had not yielded in the previous iteration. Now, we have to check whether the yielding took place in the present iteration or not, that is, we have to check if $\bar{\sigma}_e^r > \sigma_Y^{r-1}$.
   - If $\bar{\sigma}_e^r > \sigma_Y^{r-1}$, the element has yielded during this load increment. Now, we have to reduce the portion of stress that is greater than the yield value to the new yield surface. The reduction factor (R) is given by $R = \frac{\bar{\sigma}_e^r - \sigma_Y^{r-1}}{\bar{\sigma}_e^r - \bar{\sigma}^{r-1}}$, which represents the portion of stress that is to be corrected or reduced. This has been shown in Fig. 8.2.
   - If $\bar{\sigma}_e^r < \sigma_Y^{r-1}$, the element has not yielded in the present iteration and is still elastic. No correction in stresses is required, and we have $\boldsymbol{\sigma}^r = \boldsymbol{\sigma}_e^r$.

For those elements that have yielded, we calculate the portion of the total stress that satisfies the yield criterion as $\boldsymbol{\sigma}^{r-1} + (1-R)\Delta\boldsymbol{\sigma}_e^r$. The remaining portion $R\Delta\boldsymbol{\sigma}_e^r$ of the total

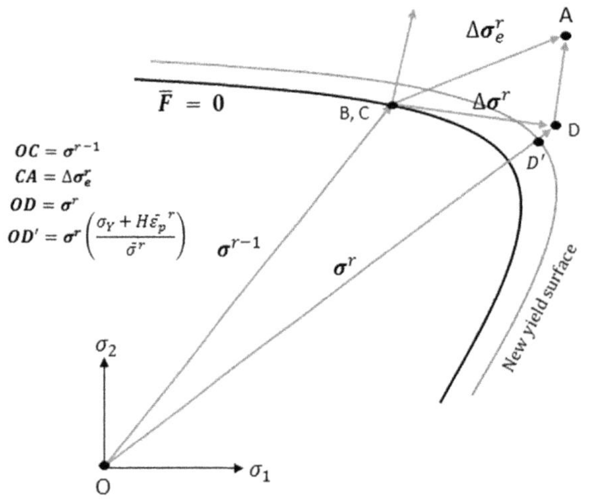

FIGURE 8.1 Plastic correction for an already yielded Gauss point.

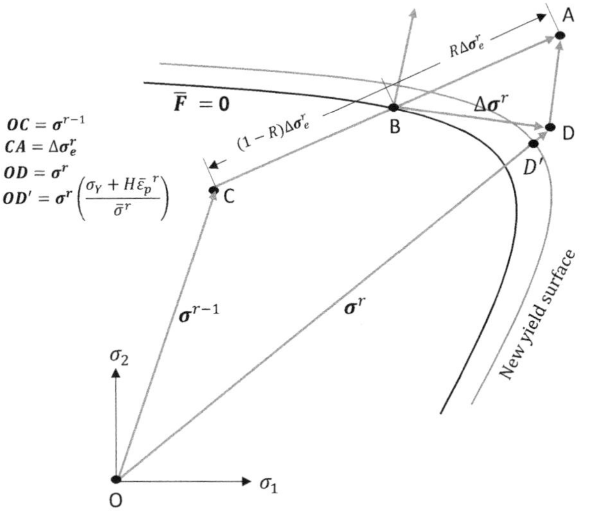

FIGURE 8.2 Incremental stresses and strains for a Gauss point after initial yielding.

stress increment must be effectively reduced by applying the appropriate plastic correction. The elastically obtained stresses (point A) must be brought onto the yield surface after allowing the plastic deformation to occur. From Fig. 8.2, we see that on loading the Gauss point from point C to point A, the stress point meets the yield surface at point B. After applying the plastic corrections, the final reduced stress for the rth iteration becomes $\boldsymbol{\sigma}^r = \boldsymbol{\sigma}^{r-1} + \Delta\boldsymbol{\sigma}_e^r - d\lambda \boldsymbol{d}_C$, where $\boldsymbol{\sigma}^r$ is the final corrected stress for the rth iteration under elastoplastic conditions.

The effective plastic strain for the rth iteration is given by $\bar{\varepsilon}_p{}^r = \bar{\varepsilon}_p{}^{r-1} + \frac{d\lambda \boldsymbol{a}^T \boldsymbol{\sigma}_f}{\bar{\sigma}_f}$, where $\boldsymbol{\sigma}_f$ is the flow stress, defined by $\boldsymbol{\sigma}_f = \boldsymbol{\sigma}^{r-1} + (1-R)\Delta\boldsymbol{\sigma}_e^r$. The flow stress corresponds to the point B on the yield surface, as shown in Fig. 8.2.

It has been observed that if the stress increment is large, the final corrected stress $\boldsymbol{\sigma}_r$, corresponding to the point D, departs from the original yield surface, which may lead to inaccurate results. Thus the size of the load increment should be small enough to avoid these discrepancies. However, appropriate scaling can be used to reduce the point D to the yield surface. The effective stress at the end of the rth iteration should be equal to the yield stress at the end of the rth iteration, that is, $\bar{\sigma}^r = \sigma_Y^r = \sigma_Y + H\bar{\varepsilon}_p{}^r$. Thus the appropriate scaling factor to bring down the point D on the yield surface can be written as $\boldsymbol{\sigma}^r = \boldsymbol{\sigma}^r\left(\frac{\sigma_Y + H\bar{\varepsilon}_p{}^r}{\bar{\sigma}^r}\right)$.

If the stress increments are large, the above procedure for scaling down the stresses to the yield surface results in inaccurate results. For larger increments, the excess stress is reduced to the yield surface in several stages, as shown in Fig. 8.3. The excess stress is divided into several parts, and each part of the excess stress is reduced to the yield surface turn by turn. The greater the number of reduction steps into which the excess stress "AB" is divided, greater is the accuracy. But the increase in the number of reduction steps increases the computational time. The excess stress to be reduced to the yield surface is divided into m equal parts such that m is the nearest integer given by $m = \left(\frac{\bar{\sigma}_e^r - \sigma_y^r}{\sigma_y}\right)8 + 1$.

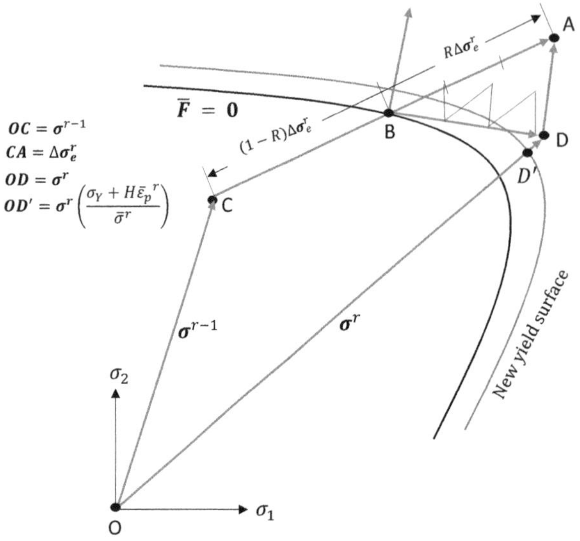

FIGURE 8.3 Representation of stepwise stress reduction.

8.4.2 Implicit integration or von Mises plasticity

8.4.2.1 The radial return method

The von Mises yield surface with a representation of the radial return method or the implicit integration method is shown in Fig. 8.4 [43]. A trial stress increment is chosen outside of the yield surface, which takes the updated stresses, $\boldsymbol{\sigma}_{t+\Delta t}^{tr}$. After trial stress increment, plastic corrector is utilized to update the stress and bring it back onto the yield surface at time $t + \Delta t$. The plane stress von Mises ellipse becomes a circle in deviatoric stress space, as shown in Fig. 8.4. Due to the normality condition, the plastic correction term is always directed toward the center of the yield surface. The technique is therefore called the radial return method. Unless specified, all the quantities are taken to be those at the end of time step $t + \Delta t$. $\boldsymbol{\sigma}_t$ represents the stress at the beginning of the time step, that is, at time t and $\boldsymbol{\sigma}$ is the stress at $t + \Delta t$.

In multiaxial form, Hooke's law can be written as

$$\Delta\boldsymbol{\sigma} = 2G\Delta\varepsilon^{e} + \lambda \mathrm{Tr}(\Delta\varepsilon^{e})\mathbf{I} \tag{8.27}$$

At the end of time step, the elastic component may be composed as

$$\varepsilon^{e} = \varepsilon_{t}^{e} + \Delta\varepsilon^{e} = \varepsilon_{t}^{e} + \Delta\varepsilon - \Delta\varepsilon^{p} \tag{8.28}$$

Therefore stress $\boldsymbol{\sigma}$ can be written as

$$\boldsymbol{\sigma} = 2G\left(\varepsilon_{t}^{e} + \Delta\varepsilon - \Delta\varepsilon^{p}\right) + \lambda \mathrm{Tr}\left(\varepsilon_{t}^{e} + \Delta\varepsilon - \Delta\varepsilon^{p}\right)\mathbf{I} \tag{8.29}$$

$$\boldsymbol{\sigma} = \underbrace{2G\left(\varepsilon_{t}^{e} + \Delta\varepsilon\right) + \lambda \mathrm{Tr}\left(\varepsilon_{t}^{e} + \Delta\varepsilon\right)\mathbf{I}}_{\text{Elastic Predictor}} - \underbrace{2G\Delta\varepsilon^{p}}_{\text{Plastic Corrector}} \tag{8.30}$$

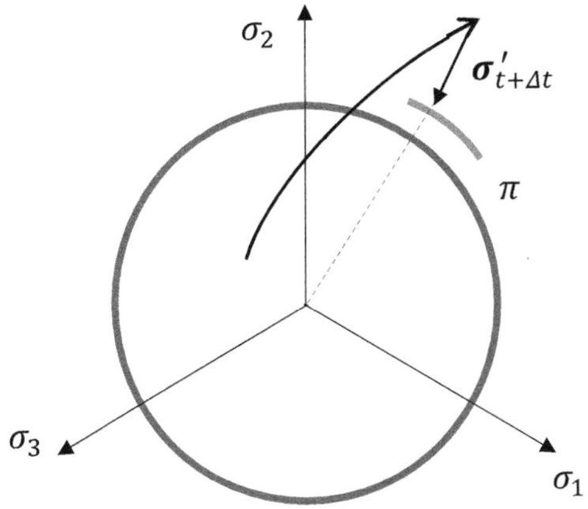

FIGURE 8.4 Principle of radial return algorithm.

Due to incompressibility condition $\mathrm{Tr}(\Delta\varepsilon^{p}) = 0$, therefore the elastic predictor or trial stress can be represented as

$$\boldsymbol{\sigma}^{tr} = 2G\left(\varepsilon_t^e + \Delta\varepsilon\right) + \lambda\,\mathrm{Tr}\left(\varepsilon_t^e + \Delta\varepsilon\right)\mathbf{I} \tag{8.31}$$

Therefore we can write $\boldsymbol{\sigma}$ as

$$\boldsymbol{\sigma} = \boldsymbol{\sigma}^{tr} - 2G\Delta\varepsilon^{p} = \boldsymbol{\sigma}^{tr} - 2G\Delta\overline{\varepsilon}^{p}\mathbf{n} \tag{8.32}$$

where $\mathbf{n} = \frac{3}{2}\frac{\sigma'}{\overline{\sigma}}$ represents the normal to the yield is surface, $\boldsymbol{\sigma}'$ and $\overline{\sigma}$ are the deviatoric and effective stress tensors. $\Delta\overline{\varepsilon}^{p}$ denotes the incremental effective plastic strain.

So above equation can be further written as

$$\boldsymbol{\sigma} = \boldsymbol{\sigma}^{tr} - 3G\Delta\overline{\varepsilon}^{p}\frac{\boldsymbol{\sigma}'}{\overline{\sigma}} \tag{8.33}$$

In terms of deviatoric and mean stress, the stresses can be expressed as

$$\boldsymbol{\sigma} = \boldsymbol{\sigma}' + \frac{1}{3}(\boldsymbol{\sigma}{:}\mathbf{I})\mathbf{I} = \boldsymbol{\sigma}^{tr} - 3G\Delta\overline{\varepsilon}^{p}\frac{\boldsymbol{\sigma}'}{\overline{\sigma}} \tag{8.34}$$

where $\mathbf{I}$ denotes the identity matrix, and we can define the operator $\boldsymbol{\sigma}{:}\mathbf{n}$ as

$$\begin{bmatrix} \sigma_{xx} & \sigma_{xy} & \sigma_{xz} \\ \sigma_{xy} & \sigma_{yy} & \sigma_{yz} \\ \sigma_{xz} & \sigma_{yz} & \sigma_{zz} \end{bmatrix} : \begin{bmatrix} n_{xx} & n_{xy} & n_{xz} \\ n_{xy} & n_{yy} & n_{yz} \\ n_{xz} & n_{yz} & n_{z} \end{bmatrix} = \begin{bmatrix} \sigma_{xx} \\ \sigma_{yy} \\ \sigma_{zz} \\ \sigma_{yz} \\ \sigma_{xz} \\ \sigma_{xy} \end{bmatrix} : \begin{bmatrix} n_{xx} \\ n_{yy} \\ n_{zz} \\ n_{yz} \\ n_{xz} \\ n_{xy} \end{bmatrix} = \boldsymbol{\sigma}\cdot\mathbf{n} \tag{8.35}$$

Such that $\sigma{:}I = \mathrm{Tr}\,(\sigma)$

Therefore we can write

$$\left(1 + 3G\frac{\Delta\overline{\varepsilon}^{p}}{\overline{\sigma}}\right)\boldsymbol{\sigma}' = \boldsymbol{\sigma}^{tr} - \frac{1}{3}(\boldsymbol{\sigma}{:}\mathbf{I})\mathbf{I} \tag{8.36}$$

where $\left(1 + 3G\frac{\Delta\overline{\varepsilon}^{p}}{\overline{\sigma}}\right)\boldsymbol{\sigma}' = \boldsymbol{\sigma}^{tr}$ is the deviatoric component of the trial stress.

After performing the contracted tensor product with itself on both sides, we get:

$$\left(1+3G\frac{\Delta\bar{\varepsilon}^{P}}{\bar{\sigma}}\right)^{2}\boldsymbol{\sigma}':\boldsymbol{\sigma}' = \boldsymbol{\sigma}^{tr'}:\boldsymbol{\sigma}^{tr'} \tag{8.37}$$

$$\left(1+3G\frac{\Delta\bar{\varepsilon}^{P}}{\bar{\sigma}}\right)\bar{\sigma} = \left(\frac{3}{2}\boldsymbol{\sigma}^{tr'}:\boldsymbol{\sigma}^{tr}\right)^{1/2} \equiv \bar{\sigma}^{tr} \tag{8.38}$$

We can finally write as

$$\bar{\sigma} + 3G\Delta\bar{\varepsilon}^{P} = \bar{\sigma}^{tr} \tag{8.39}$$

The multiaxial yield condition can be written as

$$f = \bar{\sigma} - r - \sigma_{Y}^{0} = \bar{\sigma}^{tr} - 3G\Delta\bar{\varepsilon}^{P} - r - \sigma_{Y}^{0} = 0, \tag{8.40}$$

where σ_{Y}^{0} is the initial yield stress.

Using Newton's method, this nonlinear equation in $\Delta\bar{\varepsilon}^{P}$ can be solved as

$$f + \frac{\partial f}{\partial \Delta\bar{\varepsilon}^{P}} d\Delta\bar{\varepsilon}^{P} + \ldots\ldots.. = 0$$

In the case of linear hardening, $r = H'\bar{\varepsilon}^{P}$, where H' is the hardening modulus such that, $\frac{\partial r}{\partial \Delta\bar{\varepsilon}^{P}} = \frac{\partial r}{\partial \bar{\varepsilon}^{P}} = H'$. Substituting multiaxial yield condition and hardening modulus in the above equation, we get:

$$\bar{\sigma}^{r} - 3G\Delta\bar{\varepsilon}^{P} - r - \sigma_{Y}^{0} + (-3G - H')d\Delta\bar{\varepsilon}^{P} = 0 \tag{8.41}$$

On rearranging, we can write

$$d\Delta\bar{\varepsilon}^{P} = \frac{\bar{\sigma}^{tr} - 3G\Delta\bar{\varepsilon}^{P} - r - \sigma_{Y}^{0}}{3G + H'} \tag{8.42}$$

In iterative form, the integration scheme can be written as

$$r^{(k)} = r_t + H'\Delta\bar{\varepsilon}^{P^{(k)}} \tag{8.43}$$

$$d\Delta\bar{\varepsilon}^{P} = \frac{\bar{\sigma}^{tr} - 3G\Delta\bar{\varepsilon}^{P^{(k)}} - r^{(k)} - \sigma_{Y}^{0}}{3G + H'} \tag{8.44}$$

$$\Delta\bar{\varepsilon}^{P^{(k+1)}} = \Delta\bar{\varepsilon}^{P^{(k)}} + d\Delta\bar{\varepsilon}^{P} \tag{8.45}$$

Thus we can write the plastic strain tensor increment as

$$\Delta\varepsilon^{P} = \frac{3}{2}\Delta\bar{\varepsilon}^{P}\frac{\boldsymbol{\sigma}'}{\bar{\sigma}} \equiv \frac{3}{2}\Delta\bar{\varepsilon}^{P}\frac{\boldsymbol{\sigma}^{tr'}}{\bar{\sigma}^{t}} \tag{8.46}$$

The elastic strain tensor increment can be written as

$$\Delta\varepsilon^{e} = \Delta\varepsilon - \Delta\varepsilon^{P}$$

And the stress increment can be obtained as

$$\Delta\boldsymbol{\sigma} = 2G\Delta\varepsilon^{e} + \lambda \mathrm{Tr}(\Delta\varepsilon^{e})\mathbf{I} \tag{8.47}$$

8.4.2.2 *Material Jacobian for time-independent isotropic linear hardening plasticity*

Since we know that

$$\left(1+3G\frac{\Delta\bar{\varepsilon}^{p}}{\bar{\sigma}}\right)\boldsymbol{\sigma}'=\boldsymbol{\sigma}^{tr} \quad (8.48)$$

Using the differential operator δ, the above equation can be written as

$$\left(1+3G\frac{\Delta\bar{\varepsilon}^{p}}{\bar{\sigma}}\right)\delta\boldsymbol{\sigma}'+\frac{3G}{\bar{\sigma}}\delta\Delta\bar{\varepsilon}^{p}\boldsymbol{\sigma}'-\frac{3G\Delta\bar{\varepsilon}^{p}}{\bar{\sigma}^{2}}\delta\bar{\sigma}\boldsymbol{\sigma}'=\delta\boldsymbol{\sigma}^{t'} \quad (8.49)$$

Since we know that, $\delta\bar{\sigma}+3G\delta\Delta\bar{\varepsilon}^{p}=\delta\bar{\sigma}^{tr}$. Therefore the yield condition can be written as

$$\delta\mathbf{f}=\delta\bar{\sigma}-\delta r=0 \quad (8.50)$$

Such that for linear strain hardening,

$$\delta\bar{\sigma}=\delta r=H'\delta\bar{\varepsilon}^{p}$$

$$Or\delta\bar{\sigma}=\delta\bar{\sigma}^{tr}\left(1-\frac{3G}{H'+3G}\right) \quad (8.51)$$

Now, after further evaluations and manipulations, eliminate $\delta\sigma_e, \delta\Delta\bar{\varepsilon}^{p}$, and $\Delta\bar{\varepsilon}^{p}$ above equation can be expressed as

$$\frac{\bar{\sigma}^{tr}}{\bar{\sigma}}\delta\boldsymbol{\sigma}'+\frac{\delta\bar{\sigma}^{tr}}{\bar{\sigma}^{2}}\left(\bar{\sigma}-\frac{\bar{\sigma}^{tr}}{1+(3G/H')}\right)\boldsymbol{\sigma}'=\delta\boldsymbol{\sigma}^{tr'} \quad (8.52)$$

Since we have

$$\delta\bar{\sigma}^{tr}=\delta\left(\frac{3}{2}\boldsymbol{\sigma}^{tr'}:\boldsymbol{\sigma}^{tr'}\right)^{1/2}=\frac{1}{2}\left(\frac{3}{2}\boldsymbol{\sigma}^{tr'}:\boldsymbol{\sigma}^{tr'}\right)^{-1/2}\left(\frac{3}{2}\delta\boldsymbol{\sigma}^{tr'}:\boldsymbol{\sigma}^{tr'}+\frac{3}{2}\boldsymbol{\sigma}^{tr'}:\boldsymbol{\sigma}^{tr'}\right)=\frac{3}{2\bar{\sigma}^{tr}}\boldsymbol{\sigma}^{tr'}:\boldsymbol{\sigma}^{tr'} \quad (8.53)$$

After substituting for $\boldsymbol{\sigma}'$, we obtain

$$\delta\boldsymbol{\sigma}'=\frac{3}{2}\left(\frac{1}{1+(3G/H')}-\frac{\bar{\sigma}}{\bar{\sigma}^{tr}}\right)\frac{\boldsymbol{\sigma}^{tr'}}{\bar{\sigma}^{tr}}\frac{\boldsymbol{\sigma}^{tr'}}{\bar{\sigma}^{tr}}:\delta\boldsymbol{\sigma}^{tr'}+\frac{\bar{\sigma}}{\bar{\sigma}^{tr}}\delta\boldsymbol{\sigma}^{tr'} \quad (8.54)$$

Since $Q=\frac{3}{2}\left(\frac{1}{1+(3G/H')}-\frac{\bar{\sigma}}{\bar{\sigma}^{tr}}\right)$ and $R=\frac{\bar{\sigma}}{\bar{\sigma}^{tr}}$

We can write above equation as

$$\delta\boldsymbol{\sigma}'=\left(Q\frac{\boldsymbol{\sigma}^{tr'}}{\bar{\sigma}^{tr}}\frac{\boldsymbol{\sigma}^{tr'}}{\bar{\sigma}^{tr}}+R\mathbf{I}_d\right):\delta\boldsymbol{\sigma}^{tr'} \quad (8.55)$$

where $\mathbf{I}_d$ denotes the fourth order identity tensor and $\mathbf{I}_d:\mathbf{I}=\mathbf{I}:\mathbf{I}_d=\mathbf{I}$.

Since Jacobian $(\mathbf{D}_{ep})$ is the relationship between $\delta\boldsymbol{\sigma}$ and $\delta\varepsilon$ therefore Hooke's law, we can define deviatoric trial stress in terms of the deviatoric trial strain as

$$\delta\boldsymbol{\sigma}^{tr'}=2G\delta\varepsilon^{tr'}=2G\left(\delta\varepsilon^{tr}-\frac{1}{3}\mathrm{Tr}(\varepsilon^{tr})\mathbf{I}\right)=2G\left(\delta\varepsilon^{tr}-\frac{1}{3}\mathbf{I}:\varepsilon^{tr}\right) \quad (8.56)$$

$$\text{As, } \delta\varepsilon^{tr} \equiv \delta\varepsilon$$

Therefore $\delta\boldsymbol{\sigma}^{tr'} = 2G\left(\delta\varepsilon - \frac{1}{3}\boldsymbol{\Pi}{:}\varepsilon\right)$

$$\text{Also, } \delta\boldsymbol{\sigma}^{tr'} = Q\frac{\sigma^{tr'}}{\overline{\sigma}^{tr}}\frac{\sigma^{tr'}}{\overline{\sigma}^{tr}}; \delta\varepsilon - \frac{1}{2}Q\frac{\sigma^{tr'}}{\sigma^{tr}}\frac{\sigma^{tr'}}{\overline{\sigma}^{tr}}; (\mathbf{I}; \delta\varepsilon) + 2GR\delta\varepsilon - \frac{2}{3}GR_d; (\mathbf{I}; \delta\varepsilon) \tag{8.57}$$

As $\boldsymbol{\sigma}^{tr'}$ is deviatoric, therefore second term of the RHS is zero. Therefore

$$\delta\boldsymbol{\sigma}' = Q\frac{\boldsymbol{\sigma}^{tr'}}{\overline{\sigma}^{tr}}\frac{\boldsymbol{\sigma}^{\sigma'}}{\overline{\sigma}^{tr}}{:}\delta\varepsilon + 2GR\varepsilon - \frac{2}{3}GR\mathbf{II}{:}\delta\varepsilon \tag{8.58}$$

Therefore the stress in terms of deviatoric component is given by

$$\delta\boldsymbol{\sigma} = \delta\boldsymbol{\sigma}' + \frac{1}{3}\boldsymbol{\Pi}{:}\delta\boldsymbol{\sigma} = \delta\boldsymbol{\sigma}' + K\boldsymbol{\Pi}{:}\delta\varepsilon^{e} = \delta\boldsymbol{\sigma}' + K\mathbf{I}{:}(\delta\varepsilon - \delta\varepsilon^{P}) = \delta\boldsymbol{\sigma}' + K\boldsymbol{\Pi}{:}\delta\varepsilon \tag{8.59}$$

where K represents bulk modulus.

$$\delta\boldsymbol{\sigma} = 2Q\frac{\boldsymbol{\sigma}^{tr'}}{\overline{\sigma}^{tr}}\frac{\boldsymbol{\sigma}^{tr'}}{\overline{\sigma}^{tr}}{:}\delta\varepsilon + 2GR\delta\varepsilon + \left(K - \frac{2}{3}GR\right)\mathbf{II}{:}\delta\varepsilon \tag{8.60}$$

Since the above equations provide the Jacobian or material tangent stiffness matrix $(\mathbf{D}_{ep})$, therefore expanding the terms of above equations, we get:

The last term can be put as

$$\left(K - \frac{2}{3}GR\right)\mathbf{II}\,\delta\varepsilon = \left(K - \frac{2}{3}GR\right)\mathbf{I}Tr(\delta\varepsilon) = \left(K - \frac{2}{3}GR\right)\mathbf{I}\left(\delta\varepsilon_{xx} + \delta\varepsilon_{yy} + \delta\varepsilon_{z}\right)$$
$$= \left(K - \frac{2}{3}GR\right)\begin{bmatrix} \delta\varepsilon_{xx} + \delta\varepsilon_{yy} + \delta\varepsilon_{zz} & 0 & 0 \\ 0 & \delta\varepsilon_{xx} + \delta\varepsilon_{yy} + \delta\varepsilon_{zz} & 0 \\ 0 & 0 & \delta\varepsilon_{xx} + \delta\varepsilon_{yy} + \delta\varepsilon_{z} \end{bmatrix} \tag{8.61}$$

$$\text{The second term as , } 2GR\begin{bmatrix} \delta\varepsilon_{xx} & \delta\varepsilon_{xy} & \delta\varepsilon_{xz} \\ \delta\varepsilon_{xy} & \delta\varepsilon_{yy} & \delta\varepsilon_{yz} \\ \delta\varepsilon_{xz} & \delta\varepsilon_{yz} & \delta\varepsilon_{zz} \end{bmatrix} \tag{8.62}$$

The first term:

$$\boldsymbol{\sigma}^{tr}; \delta\varepsilon = \begin{bmatrix} \sigma^{tr}_{xx} & \sigma^{tr}_{xy} & \sigma^{tr}_{xz} \\ \sigma^{tr}_{xy} & \sigma^{tr}_{yy} & \sigma^{tr}_{yz} \\ \sigma^{tr}_{yz} & \sigma^{tr}_{yz} & \sigma^{tr}_{zz} \end{bmatrix}; \begin{bmatrix} \delta\varepsilon_{zx} & \delta\varepsilon_{xy} & \delta\varepsilon_{xz} \\ \delta\varepsilon_{xy} & \delta\varepsilon_{yy} & \delta\varepsilon_{yz} \\ \delta\varepsilon_{xz} & \delta\varepsilon_{yz} & \delta\varepsilon_{zz} \end{bmatrix} \tag{8.63}$$
$$= \sigma^{tr'}_{xx}\delta\varepsilon_{xx} + \sigma^{tr'}_{yy}\delta\varepsilon_{yy} + \sigma^{tr'}_{z}\delta\varepsilon_{zz} + 2\sigma^{tr'}_{xy}\delta\varepsilon_{xy} + 2\sigma^{tr'}_{yz}\delta\varepsilon_{yz} + 2\sigma^{tr'}_{zx}\delta\varepsilon_{zx}$$

$$\delta\boldsymbol{\sigma} = 2GQ\frac{\boldsymbol{\sigma}^{tr'}}{\overline{\sigma}^{t}}\frac{1}{\overline{\sigma}^{tr}}\left(\sigma^{tr'}_{xx}\delta\varepsilon_{xx} + \sigma^{tr'}_{yy}\delta\varepsilon_{yy} + \sigma^{tr'}_{z}\delta\varepsilon_{z} + 2\sigma^{tr'}_{xy}\delta\varepsilon_{xy} + 2\sigma^{tr'}_{yz}\delta\varepsilon_{yz} + 2\sigma^{tr'}_{zx}\delta\varepsilon_{zx}\right)$$
$$= \frac{2GQ}{\overline{\sigma}^{tr^2}}\begin{bmatrix} \sigma^{tr'}_{xx} & \sigma^{tr'}_{xy} & \sigma^{tr'}_{xz} \\ \sigma^{tr'}_{xy} & \sigma^{tr'}_{yy} & \sigma^{tr}_{yz} \\ \sigma^{tr'}_{xz} & \sigma^{tr'}_{yz} & \sigma^{tr'}_{zz} \end{bmatrix}\left(\sigma^{tr'}_{xx}\delta\varepsilon_{xx} + \sigma^{tr'}_{yy}\delta\varepsilon_{yy} + \sigma^{tr'}_{zz}\delta\varepsilon_{zz} + 2\sigma^{tr'}_{xy}\delta\varepsilon_{xy} + 2\sigma^{tr'}_{yz}\delta\varepsilon_{yz} + 2\sigma^{tr'}_{zx}\delta\varepsilon_{zx}\right) \tag{8.64}$$

The material stiffness matrix $(\mathbf{D}_{ep})$ in 2-D can be written as

$$\mathbf{D}_{ep} = \begin{bmatrix} D_{11} & D_{12} & D_{13} & D_{14} \\ D_{21} & D_{22} & D_{23} & D_{24} \\ D_{31} & D_{32} & D_{33} & D_{34} \\ D_{41} & D_{42} & D_{43} & D_{44} \end{bmatrix} \tag{8.65}$$

$$\mathbf{D}_{ep} = \frac{\partial \delta \boldsymbol{\sigma}}{\partial \delta \varepsilon} = \begin{bmatrix} \frac{\partial \delta \sigma_{xx}}{\partial \delta \varepsilon_{xx}} & \frac{\partial \delta \sigma_{xx}}{\partial \delta \varepsilon_{yy}} & \frac{\partial \delta \sigma_{xx}}{\partial \delta \gamma_{xy}} & \frac{\partial \delta \sigma_{xx}}{\partial \delta \varepsilon_{z}} \\ \frac{\partial \delta \sigma_{yy}}{\partial \delta \varepsilon_{xx}} & \frac{\partial \delta \sigma_{yy}}{\partial \delta \varepsilon_{yy}} & \frac{\partial \delta \sigma_{yy}}{\partial \delta \gamma_{xy}} & \frac{\partial \delta \sigma_{yy}}{\partial \delta \varepsilon_{z}} \\ \frac{\partial \delta \sigma_{xy}}{\partial \delta \varepsilon_{xx}} & \frac{\partial \delta \sigma_{xy}}{\partial \delta \varepsilon_{yy}} & \frac{\partial \delta \sigma_{xy}}{\partial \delta \gamma_{xy}} & \frac{\partial \delta \sigma_{xy}}{\partial \delta \varepsilon_{z}} \\ \frac{\partial \delta \sigma_{z}}{\partial \delta \varepsilon_{xx}} & \frac{\partial \delta \sigma_{z}}{\partial \delta \varepsilon_{yy}} & \frac{\partial \delta \sigma_{z}}{\partial \delta \gamma_{xy}} & \frac{\partial \delta \sigma_{z}}{\partial \delta \varepsilon_{z}} \end{bmatrix} \tag{8.66}$$

Terms of $(\mathbf{D}_{ep})$ matrix can be established as

$$\begin{aligned} \delta \sigma_{xx} &= 2GQ\frac{\sigma_{xx}^{tr'}}{\overline{\sigma}^{tr^2}}\left(\sigma_{xx}^{tr'}\delta\varepsilon_{xx} + \sigma_{yy}^{tr'}\delta\varepsilon_{yy} + \sigma_{zz}^{tr'}\delta\varepsilon_{zz} + 2\sigma_{xy}^{tr'}\delta\varepsilon_{xy}\right) + 2GR\delta\varepsilon_{xx} \\ &+ \left(K - \frac{2}{3}GR\right)\left(\delta\varepsilon_{xx} + \delta\varepsilon_{yy} + \delta\varepsilon_{zz}\right) \\ \delta \sigma_{yy} &= 2GQ\frac{\sigma_{yy}^{tr'}}{\overline{\sigma}^{tr^2}}\left(\sigma_{xx}^{tr'}\delta\varepsilon_{xx} + \sigma_{yy}^{tr'}\delta\varepsilon_{yy} + \sigma_{z}^{tr'}\delta\varepsilon_{z} + 2\sigma_{xy}^{tr'}\delta\varepsilon_{xy}\right) + 2GR\delta\varepsilon_{yy} \\ &+ \left(K - \frac{2}{3}GR\right)\left(\delta\varepsilon_{xx} + \delta\varepsilon_{yy} + \delta\varepsilon_{zz}\right) \\ \delta \sigma_{xy} &= 2GQ\frac{\sigma_{xy}^{tr'}}{\overline{\sigma}^{tr^2}}\left(\sigma_{xx}^{tr'}\delta\varepsilon_{xx} + \sigma_{yy}^{tr'}\delta\varepsilon_{yy} + \sigma_{z}^{tr'}\delta\varepsilon_{zz} + 2\sigma_{xy}^{tr'}\delta\varepsilon_{xy}\right) + 2GR\delta\varepsilon_{xy} \\ \delta \sigma_{zz} &= 2GQ\frac{\sigma_{zz}^{tr'}}{\overline{\sigma}^{tr^2}}\left(\sigma_{xx}^{tr}\delta\varepsilon_{xx} + \sigma_{yy}^{tr'}\delta\varepsilon_{yy} + \sigma_{z}^{tr'}\delta\varepsilon_{z} + 2\sigma_{xy}^{tr'}\delta\varepsilon_{xy}\right) + 2GR\delta\varepsilon_{zz} \\ &+ \left(K - \frac{2}{3}GR\right)(\delta\varepsilon_{xx} + \delta\varepsilon_{yy} + \delta\varepsilon_{zz}) \end{aligned} \tag{8.67}$$

Since $\gamma_{xy} = 2\varepsilon_{xy}$ also $A = \frac{2GQ}{\overline{\sigma}^{tr2}}$ and $B = K - \frac{2}{3}GR$, therefore we can write various terms of $(\mathbf{D}_{ep})$ as

$$\begin{aligned} &D_{11} = A\sigma_{xx}^{tr'}\sigma_{xx}^{tr'} + 2GR + B, D_{12} = A\sigma_{xx}^{tr'}\sigma_{yy}^{tr'} + B, D_{13} = A\sigma_{xx}^{tr'}\sigma_{xy}^{tr'}, D_{14} = A\sigma_{xx}^{tr'}\sigma_{xz}^{tr'} + B \\ &D_{21} = A\sigma_{yy}^{tr'}\sigma_{xx}^{tr'} + B, D_{22} = A\sigma_{yy}^{tr'}\sigma_{yy}^{tr'} + 2GR + B, D_{23} = A\sigma_{yy}^{tr'}\ \sigma_{xy}^{tr'}, D_{24} = A\sigma_{yy}^{tr'}\sigma_{zz}^{tr'} + B \\ &D_{31} = A\sigma_{xy}^{tr'}\sigma_{xx}^{tr'}, D_{32} = A\sigma_{xy}^{tr'}\sigma_{yy}^{tr'}, D_{33} = A\sigma_{xy}^{tr'}\sigma_{xy}^{tr'} + 2GR, D_{34} = A\sigma_{xy}^{tr'}\sigma_{zz}^{tr'} \\ &D_{41} = A\sigma_{zz}^{tr'}\sigma_{xx}^{tr'} + B, D_{42} = A\sigma_{zz}^{tr'}\sigma_{yy}^{tr'} + B, D_{43} = A\sigma_{zz}^{tr'}\sigma_{xy}^{\theta'}, D_{44} = A\sigma_{zz}^{tr'}\sigma_{zz}^{tr'} + 2GR + B \end{aligned} \tag{8.68}$$

8.4.3 Plane stress plasticity

The strain components are prescribed in cases of plane strain and axisymmetric problems, whereas the in-plane and out-of-plane stress components are not known [45]. Eliminating the appropriate strain components from the formulation helps to obtain the particularization of 3-D plasticity for states of constrained strain. The not-vanishing stress component will be determined as an outcome of the numerical integration algorithm. Certain stress components are constrained to be zero in cases of plane stress problems. During elastoplastic problems, the treatment of plane stress constraints requires further evaluation, although it is trivial during elastic conditions. The same numerical integration algorithm as used for plane strain/axisymmetric problems can be utilized for elastoplastic problems after certain some modifications. The problem can be evaluated by three general approaches:

1. one of the methods is to include the plane stress constraint directly into the three-dimensional elastic predictor plastic corrector algorithm equations applied at the Gauss point level. Nested Newton return mapping iteration for plane stress enforcement can be utilized to implement the method.
2. with the plane stress condition added as a structural constraint at the global structural level, use of standard three-dimensional return mapping at the Gauss point level.
3. use of plane stress projected constitutive equations, where similarly to procedure of item (a), the plane stress constraint is enforced at the Gauss point level.

8.4.3.1 *Plane stress-projected plasticity model*

1. This plasticity model is defined by a set of evolution equations that
2. include the simultaneous involvement of in-plane stress and strain components.
3. are equivalent to the three-dimensional model with the added plane stress constraint.
4. In this case, the in-plane stress–strain components characterize the primary variables, which are evaluated by solving the evolution equations. The out-of-plane strains are dependent variables, determined as functions of in-plane variables used to solve the plane stress-projected model equations.

Unless specified, all the quantities are taken at the end of a time step $t+\Delta t$. So $\boldsymbol{\sigma}$ denotes the stress at time $t+\Delta t$, and $\boldsymbol{\sigma}_t$ is the stress at the beginning of the time step (at the end of time step t). Let $\mathbf{P}$ be a matrix such that, $\mathbf{P}=\frac{1}{3}\begin{bmatrix} 2 & -1 & 0 \\ -1 & 2 & 0 \\ 0 & 0 & 6 \end{bmatrix}$ the stress deviator $\mathbf{S}$ is given by

$$\mathbf{s}^{\mathrm{T}}=\left[\sigma'_{xx},\sigma'_{yy},2\sigma_{xy}\right]=\mathbf{P}\boldsymbol{\sigma} \tag{8.69}$$

Under isotropic hardening conditions, the von Mises model in terms of in-plane stress–strain components can be defined as

$$\dot{\varepsilon}=\dot{\varepsilon}^{\mathrm{e}}+\dot{\varepsilon}^{\mathrm{p}}$$

$$\boldsymbol{\sigma}=\mathbf{D}\varepsilon^{\mathrm{e}}$$

$$F=\sqrt{\frac{3}{2}\boldsymbol{\sigma}^{T}\mathbf{P}\boldsymbol{\sigma}}-\sigma_{Y}(\bar{\varepsilon}^{p}) \tag{8.70}$$

$$\varepsilon^{p}=\dot{\lambda}\frac{\partial F}{\partial \boldsymbol{\sigma}}=\dot{\lambda}\sqrt{\frac{3}{2}}\frac{\mathbf{P}\boldsymbol{\sigma}}{\sqrt{\boldsymbol{\sigma}^{T}\mathbf{P}\boldsymbol{\sigma}}} \tag{8.71}$$

$$\bar{\varepsilon}^{p}=\dot{\lambda}$$

$$\dot{\lambda}\geq 0,\quad F\leq 0,\quad \dot{\lambda}F=0$$

It is more convenient to use the squared form of the yield function for the derivation of integration algorithm. Therefore the evolution equations can be updated as

$$F=\frac{1}{2}\boldsymbol{\sigma}^{T}\mathbf{P}\boldsymbol{\sigma}-\frac{1}{3}\sigma_{Y}^{2}(\bar{\varepsilon}^{p}) \tag{8.72}$$

$$\dot{\varepsilon}^{p}=\dot{\lambda}\frac{\partial F}{\partial \boldsymbol{\sigma}}=\dot{\lambda}\mathbf{P}\boldsymbol{\sigma} \tag{8.73}$$

$$\dot{\bar{\varepsilon}}^{p}=\dot{\lambda}\sqrt{\frac{2}{3}\boldsymbol{\sigma}^{T}\mathbf{P}\boldsymbol{\sigma}} \tag{8.74}$$

8.4.3.2 Plane stress-projected integration plasticity model

This algorithm proceeds with the computation of in-plane trial elastic strain–stress components as

$$\varepsilon^{etr}=\varepsilon_{t}^{e}+\Delta\varepsilon \tag{8.75}$$

$$\boldsymbol{\sigma}^{tr}=\mathbf{D}\varepsilon^{etr} \tag{8.76}$$

$$\bar{\varepsilon}^{p\ tr}=\bar{\varepsilon}_{t}^{p} \tag{8.77}$$

The next step is to compute trial yield function so as to check for plastic admissibility of trial state as

$$F^{tr}=\frac{1}{2}\left(\boldsymbol{\sigma}^{tr}\right)^{T}\mathbf{P}\boldsymbol{\sigma}^{tr}-\frac{1}{3}\sigma_{Y}^{2}\left(\bar{\varepsilon}^{p\ tr}\right) \tag{8.78}$$

If $F^{tr}\leq 0$ means elastic trial state is admissible, that is, trial quantities are treated as the final quantities.

Else, the return mapping algorithm needs to be utilized as depicted below.

8.4.3.3 The plane stress-projected return mapping

The implicit return mapping involves in solving the below set of equations for $\varepsilon^{e},\bar{\varepsilon}^{p}$ and $\Delta\lambda$, where $\boldsymbol{\sigma}$ is expresssed by elastic law as a function of ε^{e}:

$$\varepsilon^{e}=\boldsymbol{\varepsilon}^{etr}-\Delta\lambda\mathbf{P}\boldsymbol{\sigma} \tag{8.79}$$

$$\bar{\varepsilon}^{\mathrm{P}} = \bar{\varepsilon}_{\mathrm{t}}^{\mathrm{P}} + \Delta\lambda\sqrt{\frac{2}{3}\boldsymbol{\sigma}^{\mathrm{T}}\mathbf{P}\boldsymbol{\sigma}} \tag{8.80}$$

$$\frac{1}{2}\boldsymbol{\sigma}^{\mathrm{T}}\mathbf{P}\boldsymbol{\sigma} - \frac{1}{3}\sigma_{\mathrm{Y}}^{2}(\bar{\varepsilon}^{\mathrm{P}}) = 0 \tag{8.81}$$

It is convenient to reduce the above five variable mapping system into a single scalar nonlinear equation with incremental plastic multiplier as unknown. After further evaluation of the above equations and utilizing the inverse elastic law, we get

$$\begin{gathered}\boldsymbol{\sigma} = [\mathrm{C} + \Delta\lambda\mathbf{P}]^{-1}\mathbf{C}\boldsymbol{\sigma}^{\mathrm{tr}} \\ \frac{1}{2}\boldsymbol{\sigma}^{\mathrm{T}}\mathbf{P}\boldsymbol{\sigma} - \frac{1}{3}\sigma_{\mathrm{Y}}^{2}\left(\bar{\varepsilon}_{\mathrm{t}}^{\mathrm{P}} + \Delta\lambda\sqrt{\frac{2}{3}\boldsymbol{\sigma}^{\mathrm{T}}\mathbf{P}}\right) = 0\end{gathered} \tag{8.82}$$

where $\boldsymbol{\sigma}$ is the stress and $\Delta\lambda$ is plastic multiplier are the unknowns, and $\mathbf{C} = \mathbf{D}^{-1}$ represents the inverse elastic matrix. After further manipulations and substitutions, the return mapping for plane stress von Mises model is obtained as

$$\tilde{\mathrm{F}}(\Delta\lambda) = \frac{1}{2}\xi(\Delta\lambda) - \frac{1}{3}\sigma_{\mathrm{Y}}^{2}(\bar{\varepsilon}^{\mathrm{P}}) = 0 \tag{8.83}$$

where $\Delta\lambda$ is the only unknown of this scalar consistency nonlinear equation. Also $\xi(\Delta\lambda)$ is described as

$$\xi(\Delta\lambda) = (\boldsymbol{\sigma}^{tr})^{\mathrm{T}}\mathbf{A}^{\mathrm{T}}(\Delta\lambda)\mathbf{P}\mathbf{A}(\Delta\lambda)\boldsymbol{\sigma}^{\mathrm{tr}} \text{ with } \mathbf{A}(\Delta\lambda) = [\mathbf{C} + \Delta\lambda\mathbf{P}]^{-1}\mathbf{C} \tag{8.84}$$

Therefore in return mapping for the plane stress von Mises model, we solve the consistency equation using the Newton–Raphson algorithm. After computing $\Delta\lambda$, the other variables are obtained as

$$\begin{gathered}\boldsymbol{\sigma} = \mathbf{A}(\Delta\lambda)\boldsymbol{\sigma}^{\mathrm{tr}} \\ \varepsilon^{\mathrm{e}} = \mathbf{C}\boldsymbol{\sigma} \\ \bar{\varepsilon}^{\mathrm{P}} = \bar{\varepsilon}_{\mathrm{t}}^{\mathrm{P}} + \Delta\lambda\sqrt{\frac{2}{3}\xi(\Delta\lambda)}\end{gathered} \tag{8.85}$$

Also in-plane plastic strains are obtained as

$$\varepsilon^{\mathrm{P}} = \varepsilon_{\mathrm{t}}^{\mathrm{P}} + \Delta\lambda\mathbf{P}\gamma \tag{8.86}$$

8.4.3.4 The elastoplastic consistent tangent operator for plane stress plasticity

We can define the elastoplastic consistent tangent matrix as

$$\mathbf{D}_{\mathrm{ep}} = \frac{\mathrm{d}\boldsymbol{\sigma}}{\mathrm{d}\varepsilon} = \frac{\mathrm{d}\boldsymbol{\sigma}}{\mathrm{d}\varepsilon^{\mathrm{tr}}} \tag{8.87}$$

where $\boldsymbol{\sigma}$ is obtained from the plane stress-projected return mapping algorithm. The expression for $\mathbf{D}_{\mathrm{ep}}$, can be determined by differentiating $\varepsilon^{\mathrm{e}} = \varepsilon^{\mathrm{e\ tr}} - \Delta\lambda\mathbf{P}\boldsymbol{\sigma}$ considering elastic law as

$$d\boldsymbol{\sigma} = \mathbf{E}[\mathrm{d}\boldsymbol{\varepsilon}^{tr} - \mathrm{d}\Delta\lambda\mathbf{P}\boldsymbol{\sigma}];\ \text{here, } \mathbf{E} = [\mathrm{C} + \Delta\lambda\mathbf{P}]^{-1} \tag{8.88}$$

On differentiating plastic consistency equation $\left(\tilde{F}(\Delta\lambda) = \frac{1}{2}\xi(\Delta\lambda) - \frac{1}{3}\sigma_Y^2(\bar{\varepsilon}^p) = 0\right)$, we get

$$\begin{aligned} d\tilde{F} &= \frac{1}{2}\xi - \frac{2}{3}\sigma_Y H' \sqrt{\frac{2}{3}}\left(d\Delta\lambda\sqrt{\xi} + \frac{\Delta\lambda}{2\sqrt{\xi}}d\xi\right) \\ &= \frac{1}{2}\xi - \frac{2}{3}H'\left(\xi d\Delta\lambda + \frac{1}{2}\Delta\lambda d\xi\right) = 0 \end{aligned} \tag{8.89}$$

where $\sigma_Y = \sqrt{\frac{3}{2}\xi}$, which is valid during under plastic flow has been utilized. We can solve above equation for $d\Delta\lambda$ as,

$$d\Delta\lambda = \frac{3}{4H'\xi}\left(1 - \frac{2}{3}H'\Delta\lambda\right)d\xi \tag{8.90}$$

By using the elastic law, we can be equivalently write matrix **E** and the scalar ξ (defined previously) as

$$\xi = \left(\varepsilon^{tr}\right)^T \mathbf{EPE}\varepsilon^{tr}$$

On differentiating the above equation, we get

$$\begin{aligned} d\xi &= 2\left(\varepsilon^{tr}\right)^T \mathbf{EPE} d\varepsilon^{tr} + 2\left(\varepsilon^{tr}\right)^T d\mathbf{EPE}\varepsilon^{tr} \\ &= 2\left(\boldsymbol{\sigma}^T \mathbf{PE} d\varepsilon^{tr} - \boldsymbol{\sigma}^T \mathbf{PEPE} d\Delta\lambda\right) \end{aligned}$$

On substituting the value of $d\xi$ into $d\Delta\lambda$ and then again substituting that expression into $d\boldsymbol{\sigma} = \mathbf{E}\left[d\varepsilon^{tr} - d\Delta\lambda \mathbf{P}\boldsymbol{\sigma}\right]$ and then finally after mathematical manipulations, we obtain

$$d\boldsymbol{\sigma} = [\mathbf{E} - \alpha(\mathbf{EP}\boldsymbol{\sigma}) \otimes (\mathbf{EP}\boldsymbol{\sigma})]d\varepsilon^{tr} \tag{8.91}$$

where α is a scalar quantity written as $\alpha = \frac{1}{\boldsymbol{\sigma}^T \mathbf{PEP} + \frac{2\xi H'}{3 - 2H'\Delta\lambda}}$

Finally, the elastoplastic tangent operator explicit equation in matrix form consistent with the von Mises plane stress-projected return mapping is expressed as

$$\mathbf{D}_{ep} = \mathbf{E} - \alpha(\mathbf{EP}) \otimes (\mathbf{EP}) \tag{8.92}$$

8.5 Investigation of fatigue crack growth

8.5.1 Computation of J-integral

In the present work, the crack growth problems are simulated using plane stress conditions. To simulate the elastoplastic material response, the Ramberg–Osgood model has been employed [39]:

$$\frac{\varepsilon}{\varepsilon_y} = \frac{\sigma}{\sigma_y} + \alpha\left(\frac{\sigma}{\sigma_y}\right)^n \tag{8.93}$$

where σ_y represents the initial yield stress, and ε_y is the total strain at initial yield stress of the material. Also α denotes the strain hardening coefficient, and n is the strain hardening exponent. In the presence of plasticity, the crack tip singularity in stress field is modified to the Hutchinson–Rice–Rosengren (HRR). With HRR singularity, the displacement field can be written as

$$\boldsymbol{u} = a\varepsilon_y \left(\frac{J}{a\varepsilon_y \sigma_y I_n r} \right)^{\frac{n}{n+1}} \overline{u}(\theta, n) \tag{8.94}$$

where J denotes the J-integral, I_n represents a dimensionless parameter depends on n, and $\overline{u}$ denotes a dimensionless function of θ, n. To determine the crack growth direction during the fatigue crack propagation, it requires to evaluate the stress intensity factors for mode-I and mode-II conditions. For the determination of SIF's, a path-independent integral called J-integral is used, which can calculate the rate of change of potential energy with respect to crack length. The contour form of J-integral can be expressed as

$$J = \int_{\Omega} \left[W dx_2 - \sigma_{ij} n_j \frac{\partial u_i}{\partial x_i} \right] ds$$

Using divergence theorem, the contour integral can be represented in interaction or domain integral [48] form as

$$J = \int_A \left[\sigma_{ij} \frac{\partial u_i}{\partial x_i} - w\delta_{1j} \right] \frac{\delta q}{\delta x_j} dA \tag{8.95}$$

where $W = \int_0^{\varepsilon_{ij}} \sigma_{ij} \varepsilon_{ij}$ represents the strain energy density, σ_{ij}, u_i, and δ_{1j} denote the stress tensor, displacement field vector, and Kronecker's delta, respectively. A is the area of both top and bottom crack faces, and q is function whose value is zero along the outer boundary of A and unity along the inner boundary of A. For a cracked body with defined boundary conditions, consider an independent equilibrium state or state 1 as actual state and state 2 as the auxiliary state. In the case of state 2 or auxiliary state, the HRR singularity is utilized instead of $\sqrt{r}$ singularity of the analytical linear elastic solution. For these superposed states, J-integral can be written as

$$\begin{aligned} J^{tot} &= \int_A \left[\sigma_{ij}^{(1)} \frac{\partial u_i^{(1)}}{\partial x_1} - W^{(1)} \delta_{1j} \right] \frac{\partial q}{\partial x_j} \mathrm{dA} + \int_A \left[\sigma_{ij}^{(2)} \frac{\partial u_l^{(2)}}{\partial x_1} - W^{(2)} \delta_{1j} \right] \frac{\partial q}{\partial x_j} \mathrm{dA} \\ &+ \int_A \left[\sigma_{ij}^{(1)} \frac{\partial u_l^{(2)}}{\partial x_1} + \sigma_{ij}^{(2)} \frac{\partial u_l^{(1)}}{\partial x_1} - W^{(1,2)} \delta_{1j} \right] \frac{\partial q}{\partial x_j} \mathrm{dA} \\ &= J^{(1)} + J^{(2)} + M^{(1,2)} \end{aligned} \tag{8.96}$$

$M^{(1,2)}$ can be obtained by employing Irwin's relation for the superposed states as [49,50]

$$M^{(1,2)} = \frac{2}{E^*} \left(K_I^{(1)} K_I^{(2)} + K_{II}^{(1)} K_{II}^{(2)} \right) \tag{8.97}$$

where $E^* = E/2$ and $E^* = \frac{E}{(1-\upsilon^2)}$ for plane stress and plane strain conditions. Mode-I and mode-II SIF's can be determined easily by appropriate utilization of auxiliary state.

8.5.2 Evaluation of fatigue life

In reality, the crack grows along a curved path. In this work, it has been assumed that there is a successive linear increment of the crack propagation [51]. The crack increment size is fixed for each load step. Maximum principal stress criterion is utilized to predict the direction of crack growth, which states that the crack propagates in the direction perpendicular to the maximum principal stress. Thus the crack growth direction θ_c at crack tip is obtained by imposing the local shear stress should equal to zero, that is,

$$K_I \sin\theta_c + K_{II}(3\cos\theta_c - 1) = 0 \tag{8.98}$$

On solving, we get

$$\theta_c = 2(\tan)^{-1}\left(\frac{K_I - \sqrt{K_I{}^2 + 8K_{II}{}^2}}{4K_{II}}\right) \tag{8.99}$$

In the case of pure mode-I, the crack growth angle θ_c is zero. If $K_{II} > 0$, then $\theta_c < 0$, and vice versa. Using this criterion, the equivalent mode-I SIF can be written as

$$K_{Ieq} = K_I \cos^3\frac{\theta_c}{2} - 3K_{II}\cos^2\frac{\theta_c}{2}\sin\frac{\theta_c}{2} \tag{8.100}$$

The residual compressive stresses arise along the crack faces at minimum loads during cyclic loading, giving rise to crack closure phenomena in the vicinity of crack tip. Hence, the opening SIF is given as

$$K_{open} = (K_{Ieq})_{\min} - K_{res} \tag{8.101}$$

The crack-driving SIF, also called as effective SIF (ΔK_{Ieq}), gets reduced due to the crack closure phenomenon. For linear analysis, effective SIF can be determined as

$$\Delta K_{eff} = (K_{Ieq})_{\max} - (K_{Ieq})_{\min} \tag{8.102}$$

whereas in the case of elastoplastic conditions, the effective SIF is obtained as

$$\Delta K_{eff} = (K_{Ieq})_{\max} - (K_{Ieq})_{\min} \tag{8.103}$$

where $(K_{Ieq})_{\max}$ and $(K_{Ieq})_{\min}$ denote the equivalent SIF obtained at maximum and minimum loads.

Generalized Paris law is employed to determine the fatigue life of the cracked component. Paris law can be expressed as

$$\frac{da}{dN} = C(\Delta K_{eff})^m \tag{8.104}$$

where "a" denotes the crack length, "N" is the number of loading cycles, and "C" and "m" are the material properties or Paris law constants. The cracked specimen is said to be failed when $(K_{Ieq})_{\max} \geq K_{IC}$, where K_{Ieq} is the equivalent mode-I SIF and K_{IC} represents the fracture toughness or critical stress intensity factor of the material.

8.6 Numerical results and discussions

This section pertains to the modeling and simulation of elastoplastic problems utilizing the XFEM and the EFGM. Generalized Paris law has been employed for the assessment of the fatigue life of cracked components. Indigenous codes on EFGM have been developed in MATLAB to solve different elastoplastic crack growth problems, and the validation of the codes has been established against results already available for some benchmark problems. A rectangular plate (100mm × 200mm) containing an edge crack of initial length 15mm is considered for simulation. At the top edge of the plate, a cyclic tensile load of $\sigma_{\max} = 60\text{N/mm}$ and $\sigma_{\min} = 0\text{N/mm}$ is applied, while the bottom edge of the plate remains fixed as depicted in Fig. 8.5.

The nonlinear elastoplastic equations have been solved by employing the Newton–Raphson iteration scheme. The modeling of finite strain plasticity has been obtained through von Mises yield criteria utilized in conjunction with isotropic hardening. The Ramberg–Osgood model has been utilized to determine the elastoplastic response of the material. In the present work, plasticity algorithms such as elastic predictor and plastic corrector and radial return methods have been utilized for stress computation during the elastoplastic crack growth. The material is assumed to have a fracture toughness of (K_{IC}) of $2403\text{MPa}\sqrt{\text{mm}}$. The Paris exponent ($m$), Paris constant ($C$), and Poisson's ratio ($\mu$) of the

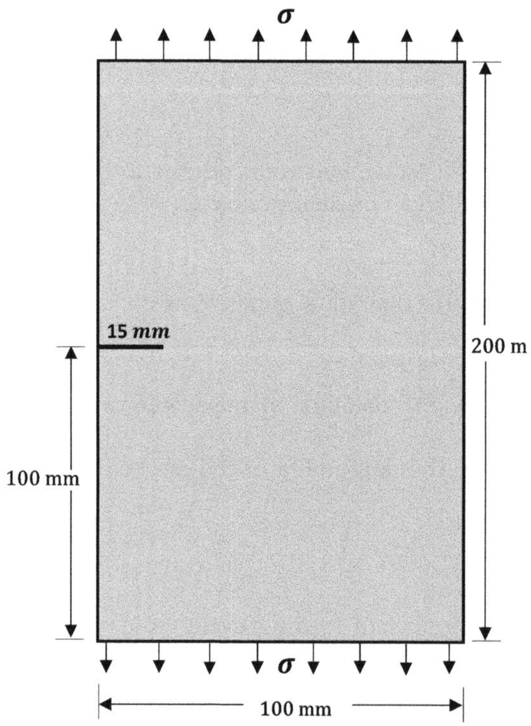

FIGURE 8.5 Edge-cracked plate with dimensions.

material are 3.32, 2.087136×10^{-13}, and 0.3. During the fatigue crack growth, a crack increment (Δa) of 2mm is given after each cycle of load until failure occurs in the plate, that is, when $K_{Ieq} > K_{IC}$. Finally, several elastoplastic crack growth problems have been solved using the XFEM and EFGM codes developed in MATLAB. Several steel alloys whose properties are depicted in Table 8.1 have been selected to determine the elastoplastic fatigue crack growth. Figs. 8.6 and 8.7 depict the domain representation of initial and final crack length by XFEM and EFGM. Fig. 8.8 denotes the fatigue life diagrams, and Fig. 8.9 represents the variation of J-integral with crack length during elastoplastic crack growth for various steel alloys mentioned in Table 8.1 by XFEM using the elastic predictor plastic corrector algorithm and radial return algorithms. Similarly, Fig. 8.10 represents the fatigue life diagrams, and Fig. 8.11 shows the variation of J-integral with crack length for various steel alloys mentioned in Table 8.1 by EFGM using the elastic predictor plastic corrector algorithm and radial return algorithms for stress computation. The comparison of fatigue life and J-integral in the case of LEFM and EPFM for alloy AISI-205 by XFEM is

TABLE 8.1 Properties of steel alloys utilized during simulation.

Material	Young's modulus (GPa)	Yield stress (GPa)	Tensile strength (GPa)	Elongation
SS-316 LN	210	230	580	0.46
AISI-303	200	240	600	0.52
AISI-201	200	310	660	0.46
AISI-205	200	460	810	0.46

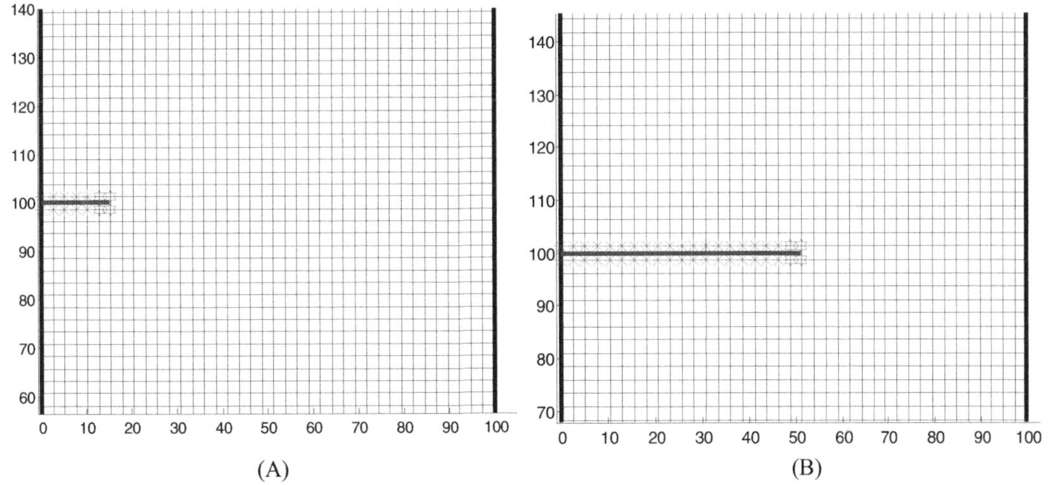

FIGURE 8.6 XFEM domain representation (zoomed view): (A) initial crack length and (B) final crack length. *XFEM*, Extended finite element method.

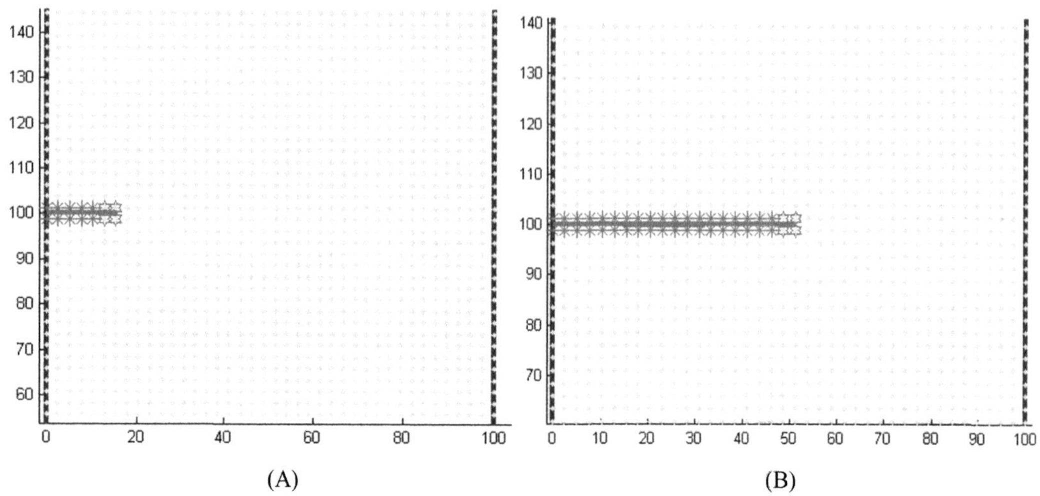

FIGURE 8.7 EFGM domain representation (zoomed view): (A) initial crack length and (B) final crack length. *EFGM*, Element-free Galerkin method.

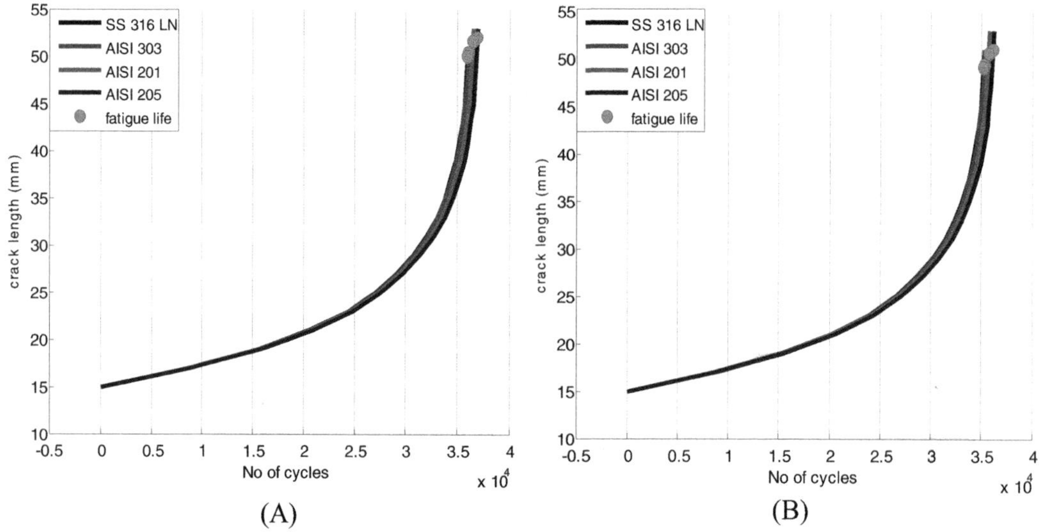

FIGURE 8.8 Fatigue life diagram for different steel alloys by XFEM: (A) elastic predictor plastic corrector algorithm and (b) radial return algorithm. *XFEM*, Extended finite element method.

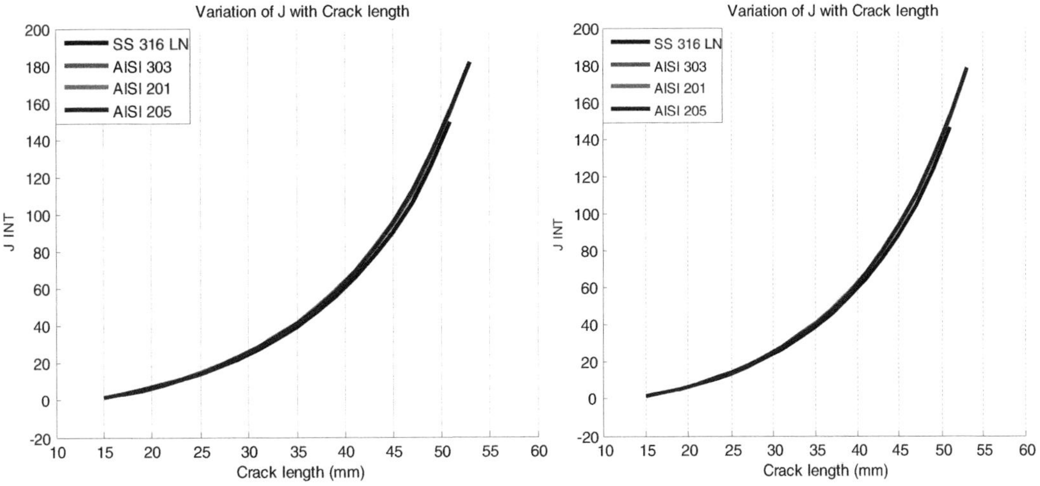

(A) (B)

FIGURE 8.9 Variation of J-integral with crack length for different steel alloys by XFEM: (A) elastic predictor plastic corrector algorithm and (b) radial return algorithm. *XFEM*, Extended finite element method.

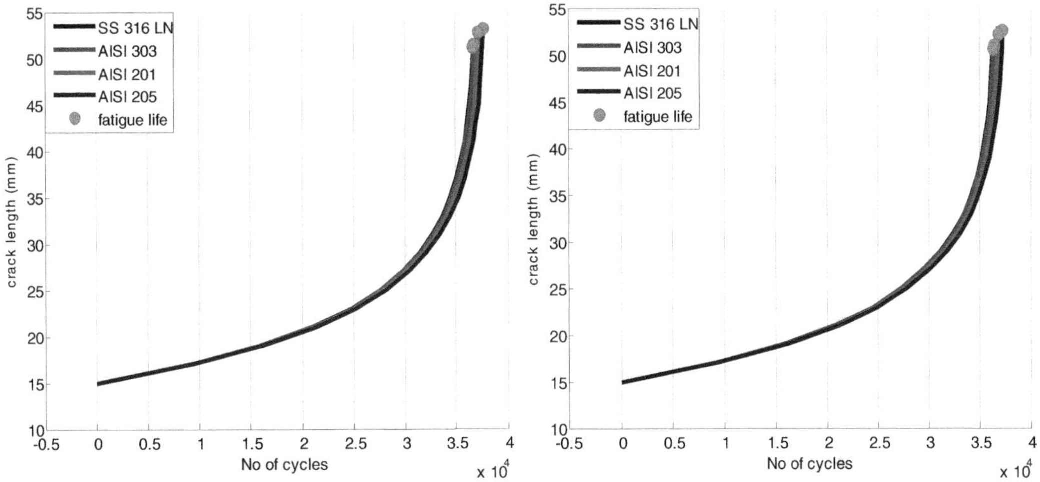

(A) (B)

FIGURE 8.10 Fatigue life diagram for different steel alloys by EFGM: (A) elastic predictor plastic corrector algorithm and (B) radial return algorithm. *EFGM*, Element-free Galerkin method.

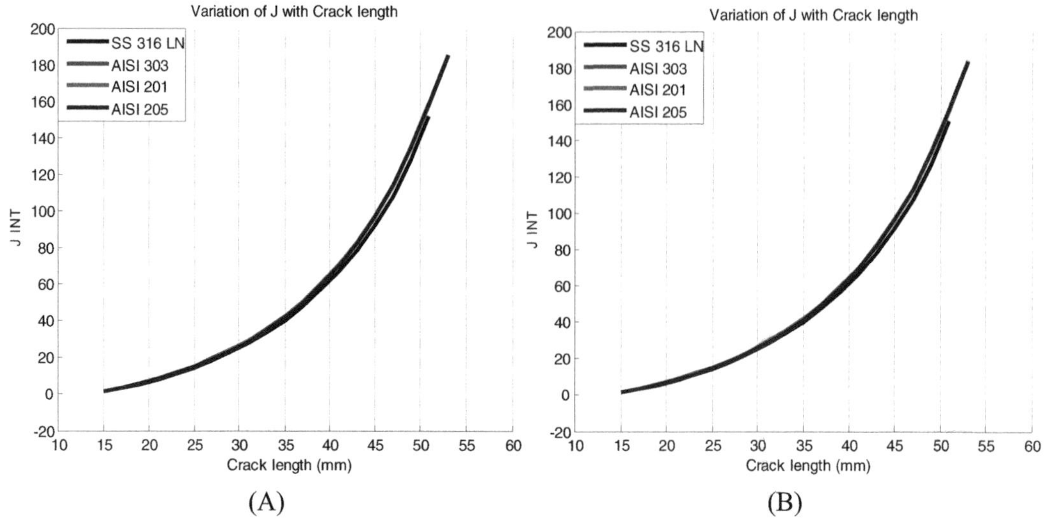

FIGURE 8.11 Variation of J-integral with crack length for different steel alloys by EFGM: (A) elastic predictor plastic corrector algorithm and (B) radial return algorithm. *EFGM*, Element-free Galerkin method.

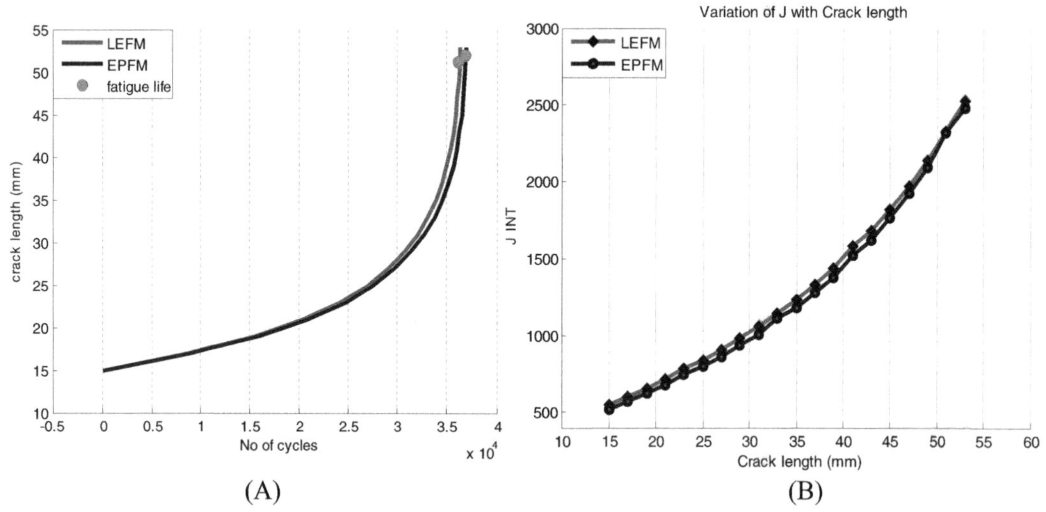

FIGURE 8.12 Comparison between LEFM and EFGM: (A) fatigue life and (B) J-integral variation. *EFGM*, Element-free Galerkin method; *LEFM*, linear elastic fracture mechanics.

represented in Fig. 8.12. The fatigue life in the case of LEFM is found as 36110 cycles, whereas in the case of EPFM, the fatigue life is found as 36878 cycles. Thus the fatigue life of the component varies by 2.08% in the case of EPFM analysis. This increase in fatigue life is attributed to the presence of residual stresses developed during unloading.

8.7 Conclusion

The present study models elastoplastic crack propagation occurring in engineering specimens. The stress computations during elastoplastic crack growth were evaluated by invoking the elastic predictor plastic corrector algorithm and radial return algorithms. Several problems of steel alloys were considered for the sake of analysis. The enriched EFGM allows to model the discontinuity accurately and efficiently independent of the mesh. A remarkable agreement was achieved between the results obtained by the XFEM and EFGM. It was concluded that these enriched techniques can model elastoplastic crack propagation very accurately and efficiently. It was also observed that the results obtained by considering the elastoplastic effects in the domain were very close to the linear elastic results as long as the size of the crack tip plastic zone remains small. If the plastic zone developed ahead of the crack tip is very large, then the linear elastic principles will not be able to govern the behavior of the cracked body accurately.

References

[1] Andersson HA. A finite-element representation of stable crack growth. Journal of the Mechanics and Physics of Solids 1973;21:337–56.

[2] Jameel A, Harmain GA. Fatigue crack growth in presence of material discontinuities by EFGM. International Journal of Fatigue 2015;81:105–16. Available from: https://doi.org/10.1016/j.ijfatigue.2015.07.021.

[3] Jameel A, Harmain GA. Modeling and numerical simulation of fatigue crack growth in cracked specimens containing material discontinuities. Strength of Materials 2016;48(2):294–307. Available from: https://doi.org/10.1007/s11223-016-9765-0.

[4] Lam PS, Kim Y, Chao YJ. The non-constant CTOD/CTOA in stable crack extension under plane-strain conditions. Engineering Fracture Mechanics 2006;73:1070–85. Available from: https://doi.org/10.1016/J.ENGFRACMECH.2005.12.008.

[5] Rice JR. A path independent integral and the approximate analysis of strain concentration by notches and cracks. Journal of Applied Mechanics ASME 1964;35:379–88. Available from: https://doi.org/10.1115/1.3601206.

[6] Lone AS, Jameel A, Harmain GA. Enriched element free Galerkin method for solving frictional contact between solid bodies. Mechanics of Advanced Materials and Structures 2022. Available from: https://doi.org/10.1080/15376494.2022.2092791.

[7] Jameel A, Harmain GA. Extended iso-geometric analysis for modeling three-dimensional cracks. Mechanics of Advanced Materials and Structures 2019;26:915–23. Available from: https://doi.org/10.1080/15376494.2018.1430275.

[8] Lone AS, Jameel A, Harmain GA. "Modelling of contact interfaces by penalty based enriched finite element method. Mechanics of Advanced Materials and Structures 2022. Available from: https://doi.org/10.1080/15376494.2022.2034075.

[9] Singh AK, Jameel A, Harmain GA. Investigations on crack tip plastic zones by the extended iso-geometric analysis. Materials Today: Proceedings 2018;5:19284–93. Available from: https://doi.org/10.1016/j.matpr.2018.06.287.

[10] Xiaozhou X, Qing Z, Hong W, Qun J. The numerical simulation of the crack elastoplastic extension based on the extended finite element method. Mathematical Problems in Engineering 2013;2013. Available from: https://doi.org/10.1155/2013/157130.

[11] Lone AS, Kanth SA, Jameel A, Harmain GA. A state of art review on the modeling of contact type nonlinearities by extended finite element method. Materials Today: Proceedings 2019;18:3462–71. Available from: https://doi.org/10.1016/j.matpr.2019.07.274.

[12] Ashraf QJ, Prasad Reddy GV, Sandhya R, Laha K, Harmain GA. Simulation of low cycle fatigue stress-strain response in 316LN stainless steel using non-linear isotropic kinematic hardening model—a comparison of

different approaches. Fatigue & Fracture of Engineering Materials & Structures 2018;41:336–47. Available from: https://doi.org/10.1111/ffe.12683.

[13] Niezgoda T, Derewońko A. Multiscale composite FEM modeling. Procedia Engineering 2009;1:209–12. Available from: https://doi.org/10.1016/j.proeng.2009.06.049.

[14] Lee WCC, Zhang M, Jia X, Cheung JTM. Finite element modeling of the contact interface between trans-tibial residual limb and prosthetic socket. Medical Engineering & Physics 2004;26:655–62. Available from: https://doi.org/10.1016/j.medengphy.2004.04.010.

[15] Partheepan G, Sehgal DK, Pandey RK. Finite element application to estimate in-service material properties using dumb-bell miniature specimen. AES-ATEMA Int Conf Ser - Adv Trends Eng Mater Their Appl 2007;369–77.

[16] Gupta V, Jameel A, Verma SK, Anand S, Anand Y. An insight on NURBS based isogeometric Analysis, its current status and involvement in mechanical applications. Archives of Computational Methods in Engineering 2022;. Available from: https://doi.org/10.1007/s11831-022-09838-0.

[17] Wen PH, Aliabadi MH, Young A. Dual boundary element methods for three-dimensional dynamic crack problems. Journal of Strain Analysis for Engineering Design 1999;34:373–94. Available from: https://doi.org/10.1177/030932479903400601.

[18] Yan X. A boundary element modeling of fatigue crack growth in a plane elastic plate. Mechanics Research Communications 2006;33:470–81. Available from: https://doi.org/10.1016/j.mechrescom.2005.06.006.

[19] Cisilino AP, Ortiz J. Boundary element analysis of three-dimensional mixed-mode cracks via the interaction integral. Computer Methods in Applied Mechanics and Engineering 2005;194:935–56. Available from: https://doi.org/10.1016/j.cma.2003.08.014.

[20] Rabczuk T, Zi G, Bordas S, Nguyen-Xuan H. A simple and robust three-dimensional cracking-particle method without enrichment. Computer Methods in Applied Mechanics and Engineering 2010;199:2437–55. Available from: https://doi.org/10.1016/j.cma.2010.03.031.

[21] Bordas SPA, Rabczuk T, Hung NX, Nguyen VP, Natarajan S, Bog T, et al. Strain smoothing in FEM and XFEM. Computers & Structures 2010;88:1419–43. Available from: https://doi.org/10.1016/j.compstruc.2008.07.006.

[22] Jameel A, Harmain GA. Effect of material irregularities on fatigue crack growth by enriched techniques. International Journal for Computational Methods in Engineering Science and Mechanics 2020;21:109–33. Available from: https://doi.org/10.1080/15502287.2020.1772902.

[23] Kanth SA, Harmain GA, Jameel A. Modeling of nonlinear crack growth in steel and aluminum alloys by the element free Galerkin method. Materials Today: Proceedings 2018;5:18805–14. Available from: https://doi.org/10.1016/j.matpr.2018.06.227.

[24] Lone AS, Jameel A, Harmain GA. A coupled finite element-element free Galerkin approach for modeling frictional contact in engineering components. Materials Today: Proceedings 2018;5:18745–54. Available from: https://doi.org/10.1016/j.matpr.2018.06.221.

[25] Pathak H, Singh A, Singh IV. Fatigue crack growth simulations of homogeneous and bi-material interfacial cracks using element free Galerkin method. Applied Mathematical Modelling 2014;38:3093–123. Available from: https://doi.org/10.1016/j.apm.2013.11.030.

[26] Jameel A, Harmain GA. Fatigue crack growth analysis of cracked specimens by the coupled finite element-element free Galerkin method. Mechanics of Advanced Materials and Structures 2019;26:1343–56. Available from: https://doi.org/10.1080/15376494.2018.1432800.

[27] Jameel A, Harmain GA. A coupled FE-IGA technique for modeling fatigue crack growth in engineering materials. Mechanics of Advanced Materials and Structures 2019;26:1764–75. Available from: https://doi.org/10.1080/15376494.2018.1446571.

[28] Gupta V, Jameel A, Verma SK, Anand S, Anand Y. Transient isogeometric heat conduction analysis of stationary fluid in a container. Part E: Journal of Process Mechanical Engineering 2022;. Available from: https://doi.org/10.1177/0954408922112.

[29] Nguyena VP, Rabczukb T, Bordas S, Duflot M. Meshless methods: a review and computer implementation aspects. Mathematics and Computers in Simulation 2008;79:763–813. Available from: https://doi.org/10.1016/j.matcom.2008.01.003.

[30] Lone AS, Kanth SA, Harmain GA, Jameel A. XFEM modeling of frictional contact between elliptical inclusions and solid bodies. Materials Today: Proceedings 2019;26:819–24. Available from: https://doi.org/10.1016/j.matpr.2019.12.424.

[31] Kumar S, Singh IV, Mishra BK. A homogenized XFEM approach to simulate fatigue crack growth problems. Computers & Structures 2015;150:1–22. Available from: https://doi.org/10.1016/j.compstruc.2014.12.008.
[32] Kanth SA, Lone AS, Harmain GA, Jameel A. Modelling of embedded and edge cracks in steel alloys by XFEM,". Materials Today: Proceedings 2020;26:814–18. Available from: https://doi.org/10.1016/j.matpr.2019.12.423.
[33] Kumar S, Singh IV, Mishra BK, Singh A. New enrichments in XFEM to model dynamic crack response of 2-D elastic solids. International Journal of Impact Engineering 2016;87:198–211. Available from: https://doi.org/10.1016/j.ijimpeng.2015.03.005.
[34] Singh IV, Mishra BK, Kumar S, Shedbale AS. Nonlinear fatigue crack growth analysis of a center crack plate by XFEM. International Journal of Advanced Materials Manufacturing and Characterization 2014;4:11–16. Available from: https://doi.org/10.11127/ijammc.2014.03.02.
[35] Sukumar N, Chopp DL, Moës N, Belytschko T. Modeling holes and inclusions by level sets in the extended finite-element method. Computer Methods in Applied Mechanics and Engineering 2001;190:6183–200. Available from: https://doi.org/10.1016/S0045-7825(01)00215-8.
[36] Belytschko T, Gracie R, Ventura G. A review of extended/generalized finite element methods for material modeling. Modelling and Simulation in Materials Science and Engineering 2009;17. Available from: https://doi.org/10.1088/0965-0393/17/4/043001.
[37] Kanth SA, Harmain GA, Jameel A. Investigation of fatigue crack growth in engineering components containing different types of material irregularities by XFEM. Mechanics of Advanced Materials and Structures 2021;0:1–39. Available from: https://doi.org/10.1080/15376494.2021.1907003.
[38] Daux C, Moës N, Dolbow J, Sukumar N, Belytschko T. Arbitrary branched and intersecting cracks with the extended finite element method. International Journal for Numerical Methods in Engineering 2000;48:1741–60. https://doi.org/10.1002/1097-0207(20000830)48:12 < 1741::AID-NME956 > 3.0.CO;2-L.
[39] Elguedj T, Gravouil A, Combescure A. Appropriate extended functions for X-FEM simulation of plastic fracture mechanics. Computer Methods in Applied Mechanics and Engineering 2006;195:501–15. Available from: https://doi.org/10.1016/j.cma.2005.02.007.
[40] Sciammarella CA, Combel O. An elasto-plastic analysis of the crack tip field in a compact tension specimen. Engineering Fracture Mechanics 1996;55:209–22. Available from: https://doi.org/10.1016/0013-7944(95)00245-6.
[41] Sheikh UA, Jameel A. Elasto-plastic large deformation analysis of bi-material components by FEM. Materials Today: Proceedings 2020;26:1795–802. Available from: https://doi.org/10.1016/j.matpr.2020.02.377.
[42] Owen DRJ, Perić D. Recent developments in the application of finite element methods to nonlinear problems. Finite Elements in Analysis and Design 1994;18:1–15. Available from: https://doi.org/10.1016/0168-874X(94)90085-X.
[43] Dunne F., Nik P. Introduction to Computational Plasticity. New York: Oxford University Press Inc; 2006.
[44] Kumar M, Singh IV, Mishra BK, Ahmad S, Rao AV, Kumar V. Mixed mode crack growth in elasto-plastic-creeping solids using XFEM. Engineering Fracture Mechanics 2018;199:489–517. Available from: https://doi.org/10.1016/J.ENGFRACMECH.2018.05.014.
[45] De Souza Neto EA, Perić D, Owen DRJ. Computational methods for plasticity: theory and applications. Wiley; 2008. Available from: https://doi.org/10.1002/9780470694626.
[46] Jameel A, Harmain GA. Large deformation in bi-material components by XIGA and coupled FE-IGA techniques. Mechanics of Advanced Materials and Structures 2020;1–23. Available from: https://doi.org/10.1080/15376494.2020.1799120.
[47] Hinton E. Finite elements in plasticity: theory and practice. Applied Ocean Research 1981;3:149. Available from: https://doi.org/10.1016/0141-1187(81)90117-6.
[48] Gupta V, Jameel A, Anand S, Anand Y. Analysis of composite plates using isogeometric analysis: a discussion. Materials Today: Proceedings 2021;44:1190–4. Available from: https://doi.org/10.1016/j.matpr.2020.11.238.
[49] Kanth SA, Lone AS, Harmain GA, Jameel A. Elasto plastic crack growth by XFEM: a review. Materials Today: Proceedings 2019;18:3472–81. Available from: https://doi.org/10.1016/j.matpr.2019.07.275.
[50] Kim JH, Paulino GH. Consistent formulations of the interaction integral method for fracture of functionally graded materials. Journal of Applied Mechanics ASME 2005;72:351–64. Available from: https://doi.org/10.1115/1.1876395.
[51] Kanth SA, Lone AS, Harmain GA, Jameel A. Estimation of crack tip plastic zones in presence of material irregularities by extended finite element method. Journal of the Brazillian Society of Mechanical Sciences and Engineering 2023;45(304):1–31. Available from: https://doi.org/10.1007/s40430-023-04235-5.

CHAPTER

9

Enriched element-free Galerkin method for modeling large elasto-plastic deformations

Ummer Amin Sheikh[1], Azher Jameel[2], G.A. Harmain[2] and Mohd Afzal Bhat[1]

[1]Department of Mechanical Engineering, Islamic University of Science and Technology, Awantipora, Jammu and Kashmir, India [2]Department of Mechanical Engineering, National Institute of Technology Srinagar, Hazratbal, Srinagar, Jammu and Kashmir, India

9.1 Introduction

Large deformation can be quite frequently seen in engineering structures such as slender beams, columns, and other engineering components subjected to mechanical loads. Large deformation problems are nonlinear in nature due to the large geometrical changes that occur during the application of load. If the material under consideration yields during large deformation, material-type nonlinearities are also introduced into the analysis of such structures. Therefore modeling large elasto-plastic deformations in engineering materials has always been of keen interest in engineering. Till date, numerous computational methodologies have been proposed to model geometric and material nonlinearities in engineering materials, such as finite element method (FEM) [1], boundary element models [2–4], extended finite element method (XFEM) [5–12], mesh-free methods [13–19], isogeometric analysis [20–27], and several other numerical techniques. Although classical FEM has always remained the most dominant computational tool for modeling different engineering problems, it faces severe mesh distortion problems while modeling geometric nonlinearities (GNL) in structural components. Frequent remeshing is required after each stage of analysis, which complicates investigations and proves to be computationally more expensive. Element-free Galerkin method (EFGM) provides an efficient computational framework for modeling large deformations in engineering structures, which is primarily

Enriched Numerical Techniques
DOI: https://doi.org/10.1016/B978-0-443-15362-4.00009-7

due to its meshless nature. Drawbacks of mesh distortion and other mesh-related problems do not occur in the EFGM. This numerical framework discretizes the domain under consideration into an array of nodes, and approximation functions are developed from knowledge of nodes only. The EFGM finds tremendous application in areas of crack growth [28–30], metal forming [31], and contact analysis [32]. In the past few decades, EFGM has also been successfully applied to analyze three-dimensional crack growth [33,34] and bi-material interface cracks [35].

This chapter reports the modeling and simulation of material and GNL in 3D bi-material structural components by invoking the use of the mesh-free method based on the enriched EFGM. Large deformations present in the bi-material structural component have been modeled by updated Lagrangian approach (ULA), in which initial unloaded configuration is chosen to be the reference configuration for investigation. Yield criterion, based on distortion energy theory, coupled with isotropic hardening considers the elasto-plastic effects present in the structural components. Finally, three numerical problems, presenting large elasto-plastic deformations in three-dimensional structural components, are solved to demonstrate the applicability and potential of EFGM in GNL occurring in bi-material structural specimens subjected to monotonic loads.

9.2 Large deformation theory

Large deformation induces GNL in structural analysis as it is accompanied with significant geometrical changes leading to change of stiffness matrices during simulation. In order to understand the basic mechanics behind large deformation, consider a point x in the undeformed state of a loaded body, which takes position $\boldsymbol{X}$ in the deformed configuration after undergoing displacement $\boldsymbol{u}$. The deformed and undeformed states of are related by the deformation gradient $\boldsymbol{F}$, which has been written as

$$\boldsymbol{F} = \frac{\partial \boldsymbol{x}}{\partial \boldsymbol{X}} = 1 + \frac{\partial \boldsymbol{u}}{\partial \boldsymbol{X}} \tag{9.1}$$

During large deformation, the state of the loaded body changes continuously during simulation. During large deformation analysis, we use different types of stress measures depending upon reference state chosen for investigation. The first stress measure that we use during analysis is Cauchy stress (σ), which may be given as the force $\boldsymbol{df}$ imposed on the current deformed area defined over the same configuration $\boldsymbol{da}$:

$$\boldsymbol{\sigma} = \frac{\boldsymbol{df}}{\boldsymbol{da}} \tag{9.2}$$

Another stress definition is the first Piola–Kirchoff stress, which is described as current force defined over original undeformed area $\boldsymbol{A}$. This stress is unsymmetrical in nature and can be written as

$$\boldsymbol{P} = \frac{\boldsymbol{df}}{\boldsymbol{dA}} \tag{9.3}$$

Second Piola–Kirchoff stress is the force defined over the undeformed area per unit undeformed area. This type of stress is symmetric and can be defined as

$$S=\frac{dF}{dA}=F^{-1}\frac{df}{dA} \tag{9.4}$$

In large deformation formulations, strains can be defined in two ways depending on reference configuration chosen for investigation. Green–Lagrange strain tensor (E) finds extensive application in total Lagrangian approaches (TLAs), whereas Almansi strain (**e**) is frequently used in the ULA approach. These two types of strains can be expressed as

$$\mathrm{E}=\frac{1}{2}\left(F^{T}F-I\right),\mathrm{e}=\frac{1}{2}\left(I-F^{T}F^{-1}\right) \tag{9.5}$$

Different stress tensors used in continuum mechanics are related to each other as

$$P=JF^{-1}\sigma=SF^{T} \tag{9.6}$$

$$\sigma=\frac{1}{J}FP=\frac{1}{J}FSF^{T} \tag{9.7}$$

$$S=JF^{-1}\sigma F^{T}=PF^{-T} \tag{9.8}$$

where J denotes determinant of deformation gradient matrix.

9.3 Introduction to element-free Galerkin method

In this numerical method, the structural component is discretized into an array of nodes, and variable approximations are constructed from the descriptions of these nodes representing given domain. The need for creating a finite element mesh does not arise in any of the meshless techniques, including EFGM. The variable approximations are developed independent of the FEM mesh, which is one of the important features of this computational tool. Meshless nature of EFGM proves to be very handy in eliminating the drawbacks associated with mesh generation. EFGM provides a strong computational tool for investigating GNL problems as compared to classical FEM, as the elemental mesh faces severe distortion during analysis, which demands remeshing of the structural component frequently during various stages of simulation. Introduction of enrichment strategies further strengthens the classical EFGM approach, which makes it more suited for modeling different discontinuities and material irregularities present in engineering specimens. EFGM constructs approximate function entirely in terms of nodes representing the domain. Need for elemental mesh is eliminated, and hence there are no issues associated with the generation of mesh. The moving least squares (MLS) approximations are used to define the variables over the engineering domain. MLS approximants can be divided into three major parts:

(1) Polynomial basis function
(2) Weight function with compact support, which defines the nodal domain of influence and introduces local character to EFGM. The weight function vanishes and is zero outside the compact support. Disks and rectangles are the commonly used compact supports.
(3) Set of coefficients that depend on the position of nodes

MLS approximation represents field variables, defined in engineering domain, in terms of polynomial basis $\mathbf{P}(\mathbf{x})$ and set of coefficients $\mathbf{a}(\mathbf{x})$. We have

$$u^h(\mathbf{x}) = \sum_{j=1}^{n} \mathrm{p}_j(\mathbf{x})\mathrm{a}_j(\mathbf{x}) = \mathbf{P}^T(\mathbf{x})\mathbf{a}(\mathbf{x}) \tag{9.9}$$

where $\mathbf{P}(\mathbf{x})$ represents a vector of polynomial basis functions defined as

$$\mathbf{P}^T(\mathbf{x}) = [1, \mathrm{x}, \mathrm{y}, \mathrm{z}, \mathrm{xy}, \mathrm{yz}, \mathrm{zx}, \ldots\ldots\ldots\ldots \mathrm{x}^k, \mathrm{y}^k, \mathrm{z}^k] \tag{9.10}$$

where k is the degree of the polynomial and coefficient matrix $\mathbf{a}(\mathbf{x})$ is given by

$$\mathbf{a}(\mathbf{x}) = [\mathrm{a}_1(\mathbf{x}), \mathrm{a}_2(\mathbf{x}), \mathrm{a}_3(\mathbf{x}), \ldots\ldots\ldots\ldots \mathrm{a}_m(\mathbf{x})] \tag{9.11}$$

To derive unknown coefficients $\mathbf{a}(\mathbf{x})$, weighted least squares sum is defined between the variable approximation $u^h(\mathbf{x})$ and nodal parameters u_i and can be described as

$$\mathrm{L}(\mathbf{x}) = s\sum_{i=1}^{n} \mathrm{w}(\mathbf{x}-\mathbf{x}_i)[\mathbf{P}^T(\mathbf{x})\mathbf{a}(\mathbf{x})-\mathrm{u}_i]^2 \tag{9.12}$$

where u_i denotes nodal parameter of node i at $\mathbf{x} = \mathbf{x}_i$, but it should be noted that these values are not equal to nodal values because $u^h(\mathbf{x})$ is not an interpolant but an approximant. To find the coefficients $\mathbf{a}(\mathbf{x})$, $\mathbf{L}(\mathbf{x})$ is to be minimized with respect to $\mathbf{a}(\mathbf{x})$, that is,

$$\frac{\partial \mathrm{L}}{\partial \mathbf{a}} = 0 \tag{9.13}$$

The above condition leads to the system of linear algebraic equations for $\mathbf{a}(\mathbf{x})$

$$\mathbf{A}(\mathbf{x})\mathbf{a}(\mathbf{x}) = \mathbf{B}(\mathbf{x})\mathbf{u} \tag{9.14}$$

where

$$\mathbf{u}^T = [\mathrm{u}_1, \mathrm{u}_2, \mathrm{u}_3, \ldots\ldots\ldots\ldots\ldots\ldots \mathrm{u}_n] \tag{9.15}$$

$$\mathbf{A} = \sum_{i=1}^{n} \mathrm{W}_i(\mathbf{x})\mathbf{P}(\mathbf{x}_i)\mathbf{P}^T(\mathbf{x}_i) \tag{9.16}$$

$$\mathbf{B} = [\mathrm{W}_1(\mathbf{x})\mathbf{P}(\mathbf{x}_1), \mathrm{W}_2(\mathbf{x})\mathbf{P}(\mathbf{x}_2), \ldots\ldots\ldots\ldots \mathrm{W}_n(\mathbf{x})\mathbf{P}(\mathbf{x}_n)] \tag{9.17}$$

Coefficient matrix $\mathbf{a}(\mathbf{x})$ is obtained as $\mathbf{a}(\mathbf{x}) = \mathbf{A}^{-1}(\mathbf{x})\mathbf{B}(\mathbf{x})\mathbf{u}$, which is defined only if matrix $\mathbf{A}$ is nonsingular. Further simplification defines the variable approximation $u(\mathbf{x})$ as

$$u^h(\mathbf{x}) = \mathbf{P}^T(\mathbf{x})\mathbf{A}^{-1}(\mathbf{x})\mathbf{B}(\mathbf{x})\mathbf{u} \tag{9.18}$$

where $\mathbf{A}(\mathbf{x})$ and $\mathbf{B}(\mathbf{x})$ are defined as

$$\mathbf{A}(\mathbf{x}) = \sum_{i=1}^{n} \mathrm{w}(\mathbf{x} - \mathbf{x}_i)\mathbf{P}(\mathbf{x}_i)\mathbf{P}^T(\mathbf{x}_i) \tag{9.19}$$

$$\mathbf{B}(\mathbf{x}) = [\mathrm{w}(\mathbf{x} - \mathbf{x}_1)\mathbf{P}(\mathbf{x}_1), \mathrm{w}(\mathbf{x} - \mathbf{x}_2)\mathbf{P}(\mathbf{x}_2), \ldots\ldots\ldots\ldots \mathrm{w}(\mathbf{x} - \mathbf{x}_n)\mathbf{P}(\mathbf{x}_n)] \tag{9.20}$$

In the above equations, n represents the nodes, the compact support of which envelopes the point $\mathbf{x}$. Weight function is represented w, and it remains zero outside the compact support. The weight function provides proper weightings to residuals at the nodes lying in the domain of influence. In terms of mesh-free shape functions, MLS approximation is expressed in standard form as

$$u^h(\mathbf{x}) = \sum_{i=1}^{n} \mathrm{N}_i(\mathbf{x})\mathrm{u}_i = \mathbf{N}^T(\mathbf{x})\mathbf{u} \tag{9.21}$$

where

$$\mathbf{N}^T(\mathbf{x}) = [\mathrm{N}_1(\mathbf{x}), \mathrm{N}_2(\mathbf{x}), \mathrm{N}_3(\mathbf{x}), \ldots\ldots\ldots \mathrm{N}_n(\mathbf{x})] \tag{9.22}$$

$$\mathbf{u}^T = [\mathrm{u}_1, \mathrm{u}_2, \mathrm{u}_3, \ldots\ldots\ldots\ldots\ldots\ldots \mathrm{u}_\mathrm{n}] \tag{9.23}$$

The mesh-free shape functions $\mathrm{N}_i(\mathbf{x})$ are defined as

$$\mathrm{N}_i(\mathbf{x}) = \mathbf{P}^T(\mathbf{x})\mathbf{A}^{-1}(\mathbf{x})\mathbf{B}(\mathbf{x}) \tag{9.24}$$

The derivatives of the mesh-free shape functions are computed as

$$\mathrm{N}_{i,x}(\mathbf{x}) = \left(\mathbf{P}^T(\mathbf{x})\mathbf{A}^{-1}(\mathbf{x})\mathbf{B}(\mathbf{x})\right)_{,x} = \mathbf{p}^T_{,x}\mathbf{A}^{-1}\mathbf{B} + \mathbf{P}^T(\mathbf{A}^{-1})_{,x}\mathbf{B} + \mathbf{P}^T\mathbf{A}^{-1}\mathbf{B}_{,x} \tag{9.25}$$

where

$$\mathbf{B}_{,x}(\mathbf{x}) = \frac{\partial \mathrm{w}}{\partial \mathbf{x}}(\mathbf{x} - \mathbf{x}_i)\mathbf{P}(\mathbf{x}_i) \tag{9.26}$$

$$(\mathbf{A}^{-1})_{,x} = -\mathbf{A}^{-1}\mathbf{A}_{,x}\mathbf{A}^{-1} \tag{9.27}$$

where

$$\mathbf{A}_{,x} = \sum_{i=1}^{n} \frac{\partial \mathrm{w}}{\partial \mathbf{x}}(\mathbf{x} - \mathbf{x}_i)\mathbf{P}(\mathbf{x}_i)\mathbf{P}^T(\mathbf{x}_i) \tag{9.28}$$

9.3.1 Choice of weight function

Selection of weight function has always remained an important decision that affects the performance of the meshless methods. One of the important features of weight functions is their positive nature and monotonic decrease with respect to distance from the center of compact support. Some of the commonly used weight functions that have found extensive applications in meshless techniques are in exponential, cubic, and quartic spline form. From a computational point of view, exponential weight functions prove to be more demanding but less sensitive to the size of compact supports. Quartic spline is the most commonly used weight function in EFGM. Domain of influence or compact support of any node should satisfy the following conditions:

- The domain of influence should be large enough so that enough neighbors are available for each nodal point, and it is ensured that the matrix **A** is invertible. If the domain of influence is too small, there may not be enough neighbors available for the given nodal point, and the matrix **A** may not be invertible.

- The support or the domain of influence should be large enough so that all the interior points of the domain, except the points lying on the boundary, have neighbors on all the four sides, so that the contribution is obtained from all sides of the sampling point. It should be kept in mind that the support of the domain should not be too large from the computational point of view.
- The domain of influence provides the local behavior to MLS-based variable approximations. Therefore compact support should be small enough to impart sufficient local character to the given MLS-based approximations.

9.3.2 Element-free Galerkin method models

To develop numerical models based on EFGM, let us assume a simple two-dimensional domain Ω, which is surrounded by the boundary Γ. For this engineering domain, equilibrium equation can be written as

$$\nabla \cdot \boldsymbol{\sigma} + \mathbf{b} = \mathbf{0} \text{ in } \Omega \tag{9.29}$$

where ∇ is the divergence operator, $\boldsymbol{\sigma}$ denotes the stress field, $\mathbf{u}$ represents displacements, and $\mathbf{b}$ denotes body forces. Imposed boundary conditions can be expressed as

$$\text{(Natural B.C)}\ \boldsymbol{\sigma} \cdot \mathbf{n} = \mathbf{t} \text{ on } \Gamma_t \tag{9.30}$$

$$\text{(Essential B.C)}\ \mathbf{u} = \overline{\mathbf{u}} \text{ on } \Gamma_u \tag{9.31}$$

where $\mathbf{t}$ and $\overline{\mathbf{u}}$ are prescribed boundary values, and $\mathbf{n}$ is unit normal to the displacement boundary. Variational form of equilibrium equation is expressed as

$$\int_{\Omega} \delta\boldsymbol{\varepsilon} : \boldsymbol{\sigma} d\Omega - \int_{\Omega} \delta\mathbf{u} : \mathbf{b} d\Omega - \int_{\Gamma_t} \delta\mathbf{u} : \mathbf{t} d\Gamma - \delta\mathbf{W}_u(\mathbf{u}, \nabla_s \boldsymbol{\lambda}) = \mathbf{0} \tag{9.32}$$

where ε represents the strain field and $\delta\mathbf{W}_u$ is added to enforce essential boundary conditions (BCs). It is well understood that the imposition of boundary conditions is quite different in EFGM as compared to that of FEM because MLS-based variable approximations do not possess the Kronecker delta property, that is, $N_i(x_j) \neq \delta_{ij}$. In view of this, several methods were developed to apply prescribed boundary conditions in meshless techniques, including EFGM. The Lagrange method has been widely used for the imposition of prescribed BCs in EFGM. Therefore we have

$$\mathbf{W}_u(\mathbf{u}, \boldsymbol{\lambda}) = \int_{\Gamma} \boldsymbol{\lambda}(\mathbf{u} - \overline{\mathbf{u}}) d\Gamma \tag{9.33}$$

$$\delta\mathbf{W}_u(\mathbf{u}, \boldsymbol{\lambda}) = \int_{\Gamma} \delta\boldsymbol{\lambda}(\mathbf{u} - \overline{\mathbf{u}}) d\Gamma + \int_{\Gamma} \delta\mathbf{u}.\boldsymbol{\lambda} d\Gamma \tag{9.34}$$

For linear elastic problems, the stress and strain tensors can be written as

$$\varepsilon = \nabla_s \mathbf{u} \tag{9.35}$$

$$\boldsymbol{\sigma} = \mathbf{D}\varepsilon \tag{9.36}$$

where ε denotes the strain, $\boldsymbol{\sigma}$ represents stresses, and $\mathbf{D}$ denotes the constitutive matrix. The Lagrangian multiplier is given as

$$\boldsymbol{\lambda}(\mathbf{x}) = \Psi_i(\mathrm{s})\lambda_i \tag{9.37}$$

$$\delta\boldsymbol{\lambda}(\mathbf{x}) = \Psi_i(\mathrm{s})\delta\lambda_i \tag{9.38}$$

where $\Psi_i(\mathrm{s})$ represents the Lagrange shape function and s denotes arc length measured along the boundary of domain. EFGM approximation $\mathrm{u}^h(\mathbf{x})$ of the field variable $\mathrm{u}(\mathbf{x})$ is written in terms of MLS shape functions as

$$\mathrm{u}^h(\mathbf{x}) = \sum_{i=1}^{n} \mathrm{N}_i(\mathbf{x})\mathrm{u}_i \tag{9.39}$$

$$\delta\mathrm{u}^h(\mathbf{x}) = \sum_{i=1}^{n} \mathrm{N}_i(\mathbf{x})\delta\mathrm{u}_i \tag{9.40}$$

Substitution of the above approximation into a suitable form of the equilibrium equation gives a set of linear algebraic equations in matrix form as

$$\begin{bmatrix} \mathbf{K} & \mathbf{G} \\ \mathbf{G}^T & 0 \end{bmatrix} \begin{Bmatrix} \mathbf{u} \\ \boldsymbol{\lambda} \end{Bmatrix} = \begin{Bmatrix} \mathbf{f} \\ \mathbf{q} \end{Bmatrix} \tag{9.41}$$

where

$$\mathbf{K}_{ij=} \int_{\Omega} \mathbf{B}^T\mathbf{D}\mathbf{B}\mathrm{d}\Omega \tag{9.42}$$

$$\mathbf{G}_{ik=} \int_{\Omega} \mathbf{N}_i\boldsymbol{\Psi}_k\mathbf{d}\Gamma \tag{9.43}$$

$$\mathbf{f}_{i=} \int_{\Gamma_\mathrm{t}} \mathbf{N}_i\mathbf{t}\mathbf{d}\Gamma_\mathrm{t} + \int_{\Omega} \mathbf{N}_i\mathbf{b}\mathrm{d}\Omega \tag{9.44}$$

$$\mathbf{q}_{k=} \int_{\Gamma_\mathrm{t}} \boldsymbol{\Psi}_k\bar{\mathbf{u}}\mathrm{d}\Gamma_\mathrm{u} \tag{9.45}$$

$$\mathbf{B}_i = \begin{bmatrix} \mathrm{N}_{i,x} & 0 \\ 0 & \mathrm{N}_{i,y} \\ \mathrm{N}_{i,y} & \mathrm{N}_{i,x} \end{bmatrix} \tag{9.46}$$

$$\boldsymbol{\Psi}_k = \begin{bmatrix} \Psi_k & 0 \\ 0 & \Psi_k \end{bmatrix} \tag{9.47}$$

$$\mathbf{D} = \frac{E}{(1+\upsilon)(1-2\upsilon)} \begin{bmatrix} 1-\upsilon & \upsilon & 0 \\ \upsilon & 1-\upsilon & 0 \\ 0 & 0 & \frac{1-2\upsilon}{2} \end{bmatrix} \text{for plane strain} \tag{9.48}$$

$$\mathbf{D} = \frac{E}{(1-\upsilon^2)} \begin{bmatrix} 1 & \upsilon & 0 \\ \upsilon & 1 & 0 \\ 0 & 0 & \frac{1-\upsilon}{2} \end{bmatrix} \text{for plane stress} \tag{9.49}$$

In the above equations, comma represents a partial derivative. E and υ are elastic modulus and Poisson's ratio, respectively. **D** is a matrix of elastic moduli, defined separately for the states of plane stress and plane strain. Ψ_k represents usual Lagrange shape functions used in classical FEM, whereas Ψ_i denotes MLS shape functions that are used in EFGM formulations.

9.4 Total Lagrangian formulation

TLA has always been a strong tool to model and simulate GNL caused by large deformations occurring in structural components. In TLA approach, a suitable form of equilibrium equation can be described in the classical form as

$$\frac{\partial P_{ij}}{\partial X_j} + b_i = 0 \tag{9.50}$$

where P_{ij} is first Piola–Kirchoff stress, b_i is the body force, X_j is the Lagrangian coordinate. As defined earlier, first Piola–Kirchoff stress is the stress calculated as force applied on the deformed state and measured over undeformed area. For a two-dimensional case, we have

$$\frac{\partial P_{11}}{\partial X_1} + \frac{\partial P_{12}}{\partial X_2} + b_1 = 0 \tag{9.51}$$

$$\frac{\partial P_{21}}{\partial X_1} + \frac{\partial P_{22}}{\partial X_2} + b_2 = 0 \tag{9.52}$$

The equilibrium equation, described above, is defined in terms of initial undeformed configuration. Green and St. Venant presented a general definition of strains that is applicable for small as well as large deformations. In terms of Green's strain tensor, linear and nonlinear components of strain can be expressed as

$$\boldsymbol{E} = \boldsymbol{E}_L + \boldsymbol{E}_{NL} = \boldsymbol{E}_L + \frac{1}{2}\mathbf{A}_\theta\theta \tag{9.53}$$

where $\boldsymbol{E}_L$ and $\boldsymbol{E}_{NL}$ denote the linear and nonlinear strains, θ is displacement gradient, and $\mathbf{A}_\theta$ is a matrix operator, containing displacement derivatives [58]. We have

$$\boldsymbol{E}_L = \begin{Bmatrix} \dfrac{\partial u}{\partial X} \\ \dfrac{\partial v}{\partial Y} \\ \dfrac{\partial v}{\partial X} + \dfrac{\partial u}{\partial Y} \end{Bmatrix} \tag{9.54}$$

$$\boldsymbol{E}_{NL} = \begin{Bmatrix} \dfrac{1}{2}\left(\frac{\partial u}{\partial X}\right)^2 + \dfrac{1}{2}\left(\frac{\partial v}{\partial X}\right)^2 \\ \dfrac{1}{2}\left(\frac{\partial u}{\partial Y}\right)^2 + \dfrac{1}{2}\left(\frac{\partial v}{\partial Y}\right)^2 \\ \dfrac{\partial u}{\partial X}\dfrac{\partial u}{\partial Y} + \dfrac{\partial v}{\partial X}\dfrac{\partial v}{\partial Y} \end{Bmatrix} \tag{9.55}$$

To derive suitable models for simulating large deformations by TLA approach, the variations form of equilibrium equation can be written as

$$\int_\Omega \delta \mathbf{F}^T \mathbf{P} \mathrm{d}\Omega - \int_\Omega \delta \mathbf{u}^T \mathbf{b} \mathrm{d}\Omega - \int_{\Gamma_t} \delta \mathbf{u}^T \mathbf{t} \mathrm{d}\Gamma_t = 0 \tag{9.56}$$

where $\mathbf{F}$ is deformation gradient tensor, the components of which are defined as $F_{ij} = \frac{\partial x_i}{\partial X_j} = \frac{\partial u_i}{\partial X_j} + \delta_{ij}$, $\mathbf{b}$ represents body forces, and $\mathbf{t}$ denotes applied traction. $\boldsymbol{x} = \begin{bmatrix} x & y \end{bmatrix}^T$ refers to the loaded and deformed configuration of the body and $\boldsymbol{X} = \begin{bmatrix} X & Y \end{bmatrix}^T$ refers to initial unloaded state of the given body. We can further write

$$\int_\Omega \delta \begin{bmatrix} \dfrac{\partial u}{\partial X} & \dfrac{\partial v}{\partial X} & \dfrac{\partial u}{\partial Y} & \dfrac{\partial v}{\partial Y} \end{bmatrix} \begin{Bmatrix} \mathrm{P}_{11} \\ \mathrm{P}_{12} \\ \mathrm{P}_{21} \\ \mathrm{P}_{22} \end{Bmatrix} \mathrm{d}\Omega - \int_\Omega \delta [u \quad v] \begin{Bmatrix} \mathrm{b}_1 \\ \mathrm{b}_2 \end{Bmatrix} \mathrm{d}\Omega - \int_{\Gamma_t} \delta [u \quad v] \begin{Bmatrix} \mathrm{t}_1 \\ \mathrm{t}_2 \end{Bmatrix} \mathrm{d}\Gamma_t = 0 \tag{9.57}$$

In enriched EFGM, the material irregularities present in structural components are modeled by modifying the classical MLS-based variable approximations with additional enrichment functions by invoking the partition of unity approach. In the case of enriched EFGM, modified variable approximation can be expressed in a suitable form as

$$\mathbf{u}(X) = \sum_i N_i(X)\mathbf{u}_i + \sum_j N_j(X)(f(X) - f(X_j))\mathbf{a}_j \quad \text{for} \quad n_i \in n \quad \text{and} \quad n_j \in n_k \tag{9.58}$$

where N_i represents MLS-based interpolation functions, $\mathbf{u}_i$ represents nodal degree of freedom, $\mathbf{a}_j$ the enriched degrees of freedom pertaining to additional enrichment functions, $f(\boldsymbol{X})$ is the corresponding enrichment function, n denotes standard nodes present in the given domain, and n_k represents the additional enriched nodes. The variable approximation, defined above, can be expressed in a more compact form as

$$\mathbf{u}(\mathbf{X}) = \sum_i N_i(\mathbf{X})\mathbf{u}_i + \sum_j \overline{N}_j(\mathbf{X})\mathbf{a}_j \tag{9.59}$$

where $\overline{N}_j = N_j(\mathbf{X})(H(\mathbf{X}) - H(\mathbf{X}_j))$. We can further express the variable approximation as

$$\begin{Bmatrix} \mathrm{u} \\ \mathrm{v} \end{Bmatrix} = \begin{bmatrix} N_i & 0 \\ 0 & N_i \end{bmatrix} \begin{Bmatrix} u_i \\ v_i \end{Bmatrix} + \begin{bmatrix} \overline{N}_j & 0 \\ 0 & \overline{N}_j \end{bmatrix} \begin{Bmatrix} a_j \\ b_j \end{Bmatrix} = \mathbf{N}\mathbf{u} + \overline{\mathbf{N}}\mathbf{a} \tag{9.60}$$

The displacement gradients can be obtained as

$$\begin{Bmatrix} \dfrac{\partial u}{\partial X} \\ \dfrac{\partial v}{\partial X} \\ \dfrac{\partial u}{\partial Y} \\ \dfrac{\partial v}{\partial Y} \end{Bmatrix} = \begin{bmatrix} \dfrac{\partial N_i}{\partial X} & 0 \\ 0 & \dfrac{\partial N_i}{\partial X} \\ \dfrac{\partial N_i}{\partial Y} & 0 \\ 0 & \dfrac{\partial N_i}{\partial Y} \end{bmatrix} \begin{Bmatrix} u_i \\ v_i \end{Bmatrix} + \begin{bmatrix} \dfrac{\partial \overline{N}_j}{\partial X} & 0 \\ 0 & \dfrac{\partial \overline{N}_j}{\partial X} \\ \dfrac{\partial \overline{N}_j}{\partial Y} & 0 \\ 0 & \dfrac{\partial \overline{N}_j}{\partial Y} \end{bmatrix} \begin{Bmatrix} a_j \\ b_j \end{Bmatrix} = \mathbf{G}\mathbf{u} + \overline{\mathbf{G}}\mathbf{a} \tag{9.61}$$

where

$$\mathbf{G} = \begin{bmatrix} \dfrac{\partial N_i}{\partial X} & 0 \\ 0 & \dfrac{\partial N_i}{\partial X} \\ \dfrac{\partial N_i}{\partial Y} & 0 \\ 0 & \dfrac{\partial N_i}{\partial Y} \end{bmatrix} \tag{9.62}$$

$$\overline{\mathbf{G}} = \begin{bmatrix} \dfrac{\partial \overline{N}_j}{\partial X} & 0 \\ 0 & \dfrac{\partial \overline{N}_j}{\partial X} \\ \dfrac{\partial \overline{N}_j}{\partial Y} & 0 \\ 0 & \dfrac{\partial \overline{N}_j}{\partial Y} \end{bmatrix} \tag{9.63}$$

We can further write

$$\left(\int_\Omega \mathbf{G}^T\mathbf{P}d\Omega - \int_\Omega \mathbf{N}^T\mathbf{b}d\Omega - \int_\Gamma \mathbf{N}^T\mathbf{t}d\Gamma\right)\delta\mathbf{u}^T + \left(\int_\Omega \overline{\mathbf{G}}^T\mathbf{P}d\Omega - \int_\Omega \overline{\mathbf{N}}^T\mathbf{b}d\Omega - \int_\Gamma \overline{\mathbf{N}}^T\mathbf{t}d\Gamma\right)\delta\mathbf{a}^T = 0 \tag{9.64}$$

The above equation can be further simplified as

$$\left(\int_\Omega \mathbf{G}^T\mathbf{P}d\Omega - \mathbf{f}\right)\delta\mathbf{u}^T + \left(\int_\Omega \overline{\mathbf{G}}^T\mathbf{P}d\Omega - \bar{\mathbf{f}}\right)\delta\mathbf{a}^T = 0 \tag{9.65}$$

where

$$\mathbf{f} = \int_\Omega \mathbf{N}^T\mathbf{b}d\Omega + \int_\Gamma \mathbf{N}^T\mathbf{t}d\Gamma \tag{9.66}$$

$$\bar{\mathbf{f}} = \int_\Omega \overline{\mathbf{N}}^T\mathbf{b}d\Omega - \int_\Gamma \overline{\mathbf{N}}^T\mathbf{t}d\Gamma \tag{9.67}$$

The incremental form of the above equation can be expressed as

$$\left(\int_\Omega \mathbf{G}^T\Delta\mathbf{P}d\Omega - \Delta\mathbf{f}\right)\delta\mathbf{u}^T + \left(\int_\Omega \overline{\mathbf{G}}^T\Delta\mathbf{P}d\Omega - \Delta\bar{\mathbf{f}}\right)\delta\mathbf{a}^T = 0 \tag{9.68}$$

where $\Delta\mathbf{P} = \begin{Bmatrix} \Delta P_{11} \\ \Delta P_{12} \\ \Delta P_{21} \\ \Delta P_{22} \end{Bmatrix}$.

As discussed earlier, first Piola–Kirchoff stress does not possess symmetry. Therefore we generally use second Piola–Kirchoff stresses, which are defined in terms of force in unloaded configuration and measured over the undeformed area. This stress tensor is symmetric, which preserves the symmetry of the numerical formulation. First Piola–Kirchoff stress is related to the second Piola–Kirchoff stress by $\mathbf{P} = \mathbf{S}\mathbf{F}^T$, which can be written in expanded form as

$$P_{11} = F_{11}S_{11} + F_{12}S_{21} \tag{9.69}$$

$$P_{12} = F_{21}S_{11} + F_{22}S_{21} \tag{9.70}$$

$$P_{21} = F_{11}S_{12} + F_{12}S_{22} \tag{9.71}$$

$$P_{22} = F_{21}S_{12} + F_{22}S_{22} \tag{9.72}$$

where $\mathbf{F}$ is the deformation gradient defined as $F_{ij} = \frac{\partial x_i}{\partial X_j} = \frac{\partial u_i}{\partial X_j} + \delta_{ij}$, and δ_{ij} is the Kronecker delta. Now, the incremental first Piola–Kirchoff $\Delta\mathbf{P}$ can be written as

$$\Delta P_{11} = F_{11}\Delta S_{11} + F_{12}\Delta S_{21} + \Delta F_{11}S_{11} + \Delta F_{12}S_{21} \tag{9.73}$$

$$\Delta P_{12} = F_{21}\Delta S_{11} + F_{22}\Delta S_{21} + \Delta F_{21}S_{11} + \Delta F_{22}S_{21} \tag{9.74}$$

$$\Delta P_{21} = F_{11}\Delta S_{12} + F_{12}\Delta S_{22} + \Delta F_{11}S_{12} + \Delta F_{12}S_{22} \tag{9.75}$$

$$\Delta P_{22} = F_{21}\Delta S_{12} + F_{22}\Delta S_{22} + \Delta F_{21}S_{12} + \Delta F_{22}S_{22} \tag{9.76}$$

We can write the above equations in matrix form as

$$\begin{Bmatrix} \Delta P_{11} \\ \Delta P_{12} \\ \Delta P_{21} \\ \Delta P_{22} \end{Bmatrix} = \begin{bmatrix} F_{11} & 0 & F_{12} \\ F_{21} & 0 & F_{22} \\ 0 & F_{12} & F_{11} \\ 0 & F_{22} & F_{21} \end{bmatrix} \begin{Bmatrix} \Delta S_{11} \\ \Delta S_{22} \\ \Delta S_{12} \end{Bmatrix} + \begin{bmatrix} S_{11} & 0 & S_{12} & 0 \\ 0 & S_{11} & 0 & S_{12} \\ S_{12} & 0 & S_{22} & 0 \\ 0 & S_{12} & 0 & S_{22} \end{bmatrix} \begin{Bmatrix} \Delta F_{11} \\ \Delta F_{21} \\ \Delta F_{12} \\ \Delta F_{22} \end{Bmatrix} \tag{9.77}$$

Thus the incremental first Piola–Kirchoff $\Delta \mathbf{P}$ can be written as

$$\Delta \mathbf{P} = \mathbf{F}^T \Delta \mathbf{S} + \boldsymbol{M}_s \Delta \mathbf{F} \tag{9.78}$$

where

$$\mathbf{F}^T = \begin{bmatrix} F_{11} & 0 & F_{12} \\ F_{21} & 0 & F_{22} \\ 0 & F_{12} & F_{11} \\ 0 & F_{22} & F_{21} \end{bmatrix} \tag{9.79}$$

$$\boldsymbol{M}_s = \begin{bmatrix} S_{11} & 0 & S_{12} & 0 \\ 0 & S_{11} & 0 & S_{12} \\ S_{12} & 0 & S_{22} & 0 \\ 0 & S_{12} & 0 & S_{22} \end{bmatrix} \tag{9.80}$$

Green's strain $\mathbf{E}$ can be expressed in terms of second Piola–Kirchoff stress as $\Delta \mathbf{S} = \mathbf{D}^{ep}\Delta \mathbf{E}$, which can be further expanded as

$$\begin{Bmatrix} \Delta S_{11} \\ \Delta S_{22} \\ \Delta S_{12} \end{Bmatrix} = \mathbf{D}^{ep} \begin{Bmatrix} \Delta E_{11} \\ \Delta E_{22} \\ 2\Delta E_{12} \end{Bmatrix} \tag{9.81}$$

The components of Green's strain $\mathbf{E}$ are written in terms of deformation gradient components as

$$2E_{11} = F_{11}F_{11} + F_{21}F_{12} - 1 \tag{9.82}$$

$$2E_{22} = F_{12}F_{12} + F_{22}F_{22} - 1 \tag{9.83}$$

$$2E_{12} = F_{11}F_{12} + F_{21}F_{22} \tag{9.84}$$

Now, incremental Green's strain can be written as

$$\begin{Bmatrix} \Delta E_{11} \\ \Delta E_{22} \\ 2\Delta E_{12} \end{Bmatrix} = \begin{bmatrix} F_{11} & F_{21} & 0 & 0 \\ 0 & 0 & F_{12} & F_{22} \\ F_{12} & F_{22} & F_{11} & F_{21} \end{bmatrix} \begin{Bmatrix} \Delta F_{11} \\ \Delta F_{21} \\ \Delta F_{12} \\ \Delta F_{22} \end{Bmatrix} = \mathbf{F} \begin{Bmatrix} \Delta F_{11} \\ \Delta F_{21} \\ \Delta F_{12} \\ \Delta F_{22} \end{Bmatrix} = \mathbf{F}\Delta\mathbf{F} \tag{9.85}$$

The deformation gradient is defined as $F_{ij} = \frac{\partial x_i}{\partial X_j} = \frac{\partial u_i}{\partial X_j} + \delta_{ij}$. Therefore the incremental form of deformation gradient is expressed as $\Delta F_{ij} = \frac{\partial \Delta u_i}{\partial X_j}$. Now, we have

$$\begin{Bmatrix} \Delta E_{11} \\ \Delta E_{22} \\ 2\Delta E_{12} \end{Bmatrix} = \mathbf{F} \begin{Bmatrix} \dfrac{\partial \Delta u}{\partial X} \\ \dfrac{\partial \Delta v}{\partial X} \\ \dfrac{\partial \Delta u}{\partial Y} \\ \dfrac{\partial \Delta v}{\partial Y} \end{Bmatrix} = \mathbf{F}\left(\mathbf{G}\Delta\mathbf{u} + \overline{\mathbf{G}}\Delta\mathbf{a}\right) \tag{9.86}$$

Thus Green's strain is further written as

$$\Delta\mathbf{E} = \mathbf{F}\mathbf{G}\Delta\mathbf{u} + \mathbf{F}\overline{\mathbf{G}}\Delta\mathbf{a} = \overline{\mathbf{B}}\Delta\mathbf{u} + \overline{\overline{\mathbf{B}}}\Delta\mathbf{a} \tag{9.87}$$

where

$$\overline{\mathbf{B}} = \mathbf{F}\mathbf{G} = \begin{bmatrix} \dfrac{\partial N_i}{\partial X}\dfrac{\partial x}{\partial X} & \dfrac{\partial N_i}{\partial X}\dfrac{\partial y}{\partial X} \\ \dfrac{\partial N_i}{\partial Y}\dfrac{\partial x}{\partial Y} & \dfrac{\partial N_i}{\partial Y}\dfrac{\partial y}{\partial Y} \\ \dfrac{\partial N_i}{\partial X}\dfrac{\partial x}{\partial Y} + \dfrac{\partial N_i}{\partial Y}\dfrac{\partial x}{\partial X} & \dfrac{\partial N_i}{\partial X}\dfrac{\partial y}{\partial Y} + \dfrac{\partial N_i}{\partial Y}\dfrac{\partial y}{\partial X} \end{bmatrix} \tag{9.88}$$

$$\overline{\overline{\mathbf{B}}} = \mathbf{F}\overline{\mathbf{G}} = \begin{bmatrix} \dfrac{\partial \overline{N}_j}{\partial X}\dfrac{\partial x}{\partial X} & \dfrac{\partial \overline{N}_j}{\partial X}\dfrac{\partial y}{\partial X} \\ \dfrac{\partial \overline{N}_j}{\partial Y}\dfrac{\partial x}{\partial Y} & \dfrac{\partial \overline{N}_j}{\partial Y}\dfrac{\partial y}{\partial Y} \\ \dfrac{\partial \overline{N}_j}{\partial X}\dfrac{\partial x}{\partial Y} + \dfrac{\partial \overline{N}_j}{\partial Y}\dfrac{\partial x}{\partial X} & \dfrac{\partial \overline{N}_j}{\partial X}\dfrac{\partial y}{\partial Y} + \dfrac{\partial \overline{N}_j}{\partial Y}\dfrac{\partial y}{\partial X} \end{bmatrix} \tag{9.89}$$

Second Piola–Kirchoff stress is further expressed as

$$\Delta\mathbf{S} = \mathbf{D}^{ep}\Delta\mathbf{E} = \mathbf{D}^{ep}\overline{\mathbf{B}}\Delta\mathbf{u} + \mathbf{D}^{ep}\overline{\overline{\mathbf{B}}}\Delta\mathbf{a} \tag{9.90}$$

We get the following relations:

$$\int_\Omega \mathbf{G}^T \Delta\mathbf{P} d\Omega = \Delta\mathbf{f} \tag{9.91}$$

$$\int_\Omega \overline{\mathbf{G}}^T \Delta\mathbf{P} d\Omega = \Delta\overline{\mathbf{f}} \tag{9.92}$$

$$\int_{\Omega} \mathbf{G}^T(\mathbf{F}^T\Delta\mathbf{S} + M_s\Delta\mathbf{F})d\Omega = \Delta\mathbf{f} \tag{9.93}$$

$$\int_{\Omega} \mathbf{G}^T\mathbf{F}^T\Delta\mathbf{S}d\Omega + \int_{\Omega} \mathbf{G}^T M_s\Delta\mathbf{F}d\Omega = \Delta\mathbf{f} \tag{9.94}$$

$$\left(\int_{\Omega} \overline{\mathbf{B}}^T\mathbf{D}^{ep}\overline{\mathbf{B}}d\Omega + \int_{\Omega} \mathbf{G}^T M_s\mathbf{G}d\Omega\right)\Delta\mathbf{u} + \left(\int_{\Omega} \overline{\mathbf{B}}^T\mathbf{D}^{ep}\overline{\overline{\mathbf{B}}}d\Omega + \int_{\Omega} \mathbf{G}^T M_s\overline{\mathbf{G}}d\Omega\right)\Delta\mathbf{a} = \Delta\mathbf{f} \tag{9.95}$$

$$\left(\int_{\Omega} \overline{\overline{\mathbf{B}}}^T\mathbf{D}^{ep}\overline{\mathbf{B}}d\Omega + \int_{\Omega} \overline{\mathbf{G}}^T M_s\mathbf{G}d\Omega\right)\Delta\mathbf{u} + \left(\int_{\Omega} \overline{\overline{\mathbf{B}}}^T\mathbf{D}^{ep}\overline{\overline{\mathbf{B}}}d\Omega + \int_{\Omega} \overline{\mathbf{G}}^T M_s\overline{\mathbf{G}}d\Omega\right)\Delta\mathbf{a} = \Delta\overline{\mathbf{f}} \tag{9.96}$$

Finally, the EFGM model for large deformations can be written in matrix form as

$$\begin{bmatrix} \int_{\Omega} \overline{\mathbf{B}}^T\mathbf{D}^{ep}\overline{\mathbf{B}}d\Omega + \int_{\Omega} \mathbf{G}^T M_s\mathbf{G}d\Omega & \int_{\Omega} \overline{\mathbf{B}}^T\mathbf{D}^{ep}\overline{\overline{\mathbf{B}}}d\Omega + \int_{\Omega} \mathbf{G}^T M_s\overline{\mathbf{G}}d\Omega \\ \int_{\Omega} \overline{\overline{\mathbf{B}}}^T\mathbf{D}^{ep}\overline{\mathbf{B}}d\Omega + \int_{\Omega} \overline{\mathbf{G}}^T M_s\mathbf{G}d\Omega & \int_{\Omega} \overline{\overline{\mathbf{B}}}^T\mathbf{D}^{ep}\overline{\overline{\mathbf{B}}}d\Omega + \int_{\Omega} \overline{\mathbf{G}}^T M_s\overline{\mathbf{G}}d\Omega \end{bmatrix} \begin{Bmatrix} \Delta\mathbf{u} \\ \Delta\mathbf{a} \end{Bmatrix} = \begin{Bmatrix} \Delta\mathbf{f} \\ \Delta\overline{\mathbf{f}} \end{Bmatrix} \tag{9.97}$$

which can be expressed in a more compact form as

$$\begin{bmatrix} \mathbf{K}^{uu} & \mathbf{K}^{ua} \\ \mathbf{K}^{au} & \mathbf{K}^{aa} \end{bmatrix} \begin{Bmatrix} \Delta\mathbf{u} \\ \Delta\mathbf{a} \end{Bmatrix} = \begin{Bmatrix} \Delta\mathbf{f} \\ \Delta\overline{\mathbf{f}} \end{Bmatrix} \tag{9.98}$$

where

$$\mathbf{K}^{\alpha\beta} = \int_{\Omega^e} (\mathbf{B}^{\alpha})^T\mathbf{D}^{ep}\mathbf{B}^{\beta}\mathrm{d}\Omega + \int_{\Omega^e} (\mathbf{G}^{\alpha})^T\mathbf{M}_s\mathbf{G}^{\beta}\mathrm{d}\Omega \quad (\alpha, \beta = u, a) \tag{9.99}$$

$$\mathbf{f}^{\alpha} = \int_{\Gamma^e} \mathbf{N}^{\alpha}\mathbf{t}\mathrm{d}\Gamma + \int_{\Omega^e} \mathbf{N}^{\alpha}\mathbf{b}\mathrm{d}\Omega \quad (\alpha = u, a) \tag{9.100}$$

where $\mathbf{D}^{ep}$ represents elasto-plastic constitutive matrix, $N^u = \mathbf{N}$, $N^a = \overline{\mathbf{N}}$, $\mathbf{B}^u = \overline{\mathbf{B}}$, $\mathbf{B}^a = \overline{\overline{\mathbf{B}}}$, $\mathbf{G}^u = \mathbf{G}$, $\mathbf{G}^a = \overline{\mathbf{G}}$.

The background cells lying near the material irregularities do not have sufficient Gauss quadrature points as required for performing numerical integration over the domain, and hence, the numerical integration may not yield accurate results. Therefore background cells located at the discontinuity are divided into triangular subdomains, the boundaries of which conform to the geometry of material irregularities. Gauss integration points obtained for subtriangular domains are used for performing numerical integration over the background cells lying near the internal material interfaces. The subtriangular domains are created only for the purpose of numerical integration, and no additional degrees of freedom are introduced because of these subdomains.

9.5 Updated Lagrangian formulation

ULA, for investigating GNL problems, uses current state of loaded body as the reference state for analysis. Here, all quantities are described with respect to latest known deformed state C_1, as shown in Fig. 9.1. The equilibrium equation is expressed with respect to the latest known C_1 configuration as

$$\int_{\Omega^1} {}_1^2S_{ij}\delta({}_1^2E_{ij})d\Omega^1 - \int_{\Omega^1} b_i\delta u_i d\Omega^1 - \int_{\Gamma_t^1} t_i\delta u_i d\Gamma_t^1 = 0 \tag{9.101}$$

where t_i represents the applied tractions, ${}_1^2S_{ij}$ is the updated second Piola–Kirchhoff (PK2) stress, and ${}_1^2E_{ij}$ represents the updated Green–Lagrange strain. Here, state "1" is the current configuration of deformable body, whereas state "2" denotes the initial unloaded and undeformed state of the structural component. Substitution of the EFGM approximation, in the above-defined equilibrium equation, yields the final set of algebraic equations for solving large deformations as $[K]\{u\} = \{f\}$, where $K = K^{mat} + K^{geo}$, such that

$$K^{mat} = \int_{\Omega} B^T D^{ep} B d\Omega \tag{9.102}$$

$$K^{geo} = \int_{\Omega} G^T M_{\sigma} G d\Omega \tag{9.103}$$

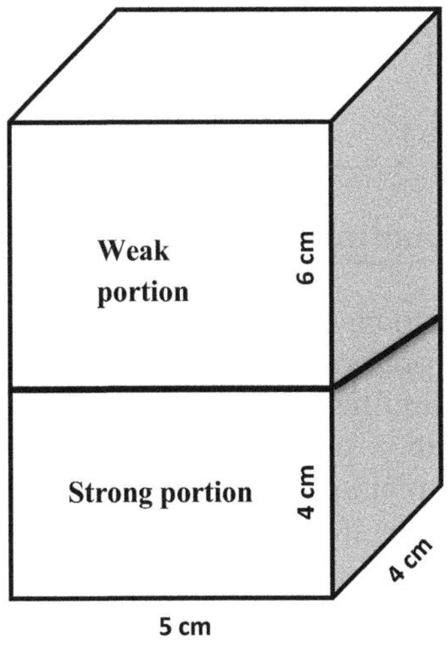

FIGURE 9.1 Three-dimensional bi-material component with a horizontal interface.

$$B = \begin{bmatrix} \frac{\partial N_i}{\partial X} & 0 & 0 \\ 0 & \frac{\partial N_i}{\partial Y} & 0 \\ 0 & 0 & \frac{\partial N_i}{\partial Z} \\ \frac{\partial N_i}{\partial Y} & \frac{\partial N_i}{\partial X} & 0 \\ 0 & \frac{\partial N_i}{\partial Z} & \frac{\partial N_i}{\partial Y} \\ \frac{\partial N_i}{\partial Z} & 0 & \frac{\partial N_i}{\partial X} \end{bmatrix} \tag{9.104}$$

$$G^T = \begin{bmatrix} \frac{\partial N_i}{\partial x} & 0 & 0 & \frac{\partial N_i}{\partial y} & 0 & 0 & \frac{\partial N_i}{\partial z} & 0 & 0 \\ 0 & \frac{\partial N_i}{\partial x} & 0 & 0 & \frac{\partial N_i}{\partial y} & 0 & 0 & \frac{\partial N_i}{\partial z} & 0 \\ 0 & 0 & \frac{\partial N_i}{\partial x} & 0 & 0 & \frac{\partial N_i}{\partial y} & 0 & 0 & \frac{\partial N_i}{\partial z} \end{bmatrix} \tag{9.105}$$

$$\boldsymbol{M}_\sigma = \begin{bmatrix} \sigma_{11}\mathbf{I}_{3\times 3} & \sigma_{12}\mathbf{I}_{3\times 3} & \sigma_{13}\mathbf{I}_{3\times 3} \\ \sigma_{21}\mathbf{I}_{3\times 3} & \sigma_{22}\mathbf{I}_{3\times 3} & \sigma_{23}\mathbf{I}_{3\times 3} \\ \sigma_{31}\mathbf{I}_{3\times 3} & \sigma_{32}\mathbf{I}_{3\times 3} & \sigma_{33}\mathbf{I}_{3\times 3} \end{bmatrix} \tag{9.106}$$

9.6 Large deformation in presence of discontinuities

The conventional finite element method has always proved to be computationally more demanding and expensive in modeling material irregularities present in structural components, as it requires mesh conformation for modeling discontinuities. Element based also suffer extreme mesh distortion while modeling large deformation problems as the aspect ratio of the elements undergoes huge changes. As already explained, large deformations induce GNL in the formulations due to large geometrical changes in structures. All mesh-related issues are eliminated in EFGM, which is a strong and efficient meshless method for carrying out structural simulations. Since no mesh is generated in EFGM, problems of mesh conformation, element distortion, and other mesh-related issues do not occur in such meshless techniques. EFGM has always proved to be a strong and efficient computational tool for modeling materials and GNL in engineering materials. GNL can be seen during various forming operations like forging and die pressing. For bi-material specimens undergoing large deformations, the final set of discrete algebraic equations or the EFGM model for analysis can be obtained as $\mathbf{Kd} = f$, where $\mathbf{d} = \begin{bmatrix} u & a \end{bmatrix}^T$. Stiffness matrix $\mathbf{K}$ and force vector f can be defined for bi-materials as

$$\mathbf{K} = \begin{bmatrix} \mathbf{K}^{uu} & \mathbf{K}^{ua} \\ \mathbf{K}^{au} & \mathbf{K}^{aa} \end{bmatrix} \text{and } f = \begin{Bmatrix} f^u \\ f^a \end{Bmatrix} \tag{9.107}$$

such that

$$\mathbf{K}^{\alpha\beta} = \int_{\Omega^e} (\mathbf{B}^\alpha)^T \mathbf{D}^{ep} \mathbf{B}^\beta \mathrm{d}\Omega + \int_{\Omega^e} (\mathbf{G}^\alpha)^T \mathbf{M}_s \mathbf{G}^\beta \mathrm{d}\Omega \ \ (\alpha, \beta = u, a) \tag{9.108}$$

where $f(x)$ represents the enrichment function that has been added to introduce the effect of discontinuities produced by different material irregularities. Bi-material interfaces are considered weak discontinuities as they produce a continuous displacement field but create a discontinuity in the first derivatives of the displacement field. In the above equations, $\boldsymbol{n}$ denotes standard nodes and $\boldsymbol{n_s}$ represents additional enriched nodes present in the domain.

In the case of crack growth analysis, the effect of large deformation is usually neglected because the displacements in most parts of the domain are very small due to the presence of crack in it. However, the material in the vicinity of the crack tip undergoes large elasto-plastic deformations, which are limited to a very small portion around the crack tip. Since the size of the elasto-plastic zone is very small, these effects are usually neglected during analysis. But as expected, the modeling of different nonlinearities near the crack tip will provide much better and accurate results, because the analysis becomes more realistic and practical. For elasto-plastic crack growth analysis, including large deformations, the final EFGM model is expressed as $\mathbf{Kd} = f$, where $\mathbf{d} = \begin{bmatrix} \mathbf{u} & \mathbf{a} & \mathbf{b}^\alpha \end{bmatrix}^T$. Stiffness and force matrices for crack modeling can be written as

$$[\mathbf{K}] = \begin{bmatrix} \mathbf{K}^{uu} & \mathbf{K}^{ua} & \mathbf{K}^{ub} \\ \mathbf{K}^{au} & \mathbf{K}^{aa} & \mathbf{K}^{ab} \\ \mathbf{K}^{bu} & \mathbf{K}^{ba} & \mathbf{K}^{bb} \end{bmatrix} \tag{9.109}$$

$$\{f\} = \left\{ f_{std} \ \ f_{jump} \ \ f_{tip} \right\}^T \tag{9.110}$$

$$\mathbf{K}^{rs} = \int_\Omega (\mathbf{B}^r)^T \mathbf{D} \mathbf{B}^s d\Omega + \int_\Omega (\mathbf{G}^r)^T \mathbf{M}_s \mathbf{G}^s \mathrm{d}\Omega \quad (r, s = u, a, b) \tag{9.111}$$

$$f_{std} = \int_\Omega \mathbf{N}_{std}^T \mathbf{b} \mathrm{d}\Omega + \int_{\Gamma_t} \mathbf{N}_{std}^T \mathbf{t} \mathrm{d}\Gamma_t \tag{9.112}$$

$$f_{jump} = \int_\Omega \mathbf{N}_{jump}^T \mathbf{b} \mathrm{d}\Omega + \int_{\Gamma_t} \mathbf{N}_{jump}^T \mathbf{t} \mathrm{d}\Gamma_t \tag{9.113}$$

$$f_{tip} = \int_\Omega \mathbf{N}_{tip}^T \mathbf{b} \mathrm{d}\Omega + \int_{\Gamma_t} \mathbf{N}_{tip}^T \mathbf{t} \mathrm{d}\Gamma_t \tag{9.114}$$

such that

$$\mathbf{N}_{std} = \begin{bmatrix} N_i & 0 \\ 0 & N_i \end{bmatrix}; \quad \mathbf{N}_{jump} = \begin{bmatrix} \overline{N}_j & 0 \\ 0 & \overline{N}_j \end{bmatrix}; \quad \mathbf{N}_{tip} = \begin{bmatrix} \overline{\overline{N}}_{k\alpha} & 0 \\ 0 & \overline{\overline{N}}_{k\alpha} \end{bmatrix} \tag{9.115}$$

$$\mathbf{B}^u = \mathbf{B}_{std} = \begin{bmatrix} \frac{\partial N_i}{\partial X}\frac{\partial x}{\partial X} & \frac{\partial N_i}{\partial X}\frac{\partial y}{\partial X} \\ \frac{\partial N_i}{\partial Y}\frac{\partial x}{\partial Y} & \frac{\partial N_i}{\partial Y}\frac{\partial y}{\partial Y} \\ \frac{\partial N_i}{\partial X}\frac{\partial x}{\partial Y} + \frac{\partial N_i}{\partial Y}\frac{\partial x}{\partial X} & \frac{\partial N_i}{\partial X}\frac{\partial y}{\partial Y} + \frac{\partial N_i}{\partial Y}\frac{\partial y}{\partial X} \end{bmatrix} \tag{9.116}$$

$$\mathbf{B}^a = \mathbf{B}_{jump} = \begin{bmatrix} \frac{\partial \overline{N}_j}{\partial X}\frac{\partial x}{\partial X} & \frac{\partial \overline{N}_j}{\partial X}\frac{\partial y}{\partial X} \\ \frac{\partial \overline{N}_j}{\partial Y}\frac{\partial x}{\partial Y} & \frac{\partial \overline{N}_j}{\partial Y}\frac{\partial y}{\partial Y} \\ \frac{\partial \overline{N}_j}{\partial X}\frac{\partial x}{\partial Y} + \frac{\partial \overline{N}_j}{\partial Y}\frac{\partial x}{\partial X} & \frac{\partial \overline{N}_j}{\partial X}\frac{\partial y}{\partial Y} + \frac{\partial \overline{N}_j}{\partial Y}\frac{\partial y}{\partial X} \end{bmatrix} \tag{9.117}$$

$$\mathbf{B}^{b\alpha} = \mathbf{B}_{tip} = \begin{bmatrix} \frac{\partial \overline{\overline{N}}_{k\alpha}}{\partial X}\frac{\partial x}{\partial X} & \frac{\partial \overline{\overline{N}}_{k\alpha}}{\partial X}\frac{\partial y}{\partial X} \\ \frac{\partial \overline{\overline{N}}_{k\alpha}}{\partial Y}\frac{\partial x}{\partial Y} & \frac{\partial \overline{\overline{N}}_{k\alpha}}{\partial Y}\frac{\partial y}{\partial Y} \\ \frac{\partial \overline{\overline{N}}_{k\alpha}}{\partial X}\frac{\partial x}{\partial Y} + \frac{\partial \overline{\overline{N}}_{k\alpha}}{\partial Y}\frac{\partial x}{\partial X} & \frac{\partial \overline{\overline{N}}_{k\alpha}}{\partial X}\frac{\partial y}{\partial Y} + \frac{\partial \overline{\overline{N}}_{k\alpha}}{\partial Y}\frac{\partial y}{\partial X} \end{bmatrix} (\alpha = 1 \ to \ 4) \tag{9.118}$$

$$\mathbf{G}^u = \mathbf{G}_{std} = \begin{bmatrix} \frac{\partial N_i}{\partial X} & 0 \\ 0 & \frac{\partial N_i}{\partial X} \\ \frac{\partial N_i}{\partial Y} & 0 \\ 0 & \frac{\partial N_i}{\partial Y} \end{bmatrix}; (i = 1 \ to \ \boldsymbol{n}) \tag{9.119}$$

$$\mathbf{G}^{a} = G_{jump} = \begin{bmatrix} \dfrac{\partial \overline{N}_j}{\partial X} & 0 \\ 0 & \dfrac{\partial \overline{N}_j}{\partial X} \\ \dfrac{\partial \overline{N}_j}{\partial Y} & 0 \\ 0 & \dfrac{\partial \overline{N}_j}{\partial Y} \end{bmatrix}; \left(j = 1 \text{ to } n_s\right) \tag{9.120}$$

$$\mathbf{G}^{b\alpha} = G_{tip} = \begin{bmatrix} \dfrac{\partial \overline{\overline{N}}_{k\alpha}}{\partial X} & 0 \\ 0 & \dfrac{\partial \overline{\overline{N}}_{k\alpha}}{\partial X} \\ \dfrac{\partial \overline{\overline{N}}_{k\alpha}}{\partial Y} & 0 \\ 0 & \dfrac{\partial \overline{\overline{N}}_{k\alpha}}{\partial Y} \end{bmatrix}; (k = 1 \text{ to } n_t) \tag{9.121}$$

where n are standard nodes, n_s denotes split nodes, and n_t are tip nodes present in the domain. The enrichment strategies for crack growth analysis are available in literature. Variable approximations corresponding to split nodes are enriched with the Heaviside jump function to include the displacement jumps that occur across crack interfaces. On the contrary, displacement approximations corresponding to tip nodes are enriched with crack tip enrichment functions. The above equations for modeling discontinuities have been obtained for the TLA, where displacements, stresses, and strains are expressed with respect to the initial undeformed state of the body. The modeling of discontinuities in the ULA follows the similar procedure, where displacements, stresses, and strains are expressed in terms of the latest known deformed configuration.

9.7 Numerical results and discussions

Mesh-free method based of EFGM has been invoked to solve large elasto-plastic deformations in three-dimensional bi-material structural components. The large deformations occurring in structural components have been investigated by using ULA, in which initial undeformed state is chosen to be the reference configuration for investigation. Yield criterion, based on distortion energy theory, coupled with isotropic hardening considers elasto-plastic effects present in the structural components. Stress integration methodology based on the elastic-predictor plastic-corrector algorithm is used for the accumulation of stresses after each step of simulation. Several numerical problems, presenting large elasto-plastic deformations in 3D structural components containing bi-material interfaces in the domain, are solved to demonstrate the applicability and potential of EFGM in GNL occurring in bi-material structural specimens subjected to monotonic loads.

9.7.1 Large deformation analysis with one bi-material surface

Large elasto-plastic deformation in a three-dimensional cuboidal specimen having a single horizontal bi-material interface is investigated using mesh-free approach based on EFGM. A cuboidal specimen with dimensions 5cm × 10cm × 4cm, as shown in Fig. 9.1, remains constrained at bottom face, whereas compression of 20mm is imposed on the top face of the component. In this bi-material specimen, the weaker portion has an elasto-plastic response with an elastic modulus of 70GPa, Poisson's ratio equal to 0.33, yield strength of 276MPa, and a hardening parameter equal to 3GPa. The stronger portion of the bi-material specimen exhibits elastic behavior during simulation and has Young's modulus equal to 200GPa and Poisson's ratio equal to 0.3. EFGM represents entire domain by a set of nodes as shown in Fig. 9.2. The problems of remeshing, mesh conformation, and element distortions are not seen in EFGM. The final deformed shape of given bi-material structural component corresponding to compression of 20 mm is shown in Fig. 9.3. Distributions of normal stresses (σ_{yy}) and shear stresses (σ_{xy}) across the three-dimensional bi-material specimen are shown in Figs. 9.4 and 9.5, respectively. Corresponding von Mises stresses for the compaction of 12 mm are depicted in Fig. 9.6.

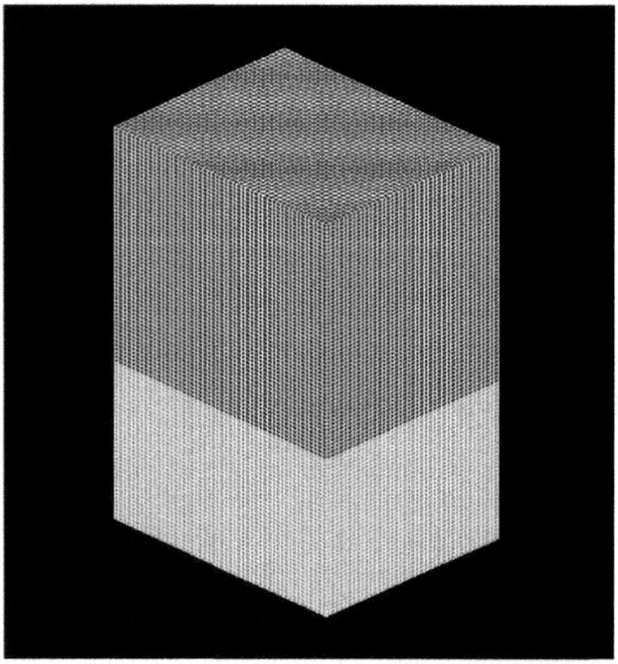

FIGURE 9.2 Representation of domain in EFGM. *EFGM*, Element-free Galerkin method.

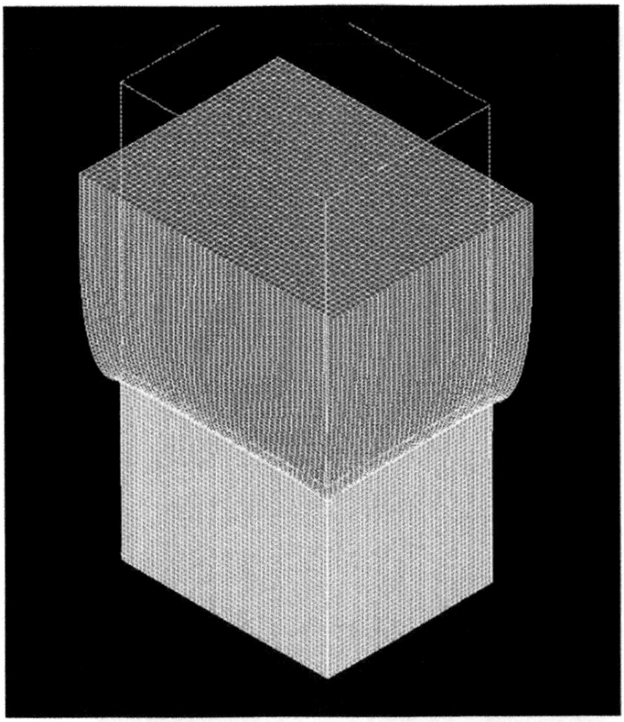

FIGURE 9.3 Deformed configuration obtained by EFGM for a compaction of 20 mm. *EFGM,* Element-free Galerkin method.

Simulations performed in current study clearly show that EFGM provides an efficient numerical framework for investigating large elasto-plastic deformations in three-dimensional structural components.

9.7.2 Large deformation analysis with two bi-material surfaces

This example reports the investigation of large elasto-plastic deformations in the 3D cuboidal specimen containing two bi-material surfaces in the domain, as can be seen in Fig. 9.7. The dimensions of cuboidal specimen and the prescribed boundary conditions are assumed to be the same as discussed in previous problem. A compressive displacement of 20mm is applied at the top face of the component. The material properties of two portions of a given structural component are also assumed to be the same as described in the previous example. The simulations are performed in EFGM by using a nodal distribution of $51 \times 41 \times 101$ nodes as shown in Fig. 9.8. The final deformed configuration of the given structural component, with two internal interfaces, corresponding to compression of 20 mm is shown in Fig. 9.9. Distributions of normal stresses (σ_{yy}) and shear stresses (σ_{xy}) across the three-dimensional bi-material specimen are shown in Figs. 9.10 and 9.11, respectively. Corresponding von Mises stresses for the compaction of 12 mm are shown in Fig. 9.12. Simulations performed in the current study clearly show that EFGM provides an efficient numerical framework for investigating large elasto-plastic deformations in three-dimensional structural components.

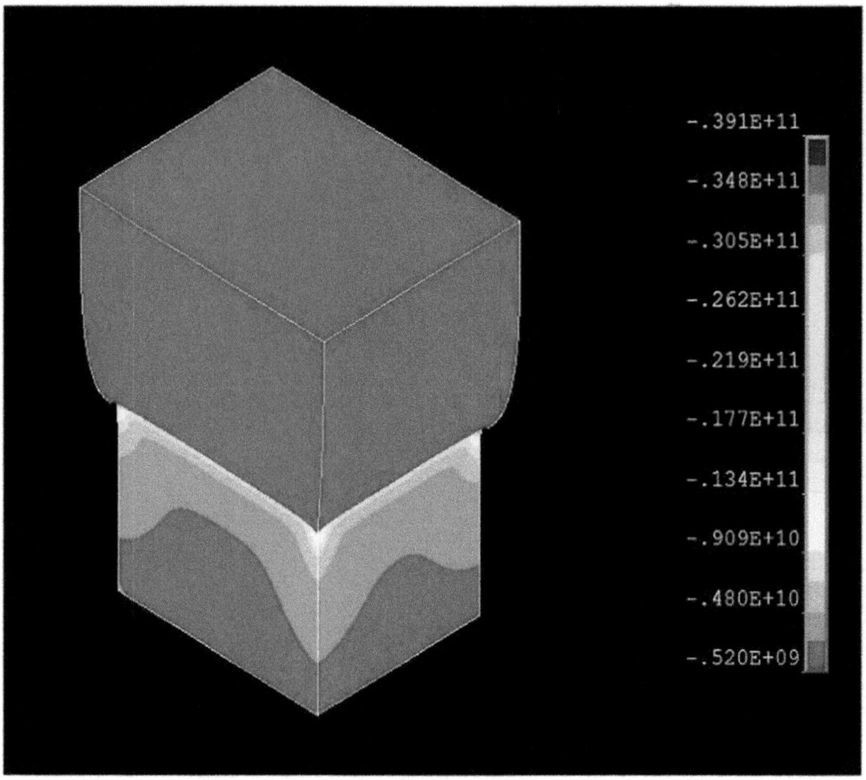

FIGURE 9.4 Normal stress (σ_{yy}) contours obtained by EFGM. *EFGM*, Element-free Galerkin method.

9.7.3 Large deformation with a spherical interface

This example reports the large elasto-plastic deformations in a three-dimensional cuboidal specimen with spherical bi-material interfaces present in the domain, as shown in Fig. 9.13. The dimensions of the cuboidal specimen and the prescribed boundary conditions are assumed to be the same as discussed in previous problem. A compressive displacement of 12mm is applied at the top face of the component. The material properties of two portions of the given structural component are also assumed to be the same as described in the previous example. Simulations have been performed in EFGM by using a nodal distribution of $51 \times 41 \times 101$ nodes as shown in Fig. 9.14. The final deformed configuration of the given structural component, with a spherical interface in the domain, corresponding to the compaction of 12 mm, is shown in Fig. 9.15. The distributions of normal stresses (σ_{yy}) and shear stresses (σ_{xy}) across the three-dimensional bi-material specimen are shown in Figs. 9.16 and 9.17, respectively. Corresponding von Mises stresses for the compaction of 12 mm are shown in Fig. 9.18. Simulations performed in current study clearly show that EFGM provides an efficient numerical framework for investigating large elasto-plastic deformations in three-dimensional structural components.

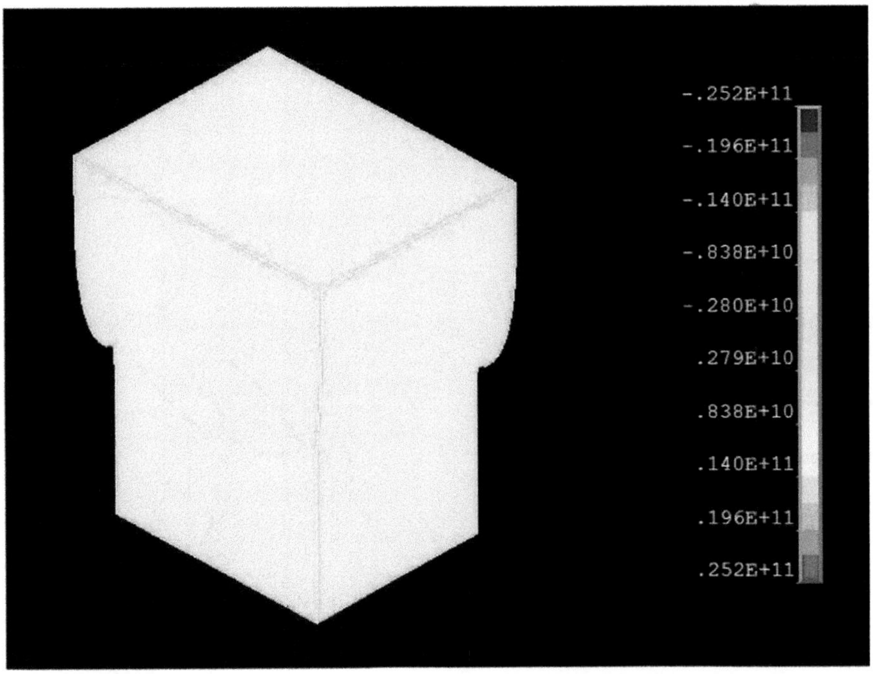

FIGURE 9.5 Shear stress (σ_{xy}) contours obtained by EFGM. *EFGM*, Element-free Galerkin method.

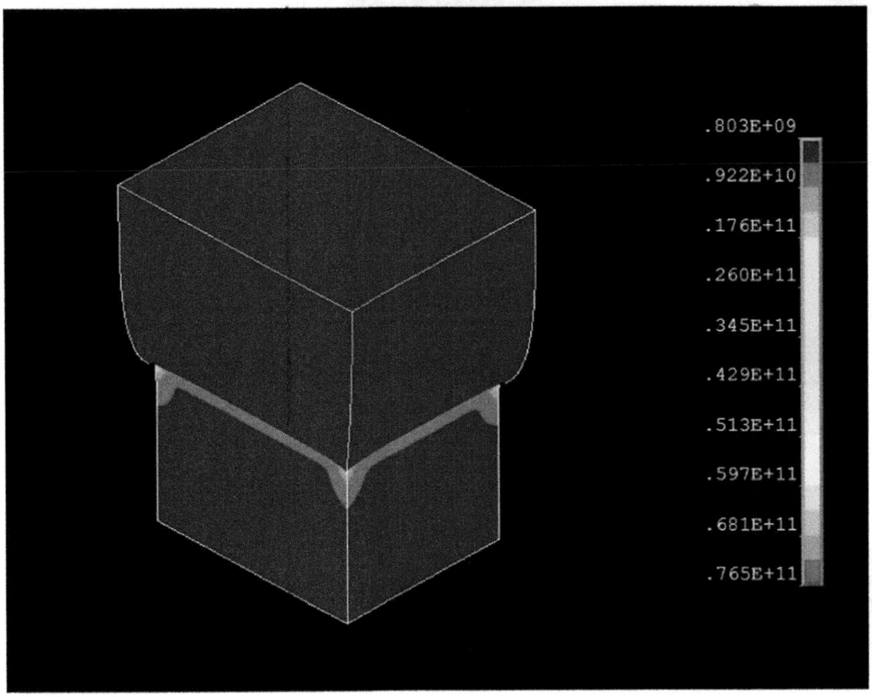

FIGURE 9.6 von Mises stress contours obtained by EFGM. *EFGM*, Element-free Galerkin method.

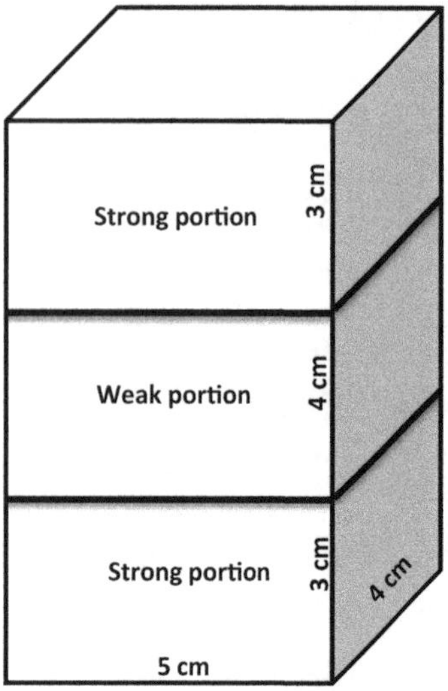

FIGURE 9.7 Three-dimensional bi-material sample with two horizontal interfaces.

FIGURE 9.8 Representation of domain in EFGM. *EFGM*, Element-free Galerkin method.

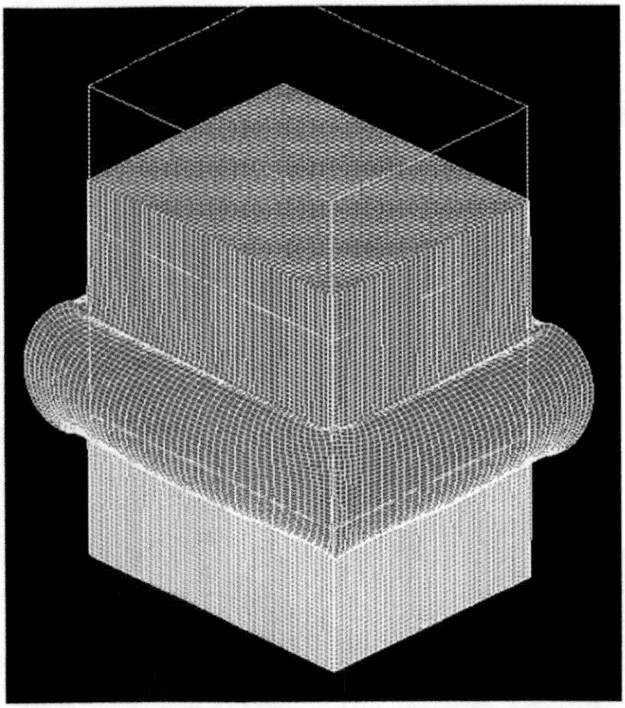

FIGURE 9.9 Deformed configuration obtained by EFGM for a compaction of 20 mm. *EFGM*, Element-free Galerkin method.

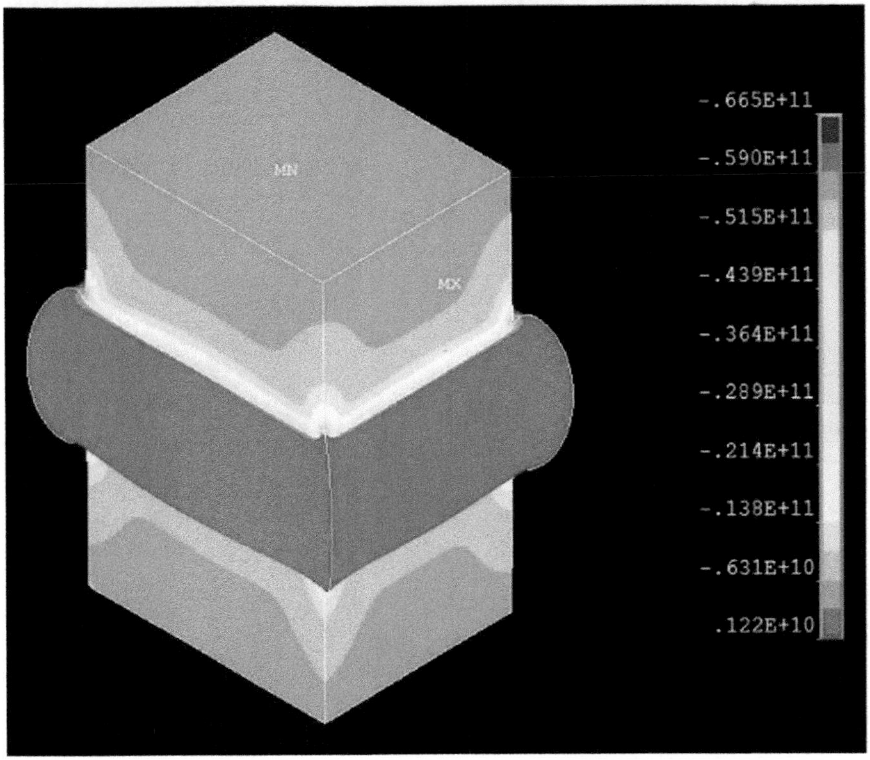

FIGURE 9.10 Normal stress (σ_{yy}) contours obtained by EFGM. *EFGM*, Element-free Galerkin method.

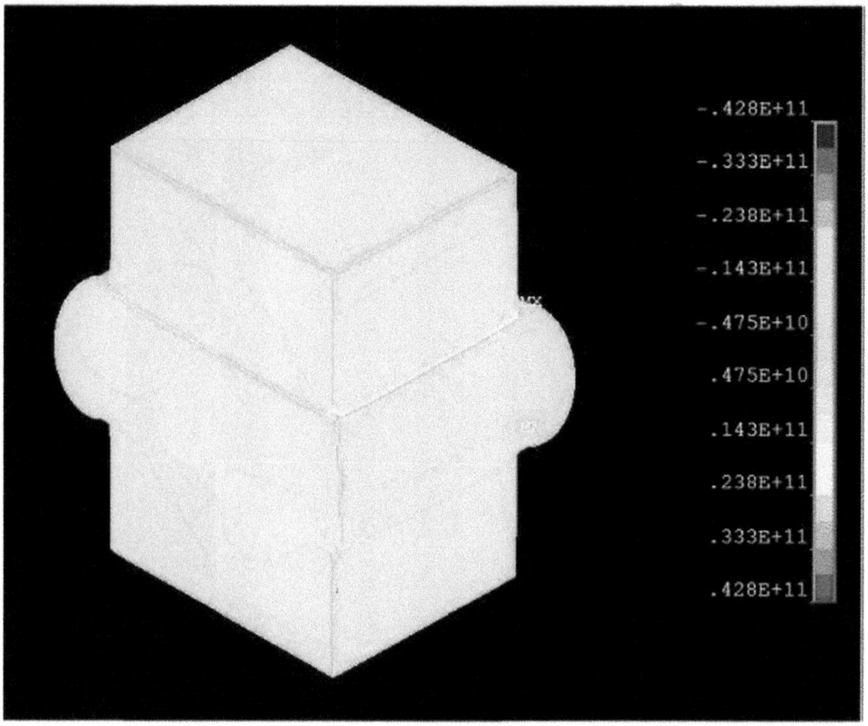

FIGURE 9.11 Shear stress (σ_{xy}) contours obtained by EFGM. *EFGM*, Element-free Galerkin method.

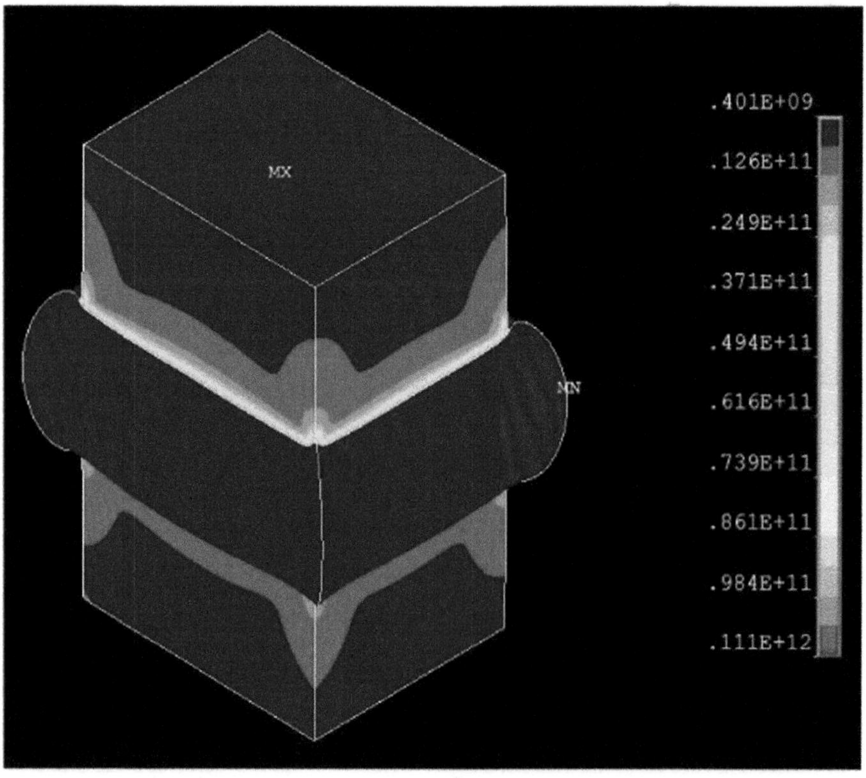

FIGURE 9.12 von Mises stress contours obtained by EFGM. *EFGM*, Element-free Galerkin method.

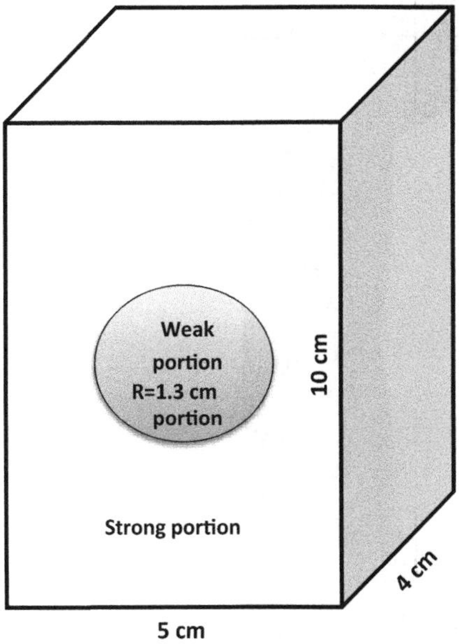

FIGURE 9.13 Three-dimensional bi-material component with spherical interface.

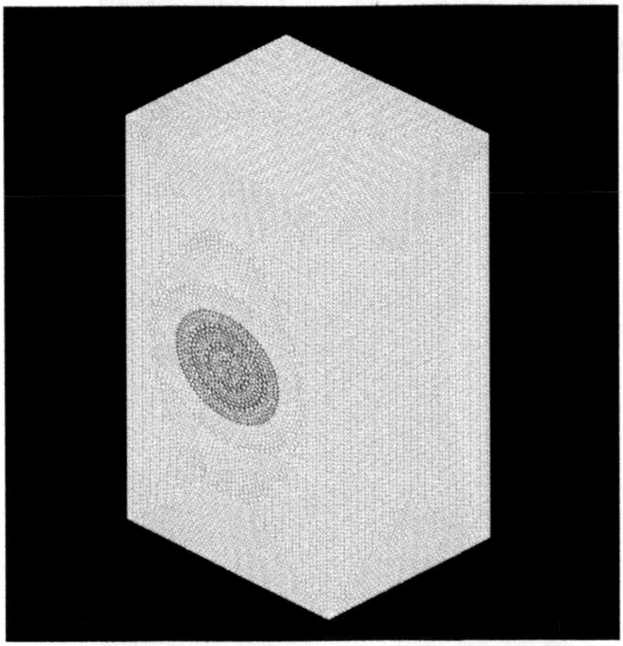

FIGURE 9.14 EFGM domain representation. *EFGM*, Element-free Galerkin method.

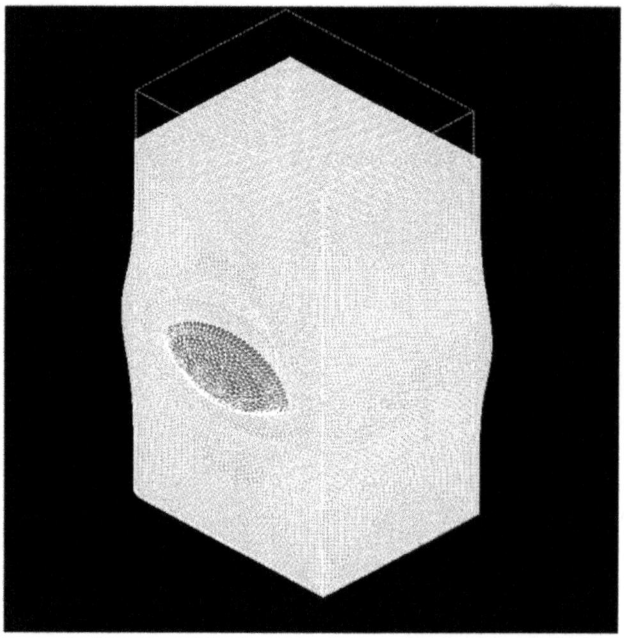

FIGURE 9.15 Deformed configuration for the bi-material component with spherical interface.

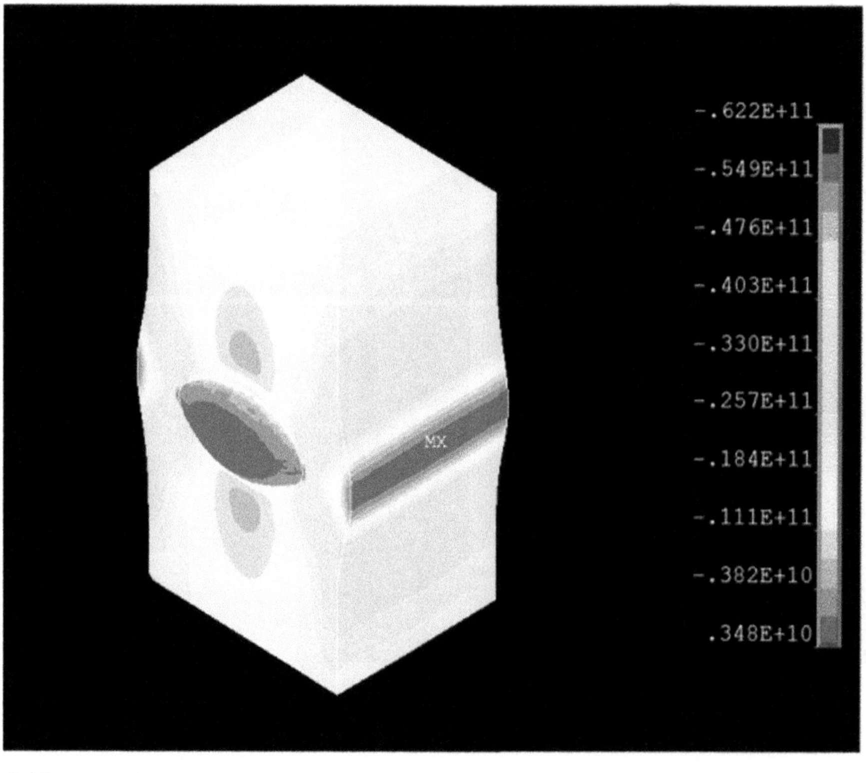

FIGURE 9.16 Normal stress (σ_{yy}) contours for the bi-material component with spherical interface.

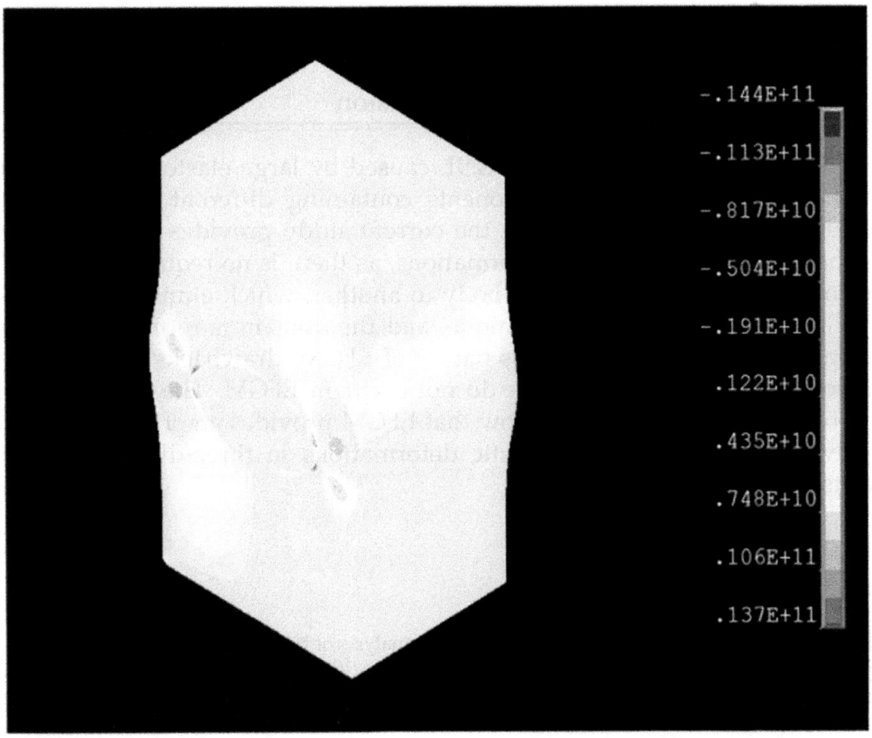

FIGURE 9.17 Shear stress (σ_{xy}) contours for the bi-material component with spherical interface.

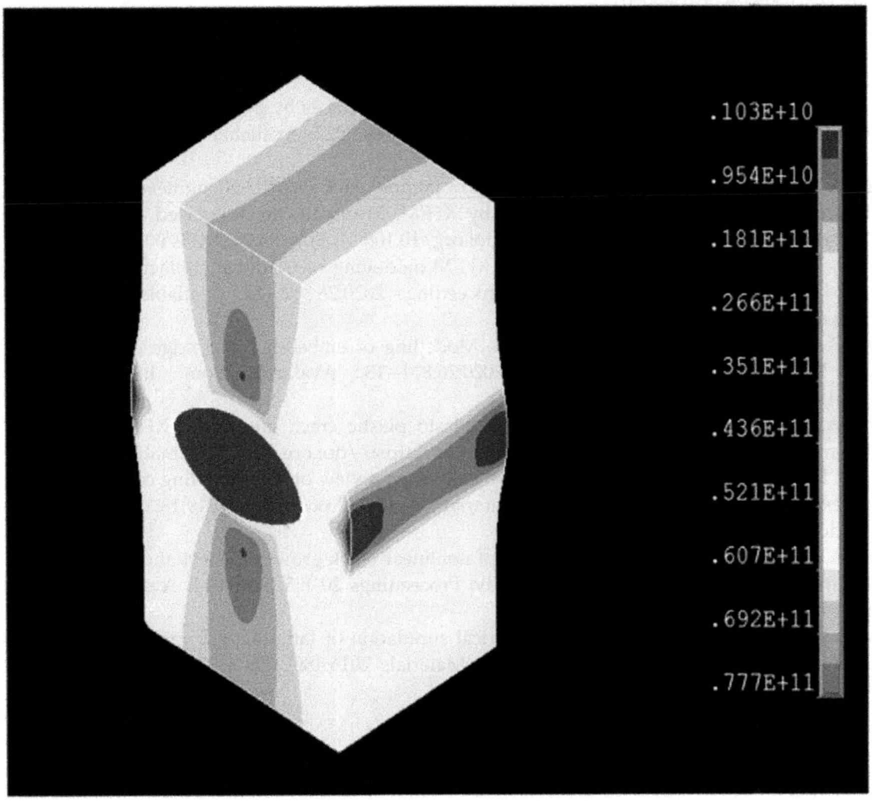

FIGURE 9.18 von Mises stress contours for the bi-material component with spherical interface.

9.8 Conclusion

The EFGM has been used to model GNL caused by large elasto-plastic deformations in three-dimensional structural components containing different types of bi-material interfaces in the domain. ULA used in the current study provides an accurate and efficient approach for modeling large deformations, as there is no requirement for mapping the data from one state of deformable body to another, which eliminated the associated errors. EFGM is a pure meshless technique, and the domain is represented by an array of nodes only. Because of the meshless nature of EFGM, drawbacks of remeshing, mesh conformation, and element distortions do not occur in EFGM. The numerical tests performed in the current study clearly show that EFGM provides an efficient computational tool for investigating large elasto-plastic deformations in three-dimensional structural components.

References

[1] Sheikh UA, Jameel A. Elasto-plastic large deformation analysis of bi-material components by FEM. Materials Today: Proceedings 2020;26:1795–802. Available from: https://doi.org/10.1016/j.matpr.2020.02.377.

[2] Aliabadi MH, Brebbia CA. Boundary element formulations in fracture mechanics: a review. Transactions on Engineering Sciences 1998;17:589–98.

[3] Wen PH, Aliabadi MH, Young A. Dual boundary element methods for three-dimensional dynamic crack problems. Journal of Strain Analysis for Engineering Design 1999;34:373–94. Available from: https://doi.org/10.1177/030932479903400601.

[4] Leonel ED, Venturini WS. Non-linear boundary element formulation applied to contact analysis using tangent operator. Engineering Analysis with Boundary Elements 2011;35:1237–47. Available from: https://doi.org/10.1016/j.enganabound.2011.06.005.

[5] Lone AS, Jameel A, Harmain GA. Modelling of contact interfaces by penalty based enriched finite element method. Mechanics of Advanced Materials and Structures 2022. Available from: https://doi.org/10.1080/15376494.2022.2034075.

[6] Kanth SA, Jameel A, Harmain GA. Investigation of fatigue crack growth in engineering components containing different types of material irregularities by XFEM. Mechanics of Advanced Materials and Structures 2022;29(24):3570–87. Available from: https://doi.org/10.1080/15376494.2021.1907003.

[7] Lone AS, Kanth SA, Jameel A, Harmain GA. XFEM modelling of frictional contact between elliptical inclusions and solid bodies. Materials Today: Proceedings 2020;26:819–24. Available from: https://doi.org/10.1016/j.matpr.2019.12.424.

[8] Kanth SA, Lone AS, Harmain GA, Jameel A. Modelling of embedded and edge cracks in steel alloys by XFEM,. Materials Today: Proceedings 2020;26:814–18. Available from: https://doi.org/10.1016/j.matpr.2019.12.423.

[9] Kanth SA, Lone AS, Harmain GA, Jameel A. Elasto plastic crack growth by XFEM: a review. Materials Today: Proceedings 2019;18:3472–81. Available from: https://doi.org/10.1016/j.matpr.2019.07.275.

[10] Lone AS, Kanth SA, Jameel A, Harmain GA. A state of art review on the modeling of contact type nonlinearities by extended finite element method. Materials Today: Proceedings 2019;18:3462–71. Available from: https://doi.org/10.1016/j.matpr.2019.07.274.

[11] Kanth SA, Harmain GA, Jameel A. Modeling of nonlinear crack growth in steel and aluminum alloys by the element free Galerkin method. Materials Today: Proceedings 2018;5:18805–14. Available from: https://doi.org/10.1016/j.matpr.2018.06.227.

[12] Jameel A, Harmain GA. Modeling and numerical simulation of fatigue crack growth in cracked specimens containing material discontinuities. Strength of Materials 2016;48(2):294–307. Available from: https://doi.org/10.1007/s11223-016-9765-0.

[13] Jameel A, Harmain GA. Fatigue crack growth in presence of material discontinuities by EFGM. International Journal of Fatigue 2015;81:105–16. Available from: https://doi.org/10.1016/j.ijfatigue.2015.07.021.
[14] Rao BN, Rahman S. A coupled meshless-finite element method for fracture analysis of cracks. International Journal of Pressure Vessels and Piping 2001;78:647–57. Available from: https://doi.org/10.1016/S0308-0161(01)00076-X.
[15] Eberhard P, Gaugele T. Simulation of cutting processes using mesh-free Lagrangian particle methods. Computational Mechanics 2013;51:261–78. Available from: https://doi.org/10.1007/s00466-012-0720-z.
[16] Gu YT, Wang QX, Lam KY. A meshless local Kriging method for large deformation analyses. Computer Methods in Applied Mechanics and Engineering 2007;196:1673–84. Available from: https://doi.org/10.1016/j.cma.2006.09.017.
[17] Belytschko T, Krongauz Y, Organ D, Fleming M, Krysl P. Meshless methods: An overview and recent developments. Computer Methods in Applied Mechanics and Engineering 1996;139:3–47. Available from: https://doi.org/10.1016/S0045-7825(96)01078-X.
[18] Rao BN, Rahman S. An enriched meshless method for non-linear fracture mechanics. International Journal for Numerical Methods in Engineering 2004;59:197–223. Available from: https://doi.org/10.1002/nme.868.
[19] Duflot M, Nguyen-Dang H. A meshless method with enriched weight functions for fatigue crack growth. International Journal for Numerical Methods in Engineering 2004;59:1945–61. Available from: https://doi.org/10.1002/nme.948.
[20] Gupta V, Jameel A, Verma SK, Anand S, Anand Y. An insight on NURBS based isogeometric analysis, its current status and involvement in mechanical applications. Archives of Computational Methods in Engineering 2022;. Available from: https://doi.org/10.1007/s11831-022-09838-0.
[21] Gupta V, Jameel A, Verma SK, Anand S, Anand Y. Transient isogeometric heat conduction analysis of stationary fluid in a container. Part E: Journal of Process Mechanical Engineering 2022. Available from: https://doi.org/10.1177/0954408922112.
[22] Jameel A, Harmain GA. Large deformation in bi-material components by XIGA and coupled FE-IGA techniques. Mechanics of Advanced Materials and Structures 2022;29:850–72. Available from: https://doi.org/10.1080/15376494.2020.1799120.
[23] Gupta V, Jameel A, Anand S, Anand Y. Analysis of composite plates using isogeometric analysis: a discussion. Materials Today: Proceedings 2021;44:1190–4. Available from: https://doi.org/10.1016/j.matpr.2020.11.238.
[24] Jameel A, Harmain GA. A coupled FE-IGA technique for modeling fatigue crack growth in engineering materials. Mechanics of Advanced Materials and Structures 2019;26:1764–75. Available from: https://doi.org/10.1080/15376494.2018.1446571.
[25] Jameel A, Harmain GA. Extended iso-geometric analysis for modeling three dimensional cracks. Mechanics of Advanced Materials and Structures 2019;26:915–23. Available from: https://doi.org/10.1080/15376494.2018.1430275.
[26] Singh AK, Jameel A, Harmain GA. Investigations on crack tip plastic zones by the extended iso-geometric analysis. Materials Today: Proceedings 2018;5:19284–93. Available from: https://doi.org/10.1016/j.matpr.2018.06.287.
[27] Jameel A, Harmain GA. Effect of material irregularities on fatigue crack growth by enriched techniques. International Journal for Computational Methods in Engineering Science and Mechanics 2020;21:109–33. Available from: https://doi.org/10.1080/15502287.2020.1772902.
[28] Lone AS, Jameel A, Harmain GA. A coupled finite element-element free Galerkin approach for modeling frictional contact in engineering components. Materials Today: Proceedings 2018;5:18745–54. Available from: https://doi.org/10.1016/j.matpr.2018.06.221.
[29] Jameel A, Harmain GA. Fatigue crack growth analysis of cracked specimens by the coupled finite element-element free Galerkin method. Mechanics of Advanced Materials and Structures 2019;26:1343–56. Available from: https://doi.org/10.1080/15376494.2018.1432800.
[30] Melenk JM, Babuška I. The partition of unity finite element method: basic theory and applications. Computer Methods in Applied Mechanics and Engineering 1996;139:289–314. Available from: https://doi.org/10.1016/S0045-7825(96)01087-0.

[31] Harmain GA, Jameel A, Najar FA, Masoodi JH. Large elasto-plastic deformations in bi-material components by coupled FE-EFGM. IOP Conference Series: Material Science and Engineering 2017;225:1–7. Available from: https://doi.org/10.1088/1757-899X/225/1/012295.
[32] Lone AS, Jameel A, Harmain GA. Enriched element free Galerkin method for solving frictional contact between solid bodies. Mechanics of Advanced Materials and Structures 2022;. Available from: https://doi.org/10.1080/15376494.2022.2092791.
[33] Sukumar N, Moran B, Black T, Belytschko T. An element-free Galerkin method for the three-dimensional fracture mechanics. Computational Mechanics 1997;20:170–5.
[34] Brighenti R. Application of the element-free Galerkin meshless method to 3-D fracture mechanics problems. Engineering Fracture Mechanics 2005;72:2808–20.
[35] Bordas S, Conley JG, Moran B, Gray J, Nichols E. A simulation based design paradigm for complex cast components. Engineering with Computers 2007;23:25–37.

CHAPTER

10

Modeling of Hertzian contact problems by enriched element-free Galerkin method

Aazim Shafi Lone, Showkat Ahmad Kanth, G.A. Harmain and Ishfaq Amin Maekai

Department of Mechanical Engineering, National Institute of Technology Srinagar, Hazratbal, Srinagar, Jammu and Kashmir, India

10.1 Introduction

For most mating parts in mechanical engineering and in the field of tribology, the stress is described by Hertzian contact stress. Most of the time, Hertzian stress may not be significant but, in some cases, may cause serious problems if not taken into account. Application of load causes a local effect, which in most solid body analyses is not taken into account. However, in the case of curved surfaces in contact, the special stress field around the zone of contact is to be considered during analysis and in determining the operating life of components such as rolling element bearings and gears [1]. The curved surface in contact can be classified into two categories, that is, counterformal and conformal [2]. When in the contact region, the radii of curvature are comparatively larger than the area of contact between two contacting surfaces, then such problems are classified under counterformal problems. In cases involving conformal problems, it is pertinent to observe that the contact area holds substantial proportions in contrast to the radii of curvature between the surfaces undergoing contact. Hertzian problems are those counterformal problems in which contacting surfaces in contact zone can be approximated by quadratic functions. Non-Hertzian problems include all conformal problems and those counterformal problems where quadratic approximation is invalid. Theoretically, the area of contact is a line contact in the case of two parallel cylinders, whereas in the case of two spheres, it is a point contact. In reality, due to elastic deformation, a small area of contact is developed, and the stress developed, called Hertzian stress, is limited to this small region.

Enriched Numerical Techniques
DOI: https://doi.org/10.1016/B978-0-443-15362-4.00017-6

In linear elasticity, the theory given by Hertz in 1882 stands as a landmark for two nonconformal bodies having normal frictionless contact [3]. Hertz formation is essentially nonlinear due to the presence of moving boundary and is based on linear kinematics. Similar problems with adhesive contact were analyzed by Spence [4], Mossakovskii [5], and later on by Spence [6] for finite friction. The introduction of finite friction contributes to heightened nonlinearity, particularly when partial slip manifests, necessitating the identification of stick–slip boundaries [7]. Additionally, Spence demonstrated that in the context of polynomial shapes, the stick–slip contour remains unaffected by contact profile alterations during monotonic loading [8]. Extensive research endeavors have been dedicated to the analysis of contact involving unconventional profiles, including scenarios such as flat indenters with blunted cones and rounded edges [9]. In elastic contact mechanics, Hertzian contact solution has wide application such as fatigue life determination for gears, bearings, and geotechnical applications.

For solving contact among complex geometries, the finite element method has been used, which can handle different problems subjected to varying loading, geometries, and material conditions with or without friction [10–14]. Gladwell [15] and Muskhelishvili [16] used the integral equation method in closed form to solve various contact problems [17,18]. Another avenue for tackling contact problems is through the formulation of mathematical programing problems centered on constrained minimization of either total or complementary potential energy. Solution to these problems can be obtained by either employing quadratic programing techniques or incremental linear programing [19]. Chandrasekaran et al. [20] used pseudoequilibrium configuration with geometric constraints imposed to solve nonlinear problems. Franke et al. [21] solved the Hertzian contact problem using classical h-, p-, hp-, and rp-versions with locally and uniformly refined meshes and compared solution with analytical solution.

Numerous instances in solid mechanics involve the interaction and contact of two separate bodies. Some of the practical problems in which contact can be observed include metal-forming operations, crack propagation, and drilling pile. Contact problems are highly nonlinear and complex because the contact tractions as well as the area of contact are unknown. Both tangential and normal stresses emerge within the zone of contact, underscoring the crucial significance of accounting for these supplementary stress components formed during contact. In different forming processes, the contact stresses significantly impact the workpiece's quality. Hence, the contact problems have been given much attention in the recent years from both numerical and experimental point of view. The extended finite element method [22,23], the coupled finite element–element-free Galerkin method (EFGM) [24–26], and the EFGM [27–29] are enhanced techniques that give a better framework for modeling contact problems than the traditional finite element method. Such enriched techniques stand out for their capacity to break free from mesh or grid limitations, enabling modeling to operate independently while sidestepping the need for grid conformity to the contact interface's geometry. Modeling irregularly shaped and evolving discontinuities becomes challenging and computationally costly when using the conventional finite element method, as it necessitates conformal meshing. Modeling large deformation contact problems using the finite element method and the extended finite element method often results in significant mesh distortion. This distortion poses challenges for conducting accurate large deformation analysis and necessitates domain remeshing after each simulation stage. Due to its meshless character, the element-free Galerkin approach outperforms the finite element method in solving massive deformation issues. When dealing with bi-material specimens experiencing significant deformations confined to specific regions of the domain, the coupled finite element–EFGM proves to be a highly efficient approach for modeling these large deformation phenomena. Hertzian contact problems have been solved using various techniques such as the virtual element method [30], the finite element method [31], the coupled FE-EFG method [32], couple-stress elasticity [33], the extended finite element method [34,35], and the frictional mortar contact approach [36].

10.2 Hertzian classical theory

As per Hertz's classical theory of contact, no tension force exists within the contact area, and the nature of contact is nonadhesive type. Certain assumptions were assumed while analyzing Hertzian contact problems. These include the following:

- Due to the substantial difference between the characteristic radius of the contacting bodies and the contact area, the approximation of each body as an elastic half-space is utilized.
- Strains developed are within elastic limit and have small magnitude.
- Contact between two bodies is frictionless.
- The contacting surface is nonconformal and continuous.

This theory is used to find indentation depth and contact area for elastic bodies in contact with simple geometry. The commonly used solution for different geometries of contacting surfaces that is available in literature has been discussed.

10.2.1 Contact between two spheres

Circular area of contact with radius equal to a is developed when two spheres of radii R_1 and R_2 are brought in contact as shown in Fig. 10.1. At the center of circular contact area, maximum contact pressure is expressed as

$$P_{\max} = \frac{3F}{2\pi a^2} \tag{10.1}$$

Here, radius of contact area is expressed as

$$a = \sqrt[3]{\frac{3F\left(\frac{1-v_1^2}{E_1} + \frac{1-v_2^2}{E_2}\right)}{4\left(\frac{1}{R_1} + \frac{1}{R_2}\right)}} \tag{10.2}$$

where moduli of elasticity are E_1 and E_2; Poisson's ratios are v_1 and v_2 for spheres 1 and 2, respectively. The depth of indentation is expressed as

$$d = \frac{a^2}{R} = \sqrt[3]{\frac{9F^2}{16R_{eff}E_{eff}^2}} \tag{10.3}$$

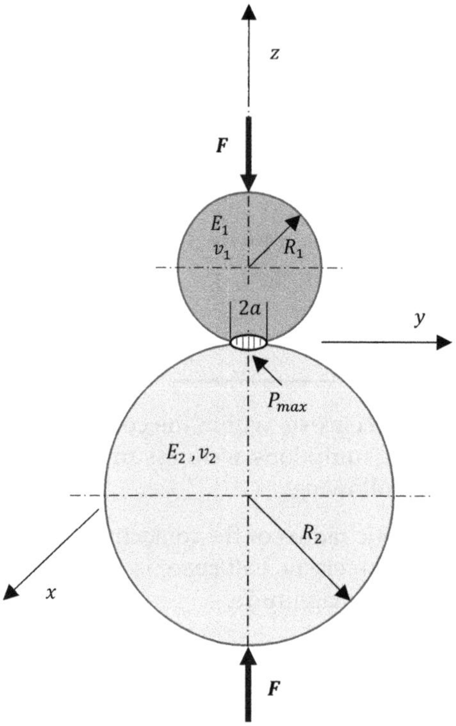

FIGURE 10.1 Two spheres in contact.

where effective radius R_{eff} and effective elastic moduli E_{eff} are given as

$$\frac{1}{R_{eff}} = \frac{1}{R_1} + \frac{1}{R_2} \tag{10.4}$$

$$\frac{1}{E_{eff}} = \frac{1}{2} * \left(\frac{1 - v_1^2}{E_1} + \frac{1 - v_2^2}{E_2}\right) \tag{10.5}$$

10.2.2 Contact between an elastic half-space and a sphere

Elastic sphere having radius R is in contact with elastic half-space as shown in Fig. 10.2, making an indent of depth u. The contact area developed is having radius given by

$$a \cong \sqrt[3]{\left(\frac{3\,R\,F}{2\,E_{eff}}\right)} \tag{10.6}$$

The indentation depth "u" is expressed as

$$u \cong \sqrt[3]{\left(\frac{2F^2}{R\,E_{eff}^2}\right)} \tag{10.7}$$

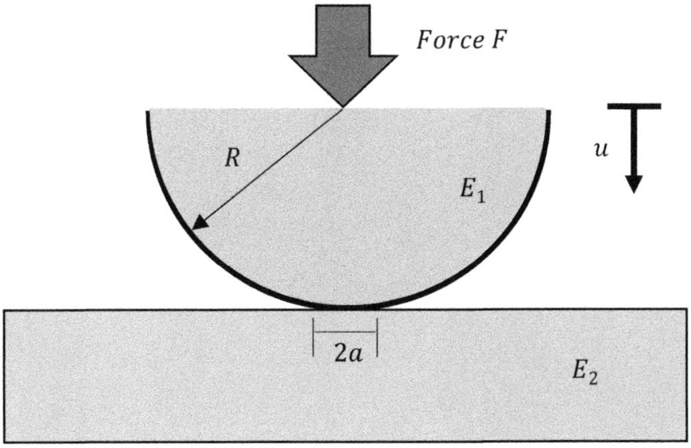

FIGURE 10.2 Contact stress analysis: elastic half-space–sphere interaction.

Stiffness, stress, and maximum shear stress are expressed as

$$K = \frac{dF}{du} \cong \sqrt[3]{R\, E_{eff}^2 F} \tag{10.8}$$

$$(\sigma_c)_{max} \cong \frac{3F}{2\pi a^2} = 0.4\sqrt[3]{\frac{E_{eff}^2 F}{R^2}} = \frac{0.4K}{R} \tag{10.9}$$

$$(\tau)_{max} \cong \frac{(\sigma_c)_{max}}{3} \tag{10.10}$$

10.2.3 Two cylinders in contact with parallel axis

In the case of two cylinders with radii $R_1$and R_2 in contact with a parallel axis, the rectangular contact area with semielliptical pressure distribution is illustrated in Fig. 10.3. In this case, the force is linearly proportional to indentation depth. The width "$2b$" of rectangular contact area developed between two cylinders with a length of contact equal to "L" is expressed as

$$b = \sqrt{\frac{4F\left(\frac{1-v_1^2}{E_1} + \frac{1-v_2^2}{E_2}\right)}{\pi L\left(\frac{1}{R_1} + \frac{1}{R_2}\right)}} \tag{10.11}$$

where two cylinders have moduli of elasticity equal to E_1 and E_2 with Poisson's ratio v_1 and v_2, respectively. The maximum pressure along the center line of rectangular area of contact is expressed as

$$P_{max} = \frac{2F}{\pi bL} \tag{10.12}$$

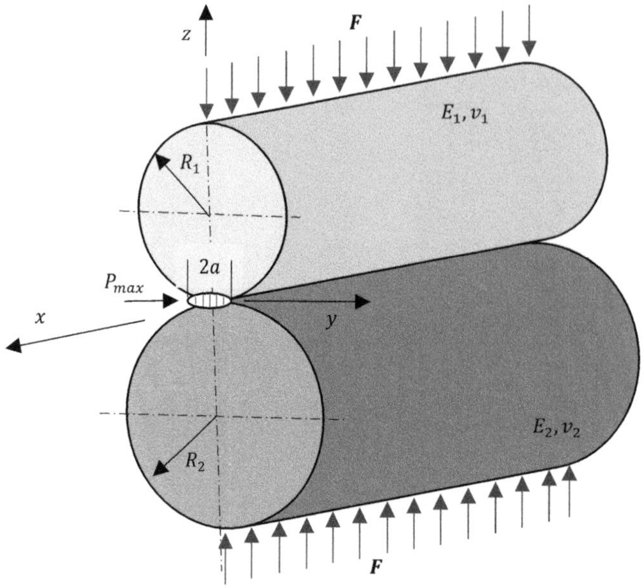

FIGURE 10.3 Two cylinders in contact.

10.3 Continuum model of contact friction

Extensive research efforts have been dedicated to the exploration of contact phenomena, driven by its intricate nature and profound significance within various engineering domains. On contact surface between two bodies, the nature of frictional forces developed is extremely complex and is governed by various factors [37], such as response to normal forces on the contact interface, thermal and inertia effects, history of loading, roughness of contacting interfaces, presence of lubricants, composition of interfaces, frequency of contact, and wear of interface materials. As such, friction emerges as a multifaceted outcome, stemming from a blend of intricate chemical and mechanical processes. Its nuanced characteristics cannot be adequately grasped through isolated, simplistic experiments but require a comprehensive understanding of the complex interplay between these factors. For efficient computational methods to solve nonlinear contact problems, a good understanding of phenomena and mathematical models involved is essential.

Curnier [38] in 1984 proposed the plasticity theory of friction, which was later extended by Laursen and Simo [39] to investigate dynamic sliding friction in the finite element framework. Dynamic sliding friction is one of the broad and essential classes of truly dynamic problems that take account of dynamic sliding effect, frictional damping effect, chattering effect, stick–slip motion effect, and other effects. The frictional forces developed in dynamic sliding friction along the contact interface are dependent upon the relative sliding velocity of one surface over another. The constitutive model for contact friction provides a theoretical framework to describe the interaction between two bodies. Frictional contact dilemmas are categorized into two main classes: one-body contact problems, often referred to as the Signorini problem, and two-body contact problems. The former entails contact between a deformable body and a rigid body, whereas the latter involves deformable bodies in contact. Frictional contact can be divided into three categories: dynamic sliding friction, quasi-static dry friction, and wear. Dynamic sliding friction encompasses frictional forces that are contingent upon the relative velocity of sliding between the contact surfaces. On the other hand, quasi-static dry friction materializes when two meticulously polished surfaces, gradually brought into contact, attain a state of static equilibrium. Wear occurs when interfaces intrude upon one another, leading to notable degradation [37]. The movement between master and slave bodies is delineated by kinematic parameters, while the establishment of their interaction is dictated by static variables. The relative motion between master and slave bodies is characterized by kinematic variables, while the enforcement of their interaction is governed by static variables. To elucidate the relative displacement, a designated reference point "M" is selected on the master body, and the corresponding portion "S" on the slave body in relation to it is depicted in Fig. 10.4. For any given point, denoted as a^s, on the slave body and represented by the coordinate x^s, there is a corresponding counterpart point, a^m on the master body such that [40]

$$x^m = x(a^m) \in argmin\{\|z - x^s\| |z \in x(\Gamma_c^m)\} \qquad \forall a^s \in \Gamma_c^s \tag{10.13}$$

The unit vector $\boldsymbol{n}$, representing the outward normal at point a^m on the master body, is denoted as $n^m(x^m)$ and is expressed within the associated local coordinate frame. Furthermore, the normal contact separation along the contact surface Γ_c^s is established as follows:

$$g_N(a^s) = \boldsymbol{n}\{x(a^s) - x(a^m(a^s))\} \quad \forall a^s \in \Gamma_c^s \tag{10.14}$$

Relative velocities can be obtained as

$$\dot{\delta}(a^s, t) = \dot{x}(a^s, t) - \dot{x}(a^m, t) \tag{10.15}$$

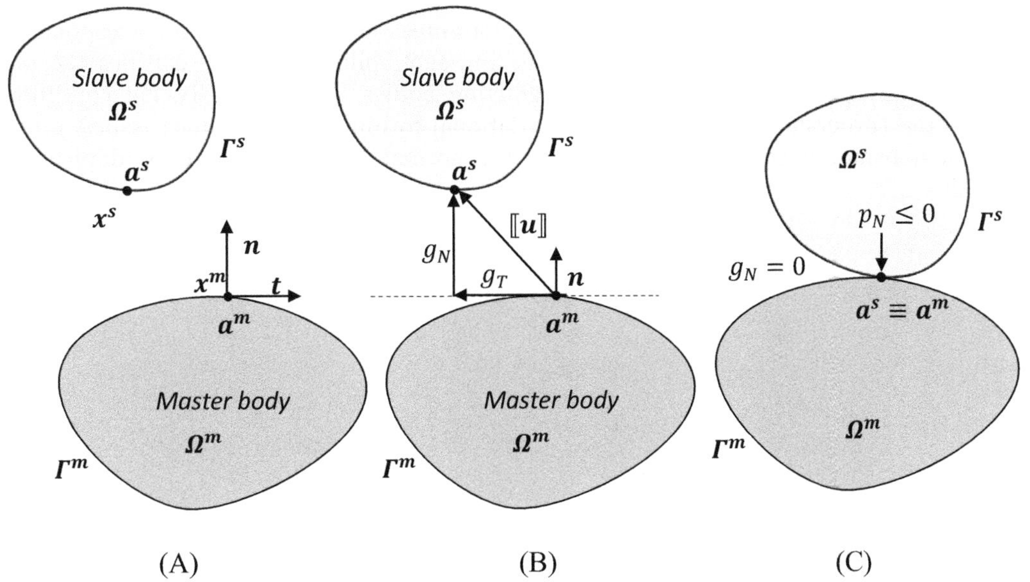

FIGURE 10.4 Depiction of slave and master bodies under different scenarios: (A) no contact, (B) normal and tangential gap, (C) contact condition.

The tangential relative velocities $(\dot{\delta}_t)$ are acquired by projecting the relative velocities "$\dot{\delta}$" onto the tangent vector space $x(\Gamma_c^m)$ at x^m. Relative displacement of point a^s on the slave body surface Γ^s to the point x^m on master body surface Γ^m is expressed as

$$[\![\boldsymbol{u}]\!] = \boldsymbol{u}^s - \boldsymbol{u}^m \tag{10.16}$$

where $\boldsymbol{u}^m$ represents displacement of point x^m, and $\boldsymbol{u}^s$ represents displacement of point a^s. Normal gap function is expressed as

$$g_N = [\![\boldsymbol{u}]\!] . \boldsymbol{n} \tag{10.17}$$

and tangential gap function is expressed as

$$g_T = (\boldsymbol{I} - \boldsymbol{n} \otimes \boldsymbol{n})[\![\boldsymbol{u}]\!] \tag{10.18}$$

In the absence of contact between two bodies, the separation gap between them maintains a value greater than zero, and the contact pressure remains at zero. Under these circumstances, the parameters are as follows: $g_N > 0$ and $p_N = 0$. When the two bodies come into contact, the gap between them diminishes to zero, resulting in a negative contact pressure. In such instances, the parameters are defined as $g_N = 0$ and $p_N < 0$. These two scenarios establish three pivotal contact constraints referred to as the Hertz–Signorini–Moreau conditions. These encompass the complementarity condition $(g_N p_N = 0)$, the compression condition $(p_N \leq 0)$, and the impenetrability condition $(g_N \geq 0)$. The requisite contact constraints applied at the interface of the contacting bodies can be formulated in the standard Kuhn–Tucker framework, as presented in the following equation:

$$g_N \geq 0, \; p_N \leq 0, \; g_N p_N = 0 \tag{10.19}$$

The complementarity condition ensures the cessation of contact tractions in the presence of an open gap. The compression condition guarantees that a closed gap corresponds to a compressive contact pressure, whereas the impenetrability condition certifies the prevention of any penetration among the interacting bodies. The Kuhn–Tucker conditions serve as the cornerstone for tackling the variational formulation of contact issues, giving rise to a nonsmooth contact law governing the normal contact pressure, as depicted in Fig. 10.5.

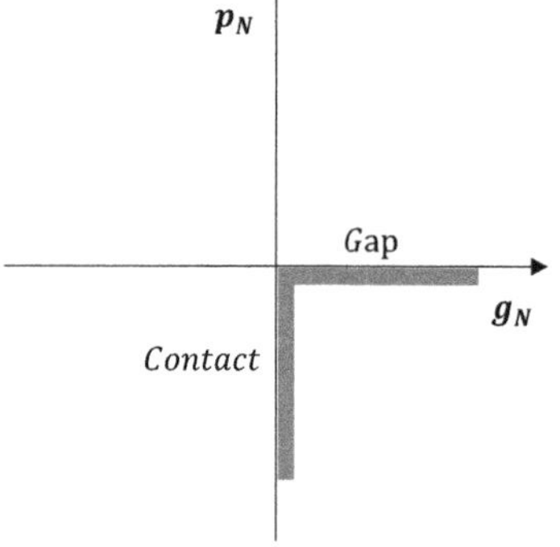

FIGURE 10.5 Unilateral contact.

10.4 Plasticity theory of friction

Frictional contact problems present considerable complexity from both theoretical and numerical standpoints. The movement occurring at the interface of contacting bodies finds its mathematical representation in the theory of friction. By establishing a parallel between plastic and frictional phenomena, the plasticity theory of friction takes shape. This conceptual framework entails the incorporation of several conditions that mirror those found in elasto-plasticity. The fundamental requirements encompass the following:

1. The stress distribution under adhering (elastic) conditions is mathematically characterized by the "adhesion (stick) law."
2. The "stick–slip law" theoretically defines the relationship between stress and stick–slip (elasto-plastic) circumstances.
3. The "wear and tear rule" provides a theoretical framework for understanding softening and hardening phenomena that occur due to sliding.
4. The "slip (yield) criterion" specifies the stress level below which relative slip motion begins.
5. The "slip (flow) principle" establishes the interplay between stress and the onset of slip motion.

The constitutive model for a nonlinear frictional contact problem can be deduced by examining the interaction between a master body and a slave body, as illustrated in Fig. 10.6. During sliding contact between these two bodies, the consideration is limited to the tangential component of contact displacement when the gap between them reduces to zero, resulting in a normal component of displacement that becomes zero. In Fig. 10.6, $\boldsymbol{n}$ signifies the outward normal unit vector on the master surface, accompanied by the normal force p_N and tangential force p_T acting on the master surface. Curnier [38] introduced the "stick–slip theory of friction," which decomposes the relative displacement at the contact interface into two distinct components.

1. Stick (or adherence) component of displacement "$[\![u^A]\!]$"
2. Slip component of displacement "$[\![u^S]\!]$"

The stick component of displacement is reversible, whereas slip component of displacement is irreversible. The elastic deformation of asperities results in the stick component of

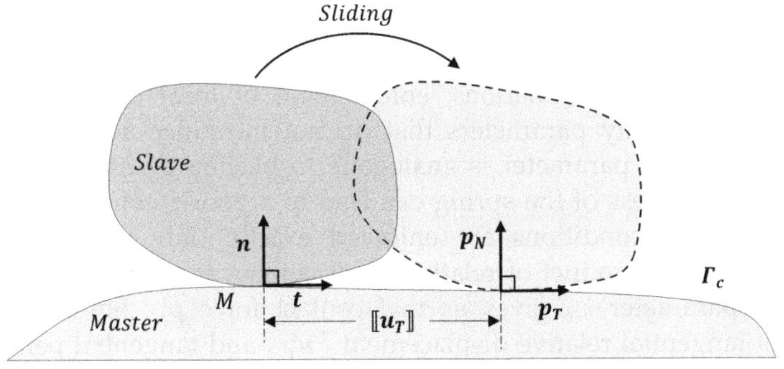

FIGURE 10.6 This figure illustrates master and slave bodies in contact as a consequence of relative tangential displacement.

displacement, whereas the partly plastic deformation of asperities results in the slip component of displacement. Thus on decomposing relative displacement, we have

$$[\![u]\!] = [\![u^A]\!] + [\![u^S]\!] = \{[\![u_N{}^A]\!] + [\![u_T{}^A]\!]\} + \{[\![u_N{}^S]\!] + [\![u_T{}^S]\!]\} = [\![u_N]\!] + [\![u_T]\!] \quad (10.20)$$

Two limitations of Curnier's plasticity theory of friction are as follows:

1. Stick and slip components are obtained by decomposing relative displacement by assuming that, throughout the measurement of relative displacement, the initial point of contact of slave body on the master surface is conserved, that is, origin M.
2. The splitting of relative displacement into tangential component and normal component is achieved by assuming that throughout the sliding process, directions of the normal vector remain nearly constant.

Considering contact velocity in place of contact displacement provides a radical remedy for these two assumptions. As a result, the tangential relative displacement can be deconstructed into two distinct components on the contact surface: an elastic portion and a plastic portion. This decomposition is analogous to the principles found in the theory of plasticity as

$$[\![u_T]\!] = [\![u_T{}^A]\!] + [\![u_T{}^S]\!] \equiv [\![u_T{}^e]\!] + [\![u_T{}^p]\!] \quad (10.21)$$

where superscripts $A, S, p,$ and e denote adherence (or stick), slip, plastic, and elastic parts of tangential displacement, respectively. Further normal displacement is written as

$$[\![u_N]\!] = ([\![u]\!] \cdot \boldsymbol{n})\boldsymbol{n} = (\boldsymbol{n} \otimes \boldsymbol{n})[\![u]\!] \quad (10.22)$$

and tangential displacements is expressed as

$$[\![u_T]\!] = ([\![u]\!] \cdot \boldsymbol{t})\boldsymbol{t} = (\boldsymbol{t} \otimes \boldsymbol{t})[\![u]\!] = (\boldsymbol{I} - \boldsymbol{n} \otimes \boldsymbol{n})[\![u]\!] \quad (10.23)$$

where $\boldsymbol{t}$ is unit tangent vector to contact.

Imposition of contact constraints into weak formulations can be achieved by various techniques such as the Lagrange multiplier approach [41–43], the penalty method [44–46], Nitsche's method [47,48], the mortar method [49–51], and the augmented Lagrange method [52,53]. In the Lagrange multiplier method, governing variational formulation is modified by introducing contact conditions as constraint in the solution. Accurate enforcement of contact constraints is achieved in the Lagrange multiplier approach, as contact forces called Lagrange multipliers are considered primary unknowns [54]. Consequently, new unknowns are introduced into the formulation as a result of a new variable p_N, which increases computational cost. In penalty formulations, enforcement of local constraints is relatively simple by introducing penalty parameters that do not introduce additional unknowns. The inclusion of the penalty parameter is analogous to placing a stiff spring on contact interface. Increasing the stiffness of the spring can lead to a poorly conditioned system of equations, and the contact conditions are enforced exactly only when the spring stiffness approaches infinity. The product of relative displacement in normal direction $[\![u_N]\!]$ and normal penalty parameter k_N gives normal contact force p_N. Similarly, the product of the elastic part of tangential relative displacement $[\![u_T^e]\!]$ and tangential penalty parameter k_T gives the elastic component of contact force in tangential direction p_T.

Total degree of freedom of the domain remains unchanged in the penalty method. The augmented Lagrange method, which inherits the benefits of both strategies, is created by combining the Lagrange multiplier method and the penalty method. Thus constitutive law governing contact loads can be encapsulated as

$$\begin{aligned} p_N &= \left(D_f^e\right)_N [\![u^e]\!] \\ p_T &= \left(D_f^e\right)_T [\![u^e]\!] \end{aligned} \tag{10.24}$$

where $\left(D_f^e\right)_N = -k_N(n \otimes n)$ denotes the normal friction elastic tensor, and $\left(D_f^e\right)_T = -k_T(t \otimes t)$ indicates the tangential friction elastic tensor. By decomposing the relative displacement of contact into distinct stick and slip components, a "slip criterion" or "friction criterion" is introduced to ascertain whether two bodies are in a stick or slip state. The stick state is formulated as $\|\dot{\delta}_t\| = 0$ and $\|p_T\| < k$, signifying that the bodies adhere to each other. Conversely, the slip state is expressed as $\|\dot{\delta}_t\| > 0$, $\|p_T\| = k$ and $\frac{p_T}{\|p_T\|} = \frac{\dot{\delta}_t}{\|\dot{\delta}_t\|}$ indicating that the bodies rub and slide in relation to one another. The friction threshold, denoted as k, is defined as $k = -\mu p_N =$ constant, where μ represents the coefficient of friction and p_N signifies the normal force. This friction criterion establishes the theoretical connection between stick–slip motion and the stress developed at the contact interface. In the case of pure friction [40], the friction criterion is expressed as $\|p_T\| \leq k$. Coulomb's friction law stands as a frequently employed criterion for slip, defined through the consideration of a slip surface denoted as F_f within the contact space. Additionally, it is posited that the coefficient of friction is contingent upon both the normal force "p_N," and the frictional work parameter, "w." As a result, the resulting slip criterion is as follows [55]:

$$F_f(p, w) = \|p_T\| - \mu_f(p_N, w)\|p_N\| - c_f \begin{cases} = 0 & slip(or \quad gap) \\ < 0 & adherence \end{cases} \tag{10.25}$$

where $\|p_T\|$ represents euclidian norm of p_T, $\mu_f = \tan\varphi_f$ is Coulomb's friction coefficient, φ_f represents interface friction angle, and c_f indicates cohesion between two contacting bodies. Fig. 10.7 shows the graph of Coulomb's friction law to show backward slip, adherence, and forward slip. Geometrically, the slip criterion is visually represented by the form

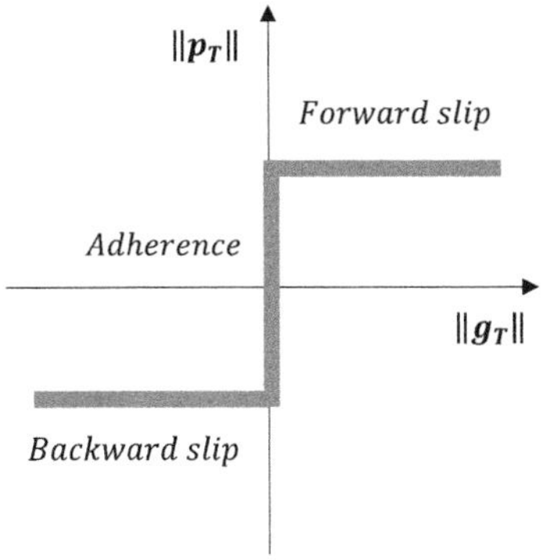

FIGURE 10.7 Nonsmooth Coulomb's friction law.

of a cone, as illustrated in Fig. 10.8, where a tangential–normal axis is affixed at the contact point. In scenarios where $\mu_f = 0$, signifying a lack of frictional force variation with load, the cone transforms into a degenerated cylinder.

The derivation of the slip rule, determining the direction of slip, employs a convex potential gradient denoted as "Z." These slip rules bear resemblance to the flow rules observed in plasticity. By substituting the slip potential "Z" with the slip criterion F_f, the slip rule becomes associated, as portrayed in Fig. 10.9. In conventional plasticity theory, the slip rule aligned with the standard criterion has demonstrated efficacy across various materials. However, adopting a slip rule derived from a friction criterion might result in the emergence of gaps or body separation at the contact interface, stemming from relative movement engendered by the associated potential where $Z = F_f$. To preclude the separation of slave bodies, a nonassociated flow rule is typically preferred [40]. In the context of isotropic frictional contact, the slip potential "Z" takes the form of a cylinder with a radius equivalent to the magnitude of the tangential force $\|\boldsymbol{p}_T\|$, as depicted in Fig. 10.9. The outward normal to the slip potential "Z" delineates the slip direction. By employing the definition of the slip rule, the tangential plastic displacement $[\![\boldsymbol{u}_T^p]\!]$ can be expressed as follows:

$$d[\![\boldsymbol{u}_T^p]\!] = d\lambda \frac{\partial Z}{\partial \boldsymbol{p}_T} = d\lambda \boldsymbol{t} \tag{10.26}$$

where $d\lambda$ is constant and represents the collinearity of the slip increment with the outward normal to the potential Z. Additionally, $\boldsymbol{t} = \frac{\boldsymbol{p}_T}{\|\boldsymbol{p}_T\|}$ denotes tangential unit vector.

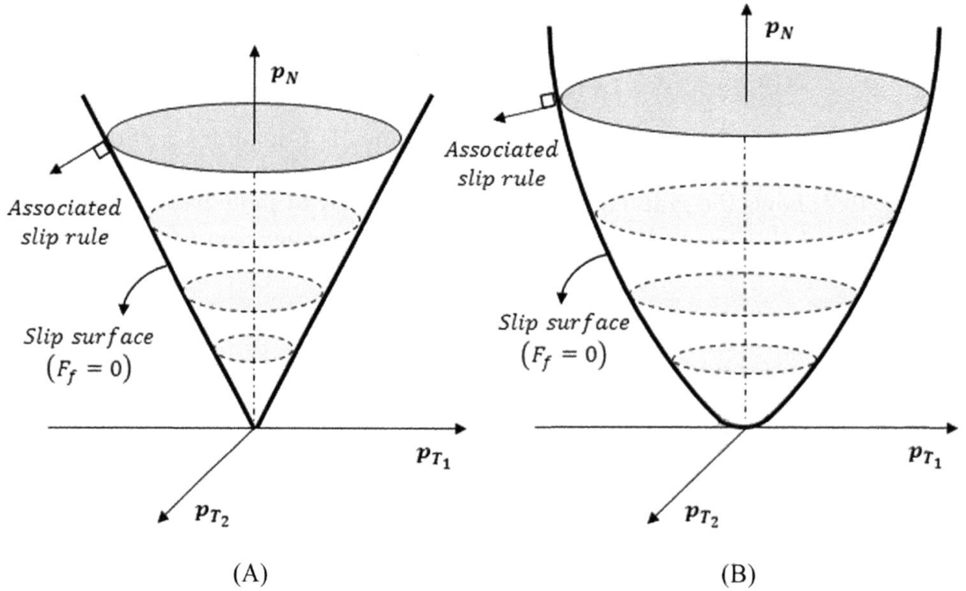

FIGURE 10.8 The slip criteria and associated slip rule: (A) Coulomb's frictional cone and (b) nonlinear slip criterion.

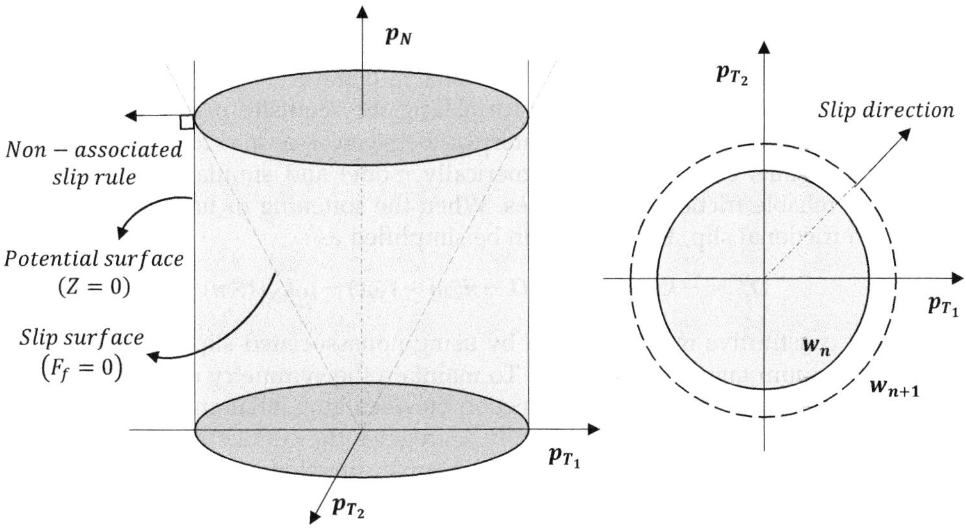

FIGURE 10.9 Nonassociated slip rule: (A) potential surface and (B) slip direction.

For a nonlinear frictional contact problem, the continuum tangent matrix similar to elasto-plasticity can be defined using consistency condition as [56]

$$\left(\frac{\partial F_f(\boldsymbol{p},w)}{\partial \boldsymbol{p}_T}\right)d\boldsymbol{p}_T+\left(\frac{\partial F_f(\boldsymbol{p},w)}{\partial \boldsymbol{p}_N}\right)d\boldsymbol{p}_N+\left(\frac{\partial F_f(\boldsymbol{p},w)}{\partial w}\right)dw=0 \tag{10.27}$$

By substituting $d\lambda$ into the constitutive law, we obtain a linearized equation of form as

$$d\boldsymbol{p}=\boldsymbol{D}_f^{ep}d[\![\boldsymbol{u}]\!] \tag{10.28}$$

where $d\boldsymbol{p}=[d\boldsymbol{p}_N d\boldsymbol{p}_T]^T$, $d[\![\boldsymbol{u}]\!]=[d[\![\boldsymbol{u}_N]\!]d[\![\boldsymbol{u}_T]\!]]^T$ and friction tangent tensor $\boldsymbol{D}_f^{ep}$ is expressed as [57]

$$\boldsymbol{D}_f^{ep}=-k_N(\boldsymbol{n}\otimes\boldsymbol{n})-k_T(\boldsymbol{I}-\boldsymbol{n}\otimes\boldsymbol{n}-\boldsymbol{t}\otimes\boldsymbol{t})-k_T\frac{1}{\beta}\left(\frac{\mu_f}{k_T}\|\boldsymbol{p}_N\|^2\left(\frac{\partial\mu_f}{\partial w}\right)\right)(\boldsymbol{t}\otimes\boldsymbol{t})-k_N\frac{1}{\beta}\left(\mu_f+\|\boldsymbol{p}_N\|\left(\frac{\partial\mu_f}{\partial\|\boldsymbol{p}_N\|}\right)\right)(\boldsymbol{t}\otimes\boldsymbol{n}) \tag{10.29}$$

The first term represents the stiffness in the normal direction, while the subsequent term signifies adhesion stiffness oriented perpendicular to the sliding direction. The third term indicates the adhesion, while the fourth term represents slip stiffness, which may soften or harden the material as it slides and

$$\beta=1+\frac{\mu_f}{k_T}\|\boldsymbol{p}_N\|^2\left(\frac{\partial\mu_f}{\partial w}\right) \tag{10.30}$$

In plasticity theory of friction, terms such as wear, tear, and force of friction are analog terms of isotropic stress, kinematic stress, and stress state, respectively, in the classical theory of plasticity. Given the constraints of available literature, there is a scarcity of comprehensive experimental data capable of furnishing the requisite precision for modeling frictional contact behavior. The intricate interplay between wear and tear forces to the sliding resistance remains challenging to numerically model and simulate, primarily due to the absence of reliable friction factor values. When the softening or hardening parameters are ignored in frictional slip, Eq. (10.29) can be simplified as

$$\boldsymbol{D}_f^{ep} = -k_N(\boldsymbol{n}\otimes\boldsymbol{n}) - k_T(\boldsymbol{I}-\boldsymbol{n}\otimes\boldsymbol{n}-\boldsymbol{t}\otimes\boldsymbol{t}) - \mu_f k_N(\boldsymbol{t}\otimes\boldsymbol{n}) \tag{10.31}$$

The contact constitutive matrix derived by using nonassociated slip rule results in nonsymmetric continuum tangent matrix $\boldsymbol{D}_f^{ep}$. To maintain the symmetry of the numerical formulation, it is possible to ignore the connection between tangential and normal traction at the contact surface. This approach artificially segregates the contact problem into distinct components: a pure contact scenario along the normal direction and a tangential frictional resistance. The constitutive relations for contact can similarly be expressed to establish a connection between relative displacement and tangential and normal contact tractions, as shown in the following equation:

$$\begin{aligned} d\boldsymbol{p}_N &= (\boldsymbol{D}_f)_N\, d[\![\boldsymbol{u}]\!] \\ d\boldsymbol{p}_T &= (\overline{\boldsymbol{D}}_f)_T\, d[\![\boldsymbol{u}]\!] \end{aligned} \tag{10.32}$$

where $(\boldsymbol{D}_f)_N = -k_N(\boldsymbol{n}\otimes\boldsymbol{n})$ represents the normal friction tensor, and $(\overline{\boldsymbol{D}}_f)_T = -\bar{k}_T(\boldsymbol{I}-\boldsymbol{n}\otimes\boldsymbol{n})$ represents the tangential friction tensor for contact surface with $\bar{k}_T$ representing tangential stiffness parameter. The simplified form of continuum tangent matrix defined in Eq. (10.31) can be written as

$$\overline{\boldsymbol{D}}_f^{ep} = -k_N(\boldsymbol{n}\otimes\boldsymbol{n}) - \bar{k}_T(\boldsymbol{I}-\boldsymbol{n}\otimes\boldsymbol{n}) \tag{10.33}$$

To define stick–slip condition, Fig. 10.10 shows stick–slip relationship that is used to obtain an appropriate variation of $\bar{k}_T$. No movement of contact surfaces occurs in stick condition, and there is a built-up of friction force up to $\|\boldsymbol{p}_T\| = \mu_f\|\boldsymbol{p}_N\|$. In stick condition, the slope of $(\|\boldsymbol{p}_T\|, \|\boldsymbol{g}_T\|)$ curve is used to directly derive stick tangential stiffness $\bar{k}_T = k_T^{stick} > 0$. Once the contacting bodies translate from stick condition to slip condition, frictional forces become constant, that is, $F_f = 0$, and corresponding slip tangential stiffness can be obtained as $\bar{k}_T = k_T^{slip} = 0$.

In the classical theory of plasticity, a numerical scheme known as an elastic-predictor or plastic-corrector can be applied to address contact problems. This approach is justified by the similarities between the classical theory of plasticity and the plasticity theory underlying friction phenomena.

10.5 Governing equation for element-free Galerkin method

The equilibrium equation for a two-dimensional (2D) problem on the domain Ω with boundary Γ is illustrated in Fig. 10.11. Displacement boundary conditions are enforced on

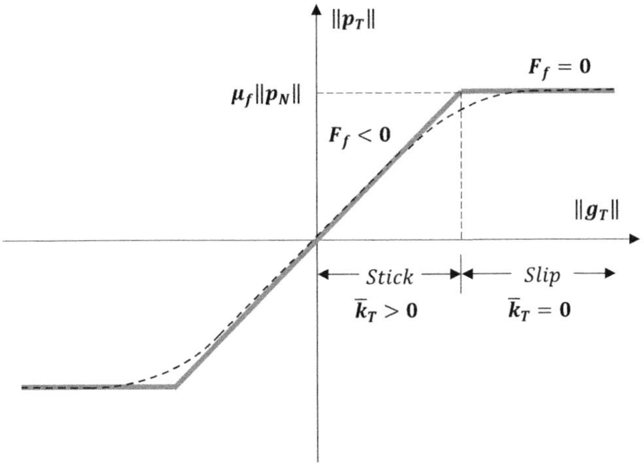

FIGURE 10.10 Frictional force relationship in slip and stick regions.

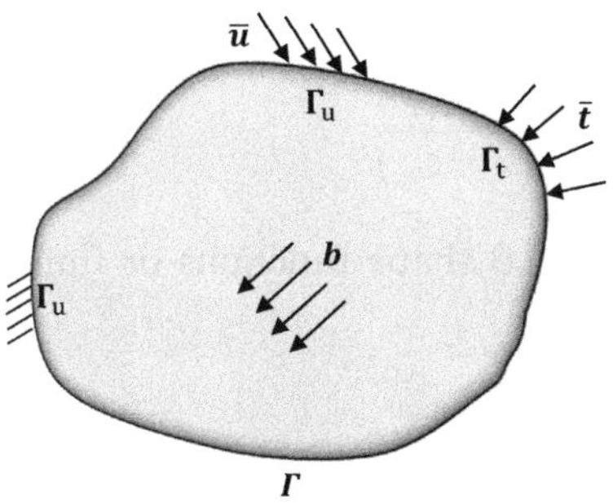

FIGURE 10.11 Illustration of a two-dimensional domain experiencing the effects of body force, along with prescribed displacements and tractions.

Γ_u, while as the tractions are applied on Γ_t. Variational form of equilibrium equation in the EFGM is expressed as

$$\int_{\Omega} \delta\varepsilon{:}\boldsymbol{\sigma}\, d\Omega - \int_{\Omega} \delta\boldsymbol{u}{:}\boldsymbol{b}\, d\Omega - \int_{\Gamma_t} \delta\boldsymbol{u}{:}\, \bar{\boldsymbol{t}}\, d\Gamma_t + \delta\boldsymbol{W}_{\boldsymbol{u}}(u, \nabla_s\lambda) = 0 \tag{10.34}$$

where ε represents a strain tensor. As moving least square (MLS) shape functions lack the fulfillment of the Kronecker delta property, the final term is incorporated to enforce boundary conditions within the element-free Galerkin methodology.

Using the Lagrange multiplier method to enforce essential boundary conditions [58], so

$$\delta\boldsymbol{W}_{\boldsymbol{u}}(\boldsymbol{u}, \lambda) = \int_{\Gamma_u} \delta\boldsymbol{\lambda}(\boldsymbol{u} - \bar{\boldsymbol{u}})d\Gamma_u + \int_{\Gamma_u} \delta\boldsymbol{u}\boldsymbol{\lambda}d\Gamma_u \tag{10.35}$$

where the term Lagrange multiplier $\lambda(x)$ is unknown along the boundary as shown in Fig. 10.12 and can be interpolated by boundary nodes as

$$\boldsymbol{\lambda}(\boldsymbol{x}) = \sum_{I}^{n_\lambda} N_I(s)\lambda_I x\Gamma_u \tag{10.36}$$

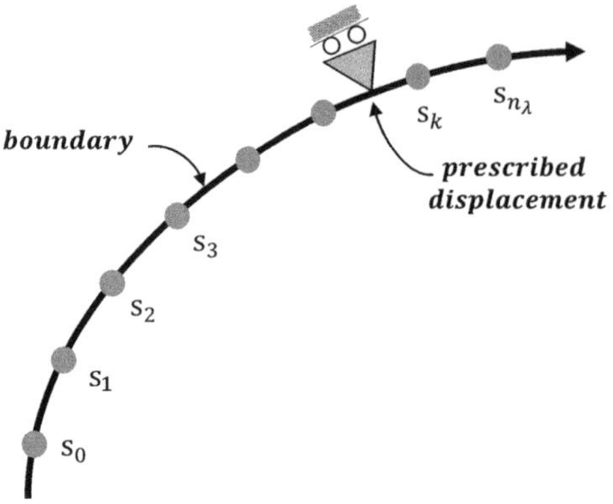

FIGURE 10.12 Local coordinate system showing nodes with applied Lagrange multiplier along an essential boundary.

where s is arc length along the boundary, n_λ are number of nodes used for interpolation. $\boldsymbol{\lambda}_I$ represents vector of discretized values of the Lagrange multiplier. Interpolant $N_I(s)$ can take any form such as Lagrangian as

$$N_I(s) = \frac{(s - s_0)(s - s_1)\cdots(s - s_{I-1})(s - s_{I+1})\left(s - s_{n_\lambda}\right)}{(s_I - s_0)(s_I - s_1)\cdots(s_I - s_{I-1})(s_I - s_{I+1})\left(s_I - s_{n_\lambda}\right)} \tag{10.37}$$

Other interpolants used for approximation can be either MLS shape functions or finite element shape function on boundary.

The equilibrium equation in weak form is stated as follows:

$$\int_\Omega \delta\boldsymbol{\varepsilon}^T\boldsymbol{\sigma}\mathrm{d}\Omega - \int_\Omega \delta\boldsymbol{u}^T\boldsymbol{b}\mathrm{d}\Omega - \int_{\Gamma_t} \delta\boldsymbol{u}^T\overline{\boldsymbol{t}}\mathrm{d}\Gamma_t + \int_{\Gamma_u} \delta\boldsymbol{\lambda}^T(\boldsymbol{u} - \overline{\boldsymbol{u}})\mathrm{d}\Gamma_u + \int_{\Gamma_u} \delta\boldsymbol{u}^T\boldsymbol{\lambda}\mathrm{d}\Gamma_u = 0 \tag{10.38}$$

The terms with the Lagrangian multiplier contribute to nodal forces and global stiffness matrix. The approximate solution in EFGM is constructed entirely in terms of nodes and hence eliminates all mesh-related drawbacks that occur in FEM. The EFGM approximation of the field variable and its variations can be written as

$$\boldsymbol{u}^h(\boldsymbol{x}) = \sum_{i=1}^{n} \Psi_i(\boldsymbol{x})\boldsymbol{u}_i = \boldsymbol{\Psi}\boldsymbol{u}$$

$$\delta\boldsymbol{u}^h(\boldsymbol{x}) = \sum_{i=1}^{n} \Psi_i(\boldsymbol{x})\delta\boldsymbol{u}_i = \boldsymbol{\Psi}\delta\boldsymbol{u} \tag{10.39}$$

where $\boldsymbol{\Psi}$ signifies the MLS shape functions utilized in EFGM, and n indicates the count of nodes within the compact support of the evaluating point being examined. Strain tensor in terms of enrichment and standard DOF is expressed as

$$\varepsilon(\boldsymbol{x}) = \sum_{i\in\Omega} \boldsymbol{B}_i(\boldsymbol{x})\boldsymbol{u}_i + \sum_{j\in\Omega_\Gamma} \overline{\boldsymbol{B}}_j(\boldsymbol{x})\boldsymbol{a}_j = \boldsymbol{B}\boldsymbol{u} + \overline{\boldsymbol{B}}\boldsymbol{a} \tag{10.40}$$

where $\boldsymbol{B}$ and $\overline{\boldsymbol{B}}$ represents spatial derivatives of standard and enriched EFGM shape functions, respectively. Substituting strain field and approximate displacement in variational equilibrium results

$$\delta \boldsymbol{u}^T\left\{\int_\Omega \boldsymbol{B}^T \boldsymbol{C}_e \boldsymbol{B}\boldsymbol{u}\, d\Omega - \int_\Omega \boldsymbol{\Psi}^T \boldsymbol{b}\, d\Omega - \int_{\Gamma_t} \boldsymbol{\Psi}^T \bar{\boldsymbol{t}}\, d\Gamma\right\} + \int_{\Gamma_u} (\boldsymbol{\Psi}\boldsymbol{u} - \bar{\boldsymbol{u}})^T \boldsymbol{N}\delta\boldsymbol{\lambda}\, d\Gamma_u + \int_{\Gamma_u} \boldsymbol{\Psi}^T \delta\boldsymbol{u}^T \boldsymbol{N}\boldsymbol{\lambda}\, d\Gamma_u = 0 \tag{10.41}$$

The aforementioned equilibrium equation can also be expressed as follows:

$$\delta \boldsymbol{u}^T\left[\boldsymbol{K}\boldsymbol{u} - \boldsymbol{f} + \boldsymbol{G}\boldsymbol{\lambda}\right] + \delta\boldsymbol{\lambda}^T\left[\boldsymbol{G}^T\boldsymbol{u} - \boldsymbol{q}\right] = \boldsymbol{0} \tag{10.42}$$

The discrete system of equations in matrix form can written as

$$\begin{bmatrix} \boldsymbol{K} & \boldsymbol{G} \\ \boldsymbol{G}^T & 0 \end{bmatrix}\begin{Bmatrix} \boldsymbol{u} \\ \lambda \end{Bmatrix} = \begin{Bmatrix} \boldsymbol{f} \\ \boldsymbol{q} \end{Bmatrix} \tag{10.43}$$

where

$$\boldsymbol{K} = \int_\Omega \boldsymbol{B}^T \boldsymbol{C}_e \boldsymbol{B} d\Omega;\ \boldsymbol{f} = \int_\Omega \boldsymbol{\Psi}^T \boldsymbol{b} d\Omega + \int_{\Gamma_t} \boldsymbol{\Psi}^T \boldsymbol{t} d\Gamma_t \tag{10.44}$$

$$\boldsymbol{G} = \int_{\Gamma_u} \boldsymbol{\Psi}^T \boldsymbol{N} d\Gamma_u;\ \boldsymbol{q} = \int_{\Gamma_u} \boldsymbol{N}^T \bar{\boldsymbol{u}} d\Gamma_u;\ \boldsymbol{B} = \begin{bmatrix} \Psi_{i,x} & 0 \\ 0 & \Psi_{i,y} \\ \Psi_{i,y} & \Psi_{i,x} \end{bmatrix} \tag{10.45}$$

10.6 Lagrangian multiplier for contact problems

In the contact problems, by minimizing the total potential energy, the equilibrium equation in weak form can be obtained, as shown by the following expression [59–67]:

$$\delta\Pi = \int_\Omega \delta\boldsymbol{\varepsilon}^T \boldsymbol{\sigma} d\Omega - \int_{\Gamma_t} \delta\boldsymbol{u}^T \bar{\boldsymbol{t}} d\Gamma - \int_\Omega \delta\boldsymbol{u}^T \boldsymbol{b} d\Omega + \int_{\Gamma_c} [\![\delta\boldsymbol{u}]\!]^T \boldsymbol{p}_c d\Gamma = 0 \tag{10.46}$$

where ε is strain tensor, $\boldsymbol{\sigma}$ is stress tensor, $\bar{\boldsymbol{t}}$ is traction, and $\boldsymbol{b}$ is body force. Last term is included to incorporate the contact constitutive model and can be written as

$$\int_{\Gamma_c} [\![\delta\boldsymbol{u}]\!]^T \boldsymbol{p}_c d\Gamma = \int_{\Gamma_c} g_N\, p_N\, d\Gamma + \int_{\Gamma_c} [\![\delta u_T]\!]^T \boldsymbol{p}_T\, d\Gamma \tag{10.47}$$

where $\int_{\Gamma_c} g_N\, p_N\, d\Gamma$ represents normal component, and $\int_{\Gamma_c} [\![\delta u_T]\!]^T \boldsymbol{p}_T d\Gamma$ represents tangential component. Normal component will vanish as normal displacement jump is zero on the contact interface. The aforementioned equilibrium equation can also be expressed in its weak form as follows:

$$\int_\Omega \delta\boldsymbol{\varepsilon}^T \boldsymbol{\sigma} d\Omega - \int_{\Gamma_t} \delta\boldsymbol{u}^T \bar{\boldsymbol{t}} d\Gamma - \int_\Omega \delta\boldsymbol{u}^T \boldsymbol{b} d\Omega + \int_{\Gamma_c} [\![\delta u_T]\!]^T \boldsymbol{p}_T d\Gamma = 0 \tag{10.48}$$

Exact enforcement of contact constraints can be achieved by using the Lagrange multiplier method, which introduces additional unknowns, usually referred to as the Lagrange multiplier. For imposing contact constraints, constraint functional Π^{con} in terms of system potential energy is defined as

$$\Pi^{con} = \int_{\Gamma_c} \lambda_N g_N(\boldsymbol{u}) d\Gamma \tag{10.49}$$

where λ_N is the new unknown Lagrange multiplier. Lagrange functional $\overline{\Pi}$ is defined by combining constraint functional Π^{con} with systems potential energy Π as

$$\overline{\Pi} = \Pi + \Pi^{con} \tag{10.50}$$

Minimization of the Lagrange functional can be obtained by taking the variation form of above equation, that is, $\delta\overline{\Pi} = \delta\Pi + \delta\Pi^{con}$, and the resulting equilibrium equation is

$$\int_{\Omega} \delta\boldsymbol{\varepsilon}^T \boldsymbol{\sigma}\, d\Omega - \int_{\Gamma_t} \delta\boldsymbol{u}^T \bar{\boldsymbol{t}}\, d\Gamma - \int_{\Omega} \delta\boldsymbol{u}^T \boldsymbol{b}\, d\Omega + \int_{\Gamma_c} [\![\delta\boldsymbol{u}_T]\!]^T \boldsymbol{p}_T\, d\Gamma + \int_{\Gamma_c} \delta\lambda_N\, g_N\, d\Gamma + \int_{\Gamma_c} \lambda_N\, \delta g_N\, d\Gamma = 0 \tag{10.51}$$

To obtain discrete equations, Lagrange multipliers need to be approximated on contact interface as

$$\lambda_N = \boldsymbol{N}_I\, \boldsymbol{\lambda}_{NI} \tag{10.52}$$

where $\boldsymbol{N}_I$ represents the Lagrange interpolant, and $\boldsymbol{\lambda}_{NI}$ represents the vector of discretized values of the Lagrange multiplier. The Lagrange interpolants used for approximation can be either MLS shape functions or finite element shape function on boundary.

10.7 Imposition of contact constraints using penalty method

The equilibrium equation for the contact problem can be established in weak form as follows:

$$\int_{\Omega} \delta\boldsymbol{\varepsilon}{:}\boldsymbol{\sigma}\, d\Omega - \int_{\Gamma_t} \delta\boldsymbol{u} \cdot \bar{\boldsymbol{t}}\, d\Gamma - \int_{\Omega} \delta\boldsymbol{u} \cdot \boldsymbol{b}\, d\Omega + \int_{\Gamma_c} [\![\delta\boldsymbol{u}]\!] \cdot \boldsymbol{p}_c\, d\Gamma = 0 \tag{10.53}$$

By substituting strain field and approximate displacement in the variational equilibrium equation, we obtain

$$\int_{\Omega} (\boldsymbol{B}\delta\boldsymbol{u} + \overline{\boldsymbol{B}}\delta\boldsymbol{a})^T \boldsymbol{\sigma}\, d\Omega - \int_{\Gamma_t} (\boldsymbol{\Psi}\delta\boldsymbol{u} + \overline{\boldsymbol{\Psi}}\delta\boldsymbol{a})^T \bar{\boldsymbol{t}}\, d\Gamma - \int_{\Omega} (\boldsymbol{\Psi}\delta\boldsymbol{u} + \overline{\boldsymbol{\Psi}}\delta\boldsymbol{a})^T \boldsymbol{b}\, d\Omega + \int_{\Gamma_c} \boldsymbol{\Psi}^T \delta\boldsymbol{a}^T \boldsymbol{p}_c\, d\Gamma = 0 \tag{10.54}$$

Rearranging the above equilibrium equation yields

$$\begin{aligned} &\delta\boldsymbol{u}^T \left\{ \int_{\Omega} \boldsymbol{B}^T \boldsymbol{\sigma}\, d\Omega - \int_{\Gamma_t} \boldsymbol{\Psi}^T \bar{\boldsymbol{t}}\, d\Gamma - \int_{\Omega} \boldsymbol{\Psi}^T \boldsymbol{b}\, d\Omega \right\} \\ &+ \delta\boldsymbol{a}^T \left\{ \int_{\Omega} \overline{\boldsymbol{B}}^T \boldsymbol{\sigma}\, d\Omega + \int_{\Gamma_c} \boldsymbol{\Psi}^T \boldsymbol{p}_c\, d\Gamma - \int_{\Gamma_t} \overline{\boldsymbol{\Psi}}^T \bar{\boldsymbol{t}}\, d\Gamma - \int_{\Omega} \overline{\boldsymbol{\Psi}}^T \boldsymbol{b}\, d\Omega \right\} = 0 \end{aligned} \tag{10.55}$$

The following set of equations can be obtained for values of δu^T and δa^T as

$$\bar{\zeta}_1 \equiv \int_\Omega \boldsymbol{B}^T \sigma d\Omega - \int_{\Gamma_t} \boldsymbol{\Psi}^T \bar{\boldsymbol{t}}\, d\Gamma - \int_\Omega \boldsymbol{\Psi}^T \boldsymbol{b} d\Omega = 0 \tag{10.56}$$

$$\bar{\zeta}_2 \equiv \int_\Omega \bar{\boldsymbol{B}}^T \sigma d\Omega + \int_{\Gamma_c} \boldsymbol{\Psi}^T \boldsymbol{p}_c\, d\Gamma - \int_{\Gamma_t} \bar{\boldsymbol{\Psi}}^T \bar{\boldsymbol{t}} d\Gamma - \int_\Omega \bar{\boldsymbol{\Psi}}^T \boldsymbol{b} d\Omega = 0 \tag{10.57}$$

in which

$$\boldsymbol{F}_{\text{int}}^{std} = \int_\Omega \boldsymbol{B}^T \sigma d\Omega;\ \boldsymbol{F}_{ext}^{std} = \int_{\Gamma_t} \boldsymbol{\Psi}^T \bar{\boldsymbol{t}}\, d\Gamma + \int_\Omega \boldsymbol{\Psi}^T \boldsymbol{b} d\Omega;\ \boldsymbol{F}_{\text{int}}^{enr} = \int_\Omega \bar{\boldsymbol{B}}^T \sigma d\Omega$$

$$\boldsymbol{F}_{ext}^{enr} = \int_{\Gamma_t} \bar{\boldsymbol{\Psi}}^T \bar{\boldsymbol{t}}\, d\Gamma + \int_\Omega \bar{\boldsymbol{\Psi}}^T \boldsymbol{b} d\Omega;\ \boldsymbol{f}_{\text{int}}^{con} = \int_{\Gamma_c} \boldsymbol{\Psi}^T \boldsymbol{p}_c\, d\Gamma \tag{10.58}$$

The normal and tangential displacements at the interface of contact between two bodies can be approximated in the following manner:

$$u_N(\boldsymbol{x}) = \sum_i \Psi_i(\boldsymbol{x}) u_{Ni} + \sum_j \Psi_j(\boldsymbol{x}) \left(H(\boldsymbol{x}) - H(\boldsymbol{x}_j) \right) a_{Nj} \tag{10.59}$$

$$u_T(\boldsymbol{x}) = \sum_i \Psi_i(\boldsymbol{x}) u_{Ti} + \sum_j \Psi_j(\boldsymbol{x}) \left(H(\boldsymbol{x}) - H(\boldsymbol{x}_j) \right) a_{Tj} \tag{10.60}$$

where the normal and tangential displacements at the contact interface are denoted by u_N and u_T, respectively. a_N stands for enhanced degrees of freedom in the normal direction, and a_T for enriched degrees of freedom in the tangential direction. Let $\boldsymbol{t} = \begin{bmatrix} t_1 & t_2 \end{bmatrix}^T$ be the unit tangential vector along the contact interface and $\boldsymbol{n} = \begin{bmatrix} n_1 & n_2 \end{bmatrix}^T$ be the unit normal vector. The aforementioned equation can be expressed in its simplest form:

$$\begin{Bmatrix} u_N \\ u_T \end{Bmatrix} = \begin{bmatrix} \Psi_i & 0 \\ 0 & \Psi_i \end{bmatrix} \begin{Bmatrix} u_{Ni} \\ u_{Ti} \end{Bmatrix} + \begin{bmatrix} \bar{\Psi}_j & 0 \\ 0 & \bar{\Psi}_j \end{bmatrix} \begin{Bmatrix} a_{Nj} \\ a_{Tj} \end{Bmatrix} \tag{10.61}$$

Based on global values of $\boldsymbol{u}_i$ and $\boldsymbol{a}_j$, we can write

$$\begin{Bmatrix} u_{Ni} \\ u_{Ti} \end{Bmatrix} = \begin{bmatrix} n_1 & n_2 \\ t_1 & t_2 \end{bmatrix} \begin{Bmatrix} u_i \\ v_i \end{Bmatrix} \tag{10.62}$$

$$\begin{Bmatrix} a_{Nj} \\ a_{Tj} \end{Bmatrix} = \begin{bmatrix} n_1 & n_2 \\ t_1 & t_2 \end{bmatrix} \begin{Bmatrix} a_j \\ b_j \end{Bmatrix} \tag{10.63}$$

Eq. (10.61) can now be expressed in the following form:

$$\begin{Bmatrix} u_N \\ u_T \end{Bmatrix} = \begin{bmatrix} \Psi_i & 0 \\ 0 & \Psi_i \end{bmatrix} \begin{bmatrix} n_1 & n_2 \\ t_1 & t_2 \end{bmatrix} \begin{Bmatrix} u_i \\ v_i \end{Bmatrix} + \begin{bmatrix} \bar{\Psi}_j & 0 \\ 0 & \bar{\Psi}_j \end{bmatrix} \begin{bmatrix} n_1 & n_2 \\ t_1 & t_2 \end{bmatrix} \begin{Bmatrix} a_j \\ b_j \end{Bmatrix} \tag{10.64}$$

The directional shape functions $\boldsymbol{N}^{con}$ and $\bar{\boldsymbol{N}}^{con}$ for standard and enriched degrees of freedom in normal and tangential directions on contact interface can be obtained as

$$\boldsymbol{\Psi}^{con} = \begin{bmatrix} n_1 & n_2 \\ t_1 & t_2 \end{bmatrix} \begin{bmatrix} \Psi_i & 0 \\ 0 & \Psi_i \end{bmatrix} \tag{10.65}$$

$$\overline{\boldsymbol{\Psi}}^{con} = \begin{bmatrix} n_1 & n_2 \\ t_1 & t_2 \end{bmatrix} \begin{bmatrix} \overline{\Psi}_j & 0 \\ 0 & \overline{\Psi}_j \end{bmatrix} \tag{10.66}$$

The displacement field along the contact surface can be expressed utilizing directional shape functions, as follows:

$$\boldsymbol{u}_C(\boldsymbol{x}) = \boldsymbol{\Psi}^{con}\boldsymbol{u} + \overline{\boldsymbol{\Psi}}^{con}\boldsymbol{a} \tag{10.67}$$

where $\boldsymbol{u}_C(\boldsymbol{x}) = [u_N \; u_T]^T$. The relative displacement at any point along the contact surface can be written as [68]

$$d\boldsymbol{u}_C = \boldsymbol{u}_{top} - \boldsymbol{u}_{bot} = \left(\boldsymbol{\Psi}^{con}\boldsymbol{u} + \overline{\boldsymbol{\Psi}}^{con}\boldsymbol{a}\right)_{top} - \left(\boldsymbol{\Psi}^{con}\boldsymbol{u} + \overline{\boldsymbol{\Psi}}^{con}\boldsymbol{a}\right)_{bot} = \left(\boldsymbol{\Psi}^{con}_{top} - \boldsymbol{\Psi}^{con}_{bot}\right)d\boldsymbol{u} + \left(\overline{\boldsymbol{\Psi}}^{con}_{top} - \overline{\boldsymbol{\Psi}}^{con}_{bot}\right)d\boldsymbol{a} \tag{10.68}$$

In the above equation, $\boldsymbol{\Psi}^{con}_{top} = \boldsymbol{\Psi}^{con}_{bot}$ and $\overline{\boldsymbol{\Psi}}^{con}_{top} = -\overline{\boldsymbol{\Psi}}^{con}_{bot}$. Thus we have

$$d\boldsymbol{u}_C = \boldsymbol{u}_{top} - \boldsymbol{u}_{bot} = 2\overline{\boldsymbol{\Psi}}^{con}_{top}d\boldsymbol{a} = -2\overline{\boldsymbol{\Psi}}^{con}_{bot}d\boldsymbol{a} \tag{10.69}$$

Contact strain matrix $\boldsymbol{B}^{con}$ can be further expressed [68] as $\boldsymbol{B}^{con} = 2\overline{\boldsymbol{\Psi}}^{con}_{top} = -2\boldsymbol{\Psi}^{con}_{bot}$. Consequently, the contact stiffness matrix becomes

$$\boldsymbol{K}^{con} = \int_\Omega (\boldsymbol{B}^{con})^T \boldsymbol{D}^{ep}_f (\boldsymbol{B}^{con}) d\Omega \tag{10.70}$$

Nonlinearities within equilibrium equations emerge from both material and contact nonlinearities. To address these intricate sets of nonlinear equations, the Newton–Raphson method has been employed. This method involves establishing a linearized system of equations, as depicted in the following equation:

$$\begin{bmatrix} \boldsymbol{K}_{uu} & \boldsymbol{K}_{ua} \\ \boldsymbol{K}_{au} & \boldsymbol{K}_{aa} \end{bmatrix} \begin{Bmatrix} d\boldsymbol{u} \\ d\boldsymbol{a} \end{Bmatrix} = -\begin{Bmatrix} \overline{\zeta}_1 \\ \overline{\zeta}_2 \end{Bmatrix} \tag{10.71}$$

The vectors $\overline{\zeta}_1$ and $\overline{\zeta}_2$ are calculated using initial values, and stiffness matrix components can be obtained as

$$\boldsymbol{K}_{uu} = \frac{\partial \overline{\zeta}_1}{\partial \boldsymbol{u}} = \int_\Omega \boldsymbol{B}^T \frac{\partial \boldsymbol{\sigma}}{\partial \boldsymbol{u}} d\Omega; \; \boldsymbol{K}_{uu} = \frac{\partial \overline{\zeta}_1}{\partial \boldsymbol{u}} = \int_\Omega \boldsymbol{B}^T \frac{\partial \boldsymbol{\sigma}}{\partial \boldsymbol{u}} d\Omega;$$

$$\boldsymbol{K}_{au} = \frac{\partial \overline{\zeta}_2}{\partial \boldsymbol{u}} = \int_\Omega \overline{\boldsymbol{B}}^T \frac{\partial \boldsymbol{\sigma}}{\partial \boldsymbol{u}} d\Omega;$$

$$\boldsymbol{K}_{aa} = \frac{\partial \overline{\zeta}_2}{\partial \boldsymbol{a}} = \int_\Omega \overline{\boldsymbol{B}}^T \frac{\partial \boldsymbol{\sigma}}{\partial \boldsymbol{a}} d\Omega + \int_{\Gamma_c} \boldsymbol{\Psi}^T \frac{\partial \boldsymbol{p}_c}{\partial \boldsymbol{a}} d\Gamma = \int_\Omega \overline{\boldsymbol{B}}^T \frac{\partial \boldsymbol{\sigma}}{\partial \boldsymbol{a}} d\Omega + \int_{\Gamma_c} \boldsymbol{\Psi}^T \boldsymbol{D}^{ep}_f \boldsymbol{\Psi} \frac{\partial \boldsymbol{a}}{\partial \boldsymbol{a}} d\Gamma; \tag{10.72}$$

Thus the stiffness matrix is composed of

$$\boldsymbol{K} = \begin{bmatrix} \int_\Omega \boldsymbol{B}^T \boldsymbol{C}_e \boldsymbol{B} d\Omega & \int_\Omega \boldsymbol{B}^T \boldsymbol{C}_e \overline{\boldsymbol{B}} d\Omega \\ \int_\Omega \overline{\boldsymbol{B}}^T \boldsymbol{C}_e \boldsymbol{B} d\Omega & \int_\Omega \overline{\boldsymbol{B}}^T \boldsymbol{C}_e \overline{\boldsymbol{B}} d\Omega + \int_{\Gamma_c} \boldsymbol{\Psi}^T \boldsymbol{D}^{ep}_f \boldsymbol{\Psi} d\Gamma \end{bmatrix} = \begin{bmatrix} \boldsymbol{K}_{uu} & \boldsymbol{K}_{ua} \\ \boldsymbol{K}_{au} & \boldsymbol{K}_{aa} + \boldsymbol{K}^{con} \end{bmatrix} \tag{10.73}$$

where the contact force vector is denoted as f_{int}^{con}, and the contact stiffness matrix is represented as K^{con}, both of which are integrated into the EFGM to incorporate contact behavior. Notably, in the normal direction, the magnitude of the penalty parameter k_N significantly influences the accuracy of enforcing contact constraints. A higher value of the penalty parameter results in a more precise imposition of contact constraints. However, choosing an excessively large penalty parameter can lead to ill-conditioning of the stiffness matrix. Hence, the selection of the penalty parameter value becomes a pivotal factor in determining the efficacy of the penalty method.

10.8 Analysis of the classical Hertzian contact problem: cylinder–rectangular block interaction

The analysis delves into the classical Hertzian problem, encompassing frictionless contact between a cylinder and a rectangular block. This investigation utilizes the EFGM for comprehensive scrutiny. The outcomes derived from this approach are then juxtaposed against the analytical solution documented in existing literature [1]. In the presented scenario, a cylindrical block with a radius of 50mm interacts with a rectangular block, each of its sides measuring 200mm. A vertical load of 5kN is applied atop the cylindrical block, as depicted in Fig. 10.13. It is postulated that the cylindrical block possesses Young's modulus of 70GPa and Poisson's ratio of 0.3. Conversely, the rectangular block is characterized by Young's modulus of 210GPa and Poisson's ratio of 0.3.

Grid independence tests were conducted using nodal distributions of 600, 1350, 2400, 3750, 5400, 7350, 9600, 12,150, and 15,000, with the aim of determining the optimal nodal arrangement for solving the Hertzian contact problem. Notably, the simulations

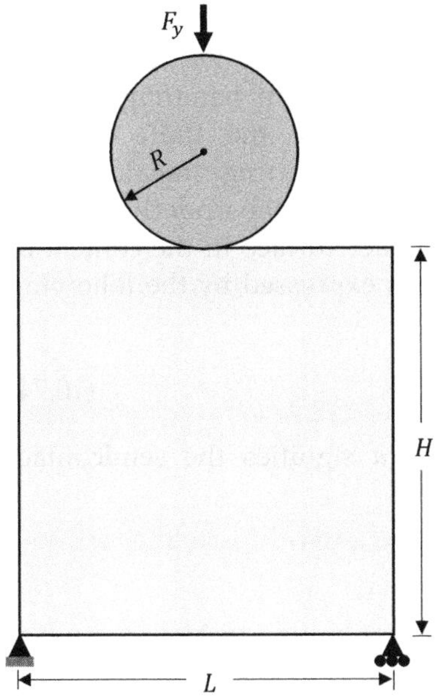

FIGURE 10.13 Hertzian contact problem: cylinder–rectangular block interaction.

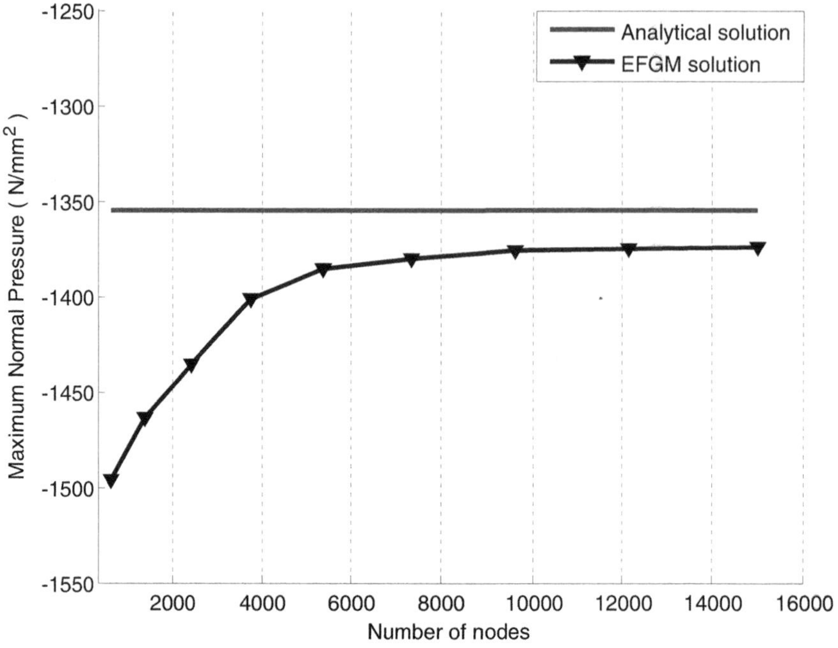

FIGURE 10.14 Grid Independence test for solving Hertzian contact problem.

reached grid independence upon reaching a nodal distribution of 9600 node 9, as indicated in Fig. 10.14. Based on this observation, a uniform nodal distribution of 80×120 was selected for analysis within the EFGM framework, as illustrated in Fig. 10.15. EFGM discretizes the entire domain into a uniform set of nodes, utilizing MLS approximations to effectively estimate the displacement field across the domain. The chosen values for the normal and tangential penalty parameters are consistent with those in Ref. [32]. It has been ascertained that enriched techniques exhibit greater efficiency in handling various types of discontinuities compared to conventional methods like the finite element approach. Remarkably, the absence of discontinuity considerations during domain discretization results in independent modeling of these discontinuities from the underlying grid.

The spatial variation of contact pressure (P) across the contact surface in the context of Hertzian contact between a rectangular block and a cylinder is expressed by the following equation:

$$P = P_{\max}\sqrt{1 - \left(\frac{x}{a}\right)^2} \tag{10.74}$$

where $P_{\max}$ represents the peak contact pressure, whereas a signifies the semicontact width, defined as

$$P_{\max} = \sqrt{\frac{F_y E^*}{\pi R^*}}$$

$$a = \sqrt{\frac{4F_y R^*}{\pi E^*}} \tag{10.75}$$

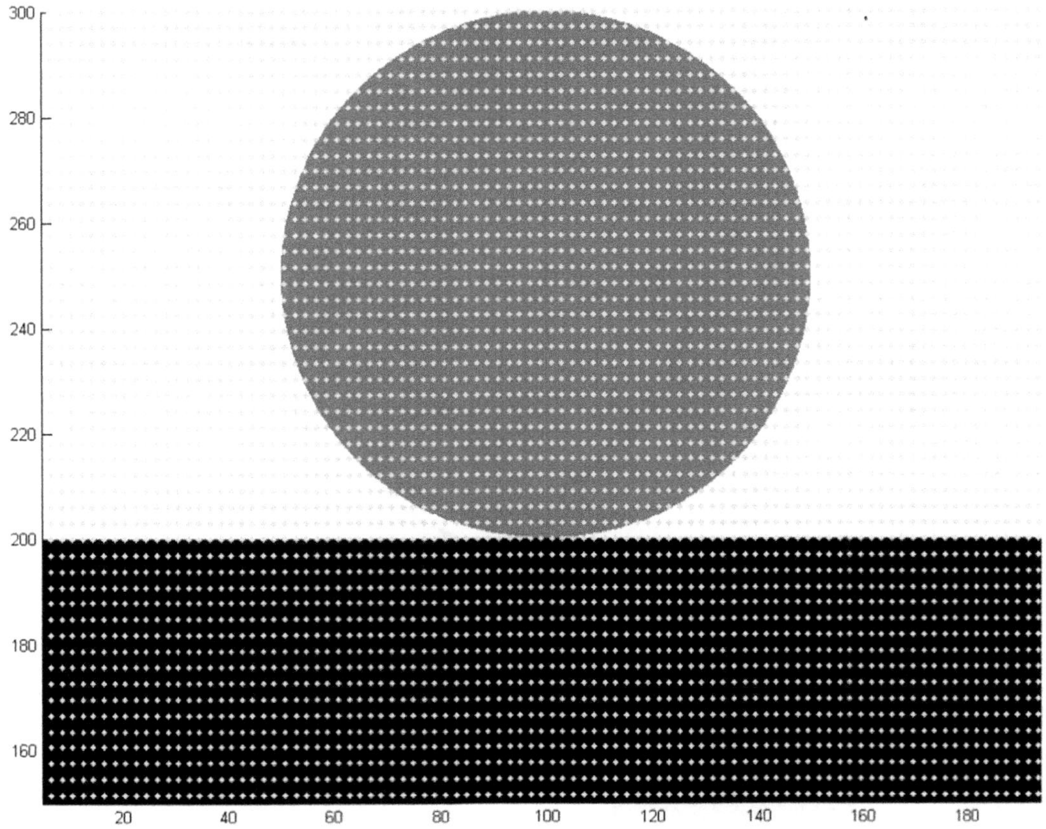

FIGURE 10.15 EFGM nodal distribution of 80 × 120 nodes in the domain representation of the Hertzian cylinder on the block (zoomed view). *EFGM*, Element-free Galerkin method.

Effective material parameter E^* and combined radius R^* are defined as

$$E^* = \frac{E_1 E_1}{E_1(1-\vartheta_2^2) + E_2(1-\vartheta_1^2)}$$

$$R^* = \lim_{R_2 \to \infty} \frac{R_1 R_2}{R_1 + R_2} = R_1 \tag{10.76}$$

The graphical representation of the normal contact pressure variation across the length of the contact surface is depicted in Fig. 10.16. The outcomes yielded by the application of EFGM exhibit a striking concordance with the analytical solution documented in existing literature. This congruence serves to underscore the precision and reliability of the developed MATLAB® codes. Additionally, Fig. 10.17 illustrates the Hertzian stress contour, providing a magnified perspective of the contact zone. The obtained results effectively showcase the capability and adaptability of EFGM in accurately modeling diverse contact interfaces involving solid bodies.

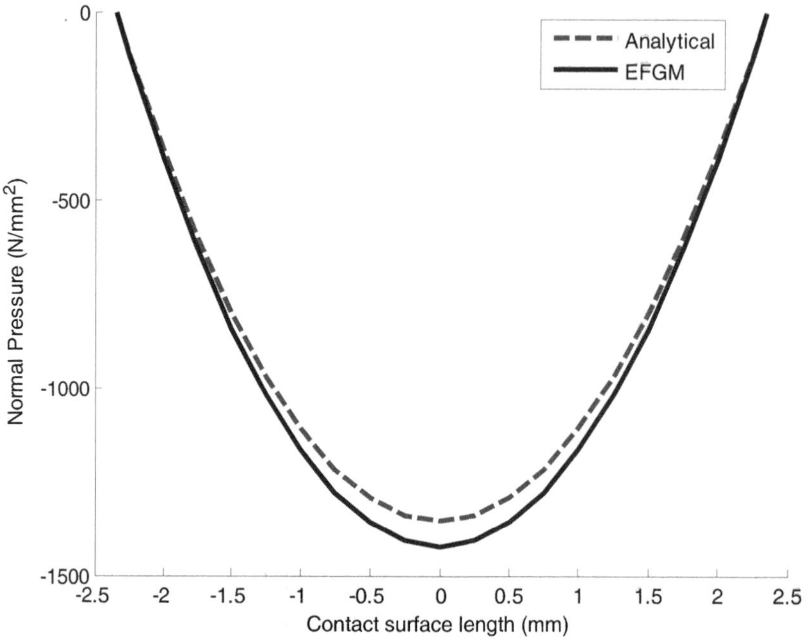

FIGURE 10.16 The depiction illustrates the change in normal pressure along the length of the contact surface. Analytical results are represented by dashed lines, while the outcomes derived from EFGM are presented as solid lines. *EFGM*, Element-free Galerkin method.

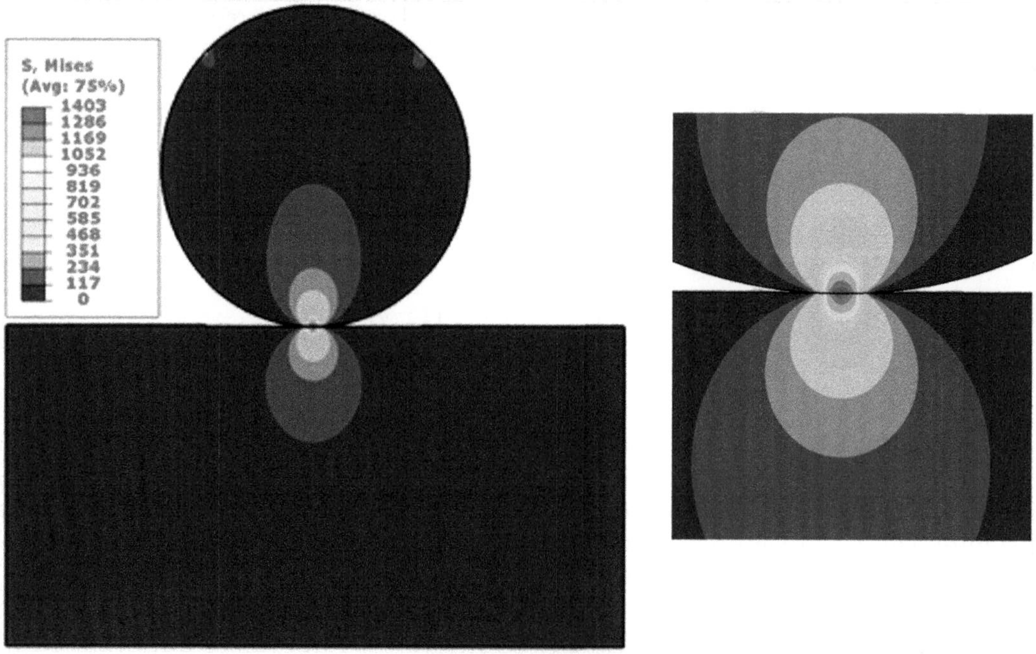

FIGURE 10.17 Hertzian stress contour with zoomed view of contact zone.

10.9 Conclusion

This book chapter presents an investigation into Hertzian contact problems using the enriched EFGM. Hertzian contact problems have long been a focal point in contact mechanics to comprehend the behavior of interacting bodies. The enriched EFGM offers a robust numerical approach to model diverse contact interfaces, as it does not involve different types of discontinuities like contact surfaces during domain discretization. This mitigates mesh-related challenges like distortion and conformal meshing. Various numerical methods exist for enforcing boundary conditions, including Lagrangian-based and penalty approaches. In this chapter, contact constraints are implemented using the penalty method. The practical significance of Hertz contact stress theory is evident, and this work delves into the elastic stress fields generated by various geometries of contact bodies, such as spheres and cylinders.

References

[1] Johnson KL. Contact Mechanics 1985;. Available from: https://doi.org/10.1201/9781003017240-4.

[2] Wriggers P. Computational Contact Mechanics 2006;. Available from: https://doi.org/10.1007/978-3-540-32609-0.

[3] Hertz H. Ueber die Verdunstung der Flüssigkeiten, insbesondere des Quecksilbers, im luftleeren Raume. Annals of Physics 1882;253. Available from: https://doi.org/10.1002/andp.18822531002.

[4] Self similar. Solutions to adhesive contact problems with incremental loading. Proceedings of the Royal Society of London. Series A, Mathematical and Physical Sciences 1968;305. Available from: https://doi.org/10.1098/rspa.1968.0105.

[5] Mossakovskii VI. Compression of elastic bodies under conditions of adhesion (Axisymmetric case). Journal of Applied Mathematics and Mechanics 1963;27. Available from: https://doi.org/10.1016/0021-8928(63)90150-3.

[6] Spence DA. The hertz contact problem with finite friction. Journal of Elasticity 1975;5. Available from: https://doi.org/10.1007/BF00126993.

[7] Storåkers B, Elaguine D. Hertz contact at finite friction and arbitrary profiles. Journal of the Mechanics and Physics of Solids 2005;53. Available from: https://doi.org/10.1016/j.jmps.2004.11.009.

[8] Spence DA. Similarity considerations for contact between dissimilar elastic bodies. Mech. contact between Deform. bodies 1975;. Available from: https://doi.org/10.1007/978-94-011-8137-2_5.

[9] Barber JR, Ciavarella M. Contact mechanics. International Journal of Solids and Structures 2000;37:29–43. Available from: https://doi.org/10.1016/S0020-7683(99)00075-X.

[10] Bathe K-J, Chaudhary A. A solution method for planar and axisymmetric contact problems. International Journal for Numerical Methods in Engineering 1985;21. Available from: https://doi.org/10.1002/nme.1620210107.

[11] Chan SK, Tuba IS. A finite element method for contact problems of solid bodies-Part I. Theory and validation. International Journal of Mechanical Sciences 1971;13. Available from: https://doi.org/10.1016/0020-7403(71)90032-4.

[12] Francavilla A, Zienkiewicz OC. A note on numerical computation of elastic contact problems. International Journal for Numerical Methods in Engineering 1975;9:913–24. Available from: https://doi.org/10.1002/nme.1620090410.

[13] Oden JT, Kikuchi N. Finite element methods for constrained problems in elasticity. International Journal for Numerical Methods in Engineering 1982;18. Available from: https://doi.org/10.1002/nme.1620180507.

[14] Oden JT, Pires EB. Numerical analysis of certain contact problems in elasticity with non-classical friction laws. Computers & Structures 1983;16. Available from: https://doi.org/10.1016/0045-7949(83)90187-6.

[15] Gladwell GML. *Contact problems in the classical theory of elasticity*. Springer Science & Business Media; 1980.

[16] Lowengrub M, Noordhoff NIM. Some basic problems of the mathematical theory of elasticity. The American Mathematical Monthly: The Official Journal of the Mathematical Association of America 1967;74. Available from: https://doi.org/10.2307/2314307.
[17] Kalker JJ. The computation of three-dimensional rolling contact with dry friction. International Journal for Numerical Methods in Engineering 1979;14. Available from: https://doi.org/10.1002/nme.1620140904.
[18] Nayak L, Johnson KL. Pressure between elastic bodies having a slender area of contact and arbitrary profiles. International Journal of Mechanical Sciences 1979;21. Available from: https://doi.org/10.1016/0020-7403(79)90067-5.
[19] Torstenfelt B. Contact problems with friction in general purpose finite element computer programs. Computers & Structures 1983;16. Available from: https://doi.org/10.1016/0045-7949(83)90188-8.
[20] Chandrasekaran N, Haisler WE, Goforth RE. Finite element analysis of Hertz contact problem with friction. Finite Elements in Analysis and Design 1987;3. Available from: https://doi.org/10.1016/0168-874X(87)90032-1.
[21] Franke D, Düster A, Nübel V, Rank E. A comparison of the h-, p-, hp-, and rp-version of the FEM for the solution of the 2D Hertzian contact problem. Computational Mechanics 2010;45. Available from: https://doi.org/10.1007/s00466-009-0464-6.
[22] Kanth SA, Lone AS, Harmain GA, Jameel A. Elasto plastic crack growth by XFEM: a review. Materials Today: Proceedings 2019;18:3472–81. Available from: https://doi.org/10.1016/j.matpr.2019.07.275.
[23] Kanth SA, Harmain GA, Jameel A. Assessment of fatigue life in presence of different hole geometries by X-FEM. Iranian Journal of Science and Technology, Transactions of Mechanical Engineering 2022;. Available from: https://doi.org/10.1007/s40997-022-00569-y.
[24] Lone AS, Jameel A, Harmain GA. A coupled finite element-element free Galerkin approach for modeling frictional contact in engineering components. Materials Today: Proceedings 2018;5:18745–54. Available from: https://doi.org/10.1016/j.matpr.2018.06.221.
[25] Jameel A, Harmain GA. Fatigue crack growth analysis of cracked specimens by the coupled finite element-element free Galerkin method. Mechanics of Advanced Materials and Structures 2019;26:1343–56. Available from: https://doi.org/10.1080/15376494.2018.1432800.
[26] Jameel A, Harmain GA. Large deformation in bi-material components by XIGA and coupled FE-IGA techniques. Mechanics of Advanced Materials and Structures 2020;. Available from: https://doi.org/10.1080/15376494.2020.1799120.
[27] Lone AS, Harmain GA, Jameel A. Enriched element free Galerkin method for solving frictional contact between solid bodies. Mechanics of Advanced Materials and Structures 2022;. Available from: https://doi.org/10.1080/15376494.2022.2092791.
[28] Lone AS, Kanth SA, Jameel A, Harmain GA. A state of art review on the modeling of Contact type nonlinearities by extended finite element method. Materials Today: Proceedings 2019;18:3462–71. Available from: https://doi.org/10.1016/j.matpr.2019.07.274.
[29] Kanth SA, Lone AS, Harmain GA, Jameel A. Modeling of embedded and edge cracks in steel alloys by XFEM. Materials Today: Proceedings 2020;26:814–18. Available from: https://doi.org/10.1016/j.matpr.2019.12.423.
[30] Wriggers P, Rust WT, Reddy BD. A virtual element method for contact. Computational Mechanics 2016;58. Available from: https://doi.org/10.1007/s00466-016-1331-x.
[31] Dag S, Guler MA, Yildirim B, Ozatag AC. Frictional Hertzian contact between a laterally graded elastic medium and a rigid circular stamp. Acta Mechanica 2013;224. Available from: https://doi.org/10.1007/s00707-013-0844-z.
[32] Chehel Amirani M, Nemati N. Simulation of two dimensional unilateral contact using a coupled FE/EFG method. Engineering Analysis with Boundary Elements 2011;35:96–104. Available from: https://doi.org/10.1016/j.enganabound.2010.05.007.
[33] Gourgiotis PA, Zisis T, Giannakopoulos AE, Georgiadis HG. The Hertz contact problem in couple-stress elasticity. International Journal of Solids and Structures 2019;168. Available from: https://doi.org/10.1016/j.ijsolstr.2019.03.032.
[34] Lone AS, Harmain GA, Jameel A. Modeling of contact interfaces by penalty based enriched finite element method. Mechanics of Advanced Materials and Structures 2022;1–19. Available from: https://doi.org/10.1080/15376494.2022.2034075.

[35] Lone AS, Kanth SA, Harmain GA, Jameel A. XFEM modeling of frictional contact between elliptical inclusions and solid bodies. Materials Today: Proceedings 2020;26:819–24. Available from: https://doi.org/10.1016/j.matpr.2019.12.424.
[36] Doca T, Andrade Pires FM, Cesar De Sa JMA. A frictional mortar contact approach for the analysis of large inelastic deformation problems. International Journal of Solids and Structures 2014;51:1697–715. Available from: https://doi.org/10.1016/j.ijsolstr.2014.01.013.
[37] Oden JT, Martins JAC. Models and computational methods for dynamic friction phenomena. Computer Methods in Applied Mechanics and Engineering 1985;52:527–634. Available from: https://doi.org/10.1016/0045-7825(85)90009-X.
[38] Curnier A. A theory of friction. International Journal of Solids and Structures 1984;20. Available from: https://doi.org/10.1016/0020-7683(84)90021-0.
[39] Laursen TA, Simo JC. A continuum-based finite element formulation for the implicit solution of multibody, large deformation-frictional contact problems. International Journal for Numerical Methods in Engineering 1993;36. Available from: https://doi.org/10.1002/nme.1620362005.
[40] Alart P, Curnier A. A mixed formulation for frictional contact problems prone to Newton like solution methods. Computer Methods in Applied Mechanics and Engineering 1991;92:353–75. Available from: https://doi.org/10.1016/0045-7825(91)90022-X.
[41] Papadopoulos P, Solberg JM. A Lagrange multiplier method for the finite element solution of frictionless contact problems. Mathematical and Computer Modelling 1998;28:373–84. Available from: https://doi.org/10.1016/S0895-7177(98)00128-9.
[42] Belytschko T, Neal MO. Contact-impact by the pinball algorithm with penalty and Lagrangian methods. International Journal for Numerical Methods in Engineering 1991;31. Available from: https://doi.org/10.1002/nme.1620310309.
[43] Baillet L, Sassi T. Finite element method with Lagrange multipliers for contact problems with friction. Comptes Rendus Mathematique 2002;334. Available from: https://doi.org/10.1016/S1631-073X(02)02356-7.
[44] Huněk I. On a penalty formulation for contact-impact problems. Computers & Structures 1993;48:193–203. Available from: https://doi.org/10.1016/0045-7949(93)90412-7.
[45] Weyler R, Oliver J, Sain T, Cante JC. On the contact domain method: a comparison of penalty and Lagrange multiplier implementations. Computer Methods in Applied Mechanics and Engineering 2012;205–8. Available from: https://doi.org/10.1016/j.cma.2011.01.011.
[46] Chouly F, Hild P. On convergence of the penalty method for unilateral contact problems. Applied Numerical Mathematics 2013;65. Available from: https://doi.org/10.1016/j.apnum.2012.10.003.
[47] Gupta V, Jameel A, Verma SK, Anand S, Anand Y. Transient isogeometric heat conduction analysis of stationary fluid in a container. Part E: Journal of Process Mechanical Engineering 2022;. Available from: https://doi.org/10.1177/0954408922112.
[48] Wriggers P, Zavarise G. A formulation for frictionless contact problems using a weak form introduced by Nitsche. Computational Mechanics 2008;41. Available from: https://doi.org/10.1007/s00466-007-0196-4.
[49] Fischer KA, Wriggers P. Frictionless 2D contact formulations for finite deformations based on the mortar method. Computational Mechanics 2005;36. Available from: https://doi.org/10.1007/s00466-005-0660-y.
[50] Cavalieri FJ, Cardona A. An augmented Lagrangian technique combined with a mortar algorithm for modelling mechanical contact problems. International Journal for Numerical Methods in Engineering 2013;93: 420–42. Available from: https://doi.org/10.1002/nme.4391.
[51] Jameel A, Harmain GA. A coupled FE-IGA technique for modeling fatigue crack growth in engineering materials. Mechanics of Advanced Materials and Structures 2019;26:1764–75. Available from: https://doi.org/10.1080/15376494.2018.1446571.
[52] Hirmand M, Vahab M, Khoei AR. An augmented Lagrangian contact formulation for frictional discontinuities with the extended finite element method. Finite Elements in Analysis and Design 2015;107. Available from: https://doi.org/10.1016/j.finel.2015.08.003.
[53] Pietrzak G, Curnier A. Large deformation frictional contact mechanics: continuum formulation and augmented Lagrangian treatment. Computer Methods in Applied Mechanics and Engineering 1999;177:351–81. Available from: https://doi.org/10.1016/S0045-7825(98)00388-0.
[54] Simo JC, Laursen TA. An augmented lagrangian treatment of contact problems involving friction. Computers & Structures 1992;42:97–116. Available from: https://doi.org/10.1016/0045-7949(92)90540-G.

[55] Khoei AR, Nikbakht M. An enriched finite element algorithm for numerical computation of contact friction problems. International Journal of Mechanical Sciences 2007;49:183–99. Available from: https://doi.org/10.1016/j.ijmecsci.2006.08.014.
[56] Khoei A. *Computational plasticity in powder forming processes*. Elsevier; 2010.
[57] Khoei AR, Keshavarz S, Khaloo AR. Modeling of large deformation frictional contact in powder compaction processes. Applied Mathematical Modelling 2008;32:775–801. Available from: https://doi.org/10.1016/j.apm.2007.02.017.
[58] Chen JS, Wang HP. New boundary condition treatments in meshfree computation of contact problems. Computer Methods in Applied Mechanics and Engineering 2000;187:441–68. Available from: https://doi.org/10.1016/S0045-7825(00)80004-3.
[59] Jameel A, Harmain GA. Modeling and numerical simulation of fatigue crack growth in cracked specimens containing material discontinuities. Strength of Materials 2016;48(2):294–307. Available from: https://doi.org/10.1007/s11223-016-9765-0.
[60] Gupta V, Jameel A, Verma SK, Anand S, Anand Y. An insight on NURBS based isogeometric analysis, its current status and involvement in mechanical applications. Archives of Computational Methods in Engineering 2022;. Available from: https://doi.org/10.1007/s11831-022-09838-0.
[61] Singh AK, Jameel A, Harmain GA. Investigations on crack tip plastic zones by the extended iso-geometric analysis. Materials Today: Proceedings 2018;5:19284–93. Available from: https://doi.org/10.1016/j.matpr.2018.06.287.
[62] Jameel A, Harmain GA. Extended iso-geometric analysis for modeling three dimensional cracks. Mechanics of Advanced Materials and Structures 2019;26:915–23. Available from: https://doi.org/10.1080/15376494.2018.1430275.
[63] Sheikh UA, Jameel A. Elasto-plastic large deformation analysis of bi-material components by FEM. Materials Today: Proceedings 2020;26:1795–802. Available from: https://doi.org/10.1016/j.matpr.2020.02.377.
[64] Gupta V, Jameel A, Anand S, Anand Y. Analysis of composite plates using isogeometric analysis: a discussion. Materials Today: Proceedings 2021;44:1190–4. Available from: https://doi.org/10.1016/j.matpr.2020.11.238.
[65] Jameel A, Harmain GA. Effect of material irregularities on fatigue crack growth by enriched techniques. International Journal for Computational Methods in Engineering Science and Mechanics 2020;21:109–33. Available from: https://doi.org/10.1080/15502287.2020.1772902.
[66] Jameel A, Harmain GA. Fatigue crack growth in presence of material discontinuities by EFGM. International Journal of Fatigue 2015;81:105–16. Available from: https://doi.org/10.1016/j.ijfatigue.2015.07.021.
[67] Kanth SA, Harmain GA, Jameel A. Modeling of nonlinear crack growth in steel and aluminum alloys by the element free Galerkin method. Materials Today: Proceedings 2018;5:18805–14. Available from: https://doi.org/10.1016/j.matpr.2018.06.227.
[68] Khoei AR, Biabanaki SOR, Anahid M. A lagrangian-extended finite-element method in modeling large-plasticity deformations and contact problems. International Journal of Mechanical Sciences 2009;51:384–401. Available from: https://doi.org/10.1016/j.ijmecsci.2009.03.012.

CHAPTER

11

Enriched element-free Galerkin method for large elasto-plastic deformations: basic mathematical foundations

Azher Jameel and G.A. Harmain

Department of Mechanical Engineering, National Institute of Technology Srinagar, Hazratbal, Srinagar, Jammu and Kashmir, India

11.1 Introduction

Nonlinear structural analysis has always remained an area of interest in computational mechanics. Structural nonlinearities can be broadly classified into material, geometric, and contact type nonlinearities. Material nonlinearities arise due to nonlinear stress–strain relationships, whereas geometric nonlinearities (GNL) exist because of large changes in the geometry of the structural component. Conventional finite element techniques face huge mesh distortion issues while modeling geometrically nonlinear problems, which demands remeshing of the domain after each stage of simulation. Thus conventional finite element techniques are computationally very expensive for modeling such problems. On the contrary, meshless techniques provide a potential computational tool for modeling GNL problems as the structural component is discretized into an array of nodes only. There are no elements in meshless techniques, due to which element distortion does not occur during simulation. Out of all meshless techniques, element-free Galerkin method (EFGM) has been the dominant numerical tool for investigating engineering problems. Till date, a number of computational techniques have been developed to model different engineering problems, which include boundary element techniques [1–3], extended finite element method (XFEM) [4–6], meshless techniques [7–11], models based on peridynamics [12–14], phase field techniques [15–17], and the standard finite element method [18–21].

Enriched Numerical Techniques
DOI: https://doi.org/10.1016/B978-0-443-15362-4.00008-5

The enriched EFGM eliminates all mesh-related issues and models all types of material irregularities, irrespective of nodal distribution selected for simulation [22–24]. EFGM is a strong and potential computational tool that has found extensive applications in modeling GNL in engineering structures. In enriched EFGM, the conventional moving least square approximations are modified with additional functions, known as enrichment functions, to investigate the effect of cracks, holes, contact interfaces, and other material irregularities. The discontinuities are represented by a strong numerical tool known as the level set method, which remains zero along the interface and changes its sign across the discontinuity [25–34]. Enrichment of conventional variable approximations is carried out by employing the partition of unity approaches [35–40].

The present book chapter reports the basic mathematical principles for the modeling of GNL problems that occur when an engineering structure undergoes large elasto-plastic deformations. Development of mathematical models for large deformation problems using various approaches has been presented in detail. Basic mathematical foundations for TLA and ULA approaches have been discussed. In TLA approach, the reference state remains the same during analysis and is taken as the initial unloaded state of the body. In ULA approach, reference state changes after each step of simulation and is taken as the current state of the deformable body. The introduction of enrichment strategies in standard EFGM is also reported and discussed in detail. The kinematics of large deformation, deformation gradient, and various stress–strain measures used in large deformation analysis are also discussed in detail. The current work reports the mathematical principles governing the elasto-plastic behavior of engineering materials subjected to different types of loads. Different stress integration algorithms and hardening laws used in elasto-plastic analysis have been discussed in detail.

11.2 Kinematics of large deformation

Most engineering components undergo very small strains (<0.001) during their service. The strain at yield in most of the ductile materials is about 0.2%, and this level of strain is rarely reached in engineering components during operation. However, during manufacturing operations, the deformations are very large, and strains are much bigger ($>200\%$ during various forging operations). Thus the complete and accurate analysis of large deformation problems is of great importance as such investigations ensure the design of safe and reliable structures.

11.2.1 The deformation gradient

Deformation gradient provides a complete description of deformation of a body and helps us relate the undeformed state with the deformed state of a particular body. Let us consider a body in the initial undeformed configuration or unloaded state, as shown in Fig. 11.1. When external loads act on this body, it undergoes deformation and attains the new deformed configuration in the loaded state. Generally, the deformation of any particular body consists of actual deflections and rigid body motions. These rigid body rotations

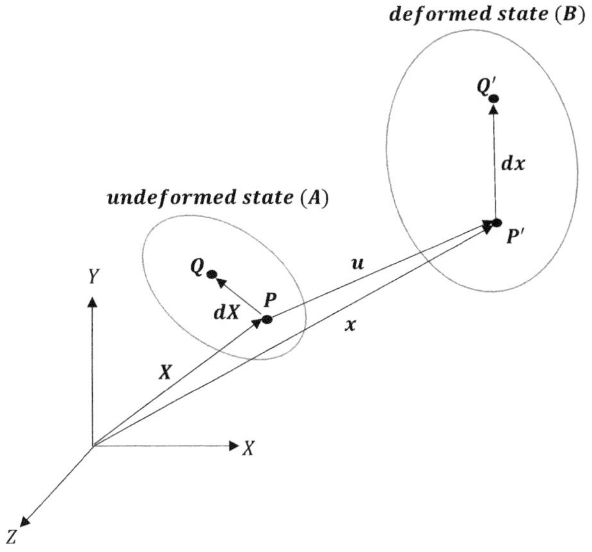

FIGURE 11.1 Two states of body before and after application of loads.

and translations do not contribute to any shape change and do not induce any stresses in the material. Now, consider a small line segment "PQ" or vector "**dX**" in the unloaded or undeformed configuration. The position of any point "P" in the undeformed state is given by vector "**X**," whereas its position in the deformed state is given by the vector "**x**." The line segment *PQ* in the undeformed configuration becomes *P′Q′* in the loaded state. The vector **dX** in the undeformed configuration is related to its deformed state **dx** by the deformation gradient '**F**' as

dx = **FdX**, which can be written as

$$\begin{Bmatrix} dx \\ dy \\ dz \end{Bmatrix} = \begin{bmatrix} F_{xx} & F_{xy} & F_{xz} \\ F_{yx} & F_{yy} & F_{yz} \\ F_{zx} & F_{zy} & F_{zz} \end{bmatrix} \begin{Bmatrix} dX \\ dY \\ dZ \end{Bmatrix} \tag{11.1}$$

This equation can be further written as

$$\begin{Bmatrix} dx \\ dy \\ dz \end{Bmatrix} = \begin{bmatrix} \frac{\partial x}{\partial X} & \frac{\partial x}{\partial Y} & \frac{\partial x}{\partial Z} \\ \frac{\partial y}{\partial X} & \frac{\partial y}{\partial Y} & \frac{\partial y}{\partial Z} \\ \frac{\partial z}{\partial X} & \frac{\partial z}{\partial Y} & \frac{\partial z}{\partial Z} \end{bmatrix} \begin{Bmatrix} dX \\ dY \\ dZ \end{Bmatrix} \tag{11.2}$$

It can be seen that the deformation gradient is actually a tensor of second order, which is defined as $\boldsymbol{F} = \frac{dx}{dX}$. The Cartesian components can be written as $F_{ij} = \frac{\partial x_i}{\partial X_j} = \delta_{ij} + \frac{\partial u_i}{\partial X_j}$, where δ represents Kronecker delta, and **u** represents displacement field. It can be proved that the determinant of the deformation gradient tensor is the ratio of the volume after deformation to the initial undeformed volume.

11.2.2 Measures of stresses and strains

An engineering structure can be under the action of different forces, which can be broadly classified into two categories. These include the boundary forces, which act at the boundary of the body, and the body forces, which act throughout the body. When these forces act on the body, it deforms, due to which stresses and strains are generated in it. During large deformation analysis, we use different types of stress measures depending upon reference state chosen for investigation. The first stress measure that we use during analysis is Cauchy stress ($\boldsymbol{\sigma}$), which may be given as the force df imposed on the current deformed area defined over the same configuration "$d\mathbf{a}$," that is, $\boldsymbol{\sigma} = \frac{df}{da}$. Cauchy stresses are symmetric and are stated to be true stresses induced in the loaded specimens. Second stress definition is the first Piola–Kirchoff stress ($\boldsymbol{P}$), which is described as force df acting on current deformed area but defined over the original undeformed area $d\boldsymbol{A}$ and can be defined as $\boldsymbol{P} = \frac{df}{dA}$. One of the drawbacks of the first Piola–Kirchoff stress lies in its unsymmetrical nature, due to which its use is limited during the analysis of structures. Another stress measure that finds extensive application is the Second Piola–Kirchoff stress ($\boldsymbol{S}$), which maintains its symmetrical nature and has been defined as the force $d\boldsymbol{F}$ acting on original undeformed area and defined over the same area "dA," that is, $\boldsymbol{S} = \frac{d\boldsymbol{F}}{dA}$. The stress tensors defined above are related to each other, and the relationships between them are given in the following equations:

$$\boldsymbol{P} = J\boldsymbol{F}^{-1}\boldsymbol{\sigma} = \boldsymbol{S}\boldsymbol{F}^{T} \tag{11.3}$$

$$\boldsymbol{\sigma} = \frac{1}{J}\boldsymbol{F}\boldsymbol{P} = \frac{1}{J}\boldsymbol{F}\boldsymbol{S}\boldsymbol{F}^{T} \tag{11.4}$$

$$\boldsymbol{S} = J\boldsymbol{F}^{-1}\boldsymbol{\sigma}\boldsymbol{F}^{-T} = \boldsymbol{P}\boldsymbol{F}^{-T} \tag{11.5}$$

where J is the determinant of the deformation gradient matrix $\boldsymbol{F}$. Now, consider the vector $\boldsymbol{dx}$ having length ds in the deformed configuration. The length of this vector can be written as

$$ds^2 = \boldsymbol{dx} \cdot \boldsymbol{dx} = (\boldsymbol{FdX}) \cdot (\boldsymbol{FdX}) = \boldsymbol{dX}^T\boldsymbol{F}^T\boldsymbol{FdX} = \boldsymbol{dX}^T\boldsymbol{CdX} \tag{11.6}$$

where $\boldsymbol{C}$ is called the left Cauchy–Green tensor, defined by $\boldsymbol{C} = \boldsymbol{F}^T\boldsymbol{F}$. Now, consider the vector $\boldsymbol{dX}$ in the deformed state with length dS, which can be obtained as

$$dS^2 = \boldsymbol{dX} \cdot \boldsymbol{dX} = \left(\boldsymbol{F}^{-1}\boldsymbol{dx}\right) \cdot \left(\boldsymbol{F}^{-1}\boldsymbol{dx}\right) = \boldsymbol{dx}^T(\boldsymbol{F}^{-1})^T\boldsymbol{F}^{-1}\boldsymbol{dx} = \boldsymbol{dx}^T\boldsymbol{B}^{-1}\boldsymbol{dx} \tag{11.7}$$

where $\boldsymbol{B}$ is the right Cauchy–Green tensor, written as $\boldsymbol{B}^{-1} = (\boldsymbol{F}^{-1})^T\boldsymbol{F}^{-1}$. We have

$$ds^2 - dS^2 = \boldsymbol{dx} \cdot \boldsymbol{dx} - \boldsymbol{dx} \cdot \boldsymbol{B}^{-1}\boldsymbol{dx} = \boldsymbol{dx} \cdot \left(\boldsymbol{I} - \boldsymbol{B}^{-1}\right)\boldsymbol{dx} \tag{11.8}$$

such that $\boldsymbol{I}$ is the identity matrix. The Cauchy–Green tensor $\boldsymbol{B}$ provides a measure of strain, but it gives nonzero components even when there is zero stretch. Therefore an alternate and more appropriate measure of strain, called the Almansi strain, is introduced. The Almansi strain "e" is zero in the case of zero stretch and is written as

$$\boldsymbol{e} = \frac{1}{2}\left(\boldsymbol{I} - \boldsymbol{B}^{-1}\right) = \frac{1}{2}\left(\boldsymbol{I} - (\boldsymbol{F}^{-1})^T\boldsymbol{F}^{-1}\right) \tag{11.9}$$

We can also write

$$ds^2 - dS^2 = \boldsymbol{dx}^T\boldsymbol{dx} - \boldsymbol{dX}^T\boldsymbol{dX} = \boldsymbol{dX}^T\left(\boldsymbol{F}^T\boldsymbol{F} - \boldsymbol{I}\right)\boldsymbol{dX} = \boldsymbol{dX}^T(\boldsymbol{C} - \boldsymbol{I})\boldsymbol{dX} \tag{11.10}$$

With the help of above equation, we define the Green–Lagrange strain tensor "E" as

$$E = \frac{1}{2}(C - I) = \frac{1}{2}\left(F^T F - I\right) \tag{11.11}$$

11.3 Large deformation analysis by element-free Galerkin method

Due to large deformations occurring in any part of the engineering component, nonlinearities are introduced in the structure, which are termed GNL. While modeling large deformation problems, the conventional finite element method suffers certain limitations in terms of mesh distortions. Extreme mesh distortion in the finite element method makes further analysis difficult, which demands remeshing of loaded domain after each stage of simulation. However, the meshless techniques like EFGM prove to be very effective for modeling such problems because of their meshless nature. Since no mesh is generated in meshless techniques, mesh distortion issues do not occur in these techniques. As discussed earlier, there are two approaches available for modeling GNL caused by large deformations. One of the approaches is the total Lagrangian approach (TLA), in which the reference state remains the same during analysis and is taken as the initial unloaded state of the body. Second approach is updated Lagrangian approach (ULA) in which the reference state changes after each step of simulation and is taken as the current state of the deformable body. Since the reference configuration remains the same in TLA, the formulation and computation become simple and easy. Furthermore, the modeling of anisotropic problems presents no difficulties in the TLA.

11.3.1 Total Lagrangian formulation

TLA has several advantages, as the reference configuration remains the same during investigation. To develop numerical models for TLA formulations, the equilibrium equation is defined with respect to the initial unloaded or undeformed state of the body as

$$\frac{\partial P_{ji}}{\partial X_j} + b_i = 0 \tag{11.12}$$

where X_j represents the initial undeformed configuration, P_{ji} denotes first Piola–Kirchhoff (PK1) stress, and b_i represents body forces. Green and St. Venant introduced the description of strains that is valid for both large and small strains. The total strain consists of the linear and nonlinear components and can be defined as $\boldsymbol{E} = \boldsymbol{E}_L + \boldsymbol{E}_{NL}$, where $\boldsymbol{E}_L$ and $\boldsymbol{E}_{NL}$ are the corresponding linear and nonlinear strain components, which are written as

$$\boldsymbol{E}_L = \left\{ \begin{array}{c} \dfrac{\partial u}{\partial X} \\[2ex] \dfrac{\partial v}{\partial Y} \\[2ex] \dfrac{\partial v}{\partial X} + \dfrac{\partial u}{\partial Y} \end{array} \right\}; \quad \boldsymbol{E}_{NL} = \left\{ \begin{array}{c} \dfrac{1}{2}\left(\frac{\partial u}{\partial X}\right)^2 + \dfrac{1}{2}\left(\frac{\partial v}{\partial X}\right)^2 \\[2ex] \dfrac{1}{2}\left(\frac{\partial u}{\partial Y}\right)^2 + \dfrac{1}{2}\left(\frac{\partial v}{\partial Y}\right)^2 \\[2ex] \dfrac{\partial u}{\partial X}\dfrac{\partial u}{\partial Y} + \dfrac{\partial v}{\partial X}\dfrac{\partial v}{\partial Y} \end{array} \right\} \tag{11.13}$$

The variational form of equilibrium equation is expressed in terms of deformation gradient as

$$\int_{\Omega} \delta\mathbf{F}^T \mathbf{P} d\Omega - \int_{\Omega} \delta\mathbf{u}^T \mathbf{b} d\Omega - \int_{\Gamma t} \delta\mathbf{u}^T \mathbf{t} d\Gamma_t = 0 \tag{11.14}$$

such that $\boldsymbol{F}$ is deformation gradient defined as $\boldsymbol{F} = [F_{11}\ F_{21}\ F_{12}\ F_{22}]^T$, $\mathbf{P}$ denotes a vector of first Piola–Kirchoff stresses given as $\boldsymbol{P} = [P_{11}\ P_{12}\ P_{21}\ P_{22}]^T$, $\boldsymbol{b}$ denotes the body forces, and $\boldsymbol{t}$ represents the applied forces. The displacement-based approximation can be described in standard form as

$$\begin{Bmatrix} u \\ v \end{Bmatrix} = \begin{bmatrix} N_i & 0 \\ 0 & N_i \end{bmatrix} \begin{Bmatrix} u_i \\ v_i \end{Bmatrix} = \mathbf{N}\boldsymbol{u}_i \tag{11.15}$$

where N_i are classical EFGM shape functions and $\boldsymbol{u}_i$ is a vector of nodal displacements. The variations in displacements can be written as $\delta\mathbf{u} = \mathbf{N}\delta\boldsymbol{u}_i$. The deformation gradient has been defined as $F_{ij} = \frac{\partial x_i}{\partial X_j} = \delta_{ij} + \frac{\partial u_i}{\partial X_j}$. Thus the variations in the deformation gradient can be obtained as

$$\delta\mathbf{F} = \begin{bmatrix} \frac{\partial \delta u_i}{\partial X} & \frac{\partial \delta v_i}{\partial X} & \frac{\partial \delta u_i}{\partial Y} & \frac{\partial \delta v_i}{\partial Y} \end{bmatrix}^T = \boldsymbol{G}\delta\boldsymbol{u}_i \tag{11.16}$$

$$\boldsymbol{G} = \begin{bmatrix} \frac{\partial N_i}{\partial X} & 0 \\ 0 & \frac{\partial N_i}{\partial X} \\ \frac{\partial N_i}{\partial Y} & 0 \\ 0 & \frac{\partial N_i}{\partial Y} \end{bmatrix} \tag{11.17}$$

After making appropriate substitutions, the equilibrium equation is further expressed as

$$\left(\int_{\Omega} \mathbf{G}^T \mathbf{P} d\Omega - \int_{\Omega} \mathbf{N}^T \mathbf{b} d\Omega - \int_{\Gamma} \mathbf{N}^T \mathbf{t} d\Gamma \right) \delta\mathbf{u}_i^T = \mathbf{0} \tag{11.18}$$

which further gives

$$\int_{\Omega} \mathbf{G}^T \mathbf{P} d\Omega - f = 0 \tag{11.19}$$

such that

$$f = \int_{\Omega} \mathbf{N}^T \mathbf{b} d\Omega + \int_{\Gamma} \mathbf{N}^T \mathbf{t} d\Gamma \tag{11.20}$$

It should be noted that the above equation calculates the force vector with respect to the initial unloaded or undeformed state of the body. Body and applied forces are calculated as $\mathbf{b} = \frac{b'}{J}$ and $\mathbf{t} = \boldsymbol{t}'\left(\frac{da}{dA}\right)$, where $\boldsymbol{b}'$ and $\boldsymbol{t}'$ are the body forces and applied tractions measured with respect to current deformed configuration and J denotes determinant of deformation gradient matrix. Incremental form of equilibrium equation is written as

$$\int_{\Omega} \mathbf{G}^T \mathbf{\Delta P} d\Omega - \Delta f = 0 \tag{11.21}$$

However, the first Piola–Kirchoff stress does not find extensive use because it is not symmetric, due to which analysis becomes difficult. So, it is advantageous to express the governing equations in terms of the second Piola–Kirchoff stress, which is symmetric. The relationships between the Piola–Kirchoff stress and the second Piola–Kirchoff stress can be expressed as

$$\mathrm{P}_{11} = F_{11}S_{11} + F_{12}S_{21} \tag{11.22}$$

$$\mathrm{P}_{21} = F_{11}S_{12} + F_{12}S_{22} \tag{11.23}$$

$$\mathrm{P}_{22} = F_{21}S_{12} + F_{22}S_{22} \tag{11.24}$$

Thus the incremental form of these equations can be expressed as

$$\Delta\mathrm{P}_{11} = F_{11}\Delta S_{11} + F_{12}\Delta S_{21} + \Delta F_{11}S_{11} + \Delta F_{12}S_{21} \tag{11.25}$$

$$\Delta\mathrm{P}_{12} = F_{21}\Delta S_{11} + F_{22}\Delta S_{21} + \Delta F_{21}S_{11} + \Delta F_{22}S_{21} \tag{11.26}$$

$$\Delta\mathrm{P}_{21} = F_{11}\Delta S_{12} + F_{12}\Delta S_{22} + \Delta F_{11}S_{12} + \Delta F_{12}S_{22} \tag{11.27}$$

$$\Delta\mathrm{P}_{22} = F_{21}\Delta S_{12} + F_{22}\Delta S_{22} + \Delta F_{21}S_{12} + \Delta F_{22}S_{22} \tag{11.28}$$

which can be represented in matrix form as

$$\Delta \boldsymbol{P} = \boldsymbol{F}_s^T \Delta \boldsymbol{S} + \boldsymbol{M}_s \Delta \boldsymbol{F} \tag{11.29}$$

such that

$$\boldsymbol{F_s} = \begin{bmatrix} F_{11} & F_{21} & 0 & 0 \\ 0 & 0 & F_{12} & F_{22} \\ F_{12} & F_{22} & F_{11} & F_{21} \end{bmatrix} \tag{11.30}$$

$$\boldsymbol{M_s} = \begin{bmatrix} S_{11} & 0 & S_{12} & 0 \\ 0 & S_{11} & 0 & S_{12} \\ S_{12} & 0 & S_{22} & 0 \\ 0 & S_{12} & 0 & S_{22} \end{bmatrix} \tag{11.31}$$

$$\Delta \boldsymbol{S} = [\Delta S_{11}\ \Delta S_{22}\ \Delta S_{12}]^T \tag{11.32}$$

$$\Delta \boldsymbol{F} = [\Delta F_{11}\ \Delta F_{21}\ \Delta F_{12}\ \Delta F_{22}]^T \tag{11.33}$$

Now, Green's strain can be related to the second Piola–Kirchoff stress as $\Delta \mathbf{S} = \mathbf{D}^{ep} \mathbf{\Delta E}$, where $\mathbf{D}^{ep}$ is the elasto-plastic constitutive matrix. Strains $\mathbf{E}$ are related to deformation gradient as

$$2E_{11} = F_{11}F_{11} + F_{21}F_{12} - 1 \tag{11.34}$$

$$2E_{22} = F_{12}F_{12} + F_{22}F_{22} - 1 \tag{11.35}$$

$$2E_{12} = F_{11}F_{12} + F_{21}F_{22} \tag{11.36}$$

Incremental form of strains can be obtained as

$$\Delta \mathbf{E} = \boldsymbol{F_s}\Delta \mathbf{F} = \boldsymbol{F_s G}\Delta \boldsymbol{u_i} = \mathbf{B}\Delta \boldsymbol{u_i} \tag{11.37}$$

The strain–displacement matrix, describing the relations between displacements and strains, has been obtained as $\boldsymbol{B} = \boldsymbol{F_s G}$, which can be obtained as

$$\mathbf{B} = \boldsymbol{F_s G} = \begin{bmatrix} \frac{\partial N_i}{\partial X}\frac{\partial x}{\partial X} & \frac{\partial N_i}{\partial X}\frac{\partial y}{\partial X} \\ \frac{\partial N_i}{\partial Y}\frac{\partial x}{\partial Y} & \frac{\partial N_i}{\partial Y}\frac{\partial y}{\partial Y} \\ \frac{\partial N_i}{\partial X}\frac{\partial x}{\partial Y} + \frac{\partial N_i}{\partial Y}\frac{\partial x}{\partial X} & \frac{\partial N_i}{\partial X}\frac{\partial y}{\partial Y} + \frac{\partial N_i}{\partial Y}\frac{\partial y}{\partial X} \end{bmatrix} \tag{11.38}$$

After appropriate substitutions, we get

$$\left(\int_\Omega \mathbf{B}^T \mathbf{D}^{ep} \mathbf{B} d\Omega + \int_\Omega \mathbf{G}^T \boldsymbol{M_s} \mathbf{G} d\Omega \right) \Delta \boldsymbol{u_i} = \Delta \boldsymbol{f} \tag{11.39}$$

Thus the final discrete system of algebraic equations or the numerical model of large deformation analysis can be written as $\boldsymbol{K\Delta u_i} = \boldsymbol{\Delta f}$, such that $= \boldsymbol{K_{mat}} + \boldsymbol{K_{geo}}$. The stiffness matrices $\boldsymbol{K_{mat}}$ and $\boldsymbol{K_{geo}}$ can be written as

$$\boldsymbol{K_{mat}} = \int_\Omega \mathbf{B}^T \mathbf{D}^{ep} \mathbf{B} d\Omega \quad ; \quad \boldsymbol{K_{geo}} = \int_\Omega \mathbf{G}^T \boldsymbol{M_s} \mathbf{G} d\Omega \tag{11.40}$$

Thus we see that the stiffness matrix has two components ($\boldsymbol{K_{mat}}$ and $\boldsymbol{K_{geo}}$) in large deformation problems. The first term $\boldsymbol{K_{mat}}$ takes the material nonlinearities into consideration, whereas the second term $\boldsymbol{K_{geo}}$ takes care of GNL. For simple linear elastic problems, there are no changes in the material properties during simulation, that is, $\mathbf{D}^{ep} = \mathbf{D}$. For such problems, there is no need to add the second term $\boldsymbol{K_{geo}}$ in the finite element model because the deformations are small.

11.3.2 Updated Lagrangian formulation

Updated Lagrangian formulation for modeling large deformation problems uses the current deformed state of the body as the reference for investigation. Hence, all parameters are expressed with respect to latest known configuration (x). In GNL problems, the entire load is applied in small increments, so that the loaded body acquires several intermediate deformed configurations before reaching the final configuration. During our analysis, we consider three equilibrium configurations of the body. The configurations are initial unloaded or undeformed state C_0, previous deformed configuration C_1, and current deformed state of the body C_2. During analysis, all quantities, including displacements, strains, and stresses, are known up to the last known C_1 configuration. The equilibrium equation of a deformable body for updated Lagrangian formulation is expressed in variational form as

$$\int_{^1\Omega} \delta\left({}_1^2E^T\right)\left({}_1^2S\right)d^1\Omega - \int_{^1\Omega} \delta\mathbf{u}^T\left({}_1^2\boldsymbol{b}\right)d^1\Omega - \int_{^1\Gamma} \delta\mathbf{u}^T\left({}_1^2\boldsymbol{t}\right)\mathrm{d}^1\Gamma = 0 \tag{11.41}$$

where ${}_1^2\boldsymbol{b}$ and ${}_1^2\boldsymbol{t}$ represent the body force vector and the applied traction vector referred to C_1 configuration, ${}_1^2E$ are updated Green–Lagrange strains, and ${}_1^2S$ denotes second Piola–Kirchoff stresses. We have ${}_1^2E = {}_1^2E_L + {}_1^2E_{NL}$, such that

$$\left({}_1^2E_{ij}\right)_L = \frac{1}{2}\left(\frac{\partial u_i}{\partial^1 x_j} + \frac{\partial u_j}{\partial^1 x_i}\right); \left({}_1^2E_{ij}\right)_{NL} = \frac{1}{2}\left(\frac{\partial u_k}{\partial^1 x_i}\frac{\partial u_k}{\partial^1 x_j}\right) \tag{11.42}$$

The updated second Piola–Kirchoff stress is expressed as ${}_1^2S = {}_1^1S + {}_1^2S$, where ${}_1^1S = {}_1^1\sigma$ denotes Cauchy stresses in C_1 configuration and ${}_1^2S$ represents updated second Piola–Kirchoff stresses. Equilibrium equation presented above can be expanded as

$$\int_{^1\Omega} \delta\left({}_1^2E^T\right)\left({}_1^1\sigma + {}_1^2S\right)d^1\Omega - \int_{^1\Omega} \delta\mathbf{u}^T\left({}_1^2\boldsymbol{b}\right)d^1\Omega - \int_{^1\Gamma} \delta\mathbf{u}^T\left({}_1^2\boldsymbol{t}\right)\mathrm{d}^1\Gamma = 0 \tag{11.43}$$

$$\int_{^1\Omega} \delta\left({}_1^2E^T\right){}_1^2S d^1\Omega + \int_{^1\Omega} \delta\left({}_1^2E^T\right){}_1^1\boldsymbol{\sigma} d^1\Omega - \bar{f} = 0 \tag{11.44}$$

such that

$$\bar{f} = \int_{^1\Omega} \delta\mathbf{u}^T\left({}_1^2\boldsymbol{b}\right)d^1\Omega - \int_{^1\Gamma} \delta\mathbf{u}^T\left({}_1^2\boldsymbol{t}\right)\mathrm{d}^1\Gamma = 0 \tag{11.45}$$

Green's strains are related to nodal displacements as $\boldsymbol{E} = \boldsymbol{B}u_i$, where $\boldsymbol{B}$ is the strain–displacement matrix. Second Piola–Kirchoff stresses are related to Green's strains as $\boldsymbol{S} = \boldsymbol{D}^{ep}\boldsymbol{E}$. Equilibrium equation is further expressed as

$$\int_{^1\Omega} \delta\left\{{}_1^2E\right\}^T {}_1^2D_{ep}\left({}_1^2E\right)d^1\Omega + \int_{^1\Omega} \delta\left\{{}_1^2E\right\}^T {}_1^1\sigma d^1\Omega - \bar{f} = 0 \tag{11.46}$$

$$\int_{^1\Omega} \delta u_i{}^T \boldsymbol{B}^T \boldsymbol{D}^{ep} \boldsymbol{B} u_i d^1\Omega + \int_{^1\Omega} \delta\left\{{}_1^2E\right\}^T {}_1^1\sigma d^1\Omega - \bar{f} = 0 \tag{11.47}$$

The linear and nonlinear components of Green's strains are

$$\boldsymbol{E}_L = \begin{Bmatrix} \dfrac{\partial u}{\partial x} \\ \dfrac{\partial v}{\partial y} \\ \dfrac{\partial v}{\partial x} + \dfrac{\partial u}{\partial y} \end{Bmatrix}; \quad \boldsymbol{E}_{NL} = \begin{Bmatrix} \dfrac{1}{2}\left(\dfrac{\partial u}{\partial x}\right)^2 + \dfrac{1}{2}\left(\dfrac{\partial v}{\partial x}\right)^2 \\ \dfrac{1}{2}\left(\dfrac{\partial u}{\partial y}\right)^2 + \dfrac{1}{2}\left(\dfrac{\partial v}{\partial y}\right)^2 \\ \dfrac{\partial u}{\partial x}\dfrac{\partial u}{\partial y} + \dfrac{\partial v}{\partial x}\dfrac{\partial v}{\partial y} \end{Bmatrix} \tag{11.48}$$

During large elasto-plastic deformations, the nonlinear strains are very large as compared to the linear strains, and we may assume that $E \approx E_{NL}$. Now, the second term of the equilibrium (Eq. 4.68) equation can be written as

$$\int_{^1\Omega} \delta\left\{{}^2_1E\right\}^T {}^1_1\sigma d^1\Omega = \int_{^1\Omega} \delta\left\{{}^2_1E_{NL}\right\}^T {}^1_1\sigma d^1\Omega \tag{11.49}$$

where

$$\delta E_{NL} = \begin{bmatrix} \dfrac{\partial \delta u}{\partial x}\dfrac{\partial u}{\partial x} & \dfrac{\partial \delta v}{\partial x}\dfrac{\partial v}{\partial x} \\ \dfrac{\partial \delta u}{\partial y}\dfrac{\partial u}{\partial y} & \dfrac{\partial \delta v}{\partial y}\dfrac{\partial v}{\partial y} \\ \dfrac{\partial \delta u}{\partial x}\dfrac{\partial u}{\partial y} + \dfrac{\partial u}{\partial x}\dfrac{\partial \delta u}{\partial y} & \dfrac{\partial \delta v}{\partial x}\dfrac{\partial v}{\partial y} + \dfrac{\partial v}{\partial x}\dfrac{\partial \delta v}{\partial y} \end{bmatrix} \tag{11.50}$$

Therefore we further write

$$\int_{^1\Omega} \delta\left\{{}^2_1E\right\}^T {}^1_1\sigma d^1\Omega = \int_{^1\Omega} \begin{Bmatrix} \frac{\partial \delta u}{\partial x} \\ \frac{\partial \delta v}{\partial x} \\ \frac{\partial \delta u}{\partial y} \\ \frac{\partial \delta u}{\partial y} \end{Bmatrix}^T \begin{bmatrix} {}^1_1\sigma_{11} & 0 & {}^1_1\sigma_{12} & 0 \\ 0 & {}^1_1\sigma_{11} & 0 & {}^1_1\sigma_{12} \\ {}^1_1\sigma_{12} & 0 & {}^1_1\sigma_{22} & 0 \\ 0 & {}^1_1\sigma_{12} & 0 & {}^1_1\sigma_{22} \end{bmatrix} \begin{Bmatrix} \dfrac{\partial u}{\partial x} \\ \dfrac{\partial v}{\partial x} \\ \dfrac{\partial u}{\partial y} \\ \dfrac{\partial v}{\partial y} \end{Bmatrix} \tag{11.51}$$

$$\int_{^1\Omega} \delta\left\{{}^2_1E\right\}^T {}^1_1\sigma d^1\Omega = \int_{^1\Omega} \delta \boldsymbol{u_i}^T \boldsymbol{G}^T \boldsymbol{M_\sigma} \boldsymbol{G} \boldsymbol{u} d^1\Omega \tag{11.52}$$

The equilibrium equation can be written as

$$\int_{^1\Omega} \delta \boldsymbol{u_i}^T \boldsymbol{B}^T \boldsymbol{D}^{ep} \boldsymbol{B} \boldsymbol{u_i} d^1\Omega + \int_{^1\Omega} \delta \boldsymbol{u_i}^T \boldsymbol{G}^T \boldsymbol{M_\sigma} \boldsymbol{G} \boldsymbol{u_i} d^1\Omega - \delta \boldsymbol{u_i}^T \boldsymbol{f} = 0 \tag{11.53}$$

Thus we obtain the final EFGM model for GNL problems in updated Lagrangian formulation as $\boldsymbol{K}\boldsymbol{u_i} = \boldsymbol{f}$, where $\boldsymbol{K}$ is stiffness matrix, and $\boldsymbol{f}$ **is** external force vector. All individual matrices in the final EFGM model are expressed as

$$\boldsymbol{K} = \boldsymbol{K_{mat}} + \boldsymbol{K_{geo}} \tag{11.54}$$

such that

$$\boldsymbol{K_{mat}} = \int_\Omega \mathbf{B}^T \mathbf{D}^{ep} \mathbf{B} d\Omega \; ; \; \boldsymbol{K_{geo}} = \int_\Omega \mathbf{G}^T \boldsymbol{M_\sigma} \mathbf{G} d\Omega \tag{11.55}$$

$$\boldsymbol{f} = \int_\Omega \mathbf{N}^T \mathbf{b} d\Omega + \int_\Gamma \mathbf{N}^T \mathbf{t} d\Gamma$$

$$\mathbf{B} = \begin{bmatrix} \frac{\partial N_i}{\partial x} & 0 \\ 0 & \frac{\partial N_i}{\partial x} \\ \frac{\partial N_i}{\partial y} & \frac{\partial N_i}{\partial x} \end{bmatrix} ; \mathbf{G} = \begin{bmatrix} \frac{\partial N_i}{\partial x} & 0 \\ 0 & \frac{\partial N_i}{\partial x} \\ \frac{\partial N_i}{\partial y} & 0 \\ 0 & \frac{\partial N_i}{\partial y} \end{bmatrix} \tag{11.56}$$

$$\boldsymbol{M}_\sigma = \begin{bmatrix} {}^1\sigma_{11} & 0 & {}^1\sigma_{12} & 0 \\ 0 & {}^1\sigma_{11} & 0 & {}^1\sigma_{12} \\ {}^1\sigma_{12} & 0 & {}^1\sigma_{22} & 0 \\ 0 & {}^1\sigma_{12} & 0 & {}^1\sigma_{22} \end{bmatrix} \tag{11.57}$$

11.3.3 Large deformation in the presence of discontinuities

The modeling of discontinuities in classical FEM requires mesh conformation, which complicates the representation of irregularly shaped and advancing discontinuities. However, such limitations do not occur in enriched techniques because the representation of discontinuities does not depend on the nodal distribution. Large deformations usually occur during various forming operations like forging and die pressing. For bi-material specimens undergoing large deformations, the final EFGM model can be obtained as $\mathbf{K}\mathbf{d} = f$, where $\mathbf{d} = \begin{bmatrix} \boldsymbol{u} & \boldsymbol{a} \end{bmatrix}^T$. Here, $\boldsymbol{u}$ represents the standard nodal displacements and $\boldsymbol{a}$ represents additional enriched nodal parameters. All individual matrices in the final EFGM model are expressed as

$$\mathbf{K} = \begin{bmatrix} \mathbf{K}^{uu} & \mathbf{K}^{ua} \\ \mathbf{K}^{au} & \mathbf{K}^{aa} \end{bmatrix} \; and \; f = \begin{Bmatrix} f^u \\ f^a \end{Bmatrix} \tag{11.58}$$

such that

$$\mathbf{K}^{\alpha\beta} = \int_{\Omega^e} (\mathbf{B}^\alpha)^T \mathbf{D}^{ep} \mathbf{B}^\beta d\Omega + \int_{\Omega^e} (\mathbf{G}^\alpha)^T \mathbf{M}_s \mathbf{G}^\beta d\Omega \quad (\alpha, \beta = u, a) \tag{11.59}$$

$$f^\alpha = \int_{\Gamma^e} \mathbf{N}^\alpha \mathbf{t} d\Gamma + \int_{\Omega^e} \mathbf{N}^\alpha \mathbf{b} d\Omega \quad (\alpha = u, a) \tag{11.60}$$

$$\mathbf{N}^u = \mathbf{N}_{std} = \begin{bmatrix} N_i & 0 \\ 0 & N_i \end{bmatrix} ; \; \mathbf{N}^a = \mathbf{N}_{enr} = \begin{bmatrix} \overline{N_J} & 0 \\ 0 & \overline{N_J} \end{bmatrix} ; \; \overline{N_J} = N_j(\boldsymbol{x})\{f(\boldsymbol{x}) - f(\boldsymbol{x}_j)\} \tag{11.61}$$

$$\mathbf{B}^u = \boldsymbol{B}_{std} = \begin{bmatrix} \frac{\partial N_i}{\partial X}\frac{\partial x}{\partial X} & \frac{\partial N_i}{\partial X}\frac{\partial y}{\partial X} \\ \frac{\partial N_i}{\partial Y}\frac{\partial x}{\partial Y} & \frac{\partial N_i}{\partial Y}\frac{\partial y}{\partial Y} \\ \frac{\partial N_i}{\partial X}\frac{\partial x}{\partial Y} + \frac{\partial N_i}{\partial Y}\frac{\partial x}{\partial X} & \frac{\partial N_i}{\partial X}\frac{\partial y}{\partial Y} + \frac{\partial N_i}{\partial Y}\frac{\partial y}{\partial X} \end{bmatrix} \tag{11.62}$$

$$\mathbf{B}^a = \boldsymbol{B}_{enr} = \begin{bmatrix} \frac{\partial \overline{N}_j}{\partial X}\frac{\partial x}{\partial X} & \frac{\partial \overline{N}_j}{\partial X}\frac{\partial y}{\partial X} \\ \frac{\partial \overline{N}_j}{\partial Y}\frac{\partial x}{\partial Y} & \frac{\partial \overline{N}_j}{\partial Y}\frac{\partial y}{\partial Y} \\ \frac{\partial \overline{N}_j}{\partial X}\frac{\partial x}{\partial Y} + \frac{\partial \overline{N}_j}{\partial Y}\frac{\partial x}{\partial X} & \frac{\partial \overline{N}_j}{\partial X}\frac{\partial y}{\partial Y} + \frac{\partial \overline{N}_j}{\partial Y}\frac{\partial y}{\partial X} \end{bmatrix} \tag{11.63}$$

$$\mathbf{G}^u = \boldsymbol{G}_{std} = \begin{bmatrix} \frac{\partial N_i}{\partial X} & 0 \\ 0 & \frac{\partial N_i}{\partial X} \\ \frac{\partial N_i}{\partial Y} & 0 \\ 0 & \frac{\partial N_i}{\partial Y} \end{bmatrix}; \; i = 1 \; to \; n \tag{11.64}$$

$$\mathbf{G}^a = \boldsymbol{G}_{enr} = \begin{bmatrix} \frac{\partial \overline{N}_j}{\partial X} & 0 \\ 0 & \frac{\partial \overline{N}_j}{\partial X} \\ \frac{\partial \overline{N}_j}{\partial Y} & 0 \\ 0 & \frac{\partial \overline{N}_j}{\partial Y} \end{bmatrix}; \; j = 1 \; to \; n_s \tag{11.65}$$

Here, $f(x)$ represents the enrichment function that introduces the effect of different material irregularities into the mathematical models. In above equations, $\boldsymbol{n}$ denotes standard or conventional nodes, and $\boldsymbol{n_s}$ are additional enriched nodes present in the domain.

11.4 Elasto-plastic analysis

Material nonlinearities are those types of nonlinearities that are seen after the material undergoes yielding, leading to elasto-plastic response or nonlinear stress–strain curves of the material. Mathematical theory of plasticity provides relationship between the stresses and strains for engineering domains exhibiting the material nonlinearities. The complete description of elasto-plastic analysis includes the following three main components:

- The elastic stress–strain relations that are applicable before the onset of yielding.
- Yield criterion defining the start of yielding, that is, stress levels beyond which material begins exhibits elasto-plastic response.
- Stress–strain relationship after the start of yielding, that is, during elasto-plastic phase.

Before the onset of yielding, the material behavior remains elastic, and Hooke's law is applicable, which describes the relations between stresses and strains as $\sigma_{ij} = D_{ijkl}\varepsilon_{kl}$, which is written in matrix form as $\boldsymbol{\sigma} = \boldsymbol{D}\boldsymbol{\varepsilon}$, where $\boldsymbol{D}$ is Hooke's matrix relating stresses to strains.

11.4.1 Elasto-plastic analysis in one dimension

Almost all materials show an initial elastic behavior up to a certain level of stress, called the yield stress, beyond which plastic deformation occurs, resulting in the reduced material stiffness. The typical elasto-plastic response for a one-dimensional situation is shown in Fig. 11.2, where the elastic modulus before the onset of yielding is E, and the elasto-plastic range is characterized by the tangent modulus E_T. After yielding, any incremental increase in the level of stress $d\sigma$ leads to the incremental change of strain $d\varepsilon$. The total change in the strain for this stress increment can be expressed as elastic part $d\varepsilon_e$ and plastic component $d\varepsilon_p$, such that

$$d\varepsilon = d\varepsilon_e + d\varepsilon_p \tag{11.66}$$

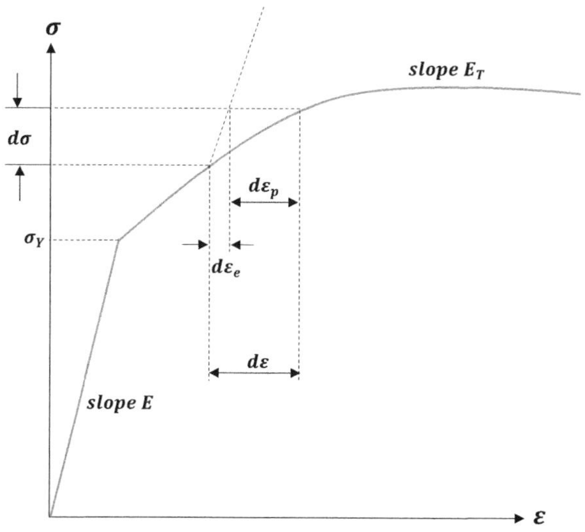

FIGURE 11.2 Elasto-Plastic response for a one-dimensional case.

$$H = \frac{d\sigma}{d\varepsilon_p} = \frac{d\sigma}{d\varepsilon - d\varepsilon_e} = \frac{E_T}{1 - E_T/E} \tag{11.67}$$

Now, we define the strain hardening parameter H as $H = \frac{d\sigma}{d\varepsilon_p}$, which can be defined as the slope of the elasto-plastic stress–strain curve after elastic strain increment is removed. The element stiffness matrix for a simple one-dimensional domain (within elastic zone) is given by

$$\boldsymbol{K}_e = \frac{EA}{L}\begin{bmatrix} 1 & -1 \\ -1 & 1 \end{bmatrix} \tag{11.68}$$

After the onset of yielding, the increase in the length will be given by $d\delta = (d\varepsilon_e + d\varepsilon_p)L$. The tangential stiffness can be written as

$$K_{ep} = \frac{dF}{d\delta} = \frac{AHd\varepsilon_p}{L(d\sigma/E + d\varepsilon_p)} = \frac{EA}{L}\left(1 - \frac{E}{E+H}\right) \tag{11.69}$$

Finally, the element stiffness matrix for modeling the elasto-plastic behavior of a body, after the onset of yielding, is given by

$$\boldsymbol{K}_{ep} = \frac{EA}{L}\left(1 - \frac{E}{E+H}\right)\begin{bmatrix} 1 & -1 \\ -1 & 1 \end{bmatrix} \tag{11.70}$$

Thus it is clear from the above equation that there is reduction in the material stiffness after yielding takes place and the material starts exhibiting the elasto-plastic behavior.

11.4.2 Elastic-predictor plastic-corrector algorithm

During elasto-plastic analysis, total load to be applied is split into small load increments. Due to the nonlinear nature of the problems, equilibrium corresponding to external and internal forces is to be established. Thus there are further iterations within each applied load step so that the condition of equilibrium is established before applying the next load step. The stresses and strains are accumulated after each load step. After certain load increments, the element may yield, and the load at which yielding begins will lie between the accumulated load of the previous cycle and the total load for the present cycle. Consequently, the element behavior becomes elasto-plastic, and the actual stresses are to be reduced to the yield surface. For this purpose, we use the elastic-predictor plastic-corrector algorithm. This algorithm predicts the values of stresses for any particular load step by assuming elastic behavior, and then plastic corrections are applied to reduce the stresses to the yield surface.

Let us consider the rth iteration of any particular load step. The elastic-predictor plastic-corrector algorithm can be applied as follows:

- Applied loads for the rth iteration of any particular load step are the residual forces f_{res}^{r-1}, obtained at the end of the $(r-1)^{th}$ iteration. For this iteration, the displacement increments are given by Δu^r, and the corresponding strain increments are given by $\Delta\varepsilon^r$.

- Assuming the linear elastic behavior, compute the incremental stresses using generalized Hooke's law. If the material did not yield during the present iteration, then the elastically obtained stresses are the true actual stresses. However, if yielding occurs during the present iteration, then the stresses obtained here have to be corrected. The stress increments are obtained by $\Delta\sigma_e^r = E\Delta\varepsilon^r$.
- Accumulate the total stress for the rth iteration, assuming the elastic behavior, as $\sigma_e^r = \sigma^{r-1} + \Delta\sigma_e^r$, where σ^{r-1} is the final corrected stress of $(r-1)^{th}$ iteration.
- Now, we have to check whether the element has yielded in the $(r-1)^{th}$ iteration or not. The yield stress for the $(r-1)^{th}$ iteration is given by $\sigma_Y^{r-1} = \sigma_Y + H\varepsilon_p{}^{r-1}$, where ε_p denotes the accumulated plastic strain and σ_Y is virgin yield stress. Thus we have to check if $\sigma^{r-1} > \sigma_Y^{r-1}$.
  - If $\sigma^{r-1} > \sigma_Y^{r-1}$, the element had yielded in the previous iteration. Now, we have to check whether $\sigma_e^r > \sigma^{r-1}$.
    - If $\sigma_e^r > \sigma^{r-1}$, the element had yielded previously and is still yielding since the stress in increasing further. In this case, all the excess stress $(\sigma_e^r - \sigma^{r-1})$ must be reduced to the yield value, as shown in Fig. 11.3. The factor R, which defines the portion of the stress that is to be reduced to the actual value, is equal to one.
    - If $\sigma_e^r < \sigma^{r-1}$, elastic unloading has taken place during this iteration. According to the plastic theory, the unloading is always elastic. In this case, σ_e^r represents the actual stress, and no modification is required, that is, $\sigma^r = \sigma_e^r$.
  - If $\sigma^{r-1} < \sigma_Y^{r-1}$, the element had not yielded in the previous iteration, as shown in Fig. 11.4. Now, we have to check whether the element has yielded in the present iteration or not, that is, we have to check if $\sigma_e^r > \sigma_Y^{r-1}$.
    - If $\sigma_e^r > \sigma_Y^{r-1}$, the element has yielded during this load increment. Here, we have to reduce stress portion that exceeds the yield stress at the elasto-plastic line.

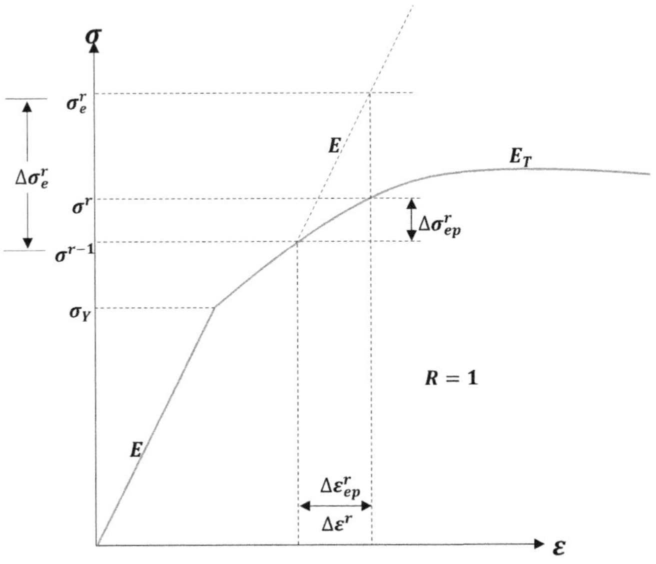

FIGURE 11.3 Incremental stresses and strains for a previously yielded material.

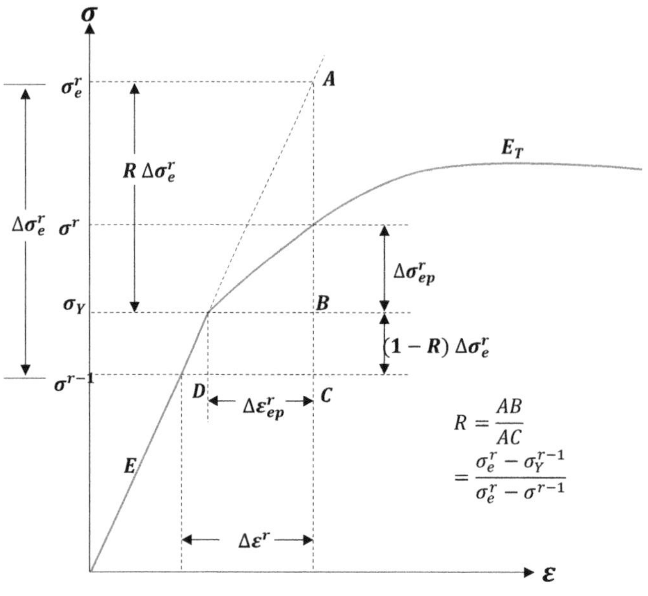

FIGURE 11.4 Incremental stresses and strains for initial yielding.

Reduction factor R is given by $R = \frac{\sigma_e^r - \sigma_Y^{r-1}}{\sigma_e^r - \sigma^{r-1}}$, which represents the portion of stress that is to be corrected or reduced.

- If $\sigma_e^r < \sigma_Y^{r-1}$, the element has not yielded in the present iteration, and is still elastic. No correction in stresses is required, and we have $\sigma^r = \sigma_e^r$.

- For those elements that have yielded, we calculate the stress increments $\Delta\sigma_{ep}{}^r$ by $\Delta\sigma_{ep}{}^r = E\left(1 - \frac{E}{E+H}\right)\Delta\varepsilon_{ep}{}^r$, where $\Delta\varepsilon_{ep}{}^r = R\Delta\varepsilon^r$. The corrected stress for the present r^{th} iteration is given by $\sigma^r = \sigma^{r-1} + (1-R)\Delta\sigma_e^r + \Delta\sigma_{ep}{}^r$. For elastic elements, we get $\sigma^r = \sigma^{r-1} + \Delta\sigma_e{}^r$.
- Now, we calculate the total plastic strain for the rth iteration as $\varepsilon_p^r = \varepsilon_p^{r-1} + \Delta\varepsilon_p^r$. The elastic component of the strain increment is given by $\Delta\varepsilon_e^r = \frac{\Delta\sigma^r}{E}$, and the plastic component of strain is given by $\Delta\varepsilon_p^r = \frac{R\Delta\varepsilon^r}{1+H/E}$.

11.4.3 General elasto-plastic analysis

Most engineering problems usually conform to plane stress or plane strain conditions. Now we present the elasto-plastic analysis of engineering components under plane stress or plane strain conditions. As discussed earlier, the mathematical theory of plasticity provides relationship between the stresses and strains for engineering domains exhibiting the material nonlinearities. The complete description of elasto-plastic analysis includes the elastic stress–strain relations that are applicable before the onset of yielding, yield criterion defining the start of yielding, and stress–strain relationship after the start of yielding, that is, during elasto-plastic phase.

11.4.3.1 The yield criteria

Selection of yield criterion is one of the most important decisions in elasto-plastic analysis, as there are several failure criteria available for estimating the failure loads in engineering materials. The yield or failure criterion is defined in a general form as

$$f(\sigma_{ij}) = K(\kappa) \tag{11.71}$$

where f is a function of stresses, K represents a material parameter, and κ is a parameter related to the hardening behavior. The structural component is said to have failed or yielded if the yield criterion is satisfied, that is, $f(\sigma_{ij}) = K(\kappa)$. Elastic stress states are obtained for $f(\sigma_{ij}) < K(\kappa)$. The states defined by $f(\sigma_{ij}) > K(\kappa)$ are not possible. It is a common practice in elasto-plastic analysis to express the yield criteria in terms of three stress invariants (J_1, J_2, J_3), which can be written as

$$J_1 = \sigma_{ii} \tag{11.72}$$

$$J_2 = \frac{1}{2}\sigma_{ij}\sigma_{ij} \tag{11.73}$$

$$J_3 = \frac{1}{3}\sigma_{ij}\sigma_{jk}\sigma_{ki} \tag{11.74}$$

It is well known that J_1 is zero for deviatoric stresses. In view of this, the yield criterion can be simplified as

$$f(J_2', J_3') = K(\kappa) \tag{11.75}$$

where J_2' and J_3' are second and third invariants of the deviatoric stresses σ_{ij}' such that $\sigma_{ij}' = \sigma_{ij} - \frac{1}{3}\delta_{ij}\sigma_{kk}$. Different yield criteria have been developed from time to time, but all of them do not conform to the experimental results. The two simplest yield criteria that show good agreement with the experimental results are the von Mises and Tresca criteria.

The Tresca yield criterion or maximum shear stress criterion predicts the yielding of a material subjected to a given set of loadings when the maximum shear stress reaches a certain critical value. If σ_1, σ_2, and σ_3 are principal stresses such that $\sigma_1 > \sigma_2 > \sigma_3$, then yielding will begin if

$$\sigma_1 - \sigma_3 = K(\kappa) \tag{11.76}$$

where K is a material parameter which is a function of the hardening parameter κ. The Tresca yield criterion shows good agreement with experimental results for ductile materials, due to which it becomes a good choice for predicting failures in ductile materials.

The von Mises yield criterion predicts the yielding of a material subjected to a given set of loadings when the second invariant of deviatoric stress tensor exceeds a critical value, which can be expressed as

$$\sqrt{J_2'} = K(\kappa) \tag{11.77}$$

such that

$$J_2' = \frac{1}{2}\sigma_{ij}'\sigma_{ij}' = \frac{1}{2}\left[\sigma_x'^2 + \sigma_y'^2 + \sigma_z'^2\right] + \tau_{xy}'^2 + \tau_{yz}'^2 + \tau_{xz}'^2 \tag{11.78}$$

Another simpler widely used version of the von Mises criterion can be written as

$$\overline{\sigma} = \sqrt{3J_2'} = K(\kappa) \tag{11.79}$$

where $\overline{\sigma}$ is von Mises stress. It has been found that the von Mises criterion fits more closely to the experimental data than the Tresca theory, but the Tresca yield criterion is simpler than the former criterion in theoretical applications. There are several other yield criteria that are available to predict failures in engineering components, such as maximum principal stress theory, maximum principal strain criterion, and strain energy density criterion. But these yield criteria do not show good agreement with the experimental data available in literature, due to which they do not find enough use and are of historical importance only.

11.4.3.2 Strain hardening

Strain hardening is an important phenomenon that can be seen in ductile materials after the onset of yielding, and it depends on the type of loading and material under consideration. In general simulations, it is seen that the yield surfaces change regularly after the commencement of yielding. This behavior has been termed strain hardening or work hardening. Three types of hardening behaviors are seen in engineering materials, which can be termed the perfectly plastic behavior and isotropic and kinematic hardening. The behavior is perfectly plastic when there is zero hardening, and yield surfaces do not undergo any change after the initiation of yielding. This type of hardening behavior does not depend on the degree of plastification and has been shown in Fig. 11.5. The yield surfaces expand uniformly without undergoing any translation in isotropic hardening, which can be seen in Fig. 11.6. For a von Mises plasticity model, the radius of the von Mises cylinder will increase, but its center will not change during hardening. Kinematic hardening represents the material behavior when the shape and orientation of yield surfaces do not change but rigid body translation of yield surfaces occurs in stress space, as shown in Fig. 11.7. The kinematic hardening produces the Bauschinger effect during cyclic loading. It has been observed that the real-life materials show mixed isotropic–kinematic hardening behavior rather than purely isotropic or purely kinematic hardening.

The case of isotropic hardening is considered here, and various mathematical foundations are presented. In order to keep track of evolving yield surfaces in isotropic hardening,

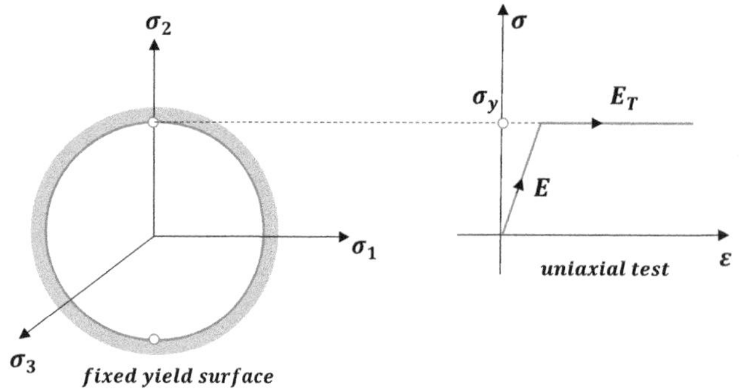

FIGURE 11.5 Representation of perfectly plastic behavior.

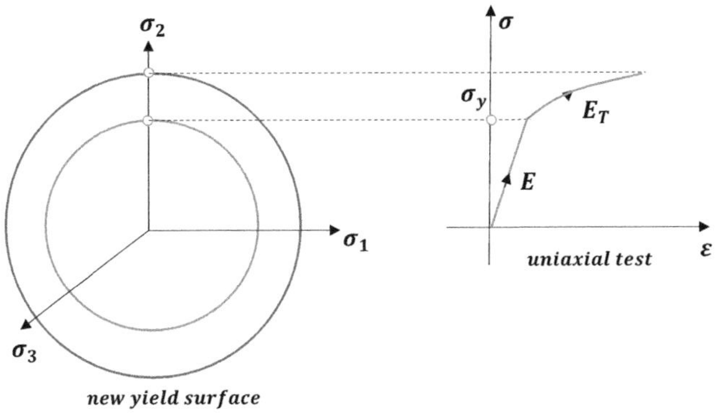

FIGURE 11.6 Representation of isotropic strain hardening behavior.

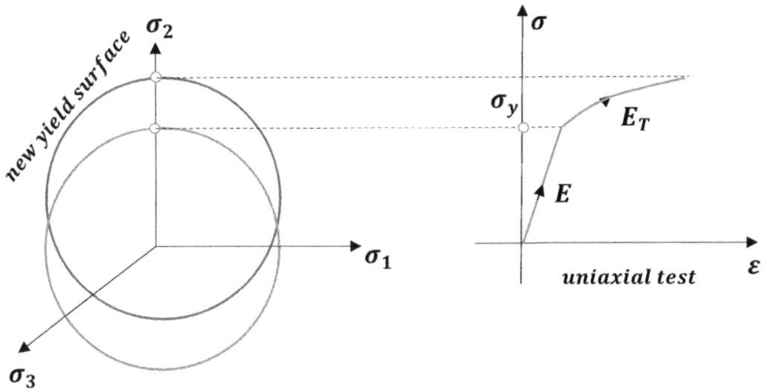

FIGURE 11.7 Representation of kinematic strain hardening behavior.

the main step is to derive the relationship between the material and hardening parameters, as defined earlier. One of the widely accepted theories is to express the hardening parameter as a function of plastic work W_p, which can be written as

$$\kappa = W_p = \int_{\Omega} \sigma_{ij} \left(d\varepsilon_{ij} \right)_p \tag{11.80}$$

where $\left(d\varepsilon_{ij} \right)_p$ denotes components of plastic strain. Another widely accepted theory is to define the hardening parameter in terms of effective plastic strain $\bar{\varepsilon}_p$, which can be written as

$$d\bar{\varepsilon}_p = \sqrt{\frac{2}{3} \left(d\varepsilon_{ij} \right)_p \left(d\varepsilon_{ij} \right)_p} \tag{11.81}$$

As we know that the plastic deformation does not depend on hydrostatic stress, we can write

$$d\bar{\varepsilon}_p = \sqrt{\frac{2}{3} \left(d\varepsilon_{ij}' \right)_p \left(d\varepsilon_{ij}' \right)_p} \tag{11.82}$$

Hardening parameter κ is obtained as

$$\kappa = \bar{\varepsilon}_p = \int d\bar{\varepsilon}_p \tag{11.83}$$

We know that when $f = K$, plastic yielding takes place. When there is an incremental change in stresses, the change in yield function can be expressed as

$$df = \frac{\partial f}{\partial \sigma_{ij}} d\sigma_{ij} \tag{11.84}$$

The above equation can be interpreted as follows:

- If the change in the yield function is less than zero, that is, $df < 0$, elastic unloading has occurred and the new stress state lies within the yield surface.
- If there is no change in yield function, that is, $df = 0$, the yield surface has not changed, and the new stress state will occur on the same yield surface. This material behavior is termed perfectly plastic behavior.
- If there is a positive change in the yield function, that is, $df > 0$, there has been an evolution in the yield surface, and the new stress state lies ahead of the existing yield surface. This material behavior is termed strain hardening.

11.4.3.3 Elasto-plastic stress–strain relations

Under elastic conditions, Hooke's law describes the stress–strain relations. After the onset of yielding, the strain increment can be expressed as

$$d\varepsilon_{ij} = \left(d\varepsilon_{ij}\right)_e + \left(d\varepsilon_{ij}\right)_p \tag{11.85}$$

where $\left(d\varepsilon_{ij}\right)_e$ and $\left(d\varepsilon_{ij}\right)_p$ are elastic and plastic strain components. The elastic strain component can be defined in terms of deviatoric and hydrostatic stresses as

$$\left(d\varepsilon_{ij}\right)_e = \frac{\partial \sigma_{ij}'}{2\mu} + \frac{1-2\nu}{E}\delta_{ij} d\sigma_{kk} \tag{11.86}$$

In order to obtain relationship between plastic strains and incremental stresses, we define a quantity called plastic potential "Q" such that plastic strains are proportional to the stress gradient of plastic potential. We can write

$$\left(d\varepsilon_{ij}\right)_p = d\lambda \frac{\partial Q}{d\sigma_{ij}} \tag{11.87}$$

where $d\lambda$ is plastic multiplier. This equation describes plastic flow after yielding and is known as flow rule. It is very difficult to obtain the general form of the plastic potential. In plasticity theories, it is a common practice to define the plastic potential in terms of the yield function, and this assumption has gained importance because a number of variational principles can be easily postulated. It further leads to the development of the associated theory of plasticity, which assumes plastic potential and yield function to be the same, that is, $Q = f$. Therefore plastic strains can be expressed as

$$\left(d\varepsilon_{ij}\right)_p = d\lambda \frac{\partial f}{d\sigma_{ij}} \tag{11.88}$$

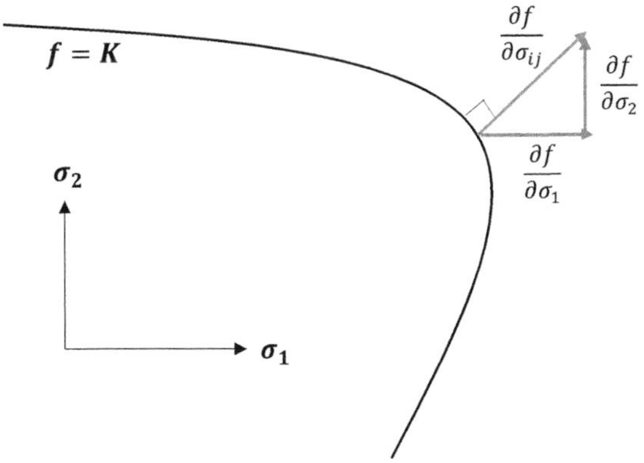

FIGURE 11.8 Geometrical representation of associated theory of plasticity.

The quantity $\frac{\partial f}{d\sigma_{ij}}$ defines a vector normal to the yield surface at the stress point, as can be seen in Fig. 11.8. If $f = J_2'$, which in fact represents a particular case, we get $\frac{\partial f}{d\sigma_{ij}} = \frac{\partial J_2'}{d\sigma_{ij}} = \sigma_{ij}'$. In this case, the flow rule becomes

$$\left(d\varepsilon_{ij}\right)_p = d\lambda\sigma_{ij}' \tag{11.89}$$

This equation is known as the Prandtl–Reuss equation, which has found extensive application in theoretical work. Complete description of stress–strain relationships for the elasto-plastic behavior of a loaded body can be expressed as

$$d\varepsilon_{ij} = \frac{\partial\sigma_{ij}'}{2\mu} + \frac{1-2\nu}{E}\delta_{ij}d\sigma_{kk} + d\lambda\frac{\partial f}{d\sigma_{ij}} \tag{11.90}$$

11.4.3.4 Ramberg–Osgood model

The evaluation of the hardening modulus H is very important in determining the elasto-plastic response of the deformable bodies. The complete elasto-plastic response for a one-dimensional case can be written as

$$\varepsilon = \varepsilon_e + \varepsilon_p = \frac{\sigma}{E} + \frac{\sigma}{H} \tag{11.91}$$

Several hardening laws have been developed from time to time to evaluate hardening modulus, out of which Ramberg–Osgood model finds extreme importance in elasto-plastic analysis. The Ramberg–Osgood model for determining the elasto-plastic response of the material can be written as

$$\frac{\varepsilon}{\varepsilon_o} = \frac{\sigma}{\sigma_o} + \alpha\left(\frac{\sigma}{\sigma_o}\right)^n \quad or \quad \varepsilon = \frac{\sigma}{E} + \frac{\alpha\sigma_o}{E}\left(\frac{\sigma}{\sigma_o}\right)^n \tag{11.92}$$

where σ_o is initial yield stress, α and n are Ramberg–Osgood parameters, and ε_o is the total strain at initial yield point. The quantity $\varepsilon_o^e = \frac{\sigma_o}{E}$ represents elastic component of strain, and the quantity $\varepsilon_o^p = \frac{\alpha\sigma_o}{E}$ denotes the plastic component of the strain at initial yield defined by the 2% offset at the yield stress, such that $\varepsilon_o^p = \frac{\alpha\sigma_o}{E} = 0.002$. Thus the first Ramberg–Osgood parameter can be obtained for any material. The second parameter of the Ramberg–Osgood model can be obtained from the knowledge of uni-axial material properties. If σ_{ut} is ultimate stress of material and ε_{ut} is strain at fracture, we can write

$$\varepsilon_{ut} = \frac{\sigma_{ut}}{E} + 0.002\left(\frac{\sigma_{ut}}{\sigma_o}\right)^n \tag{11.93}$$

which can be solved for n as

$$n = \frac{\log\frac{\left(\varepsilon_{ut} - \frac{\sigma_{ut}}{E}\right)}{0.002}}{\log\left(\frac{\sigma_{ut}}{\sigma_o}\right)} \tag{11.94}$$

Strain hardening exponent n describes the hardening behavior of material after yielding. The material behavior is linear elastic if $n = 1$ and rigid perfectly plastic if $n = \infty$. Once the Ramberg–Osgood parameters are known, the complete elasto-plastic response of the material is completely defined. Now, we can obtain the expression of hardening modulus H as

$$H = \frac{d\sigma}{d\varepsilon_p} = \frac{E}{n\alpha}\left(\frac{\sigma}{\sigma_o}\right)^{1-n} \quad where \quad \varepsilon_p = \frac{\alpha\sigma_o}{E}\left(\frac{\sigma}{\sigma_o}\right)^n \tag{11.95}$$

The Ramberg–Osgood elasto-plastic stress–strain curves for various steel and aluminum alloys are shown in Figs. 11.9 and 11.10.

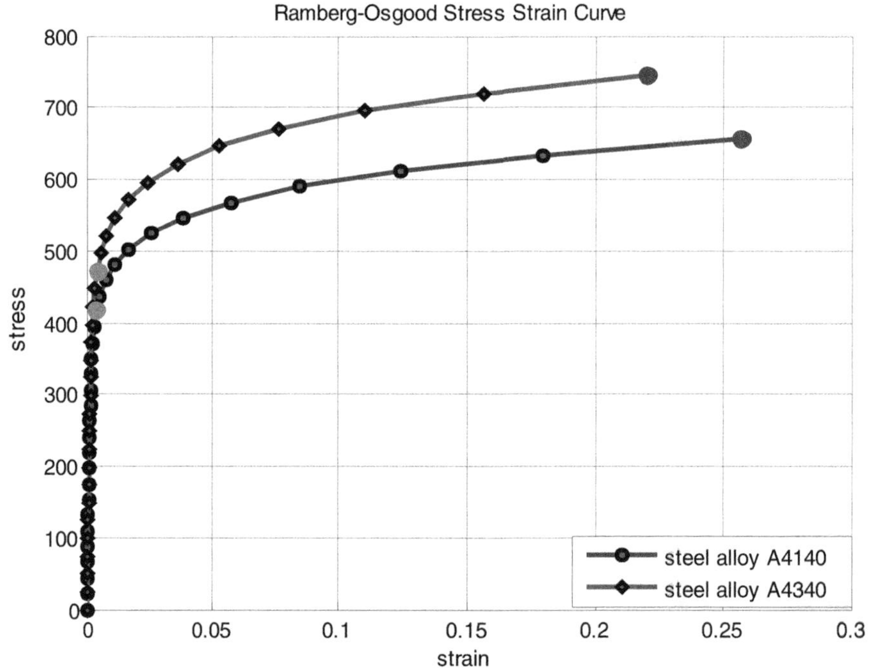

FIGURE 11.9 Ramberg–Osgood stress–strain curves for steel alloys.

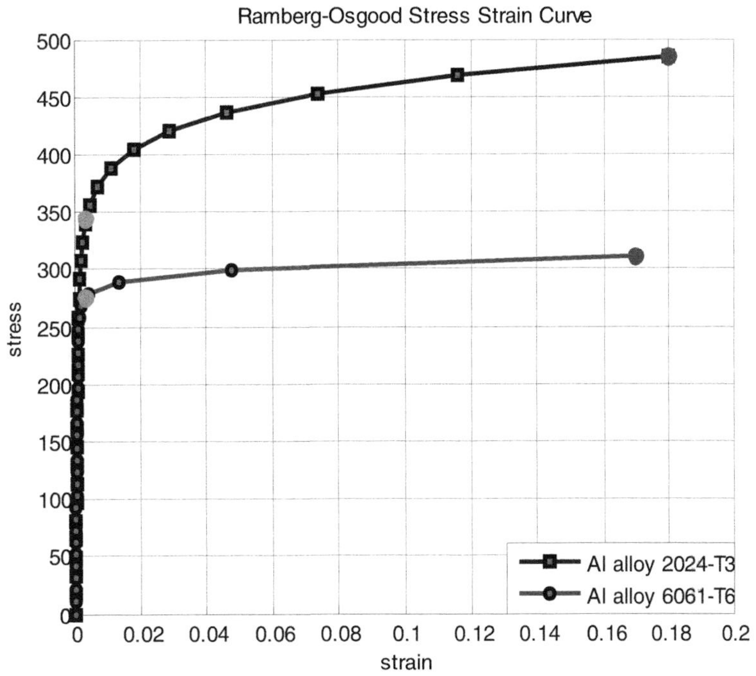

FIGURE 11.10 Ramberg–Osgood stress–strain curves for aluminum alloys.

11.5 Conclusion

The current work reports the details of mathematical formulations for modeling large elasto-plastic deformations in engineering components. TLA and ULA were discussed in details. In the case of TLA, the reference configuration of the loaded body remains the same throughout simulation, but data should be related to the initial undeformed configuration with the help of the deformation gradient tensor. On the contrary, ULA utilizes the latest known deformed configuration for analysis and hence eliminates the need for mapping data between different configurations. The EFGM provides a strong computational tool for modeling large elasto-plastic deformations occurring in structural components. Problems of mesh distortion during large deformation analysis do not occur in EFGM. Introduction of enrichment strategies makes conventional EFGM more suitable for modeling material irregularities, as the problems of mesh conformation and adaption get eliminated in enriched EFGM.

References

[1] Aliabadi MH, Brebbia CA. Boundary element formulations in fracture mechanics: a review. Transactions on Engineering Sciences 1998;17:589–98.

[2] Wen PH, Aliabadi MH, Young A. Dual boundary element methods for three-dimensional dynamic crack problems. Journal of Strain Analysis for Engineering Design 1999;34:373–94. Available from: https://doi.org/10.1177/030932479903400601.

[3] Leonel ED, Venturini WS. Non-linear boundary element formulation applied to contact analysis using tangent operator. Engineering Analysis with Boundary Elements 2011;35:1237–47. Available from: https://doi.org/10.1016/j.enganabound.2011.06.005.

[4] Spangenberger AG, Lados DA. Extended finite element modeling of fatigue crack growth microstructural mechanisms in alloys with secondary/reinforcing phases: model development and validation. Computational Mechanics 2020;. Available from: https://doi.org/10.1007/s00466-020-01921-2.

[5] Lone AS, Kanth SA, Jameel A, Harmain GA. A state of art review on the modeling of contact type nonlinearities by extended finite element method. Materials Today: Proceedings 2019;18:3462–71. Available from: https://doi.org/10.1016/j.matpr.2019.07.274.

[6] Sukumar N, Chopp DL, Moës N, Belytschko T. Modeling holes and inclusions by level sets in the extended finite-element method. Computer Methods in Applied Mechanics and Engineering 2001;190:6183–200. Available from: https://doi.org/10.1016/S0045-7825(01)00215-8.

[7] Jameel A, Harmain GA. Fatigue crack growth in presence of material discontinuities by EFGM. International Journal of Fatigue 2015;81:105–16. Available from: https://doi.org/10.1016/j.ijfatigue.2015.07.021.

[8] Lone AS, Jameel A, Harmain GA. A coupled finite element-element free Galerkin approach for modeling frictional contact in engineering components. Materials Today: Proceedings 2018;5:18745–54. Available from: https://doi.org/10.1016/j.matpr.2018.06.221.

[9] Kanth SA, Harmain GA, Jameel A. Modeling of nonlinear crack growth in steel and aluminum alloys by the element free Galerkin method. Materials Today: Proceedings 2018;5:18805–14. Available from: https://doi.org/10.1016/j.matpr.2018.06.227.

[10] Jameel A, Harmain GA. Fatigue crack growth analysis of cracked specimens by the coupled finite element-element free Galerkin method. Mechanics of Advanced Materials and Structures 2019;26:1343–56. Available from: https://doi.org/10.1080/15376494.2018.1432800.

[11] Lone AS, Jameel A, Harmain GA. Enriched element free Galerkin method for solving frictional contact between solid bodies. Mechanics of Advanced Materials and Structures 2022. Available from: https://doi.org/10.1080/15376494.2022.2092791.

[12] Lone AS, Jameel A, Harmain GA. "Modelling of contact interfaces by penalty based enriched finite element method. Mechanics of Advanced Materials and Structures 2022. Available from: https://doi.org/10.1080/15376494.2022.2034075.

[13] Jameel A, Harmain GA. Large deformation in bi-material components by XIGA and coupled FE-IGA techniques. Mechanics of Advanced Materials and Structures 2022;29:850–72. Available from: https://doi.org/10.1080/15376494.2020.1799120.

[14] Kanth SA, Jameel A, Harmain GA. Investigation of fatigue crack growth in engineering components containing different types of material irregularities by XFEM. Mechanics of Advanced Materials and Structures 2022;29(24):3570–87. Available from: https://doi.org/10.1080/15376494.2021.1907003.

[15] Gupta V, Jameel A, Anand S, Anand Y. Analysis of composite plates using isogeometric analysis: a discussion. Materials Today: Proceedings 2021;44:1190–4. Available from: https://doi.org/10.1016/j.matpr.2020.11.238.

[16] Jameel A, Harmain GA. Effect of material irregularities on fatigue crack growth by enriched techniques. International Journal for Computational Methods in Engineering Science and Mechanics 2020;21:109–33. Available from: https://doi.org/10.1080/15502287.2020.1772902.

[17] Lone AS, Kanth SA, Jameel A, Harmain GA. XFEM modelling of frictional contact between elliptical inclusions and solid bodies. Materials Today: Proceedings 2020;26:819–24. Available from: https://doi.org/10.1016/j.matpr.2019.12.424.

[18] Sheikh UA, Jameel A. Elasto-plastic large deformation analysis of bi-material components by FEM. Materials Today: Proceedings 2020;26:1795–802. Available from: https://doi.org/10.1016/j.matpr.2020.02.377.

[19] Simpson R, Trevelyan J. A partition of unity enriched dual boundary element method for accurate computations in fracture mechanics. Computer Methods in Applied Mechanics and Engineering 2011;200:1–10. Available from: https://doi.org/10.1016/j.cma.2010.06.015.

[20] Noda NA, Oda K. Numerical solutions of the singular integral equations in the crack analysis using the body force method. International Journal of Fracture 1992;58:285–304. Available from: https://doi.org/10.1007/BF00048950.

[21] Rao BN, Rahman S. A coupled meshless-finite element method for fracture analysis of cracks. International Journal of Pressure Vessels and Piping 2001;78:647–57. Available from: https://doi.org/10.1016/S0308-0161(01)00076-X.

[22] Kanth SA, Lone AS, Harmain GA, Jameel A. Modelling of embedded and edge cracks in steel alloys by XFEM. Materials Today: Proceedings 2020;26:814–18. Available from: https://doi.org/10.1016/j.matpr.2019.12.423.

[23] Kanth SA, Lone AS, Harmain GA, Jameel A. Elasto plastic crack growth by XFEM: a review. Materials Today: Proceedings 2019;18:3472–81. Available from: https://doi.org/10.1016/j.matpr.2019.07.275.

[24] Jameel A, Harmain GA. Modeling and numerical simulation of fatigue crack growth in cracked specimens containing material discontinuities. Strength of Materials 2016;48(no. 2):294–307. Available from: https://doi.org/10.1007/s11223-016-9765-0.

[25] Eberhard P, Gaugele T. Simulation of cutting processes using mesh-free Lagrangian particle methods. Computational Mechanics 2013;51:261–78. Available from: https://doi.org/10.1007/s00466-012-0720-z.

[26] De Lorenzis L, Wriggers P, Zavarise G. A mortar formulation for 3D large deformation contact using NURBS-based isogeometric analysis and the augmented Lagrangian method. Computational Mechanics 2012;49:1–20. Available from: https://doi.org/10.1007/s00466-011-0623-4.

[27] Bhardwaj G, Singh IV, Mishra BK, Bui TQ. Numerical simulation of functionally graded cracked plates using NURBS based XIGA under different loads and boundary conditions. Composite Structures 2015;126:347–59. Available from: https://doi.org/10.1016/j.compstruct.2015.02.066.

[28] Gu YT, Wang QX, Lam KY. A meshless local Kriging method for large deformation analyses. Computer Methods in Applied Mechanics and Engineering 2007;196:1673–84. Available from: https://doi.org/10.1016/j.cma.2006.09.017.

[29] Belytschko T, Krongauz Y, Organ D, Fleming M, Krysl P. Meshless methods: an overview and recent developments. Computer Methods in Applied Mechanics and Engineering 1996;139:3–47. Available from: https://doi.org/10.1016/S0045-7825(96)01078-X.

[30] Rao BN, Rahman S. An enriched meshless method for non-linear fracture mechanics. International Journal for Numerical Methods in Engineering 2004;59:197–223. Available from: https://doi.org/10.1002/nme.868.

[31] Duflot M, Nguyen-Dang H. A meshless method with enriched weight functions for fatigue crack growth. International Journal for Numerical Methods in Engineering 2004;59:1945–61. Available from: https://doi.org/10.1002/nme.948.

[32] Jameel A, Harmain GA. Extended iso-geometric analysis for modeling three-dimensional cracks. Mechanics of Advanced Materials and Structures 2019;26:915–23. Available from: https://doi.org/10.1080/15376494.2018.1430275.

[33] Singh IV, Bhardwaj G, Mishra BK. A new criterion for modeling multiple discontinuities passing through an element using XIGA. Journal of Mechanical Science and Technology 2015;29:1131–43. Available from: https://doi.org/10.1007/s12206-015-0225-8.

[34] Bhardwaj G, Singh IV, Mishra BK. Numerical simulation of plane crack problems using extended isogeometric analysis. Procedia Engineering 2013;64:661–70. Available from: https://doi.org/10.1016/j.proeng.2013.09.141.

[35] Jameel A, Harmain GA. A coupled FE-IGA technique for modeling fatigue crack growth in engineering materials. Mechanics of Advanced Materials and Structures 2019;26:1764–75. Available from: https://doi.org/10.1080/15376494.2018.1446571.

[36] Gupta V, Jameel A, Verma SK, Anand S, Anand Y. Transient isogeometric heat conduction analysis of stationary fluid in a container. Part E: Journal of Process Mechanical Engineering 2022. Available from: https://doi.org/10.1177/0954408922112.

[37] Gupta V, Jameel A, Verma SK, Anand S, Anand Y. An insight on NURBS based isogeometric analysis, its current status and involvement in mechanical applications. Archives of Computational Methods in Engineering 2022;. Available from: https://doi.org/10.1007/s11831-022-09838-0.

[38] Singh AK, Jameel A, Harmain GA. Investigations on crack tip plastic zones by the extended iso-geometric analysis. Materials Today: Proceedings 2018;5:19284–93. Available from: https://doi.org/10.1016/j.matpr.2018.06.287.

[39] Melenk JM, Babuška I. The partition of unity finite element method: basic theory and applications. Computer Methods in Applied Mechanics and Engineering 1996;139:289–314. Available from: https://doi.org/10.1016/S0045-7825(96)01087-0.

[40] Jameel A, Harmain GA. Extended iso-geometric analysis for modeling three dimensional cracks. Mechanics of Advanced Materials and Structures 2019;26:915–23. Available from: https://doi.org/10.1080/15376494.2018.1430275.

CHAPTER

12

Implementation issues in extended isogeometric analysis

Vibhushit Gupta[1], Shubham Kumar Verma[2], Sahil Thappa[2], Sanjeev Anand[2], Azher Jameel[3] and Yatheshth Anand[1]

[1]School of Mechanical Engineering, Shri Mata Vaishno Devi University, Kakryal, Katra, Jammu and Kashmir, India [2]School of Energy Management, Shri Mata Vaishno Devi University, Kakryal, Katra, Jammu and Kashmir, India [3]Department of Mechanical Engineering, National Institute of Technology Srinagar, Hazratbal, Srinagar, Jammu and Kashmir, India

12.1 Introduction

The presence of defects like voids, cracks, and inclusions in structures/components is becoming a critical issue. The development of these defects may affect the overall functioning of the components and result in reduced performance. These defects are also responsible for the reduced working life and strength capability of the structural component and, as a consequence, affect the loss of property and human lives. Thus it is becoming a compulsion to analyze the structural behavior that contains discontinuities. To perform such analysis, computational techniques can be applied to determine the response and reliability of structures containing fractural issues. The various types of techniques, that is, finite element analysis (FEA) [1,2], scaled-boundary method [3], boundary element [4,5], meshless method [6–10], extended finite element method (XFEM) [11–16], isogeometric analysis, coupled technique [7,17–19], and extended isogeometric analysis (XIGA) [18,20–24], have been employed for analyzing the fracture response in structures. However, among the discussed method, some of the methods show limitations due to continuous changes in geometries. Like in FEA, the number of remeshing during crack propagation is very high, which makes FEA a complicated and time-consuming technique. Consequently, the accuracy issues due to the low continuity interface in FEA make this technique cumbersome. Thus for analyzing crack problems, XFEM based on the partition of unity (PUM) [25] was developed, which deals with discontinuous problems very effectively. The notion of XFEM technology includes the description of the interface by using the level set method

Enriched Numerical Techniques
DOI: https://doi.org/10.1016/B978-0-443-15362-4.00012-7

[26]. However, XFEM technology generally utilizes the conventional FEA meshes as the background meshes, which does not allow the imposition of strong Dirichlet boundary conditions (BC). Thus the continuity issues are also evident in the elements containing background meshes and result in new problems relating to imposing BC.

Isogeometric analysis (IGA), a novel strategy introduced by Hughes et al., focuses on integrating design and analysis [27]. In their study, a computer-aided design (CAD) basis function, that is, nonuniform rational B-spline (NURBS), has been utilized, which shows some remarkable advantages over the conventional FEA strategy like no discretization errors as the representation of geometry is exact and higher order interelement continuity. The employment of NURBS in their study shows remarkable benefits in the accuracy of results due to its various properties like the PUM, homogeneity, affine covariance, convex hull property, and higher continuity [28]. Also, a brief comparison between the NURBS-based IGA and other numerical techniques has been presented by Gupta et al. in their study [29]. Further, the IGA technique has also been employed successfully in solving various problems like structural [30,31], fluids [32–34], thermal [35,36], thermo-elastic [37], fluid-structure [38], electromagnetics [39], electronics [40], bio-medical [41], digital image correction [42], and optimization [43,44]. In application to fractures, various researchers take IGA as a potential technique, for example, Verhoosel et al. proposed an NURBS-based technique for analyzing cohesive crack zones in which they employ a knot insertion strategy for modeling the interface [45]. Further, the modeling of various curved cracks like parabolic and circular by using NURBS was considered by Choi et al. in their study [46]. Similar to Verhoosel et al., they also take the knot insertion strategy into account for an exact representation of cracks. IGA has also been employed for checking the effect of inclined cracks, in which various refinement strategies were adopted to get better results than the coarser meshes [47].

However, the use of IGA approach for the analysis of fractures shows issues related to accuracy due to improper continuity near the crack interfaces. Thus the utilization of a higher-order basis function in IGA helps in eliminating such issues of interface element continuity. Furthermore, another limitation has been evident due to the noninterpolatory nature of NURBS (as analogous to meshless methods), which results in poor Kronecker delta properties. This issue requires special treatment of the boundary conditions. This problem in the case of NURBS was first taken by Hughes et al., in which on the control variable, they imposed essential BC by assessing the function of BC [27]. This imposition method is considered the direct method for employing essential BC and seems very efficient in the case of homogeneous boundary conditions. However, this method is not reliable in the case of nonhomogeneous boundary conditions. Also, this method has limitations due to the strong enforcement of derivatives of solution in the direction normal to a boundary. Thus various enhancement strategies were employed by various researchers, which include the utilization of the penalty method [48], Nitsche's method [49], Lagrange multiplier [50], transformation matrix [51], and augmented-Lagrangian method [52]. These methods show very satisfactory results in implementing essential BC in the case of IGA [53]. Moreover, based on this, the main focus of the present study is to provide a brief description on these enhancement strategies so the implementation of IGA can be done easily.

Furthermore, in consideration to the higher stability of results in the case of crack propagations, the concept of IGA was extended by employing local enrichment, which was named XIGA [54]. From the past decade, XIGA is considered an efficient numerical strategy for solving issues relating to fractures. The XIGA strategy includes the combination of two numerical techniques, that is, IGA and XFEM, which help in generating exact geometries and overcome issues related to remeshing during the crack propagation [54]. Also, the numerical results denoted the higher accuracy with higher order convergence rates in comparison to XFEM [54]. The utilization of higher order NURBS in XIGA results in evaluating smooth stresses with higher order continuity, which is unavailable in the case of XFEM. Various researchers have implemented XIGA for analyzing various issues of fracture problems. De Luycker et al. consider XIGA for getting robust and accurate results in the case of fracture mechanics [55]. Shojaee et al. employed XIGA for analyzing stationary cracks developed in orthotropic media [56]. Bhardwaj et al. analyze various dynamic and static fractures in the case of functionally graded plates (FGP's) by using XIGA approach [57]. Yu et al. take XIGA for analyzing the fracture behavior of FGP's under thermal buckling conditions [58]. The utilization of XIGA in solving structures under coupled loadings is also a prominent research area. Various researchers consider XIGA for analyzing fractures under coupled loadings, like Bui et al. employ NURBS-based XIGA for analyzing cracks present in magnetoelectroelastic solids (MEE) considered under coupled loadings, that is, electro, magneto, and mechanical. In their study, they illustrated scattered elastics wave propagations in MEE structures [59]. The growth of cracks in the case of functionally graded materials was investigated by Bhardwaj et al. under thermomechanical loading (TML) conditions [60]. Yadav et al. proposed XIGA-based strategy for analyzing linear elastic materials under TML by utilizing XIGA [21]. It is evident from the above investigations that the XIGA approach has been very efficient in solving fracture problems. Apart from this, it has also been noted that the utilization of the NURBS basis function in XIGA invokes some of the limitations related to watertight geometries and global refinement. These issues have been considered in the literature and various local refinement basis functions, that is, T-splines [61], Polynomial splines over hierarchical T-meshes (PHT-splines) [62], Beizer extracted NURBS [63], locally refined (LR) splines [64], and B++ splines [65], have been employed in the XIGA context. The concept on these basis functions is further discussed in the upcoming sections.

However, the issue related to the imposition of boundary conditions has also been evident in the case of NURBS-based XIGA (as similar to IGA and meshless methods) due to its inability to satisfy the Kronecker delta property. Hou et al. proposed the B++ spline-based XIGA technique for solving the issues related to the direct imposition of essential boundary conditions on weak discontinuous interfaces [66]. Further, they extend this methodology with double-layer collocation points on the crack surface instead of considering the original points and additional points [67]. The utilization of this enhanced methodology solves the issues in imposing a Dirchilet constraint boundary on the edges of crack [67]. However, the tip elements associated with control points are independent of the surface of the crack, so it is becoming difficult to impose Dirchilet BC on the whole surface of the crack. So, the focus of this study is to review enhancement strategies that help in imposing essential BC in the case of XIGA.

This study includes the explanation of the XIGA technique depending on various CAD functions and its implementation for fracture problems in Section 12.2; Section 12.3 includes the discussion on enhancement techniques for improving XIGA; consequently, fracture mechanics concept is considered in Section 12.4; the discussion on improving XIGA by utilizing enhancement strategy is presented in Section 12.5, and at last, Section 12.6 concludes the study.

12.2 Extended isogeometric analysis

The incorporation of the enrichment technique of XFEM into IGA results in a reliable numerical strategy, that is, XIGA. This approach considered the advantages of analogous mathematical foundation as XFEM. However, when compared to XFEM, the XIGA technique adopts CAD basis functions as an alternative to the Lagrange polynomial. Due to the utilization of higher-order continuity functions, smooth stresses can be obtained across the interface of the elements, which is absent in the case of XFEM due to C^0-continuity present on the Lagrange elements. In the last two decades, various CAD basis functions have been incorporated into XIGA for enhancing and improving the approach. Each CAD basis function has its benefits and limitations. Thus a brief discussion on these CAD basis functions in the context of XIGA is required, so it becomes easy for choosing the appropriate function for solving different problems. In this section, a brief discussion on implementing XIGA based on different CAD functions for solving fracture problems has been discussed.

12.2.1 Computer-aided design functions in extended isogeometric analysis

In XIGA, numerous basis functions have been adopted for geometry creation and solution approximation. The discussion on various basis functions is given as follows.

12.2.1.1 *B-splines*

The B-spline basis function is explained as the linear piecewise polynomial function that is constructed by specifying a knot vector, which is defined as $\Xi = \{\xi_1 \ldots\ldots\ldots \xi_{n+p+1}\}$. Here, n and p are defined as the count of basis functions and the order of the polynomial, respectively. The knot vector separates the parameter space into two intervals that are generally referred to as knot spans. The knot vectors are also known as patches in the CAD prevalence and include the most commonly used knot vectors as open knot vectors that have multiplicity $k = p + 1$ present in the first and last knots. Based on the order of polynomial and a knot vector, the B-spline function is defined by using the Cox–deBoor recursive formula and is presented as [28]

For $p = 0$

$$N_{i,0}(\xi) = \begin{cases} 1 & if\ \ \xi_i \leq \xi_{i+1} \\ 0 & otherwise \end{cases} \tag{12.1}$$

For $p > 0$

$$N_{i,p}(\xi) = \frac{\xi - \xi_i}{\xi_{i+p} - \xi_i} N_{i,p-1}(\xi) + \frac{\xi_{i+p+1} - \xi}{\xi_{i+p+1} - \xi_{i+1}} N_{i+1,p-1}(\xi) \tag{12.2}$$

where $N_{i,p}$ is defined as a B-spline basis function associated with a given knot vector in which the fractional value of $\frac{0}{0}$ is taken as 0. Also, the algorithms for evaluating B-splines and their depending derivatives of order can be taken from [28]. Furthermore, the utilization of the B-spline basis function exhibits some good properties, such as nonnegativity $N_{i,p} \geq \forall\xi$, linear independency, variational diminishing, the PUM, that is, $\sum_{i=1}^{p+} N_{i,p}(\xi) = 1$, convex hull, and Kronecker's delta properties. The continuity of the B-spline function is presented as C^0 at the boundaries of the patch, whereas C^{p-1} is seen at the knot point ξ. Moreover, this continuity can be decreased by utilizing the relation C^{p-1-k}. The multivariate B-spline functions are considered a tensor product of a univariate function. The relations are given as

$$N_{i,j}^{p,q}(\xi,\eta) = N_{i,p}(\xi) \times M_{j,q}(\eta) \tag{12.3}$$

$$N_{i,j,k}^{p,q,r}(\xi,\eta,\zeta) = N_{i,p}(\xi) \times M_{j,q}(\eta) \times L_{k,r}(\zeta) \tag{12.4}$$

Depending on this, the geometries using the B-spline function can be constructed by using the relation:

$$Z(\xi) = \sum_{i=1}^{n} N_i^p(\xi) C_i \tag{12.5}$$

$$Y(\xi,\eta) = \sum_{i=1}^{n} \sum_{i=1}^{m} N_{i,p}(\xi) \times M_{j,q}(\eta) C_{i,j} \tag{12.6}$$

$$X(\xi,\eta,\zeta) = \sum_{i=1}^{n} \sum_{i=1}^{m} \sum_{i=1}^{l} N_{i,p}(\xi) \times M_{j,q}(\eta) \times L_{k,r}(\zeta) C_{i,j,k} \tag{12.7}$$

where $C_i, C_{i,j}$, and $C_{i,j,k}$ represent the control points of 1D, 2D, and 3D arrays, the B-spline curve, surface, and solid can be defined by $Z(\xi)$, $Y(\xi, \eta)$, and $X(\xi, \eta, \zeta)$, respectively, and m and l show the nos. of basis function in the direction of η and ζ.

12.2.1.2 Nonuniform rational B-spline

B-splines are very much convenient and efficient for free-form modeling; however, they have limitations in the exact representation of some easy shapes like circles and ellipsoids. Thus the utilization of a generalized form of B-splines, that is, NURBS overcomes such issues and provides the ability to represent conic sections like hyperboloids, ellipsoids, spheres, and paraboloids in an exact manner. Also, the employment of NURBS provides efficient algorithms for the evaluation and refinement of the sections. A one-dimensional NURBS function is defined by considering the rationalized form of weighted B-splines function as [28]

$$R_i^p(\xi) = \frac{N_{i,p}(\xi) w_i}{\sum_{i=1}^{n} N_{i,p}(\xi) w_i} \tag{12.8}$$

In Eq. (12.8), R_i^p presents the rational function, and w_i shows the weight function associated with the vector of control points C_i. Analogously, the two-dimensional and three-dimensional basis functions can be defined as

$$R_i^p(\xi,\eta) = \frac{N_{i,p}(\xi) \times M_{j,q}(\eta) w_{i,j}}{\sum_{i=1}^{n} \sum_{j=1}^{m} N_{i,p}(\xi) \times M_{j,q}(\eta) w_{i,j}} \tag{12.9}$$

$$R_i^p(\xi,\eta,\zeta) = \frac{N_{i,p}(\xi) \times M_{j,q}(\eta) \times L_{k,r}(\zeta) w_{i,j,k}}{\sum_{i=1}^{n} \sum_{j=1}^{m} \sum_{k=1}^{l} N_{i,p}(\xi) \times M_{j,q}(\eta) \times L_{k,r}(\zeta) w_{i,j,k}} \tag{12.10}$$

where $N_{i,p}(\xi)$, $M_{j,q}(\eta)$, and $L_{k,r}(\zeta)$ are the B-spline basis functions having orders p, q, and r, respectively. Similarly, the NURBS base geometries are developed in the same manner as a B-spline geometry. For 1D, 2D, and 3D geometries, the equations are given as

$$W(\xi) = \sum_{i=1}^{n} R_i^p(\xi)C_i \tag{12.11}$$

$$V(\xi,\eta) = \sum_{i=1}^{n}\sum_{j=1}^{m} R_i^p(\xi,\eta)C_{i,j} \tag{12.12}$$

$$U(\xi,\eta,\zeta) = \sum_{i=1}^{n}\sum_{j=1}^{m}\sum_{k=1}^{l} R_i^p(\xi,\eta,\zeta)C_{i,j,k} \tag{12.13}$$

Also, it has been evident from the literature that the role of weights is very essential in the case of NURBS, as the selection of appropriate weights allows the description of the curved geometries very efficiently. Moreover, in cases of equal weights, the NURBS basis is reduced to the B-spline basis, which can happen for straight geometries. Mostly, the values of weights are certain for the simple geometries but generally are user-defined [28].

12.2.1.3 T-splines

Apart from various advantages of NURBS function in representing curved geometries, they also show deficiencies like overlaps or gaps at the intersection of surfaces, complicated mesh generations, and inability to perform local refinements due to its tensor product structure. Also, the NURBS are not sufficient in constructing single watertight geometry. The employment of T-splines instead of NURBS corrects such limitations by providing local refinements and producing watertight geometries. The T-splines are considered a generalized form of the NURBS function and can be constructed by giving proper increments in columns and rows of control points that form T-junctions. These junctions are similar to the hanging nodes present in the FEA strategy. Further, in NURBS, the basis functions and knots are not in one-to-one correspondence, as the knots presented are $n+p+1$ in number with only n basis functions. However, the identification of such points in a parametric space on which the basis functions are associated is easy and named as anchors. These anchors t_i can be defined for a N_A basis function having g degree of the curve [68]:

$$t_i = \left\{ \begin{array}{c} \xi_{i+\frac{g+1}{2}} \ \ if\ g\ is\ odd \\ \frac{1}{2}\xi_{i+\frac{g}{2}} + \xi_{i+\frac{g}{2}+1} \ \ if\ g\ is\ even \end{array} \right\} \tag{12.14}$$

The T-spline basis function is developed for every vertex in the T-mesh by using a local knot vector. The localized knot vector for a control point can be developed from a T-mesh that is defined in a parametric domain by utilizing parametric values assigned to each vertex of a T-mesh. The local knot vectors are different for even and odd degree T-splines mesh. The local bivariate basis function can be derived as

$$N_A(\xi,\eta) = N(\xi_{g-2},\xi_{g-1},\xi_g,\xi_{g+1},\xi_{g+2})(\xi) \times M(\eta_{g-2},\eta_{g-1},\eta_g,\eta_{g+1},\eta_{g+2})(\eta) \tag{12.15}$$

where $N(\xi_{g-2}, \xi_{g-1}, \xi_g, \xi_{g+1}, \xi_{g+2})(\xi)$ and $M(\eta_{g-2}, \eta_{g-1}, \eta_g, \eta_{g+1}, \eta_{g+2})(\eta)$ are the cubic univariate basis functions associated with the localized knot vectors $\xi_A = \xi_{g-2}, \xi_{g-1}, \xi_g, \xi_{g+1}, \xi_{g+2}$ and $\eta_A = \eta_{g-2}, \eta_{g-1}, \eta_g, \eta_{g+1}, \eta_{g+2}$, respectively. The value of the univariate basis function can be derived by utilizing a recurring formula, starting with the normalized (g = 0) piecewise function:

$$N^i_A(j,0) = \begin{Bmatrix} 1 \;\; if \;\; \xi^i_{A,j} \le \xi^i_A \le \xi^i_{A,j+1} \\ 0 \;\; otherwise \end{Bmatrix} \tag{12.16}$$

where $\xi^i_{A,j}$ is the jth value of knot in a local knot vector, and i ranges from 1 to parametric directions. For $g > 0$

$$N^i_A(j,p) = \frac{\xi^i_A - \xi^i_{A,j}}{\xi^i_{A,g+1} - \xi^i_{A,j}} N^i_A(j, g-1) + \frac{\xi^i_{A,j+g+1} - \xi^i_A}{\xi^i_{A,j+p+1} - \xi^i_{A,j+1}} N^i_A(j+1, p-1) \tag{12.17}$$

After the definition of the basis function, the surface based on the T-spline function is given as

$$T(\xi,\eta) = \sum\nolimits_{A=1}^{n} N_A(\xi,\eta) C_A \tag{12.18}$$

where $N_A(\xi,\eta)$ shows the basis function in association with control point C_A. Similarly, the T-spline rational surface is given as

$$\overline{T}(\xi,\eta) = \frac{N_A(\xi,\eta) w_A}{\sum\nolimits_{A=1}^{n} N_A(\xi,\eta) W_A} \tag{12.19}$$

$$\overline{T}(\xi,\eta) = \sum\nolimits_{A=1}^{n} R_A(\xi,\eta) \overline{C_A} \tag{12.20}$$

where $\overline{C_A} = \{w_A P^T_A, w_A\}^T$ shows the coordinates in the homogeneous form, w_A represents the weight of a control point, and W_A defines the cumulative weight function.

12.2.1.4 Locally refined B-splines

As already discussed, the utilization of B-splines and NURBS for local refinement is very challenging as they are developed as the tensors of various univariate splines. Thus another basis function, that is, LR B-splines are taken into consideration to reduce such limitation. This basis function is a generalized form of NURBS and B-splines such that they include various desirable properties of the NURBS and B-splines. For defining these LR B-splines, first, the concept of local knot vectors should be discussed. Hence, localized knot vectors Ξ_i can be constructed from a global knot vector as $\Xi = \{\xi_1 \ldots\ldots\ldots \xi_{n+p+1}\}$ as

$$\Xi_i = \left\{\xi_{i+j}\right\}_{j=0}^{p+1}, i = 1 \ldots\ldots\ldots n \tag{12.21}$$

Further, the B-spline basis function can be defined by using Eqs. (12.1) and (12.2), and then global functions are enriched by h-refinement without doing any changes in the geometry. As the LR B-splines are the generalized form of B-splines, some of the refinement strategies are also applied in the case of LR B-splines. Thus by inserting $\bar{\xi}$ between ξ_{i-1} and ξ_i into the knot vector, the following relation has been generated:

$$\alpha_1 = \begin{cases} 1 & \xi_{p+1} \leq \bar{\xi} \leq \xi_{p+2} \\ \dfrac{\bar{\xi} - \xi_1}{\xi_{p+1} - \xi_1} & \xi_1 \leq \bar{\xi} \leq \xi_{p+1} \end{cases} \tag{12.22}$$

$$\alpha_2 = \begin{cases} \dfrac{\xi_{p+2} - \bar{\xi}}{\xi_{p+2} - \xi_2} & \xi_2 \leq \bar{\xi} \leq \xi_{p+2} \\ 1 \, \xi_1 \leq \bar{\xi} \leq \xi_2 \end{cases} \tag{12.23}$$

and the localized knot vectors can be considered

$$\begin{aligned} \Xi &= \left[\xi_1, \xi_2 \ldots \;\ldots\; \ldots \xi_{i-1}, \xi_i \ldots \;\ldots\; \ldots \xi_{p+1}, \xi_{p+2}\right] \\ \Xi_1 &= \left[\xi_1, \xi_2 \ldots \;\ldots\; \ldots \xi_{i-1}, \bar{\xi}, \xi_i \ldots \;\ldots\; \ldots \xi_{p+1}, \xi_{p+2}\right] \\ \Xi_2 &= \left[\xi_2 \ldots \;\ldots\; \ldots \xi_{i-1}, \bar{\xi}, \xi_i \ldots \;\ldots\; \ldots \xi_{p+1}, \xi_{p+2}\right] \end{aligned} \tag{12.24}$$

the bivariate B-spline basis function that can be defined on the two localized knot vectors $\Xi = \{\xi_1, \xi_2, \xi_3 \ldots\ldots\ldots \xi_{p+2}\}$ and $H = \{\eta_1, \eta_2, \eta_3 \ldots\ldots\ldots \eta_{q+2}\}$.

$$B_{\Xi,H}(\xi, \eta) = B_\Xi(\xi) \cdot B_H(\eta) \tag{12.25}$$

Thus, the LR surface $J(\xi, \eta)$ can be defined as:

$$J(\xi, \eta) = \sum_{A=1}^{n} B_{\Xi,H}(\xi, \eta) C_A \tag{12.26}$$

12.2.1.5 PHT-splines

PHT-splines are generally called polynomial hierarchical T-meshes, which fulfills the requirement of linear independence in a basis function. This approach is also a piecewise polynomial approach due to it becomes easier to employ simpler adaptive refinements as well as integrations. In this case, the local refinements are also simpler as compared to the complex implementation of T-splines. The PHT-splines also represent C^1 interelement continuity that can be increased by doing extension in the PHT-splines [62]. The hierarchical T-meshes are considered in this spline, which has a natural nested structure. The T-meshes are started from level-0 and then moved forward to a required level as per the refinement requirements. The basis function of PHT splines, which can be derived from 0 to $k + 1$ level, is given as [69]

At Level 0

$$Y(\xi, \eta) = \sum_{i=1}^{n} R_i(\xi, \eta) C_i \tag{12.27}$$

At level k

$$Y^k(\xi, \eta) = \sum_{i=1}^{n_k} R_i^k(\xi, \eta) P_i^k \tag{12.28}$$

At level $k+1$

$$Y^{k+1}(\xi,\eta)=\sum_{i=1}^{n_{k+1}}R_i^{k+1}(\xi,\eta)P_i^{k+1} \tag{12.29}$$

All the above-discussed basis functions can be employed for enhancing the XIGA approach so that various implementation issues related to local refinements and exact representations can be resolved.

12.2.2 Fracture mechanics problem formulation

In this section, a linear elastic fracture mechanics problem is discussed so that it becomes easier to understand the concept of XIGA. The formulation is considered for a homogeneous two-dimensional domain, as most of the basis functions have been implemented for solving various 2D problems very efficiently. The domain Ω under consideration is given in Fig. 12.1, which is partitioned into displacement boundary (Γ_u), traction boundary (Γ_t), crack boundary (Γ_c), and inclusion or hole boundary (Γ_h). Depending on this, the domain when considered under elastic loading has to satisfy the equilibrium conditions [70]:

$$\nabla.\sigma+f=0 \tag{12.30}$$

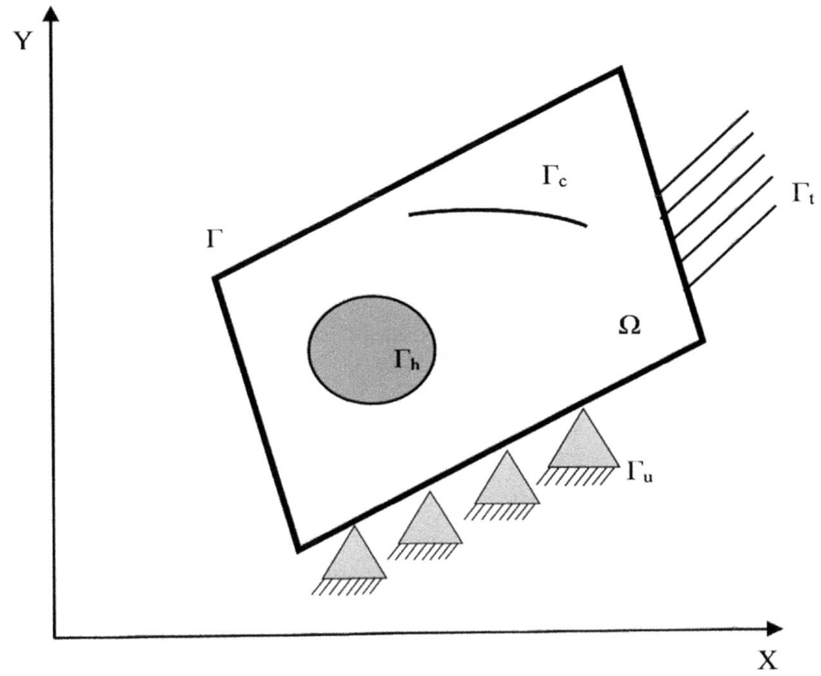

FIGURE 12.1 Engineering domain containing a crack and inclusion.

where ∇ represents the divergence operator with σ as a Cauchy stress tensor product, and f is a body force vector. The boundary conditions for the domain under consideration are explained as

$$\sigma.\hat{n} = \hat{t} \ on \ \Gamma_t \tag{12.31}$$

$$u = \bar{u} \ on \ \Gamma_u \tag{12.32}$$

$$\sigma.\hat{n} = 0 \ on \ \Gamma_c \tag{12.33}$$

The constitutive equation for a homogeneous elastic material can be derived as

$$\sigma = C \cdot \epsilon \tag{12.34}$$

This ϵ is a strain tensor, and C presents the elastic tensor. Depending on this, as per Belytschko and Black [71], the weak form equation is given as

$$\int_{\Omega} \sigma(u) \cdot \epsilon(u) d\Omega = \int_{\Omega} b \cdot u d\Omega + \int_{\Gamma} t \cdot u d\Gamma \tag{12.35}$$

Further, by inserting test and trial functions and by employing the arbitrariness of the varied control points, the equation of system in discrete form is as follows:

$$[K]\{u\} = \{f\} \tag{12.36}$$

This equation presents the external force vector as f, displacement vector u, and displacement vector by K.

12.2.3 Approximation of voids and cracks using extended isogeometric analysis

The approximation of displacement is locally enriched for analyzing discontinuities. Some additional degrees of freedom are added to the selected control points near the location of the discontinuity.

12.2.3.1 Approximation of voids using extended isogeometric analysis

The approximation of holes present in a domain can be derived from the relation presented by Bhardwaj et al. [72], which is given as

$$\mathrm{u}^{\mathrm{h}}(\xi) = \sum_{\mathrm{i}=1}^{n_{be}} \mathrm{R}_{\mathrm{i}}^{\mathrm{P}}(\xi)\mathrm{u}_{\mathrm{i}} + \sum_{l=1}^{n_z} \mathrm{R}_l^{\mathrm{P}}(\xi)\left[\chi(\xi) - \chi(\xi_{\mathrm{i}})\right]c_l \tag{12.37}$$

where u_{i} shows the parameters relates to the specified control point, n_{be} are the number of shape functions per element given as $n_{be} = (p+1) \times (q+1)$;$\mathrm{R}_{\mathrm{i}}^{\mathrm{P}}$ shows the different approximation functions that can be referred from Section 12.2.1, C_l = extra enriched degree of freedoms pertaining to $\chi(\xi)$ that takes 0 for the inner control points of the void and +1 for the outside control points and n_z are the split-enriched node elements of a void.

12.2.3.2 *Approximation of cracks using extended isogeometric analysis*

The modeling of crack tip and edge by using XIGA can be done by using the following relation [72]:

$$u^h(\xi) = \sum_{i=1}^{n_{be}} R_i^p(\xi)u_i + \sum_{j=1}^{n_s} R_j^p(\xi)\left[W(\xi) - W(\xi_i)\right]a_j + \sum_{k=1}^{n_T} R_k^p(\xi)\sum_{\alpha=1}^{4}[\beta_\alpha(\xi) - \beta_\alpha(\xi_i)]b_k^\alpha \quad (12.38)$$

where n_s = split enriched node elements of a crack, n_T = tip enriched node elements, a_j = enriched degree of freedom of split elements related to Heaviside function (HF) $W(\xi)$; b_k^α = enriched degree of freedom of a tip element relates to crack tip enrichment functions (CTEF) given by $\beta_\alpha(\xi)$. The utilization of $W(\xi)$ helps in producing discontinuity near the crack surface and ranges from -1 to $+1$ that can be written as

$$W(\xi) = \begin{cases} -1, & otherwise \\ +1, & if\ (\xi - \xi_i).n_{be} \geq 0 \end{cases} \quad (12.39)$$

The HF is efficient when an element is divided by a crack interface. Although, in the case of tip elements, a part of the element is only cut, the HF is not able to enrich that tip element. Thus CTEF is used for the enrichment of crack tip element. This enrichment function is defined by Chopp and Sukumar [73] and is explained as

$$\beta_\alpha(\xi) = \left[\sqrt{r}\cos\frac{\theta}{2}, \sqrt{r}\sin\frac{\theta}{2}, \sqrt{r}\cos\frac{\theta}{2}\sin\theta, \sqrt{r}\sin\frac{\theta}{2}\sin\theta\right] \quad (12.40)$$

where θ and r present polar coordinates at the reference line of the crack and the crack surface.

12.2.4 Selection of control points for enrichment

The enrichment of control points in XIGA is done in an analogous manner like in XFEM [74]. However, there is a small modification seen between the control points of XIGA and nodes of XFEM. In XFEM, the nodes associated with tips and faces of cracks are enriched, while in XIGA, the domain of influence is much larger; hence, the control points situated far from the discontinuity are enriched as well. The control points having a nonzero-valued basis function on the crack face or tip are generally chosen for enrichment with the HF and CTEF.

12.2.5 Extended isogeometric analysis-based discretization

The discontinuity in XIGA can be evaluated by utilizing the displacement field. Hence, the mechanical equation (Eq. 12.30) is discretized over the whole element as

$$K^{hU}u^{hU} = f^{hU} \quad (12.41)$$

where u^h represents the displacement vector, which consists of standardized as well as additional DOF's and is written as

$$u^{hU} = \begin{bmatrix} u & a^u & b_1^u & b_2^u & b_3^u & b_4^u & c^u \end{bmatrix} \quad (12.42)$$

where $u = [u_x \quad u_y]$ shows the standard DOFs, while $a^u = [a^u_x \quad a^u_y]$ and $b^u_\alpha = [b^u_{\alpha x} \quad b^u_{\alpha y}]$ are the additional DOFs in case of cracks, c^u is the extra number of degrees of freedom corresponds to an enrichment of void. f^{hU} is a force vector, and K^{hU} is element stiffness matrix in the displacement field that is formed as the following equation:

$$K^{hU} = \begin{bmatrix} K_{ij}^{uu} & K_{ij}^{ua^u} & K_{ij}^{ub^u} & K_{ij}^{uc^u} \\ K_{ij}^{a^u u} & K_{ij}^{a^u a^u} & K_{ij}^{a^u b^u} & K_{ij}^{a^u c^u} \\ K_{ij}^{b^u u} & K_{ij}^{b^u a^u} & K_{ij}^{b^u b^u} & K_{ij}^{b^u c^u} \\ K_{ij}^{c^u u} & K_{ij}^{c^u a^u} & K_{ij}^{c^u b^u} & K_{ij}^{c^u c^u} \end{bmatrix} \tag{12.43}$$

$$f^{hU} = \begin{bmatrix} f_i^u & f_i^{a^u} & f_i^{b_1^u} & f_i^{b_2^u} & f_i^{b_3^u} & f_i^{b_4^u} & f_i^{c^u} \end{bmatrix}^T \tag{12.44}$$

$$K_{ij}^{rs} = \int_{\Omega^e} (B_i^r)^T C B_j^s d\Omega \tag{12.45}$$

$$f_i^u = \int_{\Omega^e} (R_i)^T b d\Omega + \int_{\Gamma_t} (R_i)^T \hat{t} d\Gamma \tag{12.46}$$

$$f_i^{a^u} = \int_{\Omega^e} (R_i)^T H b d\Omega + \int_{\Gamma_t} (R_i)^T H \hat{t} d\Gamma \tag{12.47}$$

$$f_i^{b^u} = \int_{\Omega^e} (R_i)^T \beta_\alpha b d\Omega + \int_{\Gamma_t} (R_i)^T \beta_\alpha \hat{t} d\Gamma \tag{12.48}$$

$$f_i^{c^u} = \int_{\Omega^e} (R_i)^T \chi b d\Omega + \int_{\Gamma_t} (R_i)^T \chi \hat{t} d\Gamma \tag{12.49}$$

where R_i is the NURBS rational function, and B_i^u, $B_i^{a^u}$, $B_i^{b^u_\alpha}$, $B_i^{b^u}$, $B_i^{c^u}$ are the derivative matrices of the rational functions that are explained as

$$B_i^u = \begin{bmatrix} R_{i,X_1} & 0 \\ 0 & R_{i,X_2} \\ R_{i,X_1} & R_{i,X_2} \end{bmatrix} \tag{12.50}$$

$$B_i^{a^u} = \begin{bmatrix} R_{i,X_1} W & 0 \\ 0 & R_{i,X_2} W \\ R_{i,X_1} W & R_{i,X_2} W \end{bmatrix} \tag{12.51}$$

$$B_i^{b\alpha} = \begin{bmatrix} (R_i\beta_\alpha)_{,X_1} & 0 \\ 0 & (R_i\beta_\alpha)_{,X_2} \\ (R_i\beta_\alpha)_{,X_1} & (R_i\beta_\alpha)_{,X_1} \end{bmatrix} \tag{12.52}$$

$$B_i^b = \begin{bmatrix} B_i^{b1} & B_i^{b2} & B_i^{b3} & B_i^{b4} \end{bmatrix} \tag{12.53}$$

$$B_i^{c^u} = \begin{bmatrix} (R_i)_{,X_1}\chi & 0 \\ 0 & (R_i)_{,X_2}\chi \\ (R_i)_{,X_1}\chi & (R_i)_{,X_1}\chi \end{bmatrix} \tag{12.54}$$

where, in these equations, $r,s = u, a^u, b^u$ and c^u and $\alpha = 1, 2, 3,$ and 4.

12.3 Enhancement in extended isogeometric analysis for imposition of boundary conditions

In XIGA, some of the approximation functions like NURBS do not satisfy the Kronecker delta property, due to which it is not possible to impose the essential BC directly to the control points. This further leads to significant errors and degraded convergence rates [75]. Thus this issue requires additional treatment so that boundary conditions can be enforced easily. Several methods have been developed from time to time for implementing boundary conditions in XIGA. Some of the methods include modifications in the weak form, and some modify the basis function in order to resolve the issue. The modifications related to basis functions have been discussed in Section 12.2.1. Further, the methods utilized for modifications in the weak form are discussed in this section. The various methods are the Lagrangian multiplier method [50], the penalty method [48], Nitsche's method [49], and the augmented-Lagrangian method [52]. The implementation of these methods in the framework can be discussed as

12.3.1 Lagrange multiplier approach

In this approach, a new unknown additional vector field is considered, which is known as the Lagrange multiplier field, symbolized as λ^h. This field is given by following relation [54]:

$$\lambda^h = \begin{Bmatrix} \lambda_u \\ \lambda_v \end{Bmatrix} = \begin{bmatrix} L_1 & 0 & \ldots & L_{n_\lambda} & 0 \\ 0 & L_1 & \ldots & 0 & L_{n_\lambda} \end{bmatrix} \begin{Bmatrix} \lambda_{u1} \\ \lambda_{v1} \\ \cdot \\ \cdot \\ \cdot \\ \lambda_{un_\lambda} \\ \lambda_{vn_\lambda} \end{Bmatrix} = L(s)\lambda \tag{12.55}$$

where λ_m represents the Lagrange multiplier for $m = u$ and v that defines a set of n_λ control points located on the essential boundaries. It is also considered that the control points are not present on the essential boundaries; in that case, the nearest points to each control point are taken. L_i represents the Lagrange basis function, and s is the length of the arc presented along the essential boundary. Thus following the Lagrange multiplier, the weak form of the equation, that is, Eq. (12.35), can be modified as

$$\begin{aligned} &\int_\Omega \sigma(u).\epsilon(u)d\Omega \\ &\quad = \int_\Omega b.ud\Omega + \int_{\Gamma_t} t.ud\Gamma + \int_{\Gamma_u} \lambda(u - \overline{u})d\Gamma \\ &\quad + \int_{\Gamma_u} \lambda.ud\Gamma \end{aligned} \tag{12.56}$$

Further, the modification in the discrete form can be noticed as

$$\begin{bmatrix} K & G \\ G^T & 0 \end{bmatrix} \begin{Bmatrix} u \\ \lambda \end{Bmatrix} = \begin{Bmatrix} f \\ q \end{Bmatrix} \tag{12.57}$$

where G_{ij} shows the nodal matrix and q_i is the vector, which are defined as

$$G_{ij} = \int_{\Gamma_u} L_i^T R_j d\Gamma \tag{12.58}$$

$$q_i = \int_{\Gamma_u} L_i^T \overline{u} d\Gamma \tag{12.59}$$

where K, u, and f are defined in Eq. (12.36).

12.3.2 Penalty approach

The penalty approach is another option for easy implementation of the essential boundary condition. Also, the size of equations in this approach is unaltered, and a higher range of convergence is guaranteed when the increase in penalty parameter is done with the refinement. The penalty parameter (β) is defined as a nonnegative scalar constant and has a very high value, which directly helps in imposing the essential boundary condition with higher accuracy. The involvement of the penalty parameter in a weak form solution can be given as

$$\int_{\Omega} \sigma(u).\epsilon(u) d\Omega + \beta \int_{\Gamma_u} u d\Gamma = \int_{\Omega} b.u d\Omega + \int_{\Gamma_t} t.u d\Gamma + \beta \int_{\Gamma_u} u d\Gamma \tag{12.60}$$

The discretization of this weak form can be written as follows:

$$[K + \beta G]\{u\} = \{f + \beta q\} \tag{12.61}$$

where K and f are taken from Eq. (12.36) and G and q are considered from Eqs. (12.58) and (12.59). In the penalty approach, the condition of the stiffness matrix generally depends on the choice of penalty parameter [76]. Also, this approach does not include any additional unknown vector fields when compared with the Lagrangian multiplier method, which makes its implementation very easy. The utilization of the penalty method presents two advantages, that is, the system dimensions remain constant and the resulting matrix is symmetric and nonnegative. However, this method also shows some of the drawbacks related to the weak imposition of Dirichlet boundary conditions and poorly conditioned matrix.

12.3.3 Nitsche approach

The Nitsche approach becomes an attractive way for modification in weak formulation. The utilization of this approach provides some notable advantages like stabilized bilinear formulation and well-conditioned stiffness matrix. Also, the utilization of the Nitsche approach increases the solution accuracy and uniform convergence in the case of IGA problems, which is presented by Embar et al. [49]. The modified weak form equation by using the Nitsche method is derived by using Griebel and Schweitzer [77]:

$$\int_{\Omega}\sigma(u)\cdot\epsilon(u)d\Omega-Ч\int_{\Gamma_u}u\cdot nd\Gamma-\int_{\Gamma_u}u\cdot nd\Gamma+Ч\int_{\Gamma_u}ud\Gamma$$
$$=\int_{\Omega}b\cdot ud\Omega+\int_{\Gamma_t}t\cdot ud\Gamma+Ч\int_{\Gamma_u}ud\Gamma-Ч\int_{\Gamma_u}u\cdot nd\Gamma \tag{12.62}$$

where Ч shows the nonnegative constant scalar parameter. The left-hand side of Eq. (12.62) presents the symmetry recovery and corrective of the bilinear form provided with the higher parameter value. Further, the right-hand side terms are added to ensure consistency in the weak form. The discretized form of the equation results in a system of equations with a similar size of $\boldsymbol{K}$, which is a nonnegative and symmetric matrix provided with the condition that the value of Ч is high. Although, as analogous in the penalty method, the number of conditions of the matrix increases with the parameter, in general, very high values are not required for ensuring convergence and proper implementation.

12.4 Stress intensity factor computation

In the context of fracture mechanics, the evaluation of stress intensity factor is very essential. The stress intensity factors (SIFs) are evaluated using a domain-based interaction integral approach [78] for the mixed mode loading conditions. In this integral approach, only those field variables are chosen that are near the vicinity of the crack tip, which makes the approach more accurate for the evaluation of SIFs. Also, the use of an auxiliary field gives an additional advantage for evaluating single SIFs from a solution, which makes it a very favorable method. Now, following Singh et al. [78], the domain-based interaction integral (DBII) for the mixed load is given as

$$M^{(1,2)}=\int_A\left[\sigma_{ij}^{(1)}\frac{\partial u_i^{(2)}}{\partial x_1}+\sigma_{ij}^{(2)}\frac{\partial u_i^{(1)}}{\partial x_1}-Z^{(1,2)}\delta_{ij}\right]\frac{\partial s}{\partial x_j}dA \tag{12.63}$$

$$Z^{(1,2)}=\frac{1}{2}\left(\sigma_{ij}^{(1)}\epsilon_{ij}^{(2)}+\sigma_{ij}^{(2)}\epsilon_{ij}^{(1)}\right)=\sigma_{ij}^{(1)}\epsilon_{ij}^{(2)}=\sigma_{ij}^{(2)}\epsilon_{ij}^{(1)} \tag{12.64}$$

where "s" shows the scalar weight function (that ranges between 0 and 1), $Z^{(1,2)}$ is mutual strain energy, ϵ and σ show the strains and stresses, respectively, whereas "1" and "2" help in showing the actual and auxiliary states, respectively. Further, the $M^{(1,2)}$ and the SIFs are related as

$$M^{(1,2)}=\frac{2}{E^*}\left(K_I^{(1)}K_I^{(2)}+K_{II}^{(1)}K_{II}^{(2)}\right) \tag{12.65}$$

where $E^*=\frac{E}{(1-\upsilon^2)}$ and $E^*=E$ represents the case of plane strain and plane stress, respectively. Also, the individual SIF in mixed-mode loading can be derived by providing an appropriate choice for the auxiliary states. For the case of mode-I loading, the SIFs are derived as $K_I^{(2)}=1$ and $K_{II}^{(2)}=0$. Similarly, for mode-II loading, SIFs are evaluated as $K_I^{(2)}=0$ and $K_{II}^{(2)}=1$. Hence, mixed-mode SIFs are obtained as $K_I^{(1)}=\frac{E'M^{(1,I)}}{2}$ and $K_{II}^{(1)}=\frac{E'M^{(1,II)}}{2}$.

12.5 Discussions

A thorough study on various implementation issues and the advancements for resolving such issues has been presented in this chapter. The reviewed literature states that NURBS basis functions are very useful in representing conic sections. Ghorashi et al. considered it for straight and curved cracks, and as a result of the study, strong discontinuities were modeled very smoothly [79]. Various researchers considered NURBS-based analysis as a very crucial technique for analyzing different complicated problems. However, some implementation issues related to local refinement and watertight geometries were also evident for suitable outcomes. Thus various researchers considered the modification or enhancement in basis function for improving the strategy and for resolving such issues. Researchers like Habib and Beladi reported T-spline-based XIGA in their study, which can be used to analyze multiple defect structures very efficiently. As a result, higher accuracies were achieved in the case of fracture problems [80]. Another basis function, that is, PHT-splines, was considered by Nguyen-Thanh and Zhou in the XIGA framework for the evaluation of crack propagation near the closed discontinuities. In their study, they concluded that the results obtained were consistent with the literature results, which justifies the accuracy of the approach. Further, this strategy also resolves the issue of local refinement, which occurs during the utilization of NURBS [62]. Also, other basis functions like LR-B-splines are employed in the XIGA framework for the enhancement of the XIGA. The utilization of a different set of basis functions makes this technique very robust for easy implementation in different cases.

Literature also suggests utilization and advancement in the weak form so that issues related to the imposition of essential boundary conditions can be resolved. In this context, several researchers implemented the Lagrange multiplier, penalty, and Nitsche approach to resolve these issues. Ghorashi et al. also consider the Lagrange multiplier approach for the imposition of the boundary conditions [79]. However, the utilization of this method shows difficulty while selecting the interpolation space, which is a major shortfall of the approach. The penalty approach was further implemented for the resolvement of the such problems. This approach also shows advantages like constant dimension of the system matrix, which makes the calculation easy. Apart from the advantages, the penalty approach also presents disadvantages due to the penalty parameter effect on the essential boundary condition, that is, weakly imposition of the Dirichlet boundary condition and poorly conditioned matrix. Yin et al. employed the Nitsche approach in XIGA for analyzing static and dynamic fractures for resolving problems related to matrix symmetricity and nonnegativity [81]. The literature also suggested that the enhancement in different basis functions and improvements in the weak form can be achieved by utilizing the above deliberated techniques.

12.6 Conclusion

In this study, the discussion on various implementation issues for the case of XIGA technique is presented. Some of the key findings of this study are as follows:

1. The modified XIGA techniques are beneficial in analyzing various complex geometries, including various conic sections (circles, ellipsoids, parabolas, etc.)

2. These techniques reduce the possibilities that lead toward errors and enhance the accuracy of the result.
3. These techniques also provide us an edge over the utilization of different basis functions by enabling us to locally refine the geometry specific to the area under consideration.
4. The modification in weak form using different approaches makes the implementation of the essential boundary conditions easier.
5. These techniques can be used to analyze various fracture problems efficiently.

The above-listed remarks demonstrate that if the implementation of XIGA is made to analyze distinct discontinuities like cracks, holes, inclusions, and bi-material interfaces, it will provide suitable outcomes.

References

[1] Wilson WK, Yu IW. The use of the J-integral in thermal stress crack problems. International Journal of Fracture 1979;15(4):377–87. Available from: https://doi.org/10.1007/BF00033062.

[2] Shih CF, Moran B, Nakamura T. Energy release rate along a three-dimensional crack front in a thermally stressed body. International Journal of Fracture 1986;30(2):79–102. Available from: https://doi.org/10.1007/BF00034019.

[3] Jiang S-y, Du C-b, Ooi ET. Modelling strong and weak discontinuities with the scaled boundary finite element method through enrichment. Engineering Fracture Mechanics 2019;222:106734. Available from: https://doi.org/10.1016/j.engfracmech.2019.106734.

[4] Prasad NNV, Aliabadi MH, Rooke DP. The dual boundary element method for thermoelastic crack problems. International Journal of Fracture 1994;66(3):255–72. Available from: https://doi.org/10.1007/BF00042588.

[5] Raveendra ST, Banerjee PK, Dargush GF. Three-dimensional analysis of thermally loaded cracks. International Journal for Numerical Methods in Engineering 1993;36(11):1909–26. Available from: https://doi.org/10.1002/nme.1620361108.

[6] Jameel A, Harmain GA. Fatigue crack growth in presence of material discontinuities by EFGM. International Journal of Fatigue 2015;81:105–16. Available from: https://doi.org/10.1016/j.ijfatigue.2015.07.021.

[7] Harmain GA, Jameel A, Najar FA, Masoodi JH. Large elasto-plastic deformations in bi-material components by coupled FE-EFGM. IOP Conference Series: Materials Science and Engineering 2017;225:012295. Available from: https://doi.org/10.1088/1757-899X/225/1/012295.

[8] Lone AS, Harmain GA, Jameel A. Enriched element free Galerkin method for solving frictional contact between solid bodies. Mechanics of Advanced Materials and Structures 2022;1–19. Available from: https://doi.org/10.1080/15376494.2022.2092791.

[9] Kanth SA, Harmain GA, Jameel A. Modeling of nonlinear crack growth in steel and aluminum alloys by the element free Galerkin method. Materials Today: Proceedings 2018;5(9, Part 3):18805–14. Available from: https://doi.org/10.1016/j.matpr.2018.06.227.

[10] Lone AS, Jameel A, Harmain GA. A coupled finite element-element free Galerkin approach for modeling frictional contact in engineering components. Materials Today: Proceedings 2018;5(9, Part 3):18745–54. Available from: https://doi.org/10.1016/j.matpr.2018.06.221.

[11] Lone AS, Harmain GA, Jameel A. Modeling of contact interfaces by penalty based enriched finite element method. Mechanics of Advanced Materials and Structures 2022;1–19. Available from: https://doi.org/10.1080/15376494.2022.2034075.

[12] Kanth SA, Harmain GA, Jameel A. Investigation of fatigue crack growth in engineering components containing different types of material irregularities by XFEM. Mechanics of Advanced Materials and Structures 2022;29(24):3570–87. Available from: https://doi.org/10.1080/15376494.2021.1907003.

[13] Lone AS, Kanth SA, Harmain GA, Jameel A. XFEM modeling of frictional contact between elliptical inclusions and solid bodies. Materials Today: Proceedings 2020;26:819–24. Available from: https://doi.org/10.1016/j.matpr.2019.12.424.

[14] Kanth SA, Lone AS, Harmain GA, Jameel A. Modeling of embedded and edge cracks in steel alloys by XFEM. Materials Today: Proceedings 2020;26:814–18. Available from: https://doi.org/10.1016/j.matpr.2019.12.423.

[15] Kanth SA, Lone AS, Harmain GA, Jameel A. Elasto plastic crack growth by XFEM: a review. Materials Today: Proceedings 2019;18:3472–81. Available from: https://doi.org/10.1016/j.matpr.2019.07.275.

[16] Lone AS, Kanth SA, Jameel A, Harmain GA. A state of art review on the modeling of contact type nonlinearities by extended finite element method. Materials Today: Proceedings 2019;18:3462–71. Available from: https://doi.org/10.1016/j.matpr.2019.07.274.

[17] Jameel A, Harmain GA. Fatigue crack growth analysis of cracked specimens by the coupled finite element-element free Galerkin method. Mechanics of Advanced Materials and Structures 2019;26(16):1343–56. Available from: https://doi.org/10.1080/15376494.2018.1432800.

[18] Jameel A, Harmain GA. Large deformation in bi-material components by XIGA and coupled FE-IGA techniques. Mechanics of Advanced Materials and Structures 2022;29(6):850–72. Available from: https://doi.org/10.1080/15376494.2020.1799120.

[19] Jameel A, Harmain GA. A coupled FE-IGA technique for modeling fatigue crack growth in engineering materials. Mechanics of Advanced Materials and Structures 2019;26(21):1764–75. Available from: https://doi.org/10.1080/15376494.2018.1446571.

[20] Mohammad MS, Shahrooi S, Shishehsaz M, Hamzehei M. Fatigue crack propagation of welded steel pipeline under cyclic internal pressure by Bézier extraction based XIGA. Journal of Pipeline Systems Engineering and Practice 2022;13(2):04022001. Available from: https://doi.org/10.1061/ASCEPS.1949-1204.0000633.

[21] Yadav A, Patil RU, Singh SK, Godara RK, Bhardwaj G. A thermo-mechanical fracture analysis of linear elastic materials using XIGA. Mechanics of Advanced Materials and Structures 2022;29(12):1730–55. Available from: https://doi.org/10.1080/15376494.2020.1838006.

[22] Singh AK, Jameel A, Harmain GA. Investigations on crack tip plastic zones by the extended iso-geometric analysis. Materials Today: Proceedings 2018;5(9, Part 3):19284–93. Available from: https://doi.org/10.1016/j.matpr.2018.06.287.

[23] Jameel A, Harmain GA. Extended iso-geometric analysis for modeling three-dimensional cracks. Mechanics of Advanced Materials and Structures 2019;26(11):915–23. Available from: https://doi.org/10.1080/15376494.2018.1430275.

[24] Fathi F, Chen L, Hageman T, de Borst R. Extended isogeometric analysis of a progressively fracturing fluid-saturated porous medium. International Journal for Numerical Methods in Engineering 2022;123(8):1861–81. Available from: https://doi.org/10.1002/nme.6919.

[25] Melenk JM, Babuška I. The partition of unity finite element method: Basic theory and applications. Computer Methods in Applied Mechanics and Engineering 1996;139(1):289–314. Available from: https://doi.org/10.1016/S0045-7825(96)01087-0.

[26] Sukumar N, Chopp DL, Moës N, Belytschko T. Modeling holes and inclusions by level sets in the extended finite-element method. Computer Methods in Applied Mechanics and Engineering 2001;190(46):6183–200. Available from: https://doi.org/10.1016/S0045-7825(01)00215-8.

[27] Hughes TJR, Cottrell JA, Bazilevs Y. Isogeometric analysis: CAD, finite elements, NURBS, exact geometry and mesh refinement. Computer Methods in Applied Mechanics and Engineering 2005;194(39):4135–95. Available from: https://doi.org/10.1016/j.cma.2004.10.008.

[28] Piegl L, Tiller W. The NURBS book, 6. Springer Science & Business Media; 2012. p. 34–9.

[29] Gupta V, Jameel A, Verma SK, Anand S, Anand Y. An insight on NURBS based isogeometric analysis, its current status and involvement in mechanical applications. Archives of Computational Methods in Engineering 2022;. Available from: https://doi.org/10.1007/s11831-022-09838-0.

[30] Cottrell JA, Reali A, Bazilevs Y, Hughes TJR. Isogeometric analysis of structural vibrations. Computer Methods in Applied Mechanics and Engineering 2006;195(41):5257–96. Available from: https://doi.org/10.1016/j.cma.2005.09.027.

[31] Gupta V, Jameel A, Anand S, Anand Y. Analysis of composite plates using isogeometric analysis: a discussion. Materials Today: Proceedings 2021;44:1190–4. Available from: https://doi.org/10.1016/j.matpr.2020.11.238.

[32] Akkerman I, Bazilevs Y, Kees CE, Farthing MW. Isogeometric analysis of free-surface flow. Journal of Computational Physics 2011;230(11):4137–52. Available from: https://doi.org/10.1016/j.jcp.2010.11.044.

[33] Tagliabue A, Dedè L, Quarteroni A. Isogeometric analysis and error estimates for high order partial differential equations in fluid dynamics. Computers & Fluids 2014;102:277–303. Available from: https://doi.org/10.1016/j.compfluid.2014.07.002.

[34] Gupta V, Verma SK, Anand S, Jameel A, Anand Y. Transient isogeometric heat conduction analysis of stationary fluid in a container. Proceedings of the Institution of Mechanical Engineers, Part E: Journal of Process Mechanical Engineering 2022. Available from: https://doi.org/10.1177/09544089221125718.

[35] Duvigneau R. An introduction to isogeometric analysis with application to thermal conduction; RR-6957. INRIA 2009;28. Available from: https://hal.inria.fr/inria-0039415.

[36] Fang W, An Z, Yu T, Bui TQ. Isogeometric boundary element analysis for two-dimensional thermoelasticity with variable temperature. Engineering Analysis with Boundary Elements 2020;110:80–94. Available from: https://doi.org/10.1016/j.enganabound.2019.10.003.

[37] Dashlejeh AA. Isogeometric analysis of coupled thermo-elastodynamic problems under cyclic thermal shock. Frontiers of Structural and Civil Engineering 2019;13(2):397–405. Available from: https://doi.org/10.1007/s11709-018-0473-7.

[38] Bazilevs Y, Calo VM, Hughes TJR, Zhang Y. Isogeometric fluid-structure interaction: theory, algorithms, and computations. Computational Mechanics 2008;43(1):3–37. Available from: https://doi.org/10.1007/s00466-008-0315-x.

[39] Buffa A, Sangalli G, Vázquez R. Isogeometric methods for computational electromagnetics: B-spline and T-spline discretizations. Journal of Computational Physics 2014;257:1291–320. Available from: https://doi.org/10.1016/j.jcp.2013.08.015.

[40] Cimrman R, Novák M, Kolman R, Tůma M, Vackář J. Isogeometric analysis in electronic structure calculations. Mathematics and Computers in Simulation 2018;145:125–35. Available from: https://doi.org/10.1016/j.matcom.2016.05.011.

[41] Morganti S, Auricchio F, Benson DJ, Gambarin FI, Hartmann S, Hughes TJR, et al. Patient-specific isogeometric structural analysis of aortic valve closure. Computer Methods in Applied Mechanics and Engineering 2015;284:508–20. Available from: https://doi.org/10.1016/j.cma.2014.10.010.

[42] Kleinendorst SM, Hoefnagels JPM, Verhoosel CV, Ruybalid AP. On the use of adaptive refinement in isogeometric digitalimage correlation. International Journal for Numerical Methods in Engineering 2015;104(10):944–62. Available from: https://doi.org/10.1002/nme.4952.

[43] Hassani B, Tavakkoli SM, Moghadam NZ. Application of isogeometric analysis in structural shape optimization. Scientia Iranica 2011;18(4):846–52. Available from: https://doi.org/10.1016/j.scient.2011.07.014.

[44] Liu H, Yang D, Hao P, Zhu X. Isogeometric analysis based topology optimization design with global stress constraint. Computer Methods in Applied Mechanics and Engineering 2018;342:625–52. Available from: https://doi.org/10.1016/j.cma.2018.08.013.

[45] Verhoosel CV, Scott MA, de Borst R, Hughes TJR. An isogeometric approach to cohesive zone modeling. International Journal for Numerical Methods in Engineering 2011;87(1–5):336–60. Available from: https://doi.org/10.1002/nme.3061.

[46] Choi M-J, Cho S. Isogeometric analysis of stress intensity factors for curved crack problems. Theoretical and Applied Fracture Mechanics 2015;75:89–103. Available from: https://doi.org/10.1016/j.tafmec.2014.11.003.

[47] Khademalrasoul A. NURBS-based isogeometric analysis method application to mixed-mode computational fracture mechanics. Journal of Applied and Computational Mechanics 2019;5(2):217–30. Available from: https://doi.org/10.22055/jacm.2018.25429.1265.

[48] Chang F, Wang WQ, Liu Y, Qu YP. Isogeometric analysis: the influence of penalty coefficients in boundary condition treatments. 2015 International Conference on Computer Science and Applications (CSA). 2015. p. 213–7. Available from: https://doi.org/10.1109/CSA.2015.53.

[49] Embar A, Dolbow J, Harari I. Imposing Dirichlet boundary conditions with Nitsche's method and spline-based finite elements. International Journal for Numerical Methods in Engineering 2010;83(7):877–98. Available from: https://doi.org/10.1002/nme.2863.

[50] Shojaee S, Izadpenah E, Haeri A. Imposition of essential boundary conditions in isogeometric analysis using the Lagrange multiplier method. IUST 2012;2(2):247–71. Available from: http://ijoce.iust.ac.ir/article-1-90-en.html.

[51] Wang D, Xuan J. An improved NURBS-based isogeometric analysis with enhanced treatment of essential boundary conditions. Computer Methods in Applied Mechanics and Engineering 2010;199(37):2425–36. Available from: https://doi.org/10.1016/j.cma.2010.03.032.

[52] Apostolatos A, Bletzinger K-U, Wüchner R. Weak imposition of constraints for structural membranes in transient geometrically nonlinear isogeometric analysis on multipatch surfaces. Computer Methods in Applied Mechanics and Engineering 2019;350:938–94. Available from: https://doi.org/10.1016/j.cma.2019.01.023.
[53] Hu T, Leng Y, Gomez H. A novel method to impose boundary conditions for higher-order partial differential equations. Computer Methods in Applied Mechanics and Engineering 2022;391:114526. Available from: https://doi.org/10.1016/j.cma.2021.114526.
[54] Ghorashi SS, Valizadeh N, Mohammadi S. Extended isogeometric analysis for simulation of stationary and propagating cracks. International Journal for Numerical Methods in Engineering 2012;89(9):1069–101. Available from: https://doi.org/10.1002/nme.3277.
[55] De Luycker E, Benson DJ, Belytschko T, Bazilevs Y, Hsu MC. X-FEM in isogeometric analysis for linear fracture mechanics. International Journal for Numerical Methods in Engineering 2011;87(6):541–65. Available from: https://doi.org/10.1002/nme.3121.
[56] Shojaee S, Asgharzadeh M, Haeri A. Crack analysis in orthotropic media using combination of isogeometric analysis and extended finite element. International Journal of Applied Mechanics 2014;06(06):1450068. Available from: https://doi.org/10.1142/S1758825114500689.
[57] Bhardwaj G, Singh IV, Mishra BK, Bui TQ. Numerical simulation of functionally graded cracked plates using NURBS based XIGA under different loads and boundary conditions. Composite Structures 2015;126:347–59. Available from: https://doi.org/10.1016/j.compstruct.2015.02.066.
[58] Yu T, Bui TQ, Yin S, Doan DH, Wu CT, Do TV, et al. On the thermal buckling analysis of functionally graded plates with internal defects using extended isogeometric analysis. Composite Structures 2016;136:684–95. Available from: https://doi.org/10.1016/j.compstruct.2015.11.002.
[59] Bui TQ, Hirose S, Zhang C, Rabczuk T, Wu C-T, Saitoh T, et al. Extended isogeometric analysis for dynamic fracture in multiphase piezoelectric/piezomagnetic composites. Mechanics of Materials 2016;97:135–63. Available from: https://doi.org/10.1016/j.mechmat.2016.03.001.
[60] Bhardwaj G, Singh SK, Patil RU, Godara RK, Khanna K. Thermo-elastic analysis of cracked functionally graded materials using XIGA. Theoretical and Applied Fracture Mechanics 2021;114:103016. Available from: https://doi.org/10.1016/j.tafmec.2021.103016.
[61] Houcine SH, Belaidi I. Extended isogeometric analysis using analysis-suitable T-splines for plane crack problems. MECHANIKA 2017;23(1):11–17. Available from: https://doi.org/10.5755/j01.mech.23.1.13475.
[62] Nguyen-Thanh N, Zhou K. Extended isogeometric analysis based on PHT-splines for crack propagation near inclusions. International Journal for Numerical Methods in Engineering 2017;112(12):1777–800. Available from: https://doi.org/10.1002/nme.5581.
[63] Huang J, Nguyen-Thanh N, Zhou K. Extended isogeometric analysis based on Bézier extraction for the buckling analysis of Mindlin–Reissner plates. Acta Mechanica 2017;228(9):3077–93. Available from: https://doi.org/10.1007/s00707-017-1861-0.
[64] Gu J, Yu T, Lich LV, Nguyen T-T, Yang Y, Bui TQ. Fracture modeling with the adaptive XIGA based on locally refined B-splines. Computer Methods in Applied Mechanics and Engineering 2019;354:527–67. Available from: https://doi.org/10.1016/j.cma.2019.05.045.
[65] Jiang K, Zhu X, Hu C, Hou W, Hu P, Bordas SPA. An enhanced extended isogeometric analysis with strong imposition of essential boundary conditions for crack problems using B++ splines. Applied Mathematical Modelling 2023;116:393–414. Available from: https://doi.org/10.1016/j.apm.2022.11.032.
[66] Hou W, Jiang K, Zhu X, Shen Y, Li Y, Zhang X, et al. Extended Isogeometric Analysis with strong imposing essential boundary conditions for weak discontinuous problems using B++ splines. Computer Methods in Applied Mechanics and Engineering 2020;370:113135. Available from: https://doi.org/10.1016/j.cma.2020.113135.
[67] Hou W, Jiang K, Zhu X, Shen Y, Hu P. Extended isogeometric analysis using B++ splines for strong discontinuous problems. Computer Methods in Applied Mechanics and Engineering 2021;381:113779. Available from: https://doi.org/10.1016/j.cma.2021.113779.
[68] Singh SK, Singh IV, Mishra BK, Bhardwaj G, Bui TQ. A simple, efficient and accurate Bézier extraction based T-spline XIGA for crack simulations. Theoretical and Applied Fracture Mechanics 2017;88:74–96. Available from: https://doi.org/10.1016/j.tafmec.2016.12.002.
[69] Nguyen-Thanh N, Nguyen-Xuan H, Bordas SPA, Rabczuk T. Isogeometric analysis using polynomial splines over hierarchical T-meshes for two-dimensional elastic solids. Computer Methods in Applied Mechanics and Engineering 2011;200(21):1892–908. Available from: https://doi.org/10.1016/j.cma.2011.01.018.

[70] Duflot M. The extended finite element method in thermoelastic fracture mechanics. International Journal for Numerical Methods in Engineering 2008;74(5):827–47. Available from: https://doi.org/10.1002/nme.2197.

[71] Belytschko T, Black T. Elastic crack growth in finite elements with minimal remeshing. International Journal for Numerical Methods in Engineering 1999;45(5):601–20. Available from: https://doi.org/10.1002/(SICI)1097-0207(19990620)45:5.

[72] Bhardwaj G, Singh IV, Mishra BK. Numerical simulation of plane crack problems using extended isogeometric analysis. Procedia Engineering 2013;64:661–70. Available from: https://doi.org/10.1016/j.proeng.2013.09.141.

[73] Chopp DL, Sukumar N. Fatigue crack propagation of multiple coplanar cracks with the coupled extended finite element/fast marching method. International Journal of Engineering Science 2003;41(8):845–69. Available from: https://doi.org/10.1016/S0020-7225(02)00322-1.

[74] Bayesteh H, Afshar A, Mohammdi S. Thermo-mechanical fracture study of inhomogeneous cracked solids by the extended isogeometric analysis method. European Journal of Mechanics - A/Solids 2015;51:123–39. Available from: https://doi.org/10.1016/j.euromechsol.2014.12.004.

[75] Belytschko T, Gu L, Lu YY. Fracture and crack growth by element free Galerkin methods. Modelling and Simulation in Materials Science and Engineering 1994;2(3A):519. Available from: https://doi.org/10.1088/0965-0393/2/3A/007.

[76] Babuška I. The finite element method with Lagrangian multipliers. Numerische Mathematik 1973;20(3):179–92. Available from: https://doi.org/10.1007/BF01436561.

[77] Griebel M, Schweitzer MA. A particle-partition of unity method part V: boundary conditions. Geometric Analysis and Nonlinear Partial Differential Equations 2003;519–42. Available from: https://doi.org/10.1007/978-3-642-55627-2_27.

[78] Singh IV, Mishra BK, Bhattacharya S, Patil RU. The numerical simulation of fatigue crack growth using extended finite element method. International Journal of Fatigue 2012;36(1):109–19. Available from: https://doi.org/10.1016/j.ijfatigue.2011.08.010.

[79] Ghorashi SS, Valizadeh N, Mohammadi S, Rabczuk T. Extended isogeometric analysis of plates with curved cracks. In Proceedings of the 8th International Conference on Engineering Computational Technology. Civil-Comp Press; 2012. Available from: https://doi.org/10.4203/ccp.100.47.

[80] Habib SH, Belaidi I. Crack analysis in bimaterial interfaces using T-spline based XIGA. Journal of Theoretical and Applied Mechanics 2017;55(1):2017. Available from: https://ptmts.org.pl/jtam/index.php/jtam/article/view/3306/2598.

[81] Yin S, Yu T, Bui TQ, Zheng X, Gu S. Static and dynamic fracture analysis in elastic solids using a multiscale extended isogeometric analysis. Engineering Fracture Mechanics 2019;207:109–30. Available from: https://doi.org/10.1016/j.engfracmech.2018.12.024.

CHAPTER

13

Extended isogeometric analysis for modeling strong discontinuities

Vibhushit Gupta[1], *Shubham Kumar Verma*[2], *Sanjeev Anand*[2] *and Yatheshth Anand*[1]

[1]School of Mechanical Engineering, Shri Mata Vaishno Devi University, Kakryal, Katra, Jammu and Kashmir, India [2]School of Energy Management, Shri Mata Vaishno Devi University, Kakryal, Katra, Jammu and Kashmir, India

13.1 Introduction

The analysis of various structures as well as their components during crack presence is very essential to ensure structure safety so that catastrophic failures can be avoided. The development of these cracks/strong discontinuities is due to either of manufacturing defect or of localized damages. Also, the presence of these discontinuities will affect the working life and strength of the structures, which results in the loss of property as well as livelihood. Thus it becomes necessary to study and analyze the behavior of the structures having discontinuities. To simulate such types of problems, various numerical techniques like finite element analysis [1,2], boundary element [3], extended finite element method (XFEM) [4–9], mesh-free method [10–14], isogeometric analysis (IGA) [15,16], coupled techniques (like FEA–mesh-free [11,17] and FEA-IGA [18,19]), scaled boundary finite element [20], and extended IGA (XIGA) [18,21–24] have been developed. However, the issues related to cost association and remeshing in the case of traditional FEA make losses in the accuracy of the solution. So, the simulation of fractures has been done efficiently by utilizing the enrichment function in the standard FEA approximation, which provides an additional degree of freedom. However, in XFEM and coupled FEA–mesh-free method, the description of the geometry and the solution needs a separate basis function. Hughes et al. present the concept of IGA, which describes an unknown solution field variable and geometry with a common basis function, that is, nonuniform rational B-spline (NURBS) [25]. The utilization of the common basis function bridges a tight linkage between design and analysis by generating accurate mesh geometries with highly precise and converged results. Over the years, IGA has been

Enriched Numerical Techniques
DOI: https://doi.org/10.1016/B978-0-443-15362-4.00021-8

applied successfully for solving various problems like structures [26,27], fluids [28–30], thermal [31,32], fluid–structure interaction [33], bio-medical [34], electromagnetics [35], and many more. For the case of fractures, IGA has also proven to be a potential as well as an efficient numerical technique [19,36,37]. However, it was perceived that near the crack interfaces, the value of continuity is low due to the high multiplicity of knots. Thus such issues require the utilization of higher order basis functions instead of lower order functions.

In the past few years, the notion of IGA has been extended by considering partition of unity (PU) in IGA, which is named XIGA [38]. Various research workers used the XIGA concept to solve different sets of fracture problems. Benson et al. employed the XIGA technique for the fracture problems [39]. De Luycker et al. presented that this method gives higher convergence as well as accuracy while working on fracture problems [40]. Ghorashi et al. carried out an analysis of fractures in structures by utilizing XIGA [38]. Nguyen-Thanh et al. employed this technique for analyzing cracks presented in thin-shell structures [41]. In their study, the complexity and computation cost of this technique were reduced by neglecting the use of rotational degrees of freedom. Bhardwaj et al. utilized XIGA for analyzing crack growth under fatigue in functionally graded materials (FGMs) [42]. Jameel and Harmain solved large elasto-plastic deformation problems by using XIGA and coupled FE-IGA approach [18]. Singh and Singh implement the XIGA approach for analyzing fractures in the case of magneto–elasto–elastic FGMs. As a result, it was found that XIGA comes up as an accurate technique for solving static fracture problems [43]. Recently, Fardaghaie et al. [44] applied the *k*-refinement in XIGA to predict the fatigue life in linear elastic fracture problems. As a result of the study, it was evident that the increase in NURBS order during *k*- refinement increases the accuracy in investigating the value of stress intensity factors (SIFs) [44].

Based on the above-presented literature, it has been evident that the XIGA approach is very efficient and accurate in solving the different types of fracture problems. So, in the present study, this XIGA approach is reviewed and then implemented for solving different static crack cases. The presented study begins with Section 13.2, which includes the basic notion of the XIGA technique for modeling static cracks; Section 13.3 shows the procedure to calculate the SIFs; various numerical examples with validation is presented in Section 13.4; at last, the study is concluded in Section 13.5.

13.2 Extended isogeometric analysis

In XIGA, a refined grid structure is adopted for the existing crack areas, while a local enrichment approach is utilized for describing the geometry of discontinuities that are independent of the grid structure. In this section, a brief notion of analyzing static crack cases is presented and given as follows.

13.2.1 Basis function

It has been observed from the literature that the NURBS function is very efficient in representing complex geometries like various conic sections [16], which makes it a very advanced technology for applying it in analysis. In general, this basis function is a piecewise polynomial, which provides great precision and flexibility to a large number of modeling applications [45].

Also, NURBS are supported locally and have continuity that is directly linked to the basis. Specifically, various curved sections like ellipses and circles can be modeled very effectively by utilizing projective transformations of piecewise quadratic curves. Furthermore, this type of basis function also shows some advanced features like preserving the geometry, special mathematical properties, and free-form surface modeling. So, in this section, the basic concept of NURBS is presented as a basis function for XIGA. For defining a NURBS, the knot vector is used by a combination of one-dimensional range of nondiminshing distinct points (knots). A knot vector (Ξ) for a 1D case can be defined as $\left\{\xi_1 \ldots\ldots\ldots \xi_{n+p+1}\right\}$. In this knot vector, n shows the quantity of basis functions with p as an order of splines and ξ_i shows the value of the knot for ith index that divides B-splines into a set of subdomains generally referred to as patches. These patches consist of knot spans. Moreover, the knots with equidistant space are considered uniform knot vectors, whereas the knot spans with unequal spaces are considered nonuniform ones. In NURBS-based XIGA, the unequal spaced knots have been utilized for the modeling of geometry due to its property of interpolation, which directly helps in applying boundary conditions. Further, by using the knot vector "Ξ," the order "p," and following the Cox–deBoor recursive relation [46], a B-spline function is given as

For $p = 0$

$$N_{i,0}(\xi) = \begin{cases} 1 \; if \;\; \xi_i \le \xi_{i+1} \\ 0 \; otherwise \end{cases} \tag{13.1}$$

For $p > 0$

$$N_{i,p}(\xi) = \frac{\xi - \xi_i}{\xi_{i+p} - \xi_i} N_{i,p-1}(\xi) + \frac{\xi_{i+p+1} - \xi}{\xi_{i+p+1} - \xi_{i+1}} N_{i+1,p-1}(\xi) \tag{13.2}$$

where $N_{i,p}$ represents the B-spline basis function, and further dividing the function by weight, the one-dimensional NURBS function $R_i^p(\xi)$ is developed by and is given as

$$R_i^p(\xi) = \frac{N_{i,p}(\xi) w_i}{\sum_{i=1}^{n} N_{i,p}(\xi) w_i} \tag{13.3}$$

where weight function is shown by w_i for an ith weight function. Analogously, the NURBS two- and three-dimensional equations can be derived as

$$R_i^p(\xi, \eta) = \frac{N_{i,p}(\xi) \times M_{j,q}(\eta) w_{i,j}}{\sum_{i=1}^{n} \sum_{j=1}^{m} N_{i,p}(\xi) \times M_{j,q}(\eta) w_{i,j}} \tag{13.4}$$

$$R_i^p(\xi, \eta, \zeta) = \frac{N_{i,p}(\xi) \times M_{j,q}(\eta) \times L_{k,r}(\zeta) w_{i,j,k}}{\sum_{i=1}^{n} \sum_{j=1}^{m} \sum_{k=1}^{1} N_{i,p}(\xi) \times M_{j,q}(\eta) \times L_{k,r}(\zeta) w_{i,j,k}} \tag{13.5}$$

where $N_{i,p}(\xi)$, $M_{j,q}(\eta)$, and $L_{k,r}(\zeta)$ are the B-spline basis functions having orders p, q, and r, respectively. Based on these, the equations of spline, surface, and solid based on NURBS can be constructed as follows:

$$Z(\xi) = \sum_{i=1}^{n} R_i^p(\xi) B_i \tag{13.6}$$

$$Z(\xi, n) = \sum_{i=1}^{n} R_i^p \sum_{j=1}^{m} R_i^p(\xi, n) B_{i,j} \tag{13.7}$$

$$Z(\xi, \eta, \zeta) = \sum_{i=1}^{n}, \sum_{j=1}^{m} \sum_{k=1}^{l} R_i^p(\xi, \eta, \zeta) B_{i,j,k} \tag{13.8}$$

where coordinates of control points (CP) are presented by $B_{i,j,k}$. A NURBS basis also has some important features that can be given as

(a) It shows a form of PU, that is, $\sum_{(i=1)}^{n} R_i^p(\xi) = 1$ for all ξ.
(b) If the internal knots are not repeated, NURBS guarantees *p-1* continuous derivatives, while only $C^{p\text{-}k}$ continuity is produced for *k* multiple knots.
(c) The support of rational function $R_i^p(\xi)$ is compact and considered in the range of $\{\xi_i, \xi_{i+p+1}\}$.

whereas for 2D case, a knot span has nonzero basis function that has number $n_{en} = (p+1) \times (q+1)$. Similarly for 3D case, $n_{en} = (p+1) \times (q+1) \times (r+1)$.

Problem: Formulation

The formulation of problem is considered for a linear elastic, isotropic, and homogeneous and two-dimensional domain "Ω" depicted in Fig. 13.1. The Ω is subdivided into traction boundaries (Γ_t), displacement boundaries (Γ_u), and traction free boundaries (Γ_c).

In the domain (Ω), when subjected to mechanical loading, the solution needs to satisfy the following equilibrium conditions [47]:

$$\nabla.\sigma + f = 0 \tag{13.9}$$

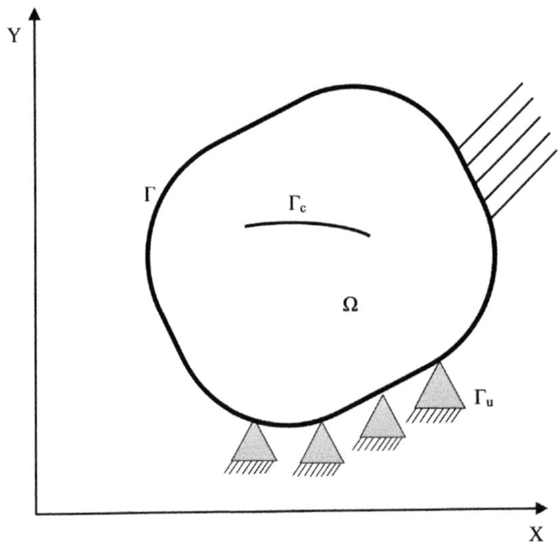

FIGURE 13.1 Problem domain considered for analysis.

In Eq. (13.8), σ, ∇, and f present a Cauchy tensor product, divergence operator, and body force vector, respectively. The value of ∇ differential operator can be given as

$$\nabla = \begin{bmatrix} \frac{\partial}{\partial X} & 0 \\ 0 & \frac{\partial}{\partial Y} \\ \frac{\partial}{\partial Y} & \frac{\partial}{\partial X} \end{bmatrix} \tag{13.10}$$

The boundary conditions (BCs) for the domain under consideration are given as

$$u = \bar{u} \; on \; \Gamma_u \tag{13.11}$$

$$\sigma.\hat{n} = 0 \; on \; \Gamma_c \tag{13.12}$$

$$\sigma.\hat{n} = \hat{t} \; on \; \Gamma_t \tag{13.13}$$

Eq. (13.8) presents the mechanical equilibrium, respectively, and shows relations of Hooke's law. Thus the fundamental relation in the case of homogeneous elastic material can be derived as

$$\sigma = C.\epsilon \tag{13.14}$$

where C presents the tensor of elasticity with the strain tensor as ϵ. Based on this, the weak form equation as per Belytschko and Black [48] can be expressed as

$$\int_{\Omega} \sigma(u){:}\epsilon(u)d\Omega = \int_{\Omega} b{:}ud\Omega + \int_{\Gamma} t{:}ud\Gamma \tag{13.15}$$

By putting the trial and test functions and utilizing the randomness of the varied CP values, the discrete system of equations formed as follows:

$$[K]\{d\} = \{f\} \tag{13.16}$$

where f, d, and K show the external force vector, displacement vector, and global stiffness matrix, respectively.

13.2.2 Extended isogeometric analysis-based approximation

By following the enrichment strategy, a local enrichment approach is considered for describing cracks that are independent of the finite element grid. In XIGA, enrichment functions are introduced into IGA for displacement approximations. Thus the approximated displacement $\left(\mathrm{u}^{\mathrm{h}}(\xi)\right)$ derived by using XIGA approach can be expressed as follows:

$$u^h(\xi) = \sum_{i=1}^{n_{be}} R_i^p(\xi)u_i + \sum_{j=1}^{n_s} R_j^p(\xi)\left[W(\xi) - W(\xi_i)\right]a_j + \sum_{k=1}^{n_T} R_k^p(\xi)\sum_{\alpha=1}^{4}[\beta_\alpha(\xi) - \beta_\alpha(\xi_i)]b_k^\alpha \tag{13.17}$$

where u_i presents the parameters relating to the specified CP, n_{be} are the amount of shape functions per element given as $n_{be} = (p+1) \times (q+1)$; R_i^p shows NURBS basis function, n_s is split-enriched node elements; a_j is enriched degrees of freedom of split elements that relates to Heaviside function $W(\xi)$; $b_k^\alpha =$ enriched degree of freedom of a tip elements relates to crack tip enrichment function (CTEF) $\beta_\alpha(\xi)$; $\beta_\alpha(\xi)$ and $W(\xi)$ present the crack tip and the Heaviside enrichment function, respectively. The main purpose of the Heaviside function $W(\xi)$ is to produce discontinuity across the surface of the crack and ranges from -1 to $+1$ that is presented as

$$\mathrm{W}(\xi) = \begin{cases} -1, otherwise \\ +1, if\ (\xi - \xi_i).n_{be} \geq 0 \end{cases} \tag{13.18}$$

The Heaviside function is efficient when an element is divided by a crack interface. However, in the case of tip elements, a portion of the element is only cut, and the Heaviside function is not able to enrich that tip element. Thus CTEF is used for enrichment of crack tip element. This enrichment function is defined by Chopp and Sukumar [49] and is explained as

$$\beta_\alpha(\xi) = \left[\sqrt{r}\cos\frac{\theta}{2}, \sqrt{r}\cos\frac{\theta}{2}\sin\theta, \sqrt{r}\sin\frac{\theta}{2}\sin\theta, \sqrt{r}\sin\frac{\theta}{2}\right] \tag{13.19}$$

where θ and r present polar coordinates at the reference line of crack and the crack surface.

13.2.3 Control point selection for enrichment

In XIGA, the number of approximation functions is equal to the number of CPs, and each approximation function is defined uniquely. Further, it is also evident that the approximation function has its own unique domain support, and these functions tend to zero on the other points of domain. As discussed in Section 2.3, the HF is utilized for enriching the CPs that have domain support intersected by the crack face, while CPs that have domain influence containing the tip of the crack are enriched by CTEF. However, for the evaluation of the tip of the crack on the enriched CPs, the parametric coordinates are calculated, and then NURBS approximation function corresponds to the evaluated parametric coordinates. The values of NURBS > 0 will specify the CPs with crack tip enrichments. Also, CPs containing HF are also calculated by adopting the similar procedure like crack tips.

The modeling of cracks with enriched CPs is shown in Fig. 13.2, in which an edge crack with intersected domains and crack tip domains is presented. It also shows the marked CP, in which two additional DOFs are added at the intersected domains and four additional DOFs at the cracked tip domain CP.

13.2.4 Integration of discontinuous elements

The integration in XIGA is done by utilizing a standard Gaussian quadrature. However, the existence of discontinuity in the integration zone reduces the reliability and exactness of the integration. Thus an efficient method is required, which alleviates this drawback and helps improve the analysis accuracy. In this case, subtriangulation strategy is adopted,

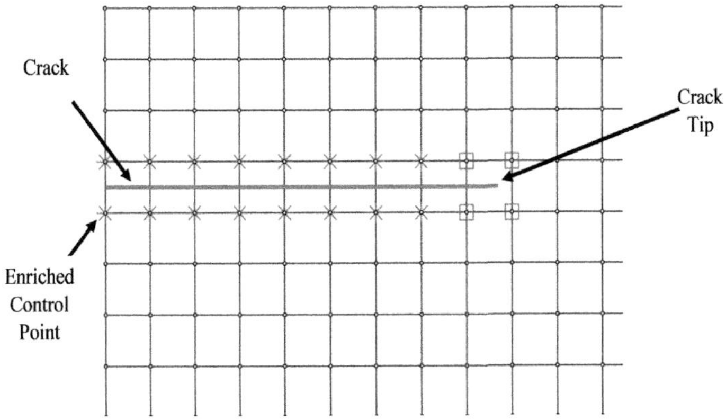

FIGURE 13.2 Crack with enriched control points.

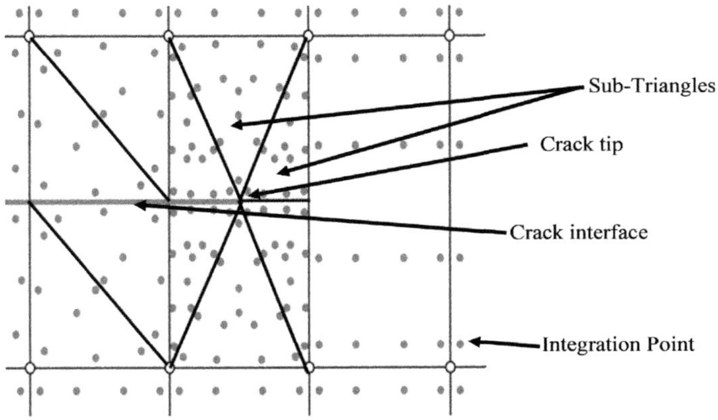

FIGURE 13.3 Subtriangulation strategy for integration of crack.

which divides the element (intersected by crack) into subtriangles at the both sides. Moreover, a higher order subtriangulation strategy is adopted for the case of crack tip element, which increases the number of triangles in that element. Fig. 13.3 shows the subtriangulation for both split elements and crack tip elements.

13.2.5 Extended isogeometric analysis crack formulation

By considering first term of Eq. (13.17), it only approximates displacement field based on conventional IGA, whereas the other remaining parts are described as enrichments for modeling a strong discontinuity. Based on this, the elemental matrices K and f can be derived by utilizing the NURBS function presented as

$$K^{hU} = \begin{bmatrix} K_{ij}^{uu} & K_{ij}^{ua^u} & K_{ij}^{ub^u} \\ K_{ij}^{a^u u} & K_{ij}^{a^u a^u} & K_{ij}^{a^u b^u} \\ K_{ij}^{b^u u} & K_{ij}^{b^u a^u} & K_{ij}^{b^u b^u} \end{bmatrix} \tag{13.20}$$

$$f^{hU} = \begin{bmatrix} f_i^u & f_i^{a^u} & f_i^{b_1^u} & f_i^{b_2^u} & f_i^{b_3^u} & f_i^{b_4^u} \end{bmatrix}^T \tag{13.21}$$

$$K_{ij}^{rs} = \int_{\Omega^e} (B_i^r)^T C B_j^s d\Omega \tag{13.22}$$

$$f_i^u = \int_{\Omega^e} (R_i)^T b d\Omega + \int_{\Gamma_t} (R_i)^T \hat{t} d\Gamma \tag{13.23}$$

$$f_i^{a^u} = \int_{\Omega^e} (R_i)^T H b d\Omega + \int_{\Gamma_t} (R_i)^T H \hat{t} d\Gamma \tag{13.24}$$

$$f_i^{b^u} = \int_{\Omega^e} (R_i)^T \beta_\alpha b d\Omega + \int_{\Gamma_t} (R_i)^T \beta_\alpha \hat{t} d\Gamma \tag{13.25}$$

where R_i is the NURBS rational function, and B_i^u, $B_i^{a^u}$, $B_i^{b_\alpha^u}$, $B_i^{b^u}$, and $B_i^{c^u}$ are the derivative matrices of the rational functions that are explained as

$$B_i^u = \begin{bmatrix} R_{i,X_1} & 0 \\ 0 & R_{i,X_2} \\ R_{i,X_1} & R_{i,X_2} \end{bmatrix} \tag{13.26}$$

$$B_i^{a^u} = \begin{bmatrix} R_{i,X_1} W & 0 \\ 0 & R_{i,X_2} W \\ R_{i,X_1} W & R_{i,X_2} W \end{bmatrix} \tag{13.27}$$

$$B_i^{b\alpha} = \begin{bmatrix} (R_i\beta_\alpha)_{,X_1} & 0 \\ 0 & (R_i\beta_\alpha)_{,X_2} \\ (R_i\beta_\alpha)_{,X_1} & (R_i\beta_\alpha)_{,X_1} \end{bmatrix} \tag{13.28}$$

$$B_i^b = \begin{bmatrix} B_i^{b1} & B_i^{b2} & B_i^{b3} & B_i^{b4} \end{bmatrix} \tag{13.29}$$

where in these equations, $\mathrm{r,s} = u, a^u$ and b^u & $\alpha = 1, 2, 3$ and 4

13.3 Evaluation of stress intensity factors

The mechanical components are mostly acted upon with the mixed loading conditions. However, the occurrence of discontinuities in the structure makes the functioning difficult under mixed loads. Thus, in this study, a domain-based interaction integral approach [50] is considered for the evaluation of SIFs under mixed mode conditions. In this domain-based strategy, only those field variables are chosen that are near the vicinity of the crack tip, which makes the approach more accurate for the evaluation of SIFs. Also, the use of auxiliary field gives an additional advantage for evaluating single SIFs from a solution, which makes it a very favorable method. Now, following Singh et al. [50], the domain-based interaction integral for the mixed load is given as

$$M^{(1,2)} = \int_A \left[\sigma_{ij}^{(1)} \frac{\partial u_i^{(2)}}{\partial x_1} + \sigma_{ij}^{(2)} \frac{\partial u_i^{(1)}}{\partial x_1} - Z^{(1,2)} \delta_{ij} \right] \frac{\partial s}{\partial x_j} dA \tag{13.30}$$

$$Z^{(1,2)} = \frac{1}{2}\left(\sigma_{ij}^{(1)} \epsilon_{ij}^{(2)} + \sigma_{ij}^{(2)} \epsilon_{ij}^{(1)} \right) = \sigma_{ij}^{(1)} \epsilon_{ij}^{(2)} = \sigma_{ij}^{(2)} \epsilon_{ij}^{(1)} \tag{13.31}$$

where "s" shows the scalar weight function (that ranges between 0 and 1), $Z^{(1,2)}$ is mutual strain energy, ϵ and σ show the strains and stresses, respectively, whereas "1" and "2" help in showing the actual and auxiliary states, respectively. Further, the $M^{(1,2)}$ and the SIFs are related as

$$M^{(1,2)} = \frac{2}{E^*}\left(K_I^{(1)} K_I^{(2)} + K_{II}^{(1)} K_{II}^{(2)} \right) \tag{13.32}$$

where $E^* = \frac{E}{(1-\upsilon^2)}$ and $E^* = E$ represent the cases of plane strain and plane stress, respectively. Also, the individual SIFs in mixed mode loading can be derived by providing an appropriate choice for the auxiliary states. For the case of mode-I loading, the SIFs are derived as $K_I^{(2)} = 1$ and $K_{II}^{(2)} = 0$. Similarly, for mode-II loading, SIFs are evaluated as $K_I^{(2)} = 0$ and $K_{II}^{(2)} = 1$. Hence, mixed mode SIFs are obtained as $K_I^{(1)} = \frac{E' M^{(1,I)}}{2}$ and $K_{II}^{(1)} = \frac{E' M^{(1,II)}}{2}$.

13.4 Numerical examples

The present study simulates the behavior of cracks under static loading conditions by using XIGA approach. The enrichment techniques model strong discontinuities independent of the mesh structure, which alleviates different grid-related problems such as grid adaptation, remeshing, and conformal meshing. The carried-out results are then checked with the exact analytical results, such that the robustness and applicability of the XIGA technique can be verified. The codes are developed in the MATLAB framework for a rectangular structure having an edge and a center crack, which are shown in Fig. 13.4. Also, the material properties of the considered structure are provided in Table 13.1.

13.4.1 Plate with static edge crack

The problem considered in this section simulates the behavior of an edge crack contained in a rectangular structural component under monotonic loads. The length and width of the structure are taken as 200 and 100 mm, respectively, with an edge crack of 30 mm. A 100 N/mm of load is applied at the top edge, while the bottom edge is kept to be fixed. The schematic diagram is presented in Fig. 13.4. With the help of XIGA, the whole domain discretizes into a set of uniform CPs. The grid independence test has been conducted for the XIGA model, and the results show that a control net of 40 × 80 CPs makes the analysis grid

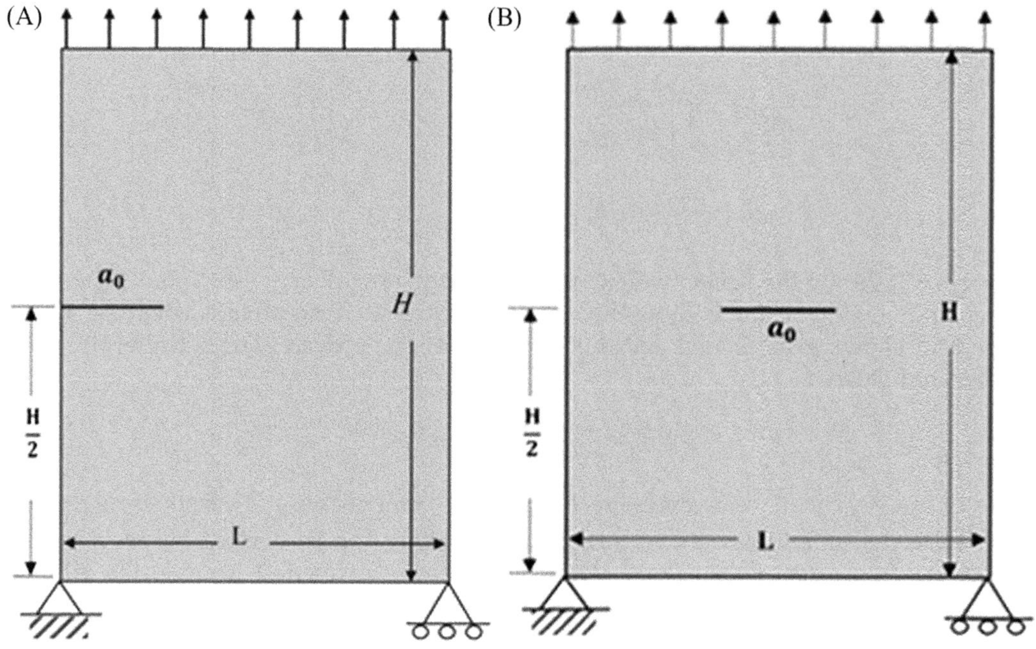

FIGURE 13.4 A rectangular plate containing: (A) Edge crack and (B) Center crack.

TABLE 13.1 Material properties of the plate structure.

Fracture toughness (K_{IC})	1897.36 MPa$\sqrt{mm}$
Poisson ratio (v)	0.3
Young's modulus of elasticity (E)	74 GPa
Yield stress (σ_Y)	240 MPa

independent. Thus the analysis is conducted with the control net of 40×80 CPs, as presented in Fig. 13.5. The NURBS basis functions have been utilized for the approximation of displacement field on the whole domain. However, the discontinuities are approximated by introducing enrichment functions in the conventional NURBS function at the crack interface. This crack with enriched CP is clearly presented by red dots in Fig. 13.5. Also, there is no need for conformal meshing as the discontinuities are modeled independently of the grid.

The numerical integration has been performed, and a total of 27,534 integration points have been developed in the entire domain. Further, by applying the boundary conditions and utilizing the XIGA methodology, the unknown field variable, that is, displacement, is approximated in the whole domain. Based on this displacement, first the stresses and strains are evaluated, which are further used for the evaluation of SIFs. The variation of SIFs with load for a linear crack is shown in Fig. 13.6. The carried-out results present that the XIGA shows a very fine agreement with the analytical results that are available in the open literature, which validates current XIGA approach.

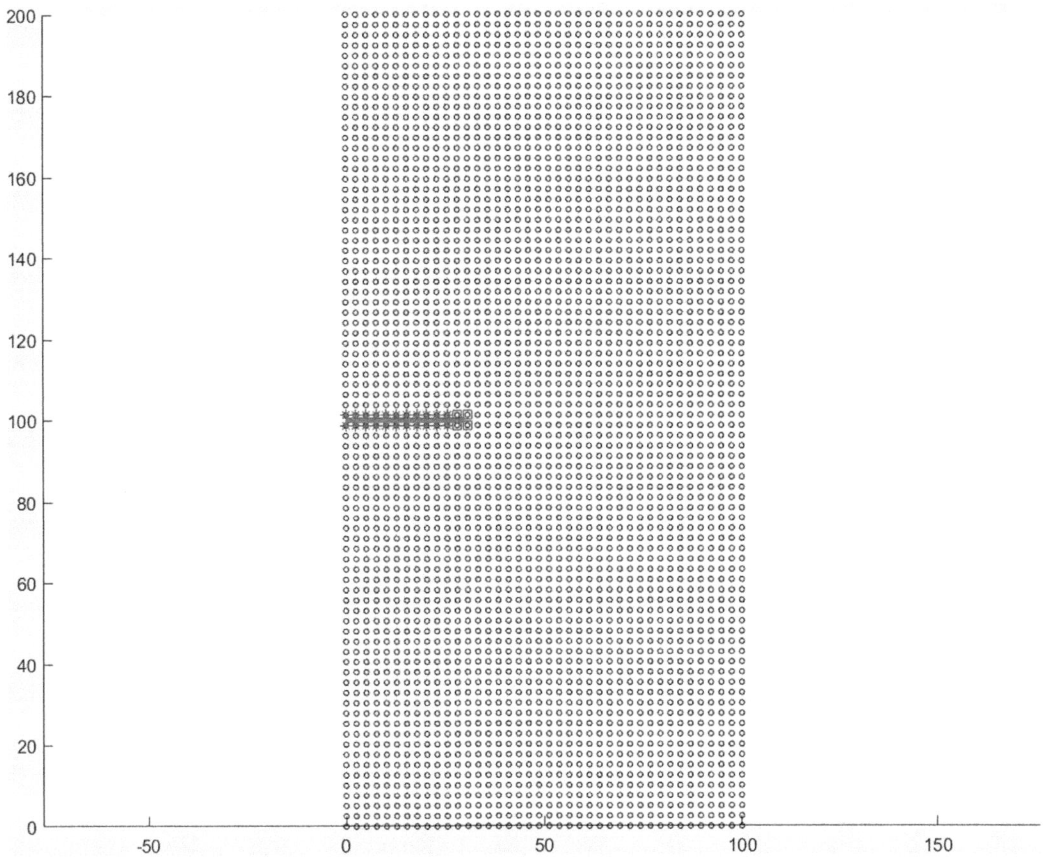

FIGURE 13.5 Meshed domain representation of edge crack plate.

13.4.2 Plate with static center crack

In this section, the behavior of the center cracked plate structure under a monotonic load is simulated. The dimensions of the plate are the same as the dimensions of the edge-cracked plate. The crack is considered at the center of the plate having 30 mm length, which is clearly presented in Fig. 13.4B. Also, at the top edge of the structure, a load of 240 N/mm is taken with fixed bottom edge. By using XIGA approach, the whole rectangular structure is discretized into a control net of 40×80 CPs, which is presented in Fig. 13.7. As already discussed, the enrichment handles the discontinuous domain very effectively. These enrichments modify the NURBS approximation by adding the appropriate enrichment functions near the discontinuity. Same material properties of the structure are considered in this case, as depicted in Table 13.1. Further, by approximating the displacement and evaluating stresses, strains, and SIFs in a similar manner of evaluation in edge-cracked plate, the solutions are checked with the analytical solutions of available literature. Also, Fig. 13.8 shows that the XIGA-evaluated values of SIFs are closely matched with the analytical values of SIFs. So, it has been evident that the XIGA analysis is very efficient and accurate for the analysis of center crack problems.

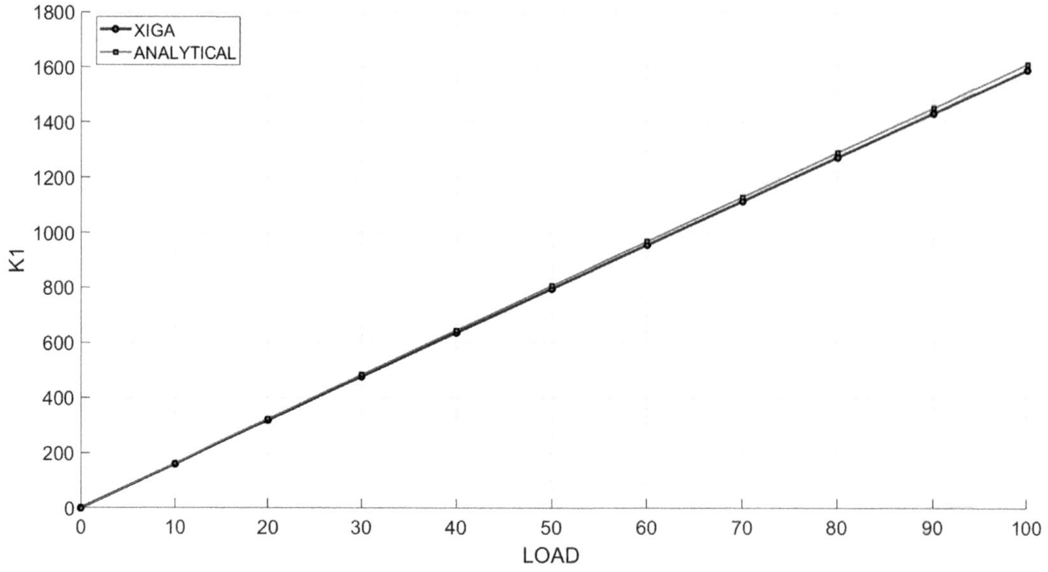

FIGURE 13.6 Variation of SIFs with Load in case of edge crack. *SIFs*, Stress intensity factors.

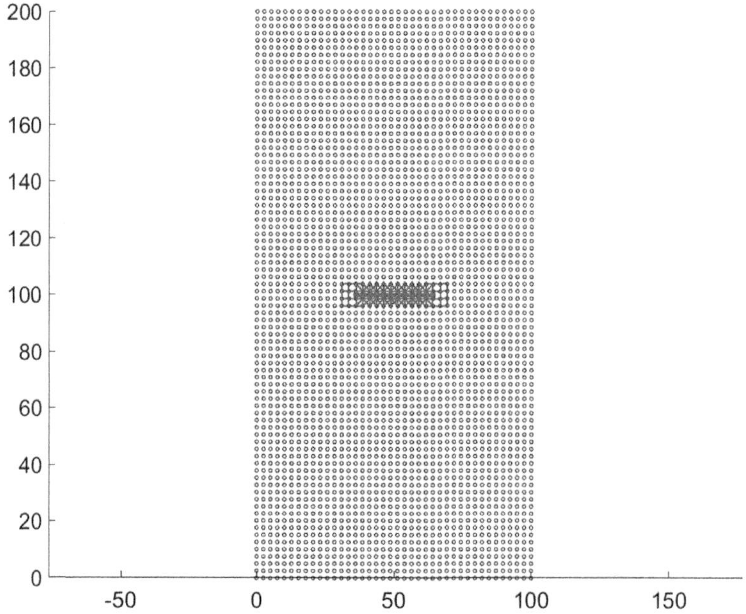

FIGURE 13.7 Meshed domain representation of center crack plate.

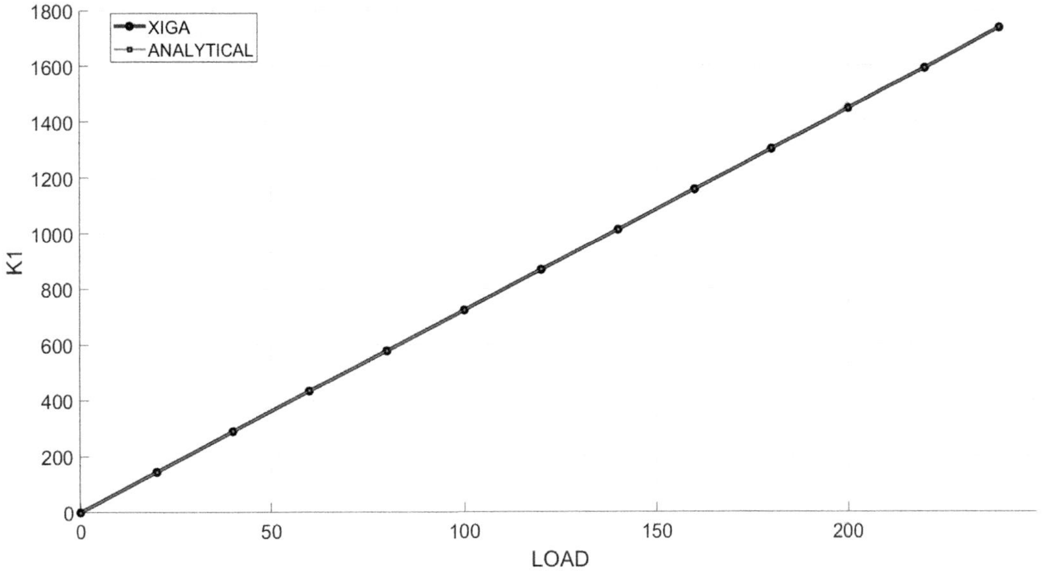

FIGURE 13.8 Variation of SIFs with Load in case of center crack. *SIFs*, Stress intensity factors.

13.5 Conclusion

The current study presents a brief discussion on implementing XIGA approach in analyzing various static crack problems. In this study, two benchmark problems, that is, edge and center cracks, are analyzed using the XIGA approach. The displacement fields in the whole domain of the problems are approximated by using NURBS functions, whereas in the crack interface and crack tip, they are approximated by modifying the NURBS function with appropriate enrichment functions. Also, the utilization of the subtriangulation model for improving the Gauss quadrature rule increased the accuracy of the analysis. The results present that SIFs obtained by XIGA are in fine agreement with analytical results of open literatures, which directly presents the high accuracy of the XIGA approach. Thus it has been concluded that XIGA approach is very efficient and accurate for solving problems of structures having strong discontinuities. Further, the present methodology can be utilized for solving problems of cracks present in the structure at different angles.

References

[1] Hansbo A, Hansbo P. A finite element method for the simulation of strong and weak discontinuities in solid mechanics. Computer Methods in Applied Mechanics and Engineering 2004;193(33):3523–40. Available from: https://doi.org/10.1016/j.cma.2003.12.041.

[2] Sheikh UA, Jameel A. Elasto-plastic large deformation analysis of bi-material components by FEM. Materials Today: Proceedings 2020;26:1795–802. Available from: https://doi.org/10.1016/j.matpr.2020.02.377.

[3] Dumont NA, Mamani EY, Cardoso ML. A boundary element implementation for fracture mechanics problems using generalised Westergaard stress functions. European Journal of Computational Mechanics 2018;27(5–6):401–24. Available from: https://doi.org/10.1080/17797179.2018.1499188.

[4] Lone AS, Harmain GA, Jameel A. Modeling of contact interfaces by penalty based enriched finite element method. Mechanics of Advanced Materials and Structures 2022;1–19. Available from: https://doi.org/10.1080/15376494.2022.2034075.

[5] Kanth SA, Harmain GA, Jameel A. Investigation of fatigue crack growth in engineering components containing different types of material irregularities by XFEM. Mechanics of Advanced Materials and Structures 2022;29(24):3570–87. Available from: https://doi.org/10.1080/15376494.2021.1907003.

[6] Lone AS, Kanth SA, Harmain GA, Jameel A. XFEM modeling of frictional contact between elliptical inclusions and solid bodies. Materials Today: Proceedings 2020;26:819–24. Available from: https://doi.org/10.1016/j.matpr.2019.12.424.

[7] Kanth SA, Lone AS, Harmain GA, Jameel A. Modeling of embedded and edge cracks in steel alloys by XFEM. Materials Today: Proceedings 2020;26:814–18. Available from: https://doi.org/10.1016/j.matpr.2019.12.423.

[8] Kanth SA, Lone AS, Harmain GA, Jameel A. Elasto plastic crack growth by XFEM: a review. Materials Today: Proceedings 2019;18:3472–81. Available from: https://doi.org/10.1016/j.matpr.2019.07.275.

[9] Lone AS, Kanth SA, Jameel A, Harmain GA. A state of art review on the modeling of contact type nonlinearities by extended finite element method. Materials Today: Proceedings 2019;18:3462–71. Available from: https://doi.org/10.1016/j.matpr.2019.07.274.

[10] Jameel A, Harmain GA. Fatigue crack growth in presence of material discontinuities by EFGM. International Journal of Fatigue 2015;81:105–16. Available from: https://doi.org/10.1016/j.ijfatigue.2015.07.021.

[11] Harmain GA, Jameel A, Najar FA, Masoodi JH. Large elasto-plastic deformations in bi-material components by coupled FE-EFGM. IOP Conference Series: Materials Science and Engineering 2017;225:012295. Available from: https://doi.org/10.1088/1757-899X/225/1/012295.

[12] Lone AS, Harmain GA, Jameel A. Enriched element free Galerkin method for solving frictional contact between solid bodies. Mechanics of Advanced Materials and Structures 2022;1–19. Available from: https://doi.org/10.1080/15376494.2022.2092791.

[13] Kanth SA, Harmain GA, Jameel A. Modeling of nonlinear crack growth in steel and aluminum alloys by the element free Galerkin method. Materials Today: Proceedings 2018;5(9, Part 3):18805–14. Available from: https://doi.org/10.1016/j.matpr.2018.06.227.

[14] Lone AS, Jameel A, Harmain GA. A coupled finite element-element free Galerkin approach for modeling frictional contact in engineering components. Materials Today: Proceedings 2018;5(9, Part 3):18745–54. Available from: https://doi.org/10.1016/j.matpr.2018.06.221.

[15] Verhoosel CV, Scott MA, de Borst R, Hughes TJR. An isogeometric approach to cohesive zone modeling. International Journal for Numerical Methods in Engineering 2011;87(1–5):336–60. Available from: https://doi.org/10.1002/nme.3061.

[16] Gupta V, Jameel A, Verma SK, Anand S, Anand Y. An insight on NURBS based isogeometric analysis, its current status and involvement in mechanical applications. Archives of Computational Methods in Engineering 2022;. Available from: https://doi.org/10.1007/s11831-022-09838-0.

[17] Jameel A, Harmain GA. Fatigue crack growth analysis of cracked specimens by the coupled finite element-element free Galerkin method. Mechanics of Advanced Materials and Structures 2019;26(16):1343–56. Available from: https://doi.org/10.1080/15376494.2018.1432800.

[18] Jameel A, Harmain GA. Large deformation in bi-material components by XIGA and coupled FE-IGA techniques. Mechanics of Advanced Materials and Structures 2022;29(6):850–72. Available from: https://doi.org/10.1080/15376494.2020.1799120.

[19] Jameel A, Harmain GA. A coupled FE-IGA technique for modeling fatigue crack growth in engineering materials. Mechanics of Advanced Materials and Structures 2019;26(21):1764–75. Available from: https://doi.org/10.1080/15376494.2018.1446571.

[20] Jiang S-Y, Du C-B, Ooi ET. Modelling strong and weak discontinuities with the scaled boundary finite element method through enrichment. Engineering Fracture Mechanics 2019;222:106734. Available from: https://doi.org/10.1016/j.engfracmech.2019.106734.

[21] Mohammad MS, Shahrooi S, Shishehsaz M, Hamzehei M. Fatigue crack propagation of welded steel pipeline under cyclic internal pressure by Bézier extraction based XIGA. Journal of Pipeline Systems Engineering and Practice 2022;13(2):04022001. Available from: https://doi.org/10.1061/(ASCEPS.1949-1204.0000633.

[22] Yadav A, Patil RU, Singh SK, Godara RK, Bhardwaj G. A thermo-mechanical fracture analysis of linear elastic materials using XIGA. Mechanics of Advanced Materials and Structures 2022;29(12):1730–55. Available from: https://doi.org/10.1080/15376494.2020.1838006.

[23] Singh AK, Jameel A, Harmain GA. Investigations on crack tip plastic zones by the extended iso-geometric analysis. Materials Today: Proceedings 2018;5(9, Part 3):19284–93. Available from: https://doi.org/10.1016/j.matpr.2018.06.287.

[24] Jameel A, Harmain GA. Extended iso-geometric analysis for modeling three-dimensional cracks. Mechanics of Advanced Materials and Structures 2019;26(11):915–23. Available from: https://doi.org/10.1080/15376494.2018.1430275.

[25] Hughes TJR, Cottrell JA, Bazilevs Y. Isogeometric analysis: CAD, finite elements, NURBS, exact geometry and mesh refinement. Computer Methods in Applied Mechanics and Engineering 2005;194(39):4135–95. Available from: https://doi.org/10.1016/j.cma.2004.10.008.

[26] Cottrell JA, Reali A, Bazilevs Y, Hughes TJR. Isogeometric analysis of structural vibrations. Computer Methods in Applied Mechanics and Engineering 2006;195(41):5257–96. Available from: https://doi.org/10.1016/j.cma.2005.09.027.

[27] Gupta V, Jameel A, Anand S, Anand Y. Analysis of composite plates using isogeometric analysis: a discussion. Materials Today: Proceedings 2021;44:1190–4. Available from: https://doi.org/10.1016/j.matpr.2020.11.238.

[28] Akkerman I, Bazilevs Y, Kees CE, Farthing MW. Isogeometric analysis of free-surface flow. Journal of Computational Physics 2011;230(11):4137–52. Available from: https://doi.org/10.1016/j.jcp.2010.11.044.

[29] Tagliabue A, Dedè L, Quarteroni A. Isogeometric Analysis and error estimates for high order partial differential equations in fluid dynamics. Computers & Fluids 2014;102:277–303. Available from: https://doi.org/10.1016/j.compfluid.2014.07.002.

[30] Gupta V, Verma SK, Anand S, Jameel A, Anand Y. Transient isogeometric heat conduction analysis of stationary fluid in a container. Proceedings of the Institution of Mechanical Engineers, Part E: Journal of Process Mechanical Engineering 2022;. Available from: https://doi.org/10.1177/09544089221125718.

[31] Duvigneau R., An introduction to isogeometric analysis with application to thermal conduction; 2009. Retrieved from https://hal.inria.fr/inria-0039415.

[32] Fang W, An Z, Yu T, Bui TQ. Isogeometric boundary element analysis for two-dimensional thermoelasticity with variable temperature. Engineering Analysis with Boundary Elements 2020;110:80–94. Available from: https://doi.org/10.1016/j.enganabound.2019.10.003.

[33] Bazilevs Y, Calo VM, Hughes TJR, Zhang Y. Isogeometric fluid-structure interaction: theory, algorithms, and computations. Computational Mechanics 2008;43(1):3–37. Available from: https://doi.org/10.1007/s00466-008-0315-x.

[34] Morganti S, Auricchio F, Benson DJ, Gambarin FI, Hartmann S, Hughes TJR, et al. Patient-specific isogeometric structural analysis of aortic valve closure. Computer Methods in Applied Mechanics and Engineering 2015;284:508–20. Available from: https://doi.org/10.1016/j.cma.2014.10.010.

[35] Buffa A, Sangalli G, Vázquez R. Isogeometric methods for computational electromagnetics: B-spline and T-spline discretizations. Journal of Computational Physics 2014;257:1291–320. Available from: https://doi.org/10.1016/j.jcp.2013.08.015.

[36] Nguyen VP, Anitescu C, Bordas SPA, Rabczuk T. Isogeometric analysis: An overview and computer implementation aspects. Mathematics and Computers in Simulation 2015;117:89–116. Available from: https://doi.org/10.1016/j.matcom.2015.05.008.

[37] Wang Y, Gao L, Qu J, Xia Z, Deng X. Isogeometric analysis based on geometric reconstruction models. Frontiers of Mechanical Engineering 2021;16(4):782–97. Available from: https://doi.org/10.1007/s11465-021-0648-0.

[38] Ghorashi SS, Valizadeh N, Mohammadi S. Extended isogeometric analysis for simulation of stationary and propagating cracks. International Journal for Numerical Methods in Engineering 2012;89(9):1069–101. Available from: https://doi.org/10.1002/nme.3277.

[39] Benson DJ, Bazilevs Y, De Luycker E, Hsu MC, Scott M, Hughes TJR, et al. A generalized finite element formulation for arbitrary basis functions: From isogeometric analysis to XFEM. International Journal for Numerical Methods in Engineering 2010;83(6):765–85. Available from: https://doi.org/10.1002/nme.2864.

[40] De Luycker E, Benson DJ, Belytschko T, Bazilevs Y, Hsu MC. X-FEM in isogeometric analysis for linear fracture mechanics. International Journal for Numerical Methods in Engineering 2011;87(6):541–65. Available from: https://doi.org/10.1002/nme.3121.

[41] Nguyen-Thanh N, Valizadeh N, Nguyen MN, Nguyen-Xuan H, Zhuang X, Areias P, et al. An extended isogeometric thin shell analysis based on Kirchhoff–Love theory. Computer Methods in Applied Mechanics and Engineering 2015;284:265–91. Available from: https://doi.org/10.1016/j.cma.2014.08.025.

[42] Bhardwaj G, Singh IV, Mishra BK. Fatigue crack growth in functionally graded material using homogenized XIGA. Composite Structures 2015;134:269–84. Available from: https://doi.org/10.1016/j.compstruct.2015.08.065.

[43] Singh SK, Singh IV. Extended isogeometric analysis for fracture in functionally graded magneto-electro-elastic material. Engineering Fracture Mechanics 2021;247:107640. Available from: https://doi.org/10.1016/j.engfracmech.2021.107640.

[44] Fardaghaie A, Shahrooi S, Shishehsaz M. The application of the extended isogeometric analysis (XIGA) with K-refinement approach for the prediction of fatigue life in linear elastic fracture mechanic. ADMT Journal 2022;15(1):29–50. Available from: https://admt.isfahan.iau.ir/article_687319.html.

[45] Khademalrasoul A, Naderi R. Local and global approaches to fracture mechanics using isogeometric analysis method. Journal of Applied and Computational Mechanics 2015;1(4):168–80. Available from: https://jacm.scu.ac.ir/article_11237.html.

[46] Piegl L, Tiller W. The NURBS book. Springer Berlin Heidelberg; 2012. Available from: https://doi.org/10.1007/978-3-642-97385-7.

[47] Rabczuk T, Song JH, Zhuang X, Anitescu C. Extended finite element and Meshfree methods. Elsevier Science; 2019. Available from: https://doi.org/10.1016/C2017-0-00659-6.

[48] Belytschko T, Black T. Elastic crack growth in finite elements with minimal remeshing. International Journal for Numerical Methods in Engineering 1999;45(5):601–20. Available from: https://doi.org/10.1002/(SICI)1097-0207(19990620)45:5.

[49] Chopp DL, Sukumar N. Fatigue crack propagation of multiple coplanar cracks with the coupled extended finite element/fast marching method. International Journal of Engineering Science 2003;41(8):845–69. Available from: https://doi.org/10.1016/S0020-7225(02)00322-1.

[50] Singh IV, Mishra BK, Bhattacharya S, Patil RU. The numerical simulation of fatigue crack growth using extended finite element method. International Journal of Fatigue 2012;36(1):109–19. Available from: https://doi.org/10.1016/j.ijfatigue.2011.08.010.

CHAPTER

14

Extended isogeometric analysis for linear elastic materials under thermomechanical loading

Vibhushit Gupta[1], Shubham Kumar Verma[2], Sanjeev Anand[2], Sahil Thappa[2], Sanjay Sharma[2], Azher Jameel[3] and Yatheshth Anand[1]

[1]School of Mechanical Engineering, Shri Mata Vaishno Devi University, Kakryal, Katra, Jammu and Kashmir, India [2]School of Energy Management, Shri Mata Vaishno Devi University, Kakryal, Katra, Jammu and Kashmir, India [3]Department of Mechanical Engineering, National Institute of Technology Srinagar, Hazratbal, Srinagar, Jammu and Kashmir, India

14.1 Introduction

The analysis of fractures becomes important for ensuring the reliability and life of the mechanical component under various loading conditions like mechanical, thermal, chemical, and electrical. In general, the direct exposure of these components in various coupled loading applications leads to failure due to the presence of fractures. A mechanical component with existing fracture discontinuities (like cracks or voids) when imperiled to the mixed loads, that is, mechanical and thermal, leads to varying stress field singularities that affect the crack propagation [1,2]. The creation of these types of situations in various components will result in the loss of property as well as human livelihood. Thus the analysis of fractures subjected to thermomechanical loading (TML) conditions is very much important for the evaluation of critical crack paths and loads so that the safety of the structures can be ensured. In addition to these parameters, the stress intensity factors (SIFs) also need to be evaluated, which plays an essential role in safety assessments of the components. To evaluate such parameters, fracture problems are generally analyzed by utilizing suitable methods. The utilization of these methods provides a detailed view of the discontinuous structures, which includes loading conditions, properties of the materials, and the

Enriched Numerical Techniques
DOI: https://doi.org/10.1016/B978-0-443-15362-4.00015-2

uncertainties that crack geometry inherits. However, in the past few decades, various researchers have utilized different numerical techniques for simulating actual geometries under realistic loading conditions. One of the most popular techniques is finite element analysis (FEA), which was utilized for evaluating SIFs of a cracked structure subjected to thermo-elastic load [3,4]. This technique was appropriately utilized for solving fracture problems under thermo-elastic load as well. Another technique named the boundary element method (BEM) was utilized by various research workers for analyzing the thermo-mechanical fracture problems in 2D space [5,6]. The analysis of fracture problems using FEA and BEM has been proven to be advantageous techniques. However, these techniques have their limitations (like the use of singular tips, conformal mesh, and remeshing) while simulating crack structures [7,8]. To overcome these limitations, various techniques, for example, extended finite element [9–13], coupled techniques [14–16], kernel particle reproducing technique [17–19], and element-free Galerkin technique [20–23] were developed. These developed techniques were proven to be beneficial in case of handling moving discontinuities like crack propagations. However, in these enhanced techniques, the geometry approximation acquaints discretization errors in the solution fields as the basis function used for describing the geometry and unknown field variable are different.

Isogeometric analysis (IGA), a novel concept, was developed by Hughes et al. for resolving such discretization issues [24]. In IGA, a standard basis function is employed for describing the geometry and finding unknown field variables, which makes IGA a precise methodology for solving various problems [25–27]. In the case of fractures, IGA has been considered an efficient and potential computational tool for solving various types of crack growth problems. Verhoosel et al. analyze cohesive cracks by developing a nonuniform rational B-spline (NURBS)-based approach in which the knot insertion method is utilized for interface modeling and inserting discontinuity directly into the solution [28]. Choi et al. modeled various geometries of cracks like circular and parabolic by utilizing the NURBS-based multipatch IGA method [29]. In this approach, a knot insertion method was employed so that the crack present in the domain is defined in an exact manner by not getting any changes in real geometry. Khademalrsoul employed the analogous methodology for analyzing the effect of various crack inclinations present in a domain. Also, in this chapter, various sets of refinement strategies were adopted that present better results than the cases with coarser meshes [30]. However, the utilization of IGA approach for crack analysis shows some issues in the accuracy of results due to improper continuity near the crack interfaces. Thus the use of higher order basis functions eliminates such types of problems.

Further, the concept of IGA was extended by introducing the partition of unity in IGA, which was entitled extended IGA (XIGA) [31]. Various research workers had utilized the XIGA concept for solving different fracture problems. De Luycker et al. incorporate IGA and XFEM for getting accurate and robust solutions in cases of fracture problems [32]. Shojaee et al. imply XIGA formulations for modeling stationary cracks presented in orthotropic media [33]. Bui analyzes various dynamic as well as static fractures in the case of piezoelectric materials by using XIGA approach [34]. Analogously, Bhardwaj et al. simulate the cracked functionally graded plates (FGPs) by implementing NURBS-based XIGA approach [35]. In their study, the composite plates under various loading and boundary conditions (BC) were analyzed by employing first-order shear deformation theory. Yu et al. implement XIGA for analyzing the fracture behavior of FGPs subjected to thermal buckling conditions [36]. Singh et al. utilized XIGA for modeling and analyzing cracks presented in an isotropic and homogeneous plate. They implemented Reddy's higher order formulations for analyzing the through-thickness cracks presented in the plates [37]. However, in these studies, the basis function employed for analysis in XIGA is NURBS, which has limitations due to its tensor product. Thus an analysis suitable for T-splines was employed by Habib and Belaidi as a basis function in XIGA for simulating discontinuities like cracks present in bi-material interfaces [38]. Similar to this local refinement strategy, another approach, including Bezier extraction in XIGA, was developed by Singh et al. and utilized for the analysis of discontinuities present in structures [39]. In these studies, it was evident that the utilization of T-splines makes the technique more accurate by employing local refinement in the domain. Also, Yadav et al. present a detailed and complete review on applying XIGA for modeling problems related to fracture mechanics. In this work, different sets of numerical tests and implementation techniques were described [40]. Also, this chapter discussed the ability of the XIGA technique in solving various fracture problems like crack behavior under dynamic loadings, elasto-plastic crack growth, and multifaceted fracture problems. Montassir et al. employed XIGA for investigating cracks present in cylinder-shaped structures [41]. Kaushik and Ghosh used the XIGA technique for analyzing the growth of cracks in case of aerospace structures, that is, laminated composite plates [42]. The growth of cracks in case of functionally graded materials was investigated by Bhardwaj et al. under TML conditions [43]. It is evident from the above studies that a lot of fracture issues have been solved by utilizing XIGA approach. However, a limited amount of work has been noted on analyzing fracture problems under mixed loading conditions. So, in the present study, a detailed review on implementing the XIGA approach for analyzing fracture problems subjected to TML conditions has been conducted.

This review begins with Section 14.2, which provides details of the XIGA technique with an emphasis on different types of basis functions and further provides a brief on the mathematical formulations of the plane strain TML problems. Section 14.3 presents procedure for the evaluation of SIFs, and Section 14.4 provides explicit culminating remarks of the presented review.

14.2 Extended isogeometric analysis

In XIGA, the analysis of discontinuous problems can be done efficiently and accurately by combining the capabilities of XFEM with IGA. Although the XFEM technique analyzes static as well as moving discontinuities present in structures, IGA is efficient in solving problems that relate to complex geometries. These techniques (XFEM and IGA) are together known as XIGA, which opens the path to analyzing these discontinuities independent of the mesh [31]. In general, XIGA is an approximation of the IGA that is enriched by the enrichment functions (partition of unity) of XFEM [44].

XIGA works with a higher rate of efficacy in performing simulation for crack propagation because, in this technique, the elements of the boundaries are not allowed to be aligned with crack surfaces, which obligates remeshing [45]. However, in XIGA, for the

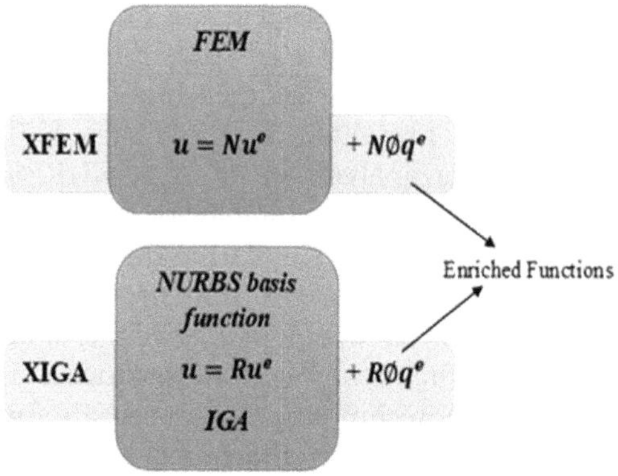

FIGURE 14.1 Basic map of numerical techniques.

simulation of discontinuities and singular fields, locally enriched functions are combined with approximations of IGA. Fig. 14.1 shows a basic map of this numerical technique.

Basically, in XIGA near the location of discontinuity, some degrees of freedom (DOFs) are inserted at specified control points of the IGA model, and for overall approximation, enrichment functions are utilized. Further, for capturing the accurate behavior of discontinuity, two enrichment functions are employed, that is, crack tip and the Heaviside function [46,47]. The selection of the enrichment function depends on the nature of the discontinuity that needs to be analyzed. The domain of control points that are bisected by the face of the crack is enriched by the Heaviside function, while the control points influenced by the tip of the crack are enriched by the crack tip enrichment function (CTEF). Also, in this technique, a standard Gauss quadrature integration is endorsed for integration over the non-enriched elements. Furthermore, a polar integration scheme is also adopted for performing integration in discontinuous elements.

14.3 Extended isogeometric analysis basis functions

As discussed in the literature, XIGA includes various types of basis functions for the approximation of the unknown solution variables. The basis functions include B-splines, NURBS, T-splines, and PHT-splines. A brief discussion on these types of functions is given in the following section.

14.3.1 B-splines

This basis function is defined as a linear piecewise polynomial function that is considered over a knot vector $\Xi = \{\xi_1 \ldots\ldots\ldots \xi_{n+p+1}\}$. In this, each knot value is taken as a real number $\mathbb{R}$; the letters "n" and "p" show the number and order of the function, respectively. The knot vectors are classified into two forms, that is, uniform and nonuniform vectors. In uniform vector, the distance between two consecutive knots remains constant; otherwise, it is a nonuniform vector. The knot vector is also known as a patch in computer-aided designing (CAD) parlance. It is seen that if the starting and ending knot vectors are repeated by $p+1$ times, it is termed an open knot vector. The B-spline functions depending on p by using the Cox–deBoor recursive formula are given as [48]For $p=0$

$$N_{i,0}(\xi) = \begin{cases} 1 & if\ \xi_i \leq \xi_{i+1} \\ 0 & otherwise \end{cases} \tag{14.1}$$

For $p > 0$

$$N_{i,p}(\xi) = \frac{\xi - \xi_i}{\xi_{i+p} - \xi_i} N_{i,p-1}(\xi) + \frac{\xi_{i+p+1} - \xi}{\xi_{i+p+1} - \xi_{i+1}} N_{i+1,p-1}(\xi) \tag{14.2}$$

B-spline function exhibits some good properties like linear independency, partition of unity, that is, $\sum_{(i=1)}^{(p+)} N_{(i,p)}(\xi) = 1$, nonnegativity $N_{i,p} \geq \forall \xi$, variational diminishing property, and Kronecker's delta properties. About the continuity of B-spline function, it shows C^0 at the boundaries of a patch, whereas C^{p-1} is seen at the knot point ξ. Also, this continuity can be decreased by using the relation C^{p-1-k}, where k shows the number of times the knots are repeated.

The multivariate B-splines functions are derived as the tensor product of a univariate function. For a bi- and trivariate case, the basis function is given as

$$N_{i,j}^{p,q}(\xi, \eta) = N_{i,p}(\xi) \times M_{j,q}(\eta) \tag{14.3}$$

$$N_{i,j,k}^{p,q,r}(\xi, \eta, \zeta) = N_{i,p}(\xi) \times M_{j,q}(\eta) \times L_{k,r}(\zeta) \tag{14.4}$$

Based on this basis function, the B-spline-based geometry can be derived as

$$Z(\xi) = \sum_{i=1}^{n} N_i^p(\xi) C_i \tag{14.5}$$

$$Y(\xi, \eta) = \sum_{i=1}^{n} \sum_{i=1}^{n} N_{i,p}(\xi) \times M_{j,q}(\eta) C_{i,j} \tag{14.6}$$

$$X(\xi, \eta, \zeta) = \sum_{i=1}^{n} \sum_{i=1}^{m} \sum_{i=1}^{l} N_{i,p}(\xi) \times M_{j,q}(\eta) \times L_{k,r}(\zeta) C_{i,j,k} \tag{14.7}$$

Here in these equations, the $C_{i,j,k}$ represents the control point volume array, and the B-spline curve, surface, and solid are defined by $Z(\xi)$, $Y(\xi, \eta)$, and $Y(\xi, \eta, \zeta)$, respectively, m and l show the nos. of basis function in the direction of η and ζ.

14.3.2 Nonuniform rational B-spline

NURBS functions are frequently utilized in CAD industries. These functions provide a high rate of accuracy and flexibility while developing and representing different geometries. Basically, NURBS are considered the generalized form of B-splines. A one-direction NURBS function is given by the rationalized form of weighted B-splines function as [48]

$$R_i^p(\xi) = \frac{N_{i,p}(\xi) w_i}{\sum_{i=l}^{n} N_{i,p}(\xi) w_i} \tag{14.8}$$

where w_i represents the weight function associated with the control point vector C_i. Similarly, for bivariate and trivariate basis functions, the function will be given as

$$R_i^p(\xi,\eta) = \frac{N_{i,p}(\xi)\times M_{j,q}(\eta)w_{i,j}}{\sum_{i=1}^{n}\sum_{j=1}^{m} N_{i,p}(\xi)\times M_{j,q}(\eta)w_{i,j}} \tag{14.9}$$

$$R_i^p(\xi,\eta,\zeta) = \frac{N_{i,p}(\xi)\times M_{j,q}(\eta)\times L_{k,r}(\zeta)w_{i,j}}{\sum_{i=1}^{n}\sum_{j=1}^{m}\sum_{k=1}^{l} N_{i,p}(\xi)\times M_{j,q}(\eta)\times L_{k,r}(\zeta)w_{i,j,k}} \tag{14.10}$$

where $N_{i,p}(\xi)M_{j,q}(\eta)$ *and* $L_{k,r}(\zeta)$ are the B-spline basis function having orders p, q, and r, respectively. Analogously, the NURBS base geometries are developed in the same manner like B-spline geometry. For 1D, 2D, and 3D geometries, the equations are given as

$$W(\xi) = \sum_{i=1}^{n} R_i^p(\xi)C_i \tag{14.11}$$

$$V(\xi,\eta) = \sum_{i=1}^{n}\sum_{j=1}^{m} R_i^p(\xi,\eta)C_{i,j} \tag{14.12}$$

$$U(\xi,\eta,\zeta) = \sum_{i=1}^{n}\sum_{j=1}^{m}\sum_{k=1}^{l} R_i^p(\xi,\eta,\zeta)C_{i,j,k} \tag{14.13}$$

14.3.3 T-spline

T-splines are a generalized form of NURBS surface that maintains all the needed properties similar to NURBS and is also considered a finer alternative to the NURBS function [39]. The employment of this type of function allows the increment of control points in columns or rows, which results in T-junctions. These junctions are analogous to the hanging nodes in FEA. In this type of basis function, they are associated with anchors. The anchors t_i for the T-spline basis function of degree g can be defined as [39]

$$t_i = \begin{Bmatrix} \xi_{i+\frac{g+1}{2}} & if \quad g \quad is \quad odd \\ \frac{1}{2}\xi_{i+\frac{g}{2}} + \xi_{i+\frac{g}{2}+1} & if \quad g \quad is \quad even \end{Bmatrix} \tag{14.14}$$

The T-spline basis function is developed for every vertex in the T-mesh by using a local knot vector. The localized knot vector for a control point can be developed from a T-mesh that is defined in a parametric domain by utilizing parametric values assigned to each vertex of a T-mesh. The local knot vectors are different for even and odd degree T-splines mesh. The local bivariate basis function can be derived as

$$N_A(\xi,\eta) = N(\xi_{g-2},\xi_{g-1},\xi_g,\xi_{g+1},\xi_{g+2})(\xi)\times M(\eta_{g-2},\eta_{g-1},\eta_g,\eta_{g+1},\eta_{g+2})(\eta) \tag{14.15}$$

where the $N(\xi_{g-2},\xi_{g-1},\xi_g,\xi_{g+1},\xi_{g+2})(\xi)$ and $M(\eta_{g-2},\eta_{g-1},\eta_g,\eta_{g+1},\eta_{g+2})(\eta)$ are the cubic univariate basis functions associated with the localized knot vectors $\xi_A = \xi_{g-2},\xi_{g-1},\xi_g,$ ξ_{g+1},ξ_{g+2} and $\eta_A = \eta_{g-2},\eta_{g-1},\eta_g,\eta_{g+1},\eta_{g+2}$, respectively. The value of univariate basis function can be derived by utilizing a recurring formula, starting with normalized ($g = 0$) piecewise function is

$$N_A^i(j,0) = \begin{Bmatrix} 1 & if\ \xi_{A,j}^i \le \xi_A^i \le \xi_{A,j+1}^i \\ 0 & otherwise \end{Bmatrix} \tag{14.16}$$

where $\xi_{A,j}^i$ is the jth value of knot in a local knot vector, and i ranges from 1 to parametric directions. For $g > 0$

$$N_A^i(j,p) = \frac{\xi_A^i - \xi_{A,j}^i}{\xi_{A,g+1}^i - \xi_{A,j}^i} N_A^i(j,g-1) + \frac{\xi_{A,j+g+1}^i - \xi_A^i}{\xi_{A,j+p+1}^i - \xi_{A,j+1}^i} N_A^i(j+1,p-1) \tag{14.17}$$

After the definition of the basis function, the surface based on T-spline function is given as

$$T(\xi,\eta) = \sum\nolimits_{A=1}^{n} N_A(\xi,\eta) C_A \tag{14.18}$$

where $N_A(\xi,\eta)$ shows the basis function in association with control point C_A. Similarly, the T-spline rational surface is given as

$$\overline{T}(\xi,\eta) = \frac{N_A(\xi,\eta) w_A}{\sum_{A=1}^{n} N_A(\xi,\eta) W_A} \tag{14.19}$$

$$\overline{T}(\xi,\eta) = \sum\nolimits_{A=1}^{n} R_A(\xi,\eta) \overline{C_A} \tag{14.20}$$

where $\overline{C_A} = \{w_A P_A^T, w_A\}^T$ shows the coordinates in the homogeneous form, w_A represents the weight of a control point, and W_A defines the cumulative weight function.

14.3.4 PHT-splines

PHT-splines are generally known as polynomial splines over hierarchical T-meshes and are the generalization of B-splines over hierarchical T-meshes that were found to be very efficient for performing local refinement. This advantageous property makes these splines considerable for implementation in various applications like geometry processing, adaptive FEA, and IGA. Also, in the case of XIGA, the PHT-splines have been utilized for performing adaptive refinements during crack propagation. The utilization of PHT-splines fulfills the linear independence requirements of a function. This approach is quite simpler as compared to the complex T-spline-based approach [49]. Also, in comparison with T-spline and hierarchical B-splines, these types of splines have C^1 continuity, which can be increased through the extension of the PHT-splines. Also, in this hierarchical structure, T-meshes are considered to have a natural nested structure. This special form of T-meshes started from T-mesh assumed at level 0 and is represented in Fig. 14.2. Further, if a level-k mesh is provided, then the level $(k+1)$ mesh is constructed by subdividing some of the cells in level-k. Every cell is further divided into four subcells by connecting the central points of the across edges in the cell. Fig. 14.2 demonstrates the procedure for the generation of hierarchical T-meshes.

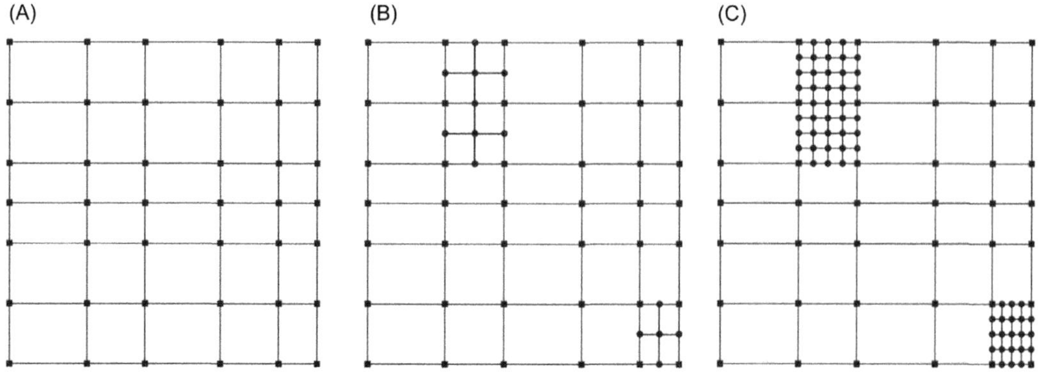

FIGURE 14.2 Process of constructing a hierarchical T-meshes: (A) initial level 0, (B) intermediate level-k, and (C) final level-$k+1$.

The basis function of PHT-splines is similar to that of NURBS at the initial level, which is given as follows. The surface based on PHT-splines can be constructed as [50]

$$Y(\xi,\eta) = \sum\nolimits_{i=1}^{n} R_i(\xi,\eta)C_i \tag{14.21}$$

At level-k, the PHT-spline basis function is modified as

$$Y^k(\xi,\eta) = \sum\nolimits_{i=1}^{nk} R_i^k(\xi,\eta)P_i^k \tag{14.22}$$

Similarly, at level-$k+1$, the PHT-spline basis function is further generated as

$$Y^{k+1}(\xi,\eta) = \sum\nolimits_{i=1}^{nk+1} R_i^{k+1}(\xi,\eta)P_i^{k+1} \tag{14.23}$$

14.4 Extended isogeometric analysis problem formulation

The current section represents the problem formulation of mixed loading conditions, that is, thermal and mechanical, in the XIGA framework. To deal with this, the approximation of these fields is done by approximating temperature and displacement field variables. In the case of temperature, two forms of cracks are investigated, that is, isothermal and adiabatic. As compared to the isothermal crack case, the adiabatic crack acquaintances various crack tip singularities. Consequently, by considering a problem domain "Ω" presented in Fig. 14.3 of a homogeneous solid under TML.

The solid "Ω" is enclosed by a boundary "Γ" which again branched into traction boundary (Γ_t), displacement boundary (Γ_u), heat flux boundary (Γ_q), and temperature boundary (Γ_T). Also, domain (Ω) has an arbitrary crack, which is presented by (Γ_c). Based on this, the governing equations based on two different loading conditions can be expressed as follows. In the domain (Ω), when subjected to mechanical loading, the solution needs to satisfy the following equilibrium conditions [51]:

$$\nabla.\sigma + f = 0 \tag{14.24}$$

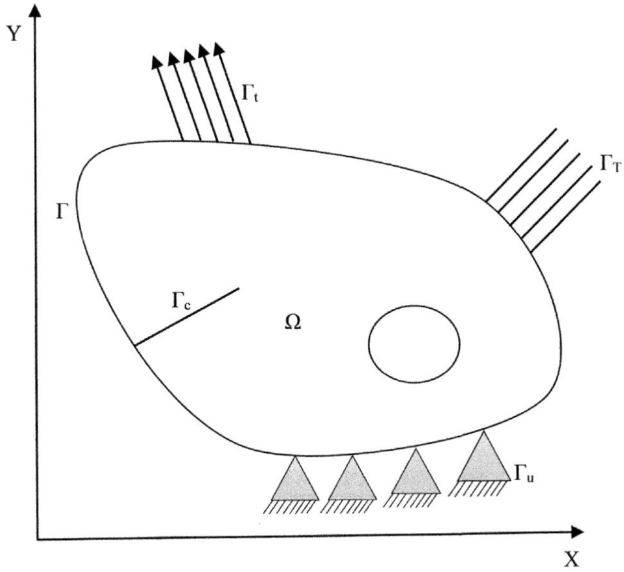

FIGURE 14.3 Problem domain under consideration.

where σ is a Cauchy tensor product, ∇ is a divergence operator, and f is a body force vector. The BCs for the domain under consideration are explained as

$$\sigma \cdot \widehat{n} = \hat{t} \ on \ \Gamma_t \tag{14.25}$$

$$\sigma \cdot \widehat{n} = o \ on \ \Gamma_c \tag{14.26}$$

$$u = \overline{u} \ on \ \Gamma_u \tag{14.27}$$

Now, in the case of a mixed linear TML crack domain, the governing equations [51] with small displacements over the domain and boundary are defined as

$$-\nabla .q + Q = 0 \tag{14.28}$$

$$q = -k\nabla T \tag{14.29}$$

$$\nabla .\sigma + f = 0 \text{ in } \Omega \tag{14.30}$$

$$\sigma = D(\epsilon - \epsilon_T) \tag{14.31}$$

Eqs. (14.30) and (14.28) present the mechanical and thermal equilibrium, respectively, and Eqs. (14.31) and (14.29) show the relations between Hooke's law and Fourier's law of conduction, respectively. The thermo-elastic tensor is given as

$$\epsilon = \nabla_s u \tag{14.32}$$

$$\epsilon_T = \alpha(T - T_0)I \tag{14.33}$$

In the aforementioned equations, Q is a heat source, q is heat flux, T is a temperature, and $\epsilon \& \epsilon_T$ represents strain and thermal strain tensors, respectively. Also, k is thermal diffusivity, with α as the expansion coefficient, D as the constitutive matrix, and I is an identity tensor.

Similarly, the associated BC [51] to these equations is presented as

$$T = \overline{T} \ on \ \Gamma_T \tag{14.34}$$

$$q \cdot n = \overline{q} \ on \ \Gamma_q \tag{14.35}$$

$$u = \overline{u} \ on \ \Gamma_u \tag{14.36}$$

In the aforementioned equations, the $\overline{T}$ and $\overline{u}$ are the field temperatures and displacement that are imposed on the boundaries Γ_T and Γ_u, respectively, whereas the heat flux $\overline{q}$ and traction vector $\hat{t}$ are applied on Γ_q and Γ_t, respectively. The problem is stated in a way such that $\Gamma_u \cup \Gamma_t = \Gamma_T \cup \Gamma_q = \Gamma$ and $\Gamma_u \cap \Gamma_t = \Gamma_T \cap \Gamma_q = \phi$. Furthermore, two cases have been considered in the upcoming sections, that is, isothermal crack ($\Gamma_c \subset \Gamma_T, \overline{T} = \overline{T_c} \ on \ \Gamma_c$) and adiabatic crack ($\Gamma_c \subset \Gamma_q, \overline{q} = o \ on \ \Gamma_c$).

14.5 Extended isogeometric analysis approximation

Firstly, in XIGA, the approximation of displacement for void and crack is formulated. Then, two cases, that is, isothermal and adiabatic cracks, are reviewed individually for evaluation of the thermal field. The following approximations are given.

14.5.1 Displacement approximation

In the parametric space of an arbitrary point ξ_i, the approximation of displacement at a particular control point for inclusions, holes, and cracks given by Bhardwaj and Singh [52] can be expressed as

$$u^h(\xi) = \sum_{i=1}^{n_{be}} R_i^p(\xi)u_i + \sum_{j=1}^{n_s} R_j^p(\xi)\left[W(\xi) - W(\xi_i)\right]a_j + \sum_{k=1}^{n_T} R_k^p(\xi)\sum_{\alpha=1}^{4}[\beta_\alpha(\xi)$$

$$- \beta_\alpha(\xi_i)]b_k^\alpha + \sum_{l=1}^{n_z} R_l^p(\xi)\left[\chi(\xi) - \chi(\xi_i)\right]c_l \tag{14.37}$$

where u_i presents the parameters relating to the specified control point, n_{be} are the number of shape functions per element given as $n_{be} = (p+1) \times (q+1)$; R_i^p shows the different approximation functions that can be referred to from Section 2.1; n_s is split-enriched node elements; n_T is tip-enriched node elements; a_j is enriched DOF of split elements related to the Heaviside function $W(\xi)$; b_k^α is enriched DOF of tip elements relates to CTEF $\beta_\alpha(\xi)$; c_l is extra enriched DOFs pertaining to $\chi(\xi)$ that takes 0 for the inner control points of the void and +1 for the outside control points; $\beta_\alpha(\xi)$ and $W(\xi)$ present the CTEF and the Heaviside enrichment function, respectively. The main purpose of the Heaviside function $W(\xi)$ is to produce discontinuity across the surface of the crack and ranges from −1 to +1, which is presented as

$$W(\xi) = \begin{cases} -1, & otherwise \\ +1, & if \ (\xi - \xi_i).n_{be} \geq 0 \end{cases} \tag{14.38}$$

The Heaviside function is efficient when an element is divided by a crack interface. However, for the case of tip elements, a portion of the element is only cut, and the Heaviside function is not able to enrich that tip element. Thus CTEFs are used for the enrichment of crack tip element. This enrichment function is defined by Chopp and Sukumar [53] and is explained as

$$\beta_\alpha(\xi) = \left[\sqrt{r}\cos\frac{\theta}{2}, \sqrt{r}\sin\frac{\theta}{2}, \sqrt{r}\cos\frac{\theta}{2}\sin\theta, \sqrt{r}\sin\frac{\theta}{2}\sin\theta\right] \tag{14.39}$$

where θ and r present polar coordinates at the reference line of crack and the crack surface.

14.5.2 Temperature approximation

The approximation of thermal field, that is, the temperature, is done in a similar manner to displacement approximation. In this, the approximation of temperature is presented for two cases of cracks, that is, adiabatic and isothermal, which are explained as follows:

14.5.2.1 Adiabatic crack

In this case, across the surface of crack, the temperature and displacement field are discontinuous, and the flux of heat is singular near the crack tip [54]. The approximation equation of discontinuous displacement is presented in Eq. (14.37). Similarly, the approximation equation for discontinuous temperature field in the case of both holes and cracks can be written as

$$T^h(\xi) = \sum_{i=1}^{n_{be}} R_i^p(\xi)T_i + \sum_{j=1}^{n_s} R_j^p(\xi)\left[W(\xi) - W(\xi_i)\right]a_j{}^T$$

$$+ \sum_{k=1}^{n_T} R_k^p(\xi)\sum_{\alpha=1}^{4}\left[\beta_\alpha(\xi) - \beta_\alpha(\xi_i)\right]b_k^{\alpha T} + \sum_{l=1}^{n_z} R_l^p(\xi)\left[\chi(\xi)\right]c_l{}^T \tag{14.40}$$

14.5.2.2 Isothermal crack

In the isothermal crack case, at crack surface, a particular temperature is defined with desired BC. Also, at the crack surface, the heat flux is discontinuous while the temperature profile remains continuous. The major difference between the isothermal and adiabatic cases is in terms of heat flux and temperature angular variation. In an isothermal crack case, the temperature varies radially to the surface of crack, while in an adiabatic crack, it is perpendicular. The distribution of temperature, in this case, is given as [51]

$$T = \frac{K_T}{k}\sqrt{\frac{2r}{\pi}}\cos\frac{\theta}{2} \tag{14.41}$$

Thus for capturing crack tip singularity, $\sqrt{r}\cos\frac{\theta}{2}$ is the enrichment function used for enriching the approximation function of temperature. The generalized approximation function for the isothermal crack case is explained as

$$T^h(\xi) = \sum_{i=1}^{n_{be}} R_i^p(\xi)T_i + \sum_{j=1}^{n_s} R_j^p(\xi)\left(\sqrt{r}\cos\frac{\theta}{2}\right)a_j + \sum_{k=1}^{n_T} R_j^p(\xi)[\chi(\xi)]b_k \tag{14.42}$$

Further, for the displacement and temperature approximations, the employment of nodal variation arbitrariness helps in finding out a group of discrete equations that can be derived as

$$\left[K^{TT}\right]\{T\} = \{F^T\} \tag{14.43}$$

$$\left[K^{UU}\right]\{u\} = \{F^u\} \tag{14.44}$$

where T and u show the nodal unknowns, K^{UU} and K^{TT} present the stiffness matrix for displacement and temperature field, respectively, and F^U and F^T are the force vectors for displacement and temperature fields.

14.6 Enrichment of control points

In the case of TML, the selection of enriched control points can be done in a similar manner like XFEM [55]. Moreover, there is little modification evident between the control points of XIGA and nodes of XFEM. In the case of XFEM, the nodes that are enriched are associated with either crack tip or crack face, whereas in XIGA, the domain under influence is much larger, so many control points located far from the crack are enriched as well. The control points that have nonzero basis function value in the crack face or tip are generally chosen for enrichment with Heaviside or CTEF function.

The modeling of cracks and voids with enriched control points is presented in Fig. 14.4, in which an edge crack with a hole has been presented for a first-order basis function. In

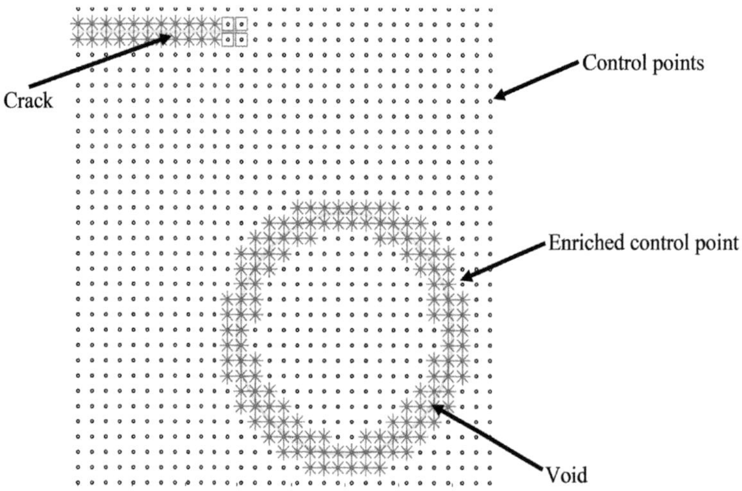

FIGURE 14.4 Domain with enriched control points.

Fig. 14.4, the crack with intersected domains, crack tip domains, and void domains is presented. Also, the enriched control points are marked with red crosses and squares. It also shows the marked control points, in which two additional DOFs are added at the intersected domains and four additional DOFs at the cracked tip domain of control points.

14.7 Change of temperature and displacement fields in between elements

It is very well known that the fields of approximation are different in the regions of enrichment and nonenrichment. A sharp change in these regions may result in substantial computational inaccuracy [55]. To overcome this, blended elements are considered in XFEM; analogously, it can be considered in the case of XIGA. In this, the knot span basis functions located in the support region of the control points are also enriched so that the blending effect is considered and discontinuous displacement fields can be avoided. It is further considered that the utilization of the Heaviside and CTEF functions does not affect the order of standardized IGA basis function; hence, this method can be utilized for the cases of transition between regions of subdomain having CTEF enrichment and subdomain with the Heaviside enrichment. It is also evident that the additional smoothness in the case of stress fields can be managed by employing more advanced transition methods. The discussed method is quite sufficient for getting the appropriate results, as the path integral method has been utilized with this method for the evaluation of SIFs [55]. The reason behind this accuracy is that, in the domain interaction integral approach, a zone near the crack tip is selected for evaluation of SIFs. This zone includes a larger number of field variables than the number of variables associated with the crack tip vicinity.

14.8 Modeling of multiple defects present in an element

In this section, the modeling of multiple defects present in a solo element is discussed. Firstly, the elements having intersections are computed and named as enriched elements. Further, additional DOFs are assigned to the intersected control points depending on the classification of the element, that is, tip or split. The number of control points linked to one element depends on the order of the basis function. The domain of enrichment does not remain fixed near the discontinuity due to the varying order. The following section considers modeling of one crack and an inclusion, one crack and a hole, one hole and an inclusion, and two holes and two inclusions as an example. The modeling of multiple discontinuities is given as follows:

14.8.1 Crack and inclusion

In this type of multiple discontinuity, a total of ten additional DOFs can be added at a control point. Fig. 14.5 shows the combination of crack with an inclusion in which additional DOFs are presented with green squares, blue squares, and orange circles. In this, eight DOFs are included due to the crack presence, whereas other two are added due to the inclusion.

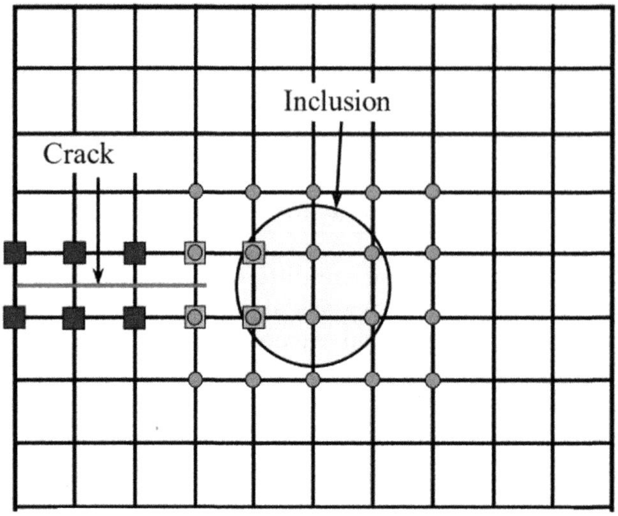

FIGURE 14.5 Representation of inclusion with a crack in a domain.

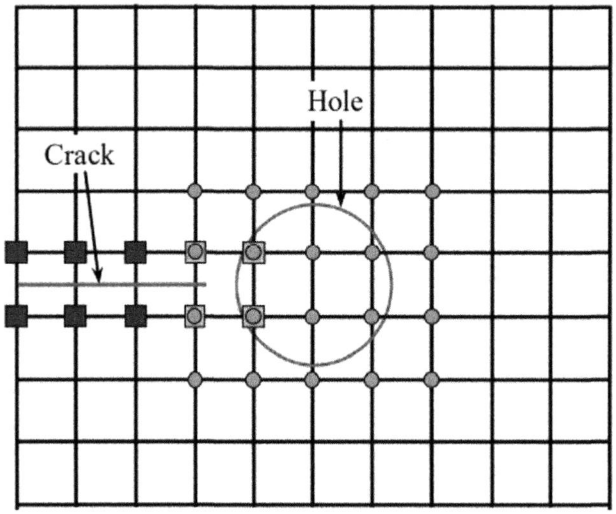

FIGURE 14.6 Representation of hole with a crack in a domain.

14.8.2 Crack with hole

The modeling method for a crack tip with a hole fleeting through an element is presented in Fig. 14.6. In this case, eight DOFs are added on control points due to the presence of crack tip, and two DOFs are included due to the presence of a hole, as similar to the crack inclusion case. An analogous procedure has been followed if an element containing multiple holes is intersected with the element that contains numerous crack tips.

14.8.3 One hole and an inclusion

In this case, the modeling of a hole with an inclusion passing through a common element is discussed. Similarly, transition control points linked to the intersected elements are identified, and further enrichment is done by adding more DOFs to them. At every control point, a total of two additional DOFs have been added to hole and inclusion, as presented in Fig. 14.7 by orange circles and green squares. Hence, at every intersected element control point, four additional DOFs are considered. The similar way can be utilized for multiple inclusions and holes passing through an element.

14.8.4 Two inclusions

Fig. 14.8 represents the two inclusions present in a domain and passing through an element. A few numbers of DOFs are added at the intersected element control point, that is, two on both inclusion 1 and inclusion 2 sides are added as the additional DOFs, represented by green squares and orange circles. Thus an accumulative of additional four DOFs are employed at every control point that has the influence of the two inclusions. An analogous procedure has been followed for the case of multiple inclusions present in a single domain.

14.8.5 Two holes

In this case, two holes presented in a single domain and passed through an element are discussed. A few numbers of DOFs are added at every control point linked with the intersected elements produced by holes. Fig. 14.9 shows the two holes and the intersected elements generated by these holes. In this, firstly, the intersected elements are figured out by utilizing the distance method, and then further, the associated control points

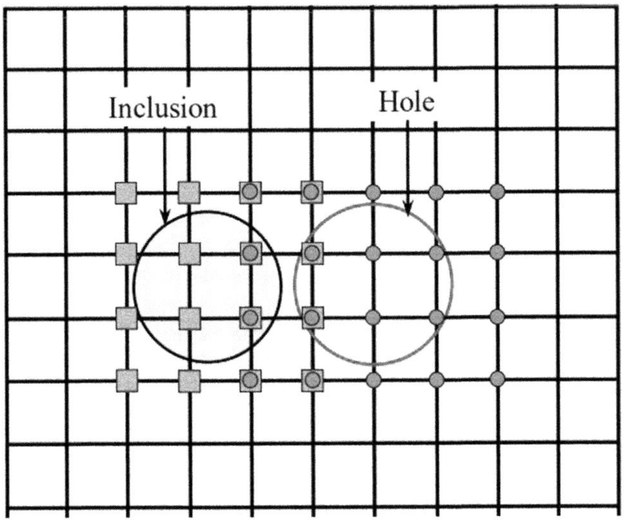

FIGURE 14.7 Representation of hole with an inclusion in a domain.

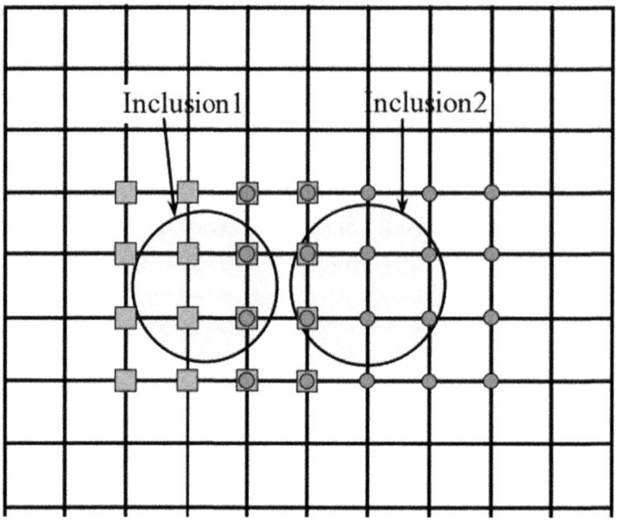

FIGURE 14.8 Representation of two inclusions in a domain.

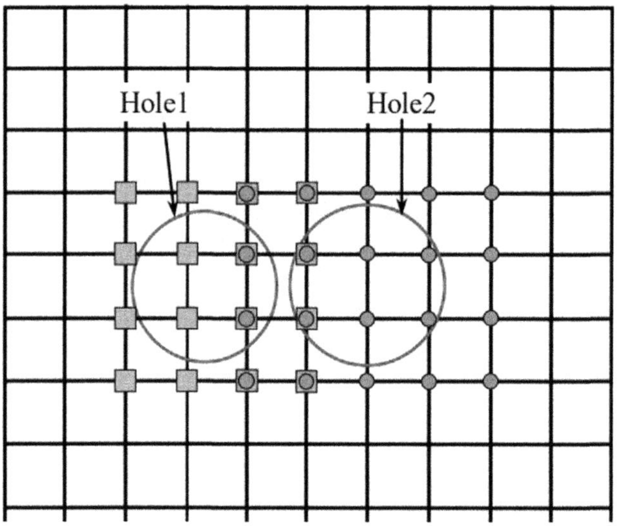

FIGURE 14.9 Representation of two holes in a domain.

with the intersected elements are identified. Two additional DOFs are employed on each control point when two holes are passed through an element, which is also presented in Fig. 14.9 by green squares and orange circles. Hence, a total of four additional DOFs are employed at every control point that gets influenced by the two holes. A similar procedure can be used in this case if there are more holes than two present in a single domain.

14.9 Integration of the elements

The integration in XIGA is done by utilizing a standard Gaussian quadrature. However, the existence of discontinuity in the integration zone reduces the reliability and exactness of the integration. Since the accuracy of the Gaussian quadrature rule has significantly declined, thus a well-known and established subtriangulation strategy can be adopted for improving the accuracy of the XIGA-based results [31]. The knot spans or elements containing a tip or crack face are divided into a set of subregions of sub-triangles, and the Gaussian rule is employed for each of the triangles. In this case of integration, a three-stage mapping is needed. At the last step of mapping, an extra step to the typical mappings of IGA is added, which shows the utilization of the Newton–Raphson algorithm for the evaluation of the position of the crack tip in the parent domain. Further, for the elements having crack faces, a similar nonlinear solver can be utilized for obtaining the coordinates of any other physical point in the parent domain.

14.10 Extended isogeometric analysis-based discretization

The discontinuity in XIGA can be derived by using the fundamental equation of displacement and temperature fields; thus the mechanical equation (Eq. 14.30) is discretized over the whole element as

$$K^{hU} u^{hU} = f^{hU} \tag{14.45}$$

were u^h represents the displacement vector, which consists of standardized as well as additional DOFs and is written as

$$u^{hU} = \begin{bmatrix} u & a^u & b_1^u & b_2^u & b_3^u & b_4^u & c^u \end{bmatrix} \tag{14.46}$$

where u $= \begin{bmatrix} u_x & u_y \end{bmatrix}$ shows the standard DOFs, while $a^u = \begin{bmatrix} a_x^u & a_y^u \end{bmatrix}$ and $b_\alpha^u = \begin{bmatrix} b_{\alpha x}^u & b_{\alpha y}^u \end{bmatrix}$ are the additional DOFs in the case of cracks, c^u is the extra number of degrees of freedom corresponds to an enrichment of void. f^{hU} is a force vector, and K^{hU} is an element stiffness matrix in the displacement field that is formed as the following:

$$K^{hU} = \begin{bmatrix} K_{ij}^{uu} & K_{ij}^{ua^u} & K_{ij}^{ub^u} & K_{ij}^{uc^u} \\ K_{ij}^{a^u u} & K_{ij}^{a^u a^u} & K_{ij}^{a^u b^u} & K_{ij}^{a^u c^u} \\ K_{ij}^{b^u u} & K_{ij}^{b^u a^u} & K_{ij}^{b^u b^u} & K_{ij}^{b^u c^u} \\ K_{ij}^{c^u u} & K_{ij}^{c^u a^u} & K_{ij}^{c^u b^u} & K_{ij}^{c^u c^u} \end{bmatrix} \tag{14.47}$$

$$f^{hU} = \begin{bmatrix} f_i^u & f_i^{a^u} & f_i^{b_1^u} & f_i^{b_2^u} & f_i^{b_3^u} & f_i^{b_4^u} & f_i^{c^u} \end{bmatrix}^T \tag{14.48}$$

$$K_{ij}^{rs} = \int_{\Omega^e} \left(B_i^r\right)^T C B_j^s d\Omega \tag{14.49}$$

$$f_i^u = \int_{\Omega^e} (R_i)^T b d\Omega + \int_{\Gamma_t} (R_i)^T \hat{t} d\Gamma \tag{14.50}$$

$$f_i^{a^u} = \int_{\Omega^e} (R_i)^T H b d\Omega + \int_{\Gamma_t} (R_i)^T H \hat{t} d\Gamma \tag{14.51}$$

$$f_i^{b^u} = \int_{\Omega^e} (R_i)^T \beta_\alpha b d\Omega + \int_{\Gamma_t} (R_i)^T \beta_\alpha \hat{t} d\Gamma \tag{14.52}$$

$$f_i^{c^u} = \int_{\Omega^e} (R_i)^T \chi b d\Omega + \int_{\Gamma_t} (R_i)^T \chi \hat{t} d\Gamma \tag{14.53}$$

where R_i is the NURBS rational function, and B_i^u, $B_i^{a^u}$, $B_i^{b_\alpha^u}$, $B_i^{b^u}$, and $B_i^{c^u}$ are the derivative matrices of the rational functions that are explained as

$$B_i^u = \begin{bmatrix} R_{i,X_1} & 0 \\ 0 & R_{i,X_2} \\ R_{i,X_1} & R_{i,X_2} \end{bmatrix} \tag{14.54}$$

$$B_i^{a^u} = \begin{bmatrix} R_{i,X_1} W & 0 \\ 0 & R_{i,X_2} W \\ R_{i,X_1} W & R_{i,X_2} W \end{bmatrix} \tag{14.55}$$

$$B_i^{b\alpha} = \begin{bmatrix} (R_i\beta_\alpha)_{,X_1} & 0 \\ 0 & (R_i\beta_\alpha)_{,X_2} \\ (R_i\beta_\alpha)_{,X_1} & (R_i\beta_\alpha)_{,X_1} \end{bmatrix} \tag{14.56}$$

$$B_i^b = \begin{bmatrix} B_i^{b1} & B_i^{b2} & B_i^{b3} & B_i^{b4} \end{bmatrix} \tag{14.57}$$

$$B_i^{c^u} = \begin{bmatrix} (R_i)_{,X_1}\chi & 0 \\ 0 & (R_i)_{,X_2}\chi \\ (R_i)_{,X_1}\chi & (R_i)_{,X_1}\chi \end{bmatrix} \tag{14.58}$$

where in these equations, $r, s = u, a^u, b^u$ and c^u and $\alpha = 1, 2, 3 \text{and} 4$. Now, XIGA formulations for different forms of temperature fields, that is, isothermal and adiabatic, are discussed. Firstly, Eq. (14.28) with adiabatic crack conditions is considered, which can be discretized over the elements as

$$K^{aTh} T^{ha} = f^{ahT} \tag{14.59}$$

Similarly, the temperature vector is defined in the same manner as the displacement vector in Eq. (14.46) and is given as

$$T^{ha} = \begin{bmatrix} T & a^T & b_1^T & b_2^T & b_3^T & b_4^T & c^T \end{bmatrix} \tag{14.60}$$

where T shows the standard DOFs, and a^T and b_1^T present the additional DOFs for cracks. In addition, for additional DOFs of void enrichment, c^T is used. Moreover, K^{aTh} and $\mathrm{f}^{\mathrm{ahT}}$

represent the elemental stiffness and force vector in the case of thermal fields that are explained as

$$K_{ij}^{aTh} = \begin{bmatrix} K_{ij}^{TT} & K_{ij}^{Ta} & K_{ij}^{Tb} & K_{ij}^{Tc} \\ K_{ij}^{aT} & K_{ij}^{aa} & K_{ij}^{ab} & K_{ij}^{ac} \\ K_{ij}^{bT} & K_{ij}^{ba} & K_{ij}^{bb} & K_{ij}^{bc} \\ K_{ij}^{cT} & K_{ij}^{ca} & K_{ij}^{cb} & K_{ij}^{cc} \end{bmatrix} \tag{14.61}$$

$$f_i^{ahT} = \begin{bmatrix} f_i^T & f_i^{a^T} & f_i^{b_1^T} & f_i^{b_2^T} & f_i^{b_3^T} & f_i^{b_4^T} & f_i^{c^T} \end{bmatrix}^T \tag{14.62}$$

In Eqs. (14.61) and (14.62), the terms have a similar meaning as defined in Eqs. (14.49)–(14.53). For obtaining the elemental stiffness matrix in the case of temperature field, the derivative matrices of the basis function are expressed as

$$B_i^T = \begin{bmatrix} R_{i,X_1} & 0 \\ 0 & R_{i,X_2} \\ R_{i,X_1} & R_{i,X_2} \end{bmatrix} \tag{14.63}$$

$$B_i^{a^T} = \begin{bmatrix} R_{i,X_1}W & 0 \\ 0 & R_{i,X_2}W \\ R_{i,X_1}W & R_{i,X_2}W \end{bmatrix} \tag{14.64}$$

$$B_i^{b^T\alpha} = \begin{bmatrix} (R_i\beta_\alpha)_{,X_1} & 0 \\ 0 & (R_i\beta_\alpha)_{,X_2} \\ (R_i\beta_\alpha)_{,X_1} & (R_i\beta_\alpha)_{,X_1} \end{bmatrix} \tag{14.65}$$

$$B_i^{b^T} = \begin{bmatrix} B_i^{b1} & B_i^{b2} & B_i^{b3} & B_i^{b4} \end{bmatrix} \tag{14.66}$$

$$B_i^{c^T} = \begin{bmatrix} (R_i)_{,X_1}\chi & 0 \\ 0 & (R_i)_{,X_2}\chi \\ (R_i)_{,X_1}\chi & (R_i)_{,X_1}\chi \end{bmatrix} \tag{14.67}$$

where $4 = 1, 2, 3$, and 4 $r, s = u, a^u, b^u$, and c^u. Secondly, the isothermal crack is considered and the thermal equation (Eq. 14.28) is discretized as

$$K^{iTh}T^h = f^{ihT} \tag{14.68}$$

the temperature vector in case of an isothermal crack is defined as

$$T^{hi} = \begin{bmatrix} T & b_1^T & c^T \end{bmatrix} \tag{14.69}$$

In the above equation, T presents the standard DOFs of the temperature field, while b_1^T and c^T show the additional DOF in the case of crack tip field and void. Similar to the adiabatic case, K^{iTh} and f^{ihT} represent the elemental stiffness and force matrixes in the case of the isothermal field, which are given as

$$K_{ij}^{iTh} = \begin{bmatrix} K_{ij}^{TT} & K_{ij}^{Tb} & K_{ij}^{Tc} \\ K_{ij}^{bT} & K_{ij}^{bb} & K_{ij}^{bc} \\ K_{ij}^{cT} & K_{ij}^{cb} & K_{ij}^{cc} \end{bmatrix} \tag{14.70}$$

$$f_i^{ihT} = \left[f_i^T \quad f_i^{b_1^T} \quad f_i^{c^T} \right]^T \tag{14.71}$$

The individual terms in the above equations have analogously been written as presented in Eqs. (14.49)–(14.53). So, for the removal of repeatability, these individual terms are discussed here again. Also, the derivative function can be obtained from Eqs. (14.63)–(14.67).

14.11 Computation of stress intensity factors

In this section of the review, discussion on SIF computation with the help of domain-based interaction integrals has been considered. In this domain-based approach, only those field variables are chosen that are not near the vicinity of the crack tip, which makes the approach more accurate for the evaluation of SIFs. Also, it is proven to be more favorable during the evaluation of single SIFs from a solution when only an auxiliary field is judiciously chosen. The domain form of interaction integral in the case of mechanical loading only is derived in [56] and is given as

$$M^{(1,2)} = \int_A \left[\sigma_{ij}^{(1)} \frac{\partial u_i^{(2)}}{\partial x_1} + \sigma_{ij}^{(2)} \frac{\partial u_i^{(1)}}{\partial x_1} - Z^{(1,2)} \delta_{ij} \right] \frac{\partial v}{\partial x_j} dA \tag{14.72}$$

$$Z^{(1,2)} = \frac{1}{2} \left(\sigma_{ij}^{(1)} \epsilon_{ij}^{(2)} + \sigma_{ij}^{(2)} \epsilon_{ij}^{(1)} \right) = \sigma_{ij}^{(1)} \epsilon_{ij}^{(2)} = \sigma_{ij}^{(2)} \epsilon_{ij}^{(1)} \tag{14.73}$$

where "v" shows the scalar weight function (that ranges between 0 and 1); $Z^{(1,2)}$ is mutual strain energy; σ and ϵ represent the stresses and strains, respectively; "1" and "2" represent the actual state and the auxiliary states, respectively

Analogously, interaction integral for TML can be written as per [57]:

$$M^{(1,2)} = \int_A \left[\sigma_{ij}^{(1)} \frac{\partial u_i^{(2)}}{\partial x_1} + \sigma_{ij}^{(2)} \frac{\partial u_i^{(1)}}{\partial x_1} - Z^{(1,2)} \delta_{ij} \right] \frac{\partial v}{\partial x_j} dA + \alpha \int_a \frac{\partial T}{\partial x_1} \sigma_{kk}^{(2)} q dA \tag{14.74}$$

Eq. (14.74) can further be modified for linear elastic problems as

$$M^{(1,2)} = \frac{2}{E^*} \left(K_I^{(1)} K_I^{(2)} + K_{II}^{(1)} K_I^{(2)} \right) \tag{14.75}$$

where $E^* = E$ and $E^* = \frac{E}{(1 - v^2)}$ show the case of plane stress and strain correspondingly. Further, single SIFs under mixed loads can be derived by taking the appropriate choice of auxiliary states. In the case of mode-I loading, the SIFs are derived as $K_I^{(2)} = 1$ and $K_{II}^{(2)} = 0$, and for mode-II as $K_I^{(2)} = 0$ and $K_{II}^{(2)} = 1$.

Further, the crack growth direction θ_c expressed by [58] and is given as

$$\theta_c = 2\tan^{-1}\frac{1}{4}\left(\frac{K_I}{K_{II}} + \sqrt{\left(\frac{K_I}{K_{II}}\right)^2 + 8}\right) \tag{14.76}$$

Thus the effective SIF value can be written as [58]

$$K_{eff} = K_I\cos^2\left(\frac{\theta_c}{2}\right) - 3K_{II}\cos^2\left(\frac{\theta_c}{2}\right)\sin\left(\frac{\theta_c}{2}\right) \tag{14.77}$$

This evaluation of K_{eff} is repeated until the satisfactory convergence is achieved.

14.12 Conclusion

The exposure of various mechanical components to mixed loading conditions results in failures of their functions due to the development of fractures. Thus the analysis of these mechanical components under various mixed loading conditions is very important. This study is dedicated to a discussion on analyzing such components using the numerical technique. Based on the advantages of representing exact geometries, the XIGA technique has been selected, and a brief review on analyzing fractures in elastic materials under TML conditions is done. Also, the utilization of different basis functions like B-splines, NURBS, T-splines, and PHT-splines opens the path for analyzing TML problems very efficiently. The procedure of approximating both thermal and mechanical field variables has been discussed with the evaluation of SIFs based on domain-based interaction integral approach. Also, an approximation of field variables using XIGA under isothermal and adiabatic crack conditions is represented. As a result, it has been evident that the XIGA approach is robust and efficient in analyzing fracture problems, and the proposed procedure can be used for evaluating mixed-mode loading conditions. It also seems that the XIGA approach is quite suitable for analyzing thermomechanical cracked domains. For future work, it is required to explore various local refinement-based XIGA implementations for analyzing fracture problems under coupled mode loading conditions.

List of acronyms

BEM boundary element method
CAD computer-aided designing
CTEF crack tip enrichment function
DOFs degree of freedoms
FEA finite element analysis
FGPs functionally graded plates
IGA isogeometric analysis
NURBS nonuniform rational B-spline
SIFs stress intensity factor
TML thermo–mechanical loading
XIGA extended isogeometric analysis

References

[1] Bouhala L, Makradi A, Belouettar S. Thermal and thermo-mechanical influence on crack propagation using an extended mesh free method. Engineering Fracture Mechanics 2012;88:35–48. Available from: https://doi.org/10.1016/j.engfracmech.2012.04.001.

[2] Pathak H, Singh A, Singh IV. Numerical simulation of thermo-mechanical crack problems using EFGM and XFEM. Paper presented at the 1st National Conference on Recent Advances in Technology and Engineering, Mangalayatan University, Aligarh, Uttar Pradesh; 2012.

[3] Wilson WK, Yu IW. The use of the J-integral in thermal stress crack problems. International Journal of Fracture 1979;15(4):377–87. Available from: https://doi.org/10.1007/BF00033062.

[4] Shih CF, Moran B, Nakamura T. Energy release rate along a three-dimensional crack front in a thermally stressed body. International Journal of Fracture 1986;30(2):79–102. Available from: https://doi.org/10.1007/BF00034019.

[5] Prasad NNV, Aliabadi MH, Rooke DP. The dual boundary element method for thermoelastic crack problems. International Journal of Fracture 1994;66(3):255–72. Available from: https://doi.org/10.1007/BF00042588.

[6] Raveendra ST, Banerjee PK, Dargush GF. Three-dimensional analysis of thermally loaded cracks. International Journal for Numerical Methods in Engineering 1993;36(11):1909–26. Available from: https://doi.org/10.1002/nme.1620361108.

[7] Kanth SA, Shafi Lone A, Harmain GA, Jameel A. Elasto plastic crack growth by XFEM: a review. Materials Today: Proceedings 2019;18:3472–81. Available from: https://doi.org/10.1016/j.matpr.2019.07.275.

[8] Lone AS, Kanth SA, Jameel A, Harmain GA. A state of art review on the modeling of Contact type Nonlinearities by Extended Finite Element method. Materials Today: Proceedings 2019;18:3462–71. Available from: https://doi.org/10.1016/j.matpr.2019.07.274.

[9] Jameel A, Harmain GA. Effect of material irregularities on fatigue crack growth by enriched techniques. International Journal for Computational Methods in Engineering Science and Mechanics 2020;21(3):109–33. Available from: https://doi.org/10.1080/15502287.2020.1772902.

[10] Kanth SA, Harmain GA, Jameel A. Investigation of fatigue crack growth in engineering components containing different types of material irregularities by XFEM. Mechanics of Advanced Materials and Structures 2022;29(24):3570–87. Available from: https://doi.org/10.1080/15376494.2021.1907003.

[11] Lone AS, Ahmad Kanth S, Harmain GA, Jameel A. XFEM modeling of frictional contact between elliptical inclusions and solid bodies. Materials Today: Proceedings 2020;26:819–24. Available from: https://doi.org/10.1016/j.matpr.2019.12.424.

[12] Kanth SA, Lone AS, Harmain GA, Jameel A. Modeling of embedded and edge cracks in steel alloys by XFEM. Materials Today: Proceedings 2020;26:814–18. Available from: https://doi.org/10.1016/j.matpr.2019.12.423.

[13] Jameel A, Harmain GA. Modeling and numerical simulation of fatigue crack growth in cracked specimens containing material discontinuities. Strength of Materials 2016;48(2):294–307. Available from: https://doi.org/10.1007/s11223-016-9765-0.

[14] Jameel A, Harmain GA. A coupled FE-IGA technique for modeling fatigue crack growth in engineering materials. Mechanics of Advanced Materials and Structures 2019;26(21):1764–75. Available from: https://doi.org/10.1080/15376494.2018.1446571.

[15] Jameel A, Harmain GA. Large deformation in bi-material components by XIGA and coupled FE-IGA techniques. Mechanics of Advanced Materials and Structures 2022;29(6):850–72. Available from: https://doi.org/10.1080/15376494.2020.1799120.

[16] Lone AS, Jameel A, Harmain GA. A coupled finite element-element free Galerkin approach for modeling frictional contact in engineering components. Materials Today: Proceedings 2018;5(9, Part 3):18745–54. Available from: https://doi.org/10.1016/j.matpr.2018.06.221.

[17] Dai M-J, Tanaka S, Sadamoto S, Yu T, Bui TQ. Advanced reproducing kernel meshfree modeling of cracked curved shells for mixed-mode stress resultant intensity factors. Engineering Fracture Mechanics 2020;233:107012. Available from: https://doi.org/10.1016/j.engfracmech.2020.107012.

[18] Sheikh UA, Jameel A. Elasto-plastic large deformation analysis of bi-material components by FEM. Materials Today: Proceedings 2020;26:1795–802. Available from: https://doi.org/10.1016/j.matpr.2020.02.377.

[19] Kanth SA, Harmain GA, Jameel A. Modeling of nonlinear crack growth in steel and aluminum alloys by the element free Galerkin method. Materials Today: Proceedings 2018;5(9, Part 3):18805–14. Available from: https://doi.org/10.1016/j.matpr.2018.06.227.

[20] Jameel A, Harmain GA. Fatigue crack growth in presence of material discontinuities by EFGM. International Journal of Fatigue 2015;81:105–16. Available from: https://doi.org/10.1016/j.ijfatigue.2015.07.021.

[21] Harmain GA, Jameel A, Najar FA, Masoodi JH. Large elasto-plastic deformations in bi-material components by coupled FE-EFGM. IOP Conference Series: Materials Science and Engineering 2017;225:012295. Available from: https://doi.org/10.1088/1757-899X/225/1/012295.

[22] Lone AS, Harmain GA, Jameel A. Enriched element free Galerkin method for solving frictional contact between solid bodies. Mechanics of Advanced Materials and Structures 2022;1–19. Available from: https://doi.org/10.1080/15376494.2022.2092791.

[23] Lone AS, Harmain GA, Jameel A. Modeling of contact interfaces by penalty based enriched finite element method. Mechanics of Advanced Materials and Structures 2022;1–19. Available from: https://doi.org/10.1080/15376494.2022.2034075.

[24] Hughes TJR, Cottrell JA, Bazilevs Y. Isogeometric analysis: CAD, finite elements, NURBS, exact geometry and mesh refinement. Computer Methods in Applied Mechanics and Engineering 2005;194(39):4135–95. Available from: https://doi.org/10.1016/j.cma.2004.10.008.

[25] Gupta V, Jameel A, Verma SK, Anand S, Anand Y. An insight on NURBS based isogeometric analysis, its current status and involvement in mechanical applications. Archives of Computational Methods in Engineering 2022;. Available from: https://doi.org/10.1007/s11831-022-09838-0.

[26] Gupta V., Verma S.K., Anand S., Jameel A., Anand Y. Transient isogeometric heat conduction analysis of stationary fluid in a container. Proceedings of the Institution of Mechanical Engineers, Part E: Journal of Process Mechanical Engineering; 2022. 09544089221125718. doi:https://doi.org/10.1177/09544089221125718.

[27] Gupta V, Jameel A, Anand S, Anand Y. Analysis of composite plates using isogeometric analysis: a discussion. Materials Today: Proceedings 2021;44:1190–4. Available from: https://doi.org/10.1016/j.matpr.2020.11.238.

[28] Verhoosel CV, Scott MA, de Borst R, Hughes TJR. An isogeometric approach to cohesive zone modeling. International Journal for Numerical Methods in Engineering 2011;87(1–5):336–60. Available from: https://doi.org/10.1002/nme.3061.

[29] Choi M-J, Cho S. Isogeometric analysis of stress intensity factors for curved crack problems. Theoretical and Applied Fracture Mechanics 2015;75:89–103. Available from: https://doi.org/10.1016/j.tafmec.2014.11.003.

[30] Khademalrasoul A. NURBS-based isogeometric analysis method application to mixed-mode computational fracture mechanics. Journal of Applied and Computational Mechanics 2019;5(2):217–30. Available from: https://doi.org/10.22055/jacm.2018.25429.1265.

[31] Ghorashi SS, Valizadeh N, Mohammadi S. Extended isogeometric analysis for simulation of stationary and propagating cracks. International Journal for Numerical Methods in Engineering 2012;89(9):1069–101. Available from: https://doi.org/10.1002/nme.3277.

[32] De Luycker E, Benson DJ, Belytschko T, Bazilevs Y, Hsu MC. X-FEM in isogeometric analysis for linear fracture mechanics. International Journal for Numerical Methods in Engineering 2011;87(6):541–65. Available from: https://doi.org/10.1002/nme.3121.

[33] Shojaee S, Asgharzadeh M, Haeri A. Crack analysis in orthotropic media using combination of isogeometric analysis and extended finite element. International Journal of Applied Mechanics 2014;06(06):1450068. Available from: https://doi.org/10.1142/S1758825114500689.

[34] Bui TQ. Extended isogeometric dynamic and static fracture analysis for cracks in piezoelectric materials using NURBS. Computer Methods in Applied Mechanics and Engineering 2015;295:470–509. Available from: https://doi.org/10.1016/j.cma.2015.07.005.

[35] Bhardwaj G, Singh IV, Mishra BK, Bui TQ. Numerical simulation of functionally graded cracked plates using NURBS based XIGA under different loads and boundary conditions. Composite Structures 2015;126:347–59. Available from: https://doi.org/10.1016/j.compstruct.2015.02.066.

[36] Yu T, Bui TQ, Yin S, Doan DH, Wu CT, Do TV, et al. On the thermal buckling analysis of functionally graded plates with internal defects using extended isogeometric analysis. Composite Structures 2016;136:684–95. Available from: https://doi.org/10.1016/j.compstruct.2015.11.002.

[37] Singh SK, Singh IV, Mishra BK, Bhardwaj G, Singh SK. Analysis of cracked plate using higher-order shear deformation theory: asymptotic crack-tip fields and XIGA implementation. Computer Methods in Applied Mechanics and Engineering 2018;336:594–639. Available from: https://doi.org/10.1016/j.cma.2018.03.009.

[38] Habib SH, Belaidi I. Crack analysis in bimaterial interfaces using T-spline based XIGA. Journal of Theoretical and Applied Mechanics 2017;55(1):55–68. Available from: https://doi.org/10.15632/jtam-pl.55.1.55.

[39] Singh SK, Singh IV, Mishra BK, Bhardwaj G, Bui TQ. A simple, efficient and accurate Bézier extraction based T-spline XIGA for crack simulations. Theoretical and Applied Fracture Mechanics 2017;88:74–96. Available from: https://doi.org/10.1016/j.tafmec.2016.12.002.

[40] Yadav A, Godara RK, Bhardwaj G. A review on XIGA method for computational fracture mechanics applications. Engineering Fracture Mechanics 2020;230:107001. Available from: https://doi.org/10.1016/j.engfracmech.2020.107001.

[41] Montassir S, Moustabchir H, Elkhalfi A, Scutaru ML, Vlase S. Fracture modelling of a cracked pressurized cylindrical structure by using extended iso-geometric analysis (X-IGA). Mathematics 2021;9(23). Available from: https://doi.org/10.3390/math9232990.

[42] Kaushik V, Ghosh A. Experimental and XIGA-CZM based Mode-II and mixed-mode interlaminar fracture model for unidirectional aerospace-grade composites. Mechanics of Materials 2021;154:103722. Available from: https://doi.org/10.1016/j.mechmat.2020.103722.

[43] Bhardwaj G, Singh SK, Patil RU, Godara RK, Khanna K. Thermo-elastic analysis of cracked functionally graded materials using XIGA. Theoretical and Applied Fracture Mechanics 2021;114:103016. Available from: https://doi.org/10.1016/j.tafmec.2021.103016.

[44] Singh AK, Jameel A, Harmain GA. Investigations on crack tip plastic zones by the extended iso-geometric analysis. Materials Today: Proceedings 2018;5(9, Part 3):19284–93. Available from: https://doi.org/10.1016/j.matpr.2018.06.287.

[45] Jameel A, Harmain GA. Extended iso-geometric analysis for modeling three-dimensional cracks. Mechanics of Advanced Materials and Structures 2019;26(11):915–23. Available from: https://doi.org/10.1080/15376494.2018.1430275.

[46] Bhardwaj G, Godara RK, Khanna K, Patil RU. A semi-homogenized extended isogeometric analysis approach for fracture in functionally graded materials containing discontinuities. Proceedings of the Institution of Mechanical Engineers, Part C: Journal of Mechanical Engineering Science 2020;234(11):2211–32. Available from: https://doi.org/10.1177/0954406220905857.

[47] Hou W, Jiang K, Zhu X, Shen Y, Li Y, Zhang X, et al. Extended Isogeometric Analysis with strong imposing essential boundary conditions for weak discontinuous problems using B++ splines. Computer Methods in Applied Mechanics and Engineering 2020;370:113135. Available from: https://doi.org/10.1016/j.cma.2020.113135.

[48] Piegl L, Tiller W. The NURBS book, 6. *Springer Science & Business Media*; 2012. p. 34–9.

[49] Nguyen-Thanh N, Zhou K. Extended isogeometric analysis based on PHT-splines for crack propagation near inclusions. International Journal for Numerical Methods in Engineering 2017;112(12):1777–800. Available from: https://doi.org/10.1002/nme.5581.

[50] Nguyen-Thanh N, Nguyen-Xuan H, Bordas SPA, Rabczuk T. Isogeometric analysis using polynomial splines over hierarchical T-meshes for two-dimensional elastic solids. Computer Methods in Applied Mechanics and Engineering 2011;200(21):1892–908. Available from: https://doi.org/10.1016/j.cma.2011.01.018.

[51] Duflot M. The extended finite element method in thermoelastic fracture mechanics. International Journal for Numerical Methods in Engineering 2008;74(5):827–47. Available from: https://doi.org/10.1002/nme.2197.

[52] Bhardwaj G, Singh IV. Fatigue crack growth analysis of a homogeneous plate in the presence of multiple defects using extended isogeometric analysis. Journal of the Brazilian Society of Mechanical Sciences and Engineering 2015;37(4):1065–82. Available from: https://doi.org/10.1007/s40430-014-0232-1.

[53] Chopp DL, Sukumar N. Fatigue crack propagation of multiple coplanar cracks with the coupled extended finite element/fast marching method. International Journal of Engineering Science 2003;41(8):845–69. Available from: https://doi.org/10.1016/S0020-7225(02)00322-1.

[54] Sih GC. On the singular character of thermal stresses near a crack tip. J Journal of Applied Mechanics 1962;29:587. Available from: https://doi.org/10.1115/1.3640612.

[55] Bayesteh H, Afshar A, Mohammdi S. Thermo-mechanical fracture study of inhomogeneous cracked solids by the extended isogeometric analysis method. European Journal of Mechanics - A/Solids 2015;51:123–39. Available from: https://doi.org/10.1016/j.euromechsol.2014.12.004.

[56] Singh IV, Mishra BK, Bhattacharya S, Patil RU. The numerical simulation of fatigue crack growth using extended finite element method. International Journal of Fatigue 2012;36(1):109–19. Available from: https://doi.org/10.1016/j.ijfatigue.2011.08.010.

[57] Banks-Sills L, Dolev O. The conservative M-integral for thermal-elastic problems. International Journal of Fracture 2004;125(1):149–70. Available from: https://doi.org/10.1023/B:FRAC.0000021065.46630.4d.

[58] Erdogan F, Sih GC. On the crack extension in plates under plane loading and transverse shear. Journal of Basic Engineering 1963;85(4):519–25. Available from: https://doi.org/10.1115/1.3656897.

CHAPTER

15

Extended isogeometric analysis for thermal absorber coatings

Sahil Thappa[1,2], Shubham Kumar Verma[2,3], Vibhushit Gupta[2], Sanjay Sharma[2], Yatheshth Anand[2] and Sanjeev Anand[2]

[1]Department of Mechanical Engineering, Maharishi Markandeshwar (Deemed to be University), Ambala, Haryana, India [2]School of Energy Management, Shri Mata Vaishno Devi University, Kakryal, Katra, Jammu and Kashmir, India [3]Built Environment Lab, Indian Institute of Technology, Gandhinagar, Gujarat, India

15.1 Introduction

This investigation presents the extended isogeometric analysis (IGA) (XIGA) for a thermal absorber coating (TAC). Actually, XIGA is an enrichment technique that enables us to investigate surfaces for possible cracks and discontinuities occurring due to continuous thermal loads. These thermal absorber surfaces are exposed to high temperatures, and it is necessary to entrap maximum available heat for the system to be efficient. To maximize the surface heat absorption, TACs are applied to the absorber surface. These coatings can enhance the efficiency of the system effectively. Many thermal absorptive coatings suitable for the purpose have been developed in recent decades, as cited in literature [1]. A study was carried out by [2], where the thermal barrier coating was analyzed for various discontinuities. The researchers tried to predict the crack propagation behavior of the coating by using an extended finite element model (XFEM). Results revealed that the unmelted nanoparticles (UNPs) accumulated the tensile stress within them. Materials with better adhesive qualities and suitable wear strength have been coined by various researchers. As per as thermal absorption is concerned, the heat involved in the process makes the coating surfaces transform their embedded properties like good adhesion, corrosion resistance, erosion resistance, and thermal stability. And cumulatively, all these factors may give rise to problems like crack/fracture propagation on the coating surface [3,4]. It is important to understand these cracks/fractures formed due to mechanical as well as thermal loading conditions. To ensure the reliability of the system,

Enriched Numerical Techniques
DOI: https://doi.org/10.1016/B978-0-443-15362-4.00018-8

discontinuities like cracks/fractures need to be minimized. Additionally, the stress intensity factor (SIF) evaluation is also necessary and plays a vital role in safety assessment for the system [5–7]. These methods are used to get precise data regarding the fractures/cracks, material properties, and discontinuities. Researchers used a number of techniques in past few years for simulation under actual/realistic environmental loading conditions. Finite element analysis (FEA) is one of the techniques to analyze SIFs in a component when it is cracked/fractured [8,9]. FEA is found useful in the case of thermo-elastic loading conditions as well as when fractures and discontinuities occur in a coated surface. The boundary element method (BEM) is also found useful for analyzing thermo-mechanical discontinuities in 2D [4,7]. These techniques have some limitations and problems like conformal mesh, problem in remeshing, and using a singular tip. Some techniques that can overcome these limitations have been developed and suggested in literature. Some of the suggested techniques are named "kernel particle reproducing technique, extended finite element technique, and element-free Galerkin technique." These techniques have proven advantageous in handling the propagating discontinuities like cracks [10–12]. Recently, the immersed boundary techniques have been very popular, which give a suitable solution to the multimaterial problems related to crack propagation. These methods can accommodate complex domains in terms of boundaries and interfaces without any need to construct a body-fitted conforming mesh. The very first method of immersed boundary was reportedly known to be introduced by Peskin [13]. The immersed finite element approach includes: fictitious domain method or embedded domain method, which helps to empty space/void in the conformal meshes by inserting the calculative or computational domain and applying specific integration techniques with Lagrange multipliers approach suggested by [14] with [15,16] suggested Nitsche's method [17–19]. Actually, enrichment techniques help in generating conforming approximation meshes by encapsulating the discontinuous behaviors within the formed mesh structure. Although the originally developed approaches represent moving fronts and crack propagation, they have been extended to tackle various types of problems at the interface with strong as well as weak discontinuities. However, among these suggested techniques, noticeable techniques are the partition of unity method (written as PUM in short form) developed by Babuška and Melenk [20], the generalized extended finite element method (written as GFEM) as introduced by Strouboulis et al. [21,22], the extended (X) finite element method (written as XFEM) as proposed by Moes et al. [23,24], and the IGFEM (elaboratively written as interface-enriched generalized finite element method) introduced by [25]. These are a few methods/techniques that make the crack propagation easy to analyze.

In past few decades, protective properties of certain mixtures or individual materials, such as TiO_2, SiO_2, and Al_2O_3, have been studied, and the electrochemical calculations were the basis of the calculation of protective properties of these coatings. These materials possess high chemical and thermal stability, which makes them suitable for the purpose, but sooner, smaller defects (cracks) appear in the coating/sol–gel or applied layer, and then these coatings are no longer capable of protecting the surface from the degraded patch/area. These degradations are not easy to avoid and can result in energy losses. These coatings provide the additional properties, such as higher absorptivity, better chemical and thermal stability, along with wear resistance, and reduced emissivity from the absorber surfaces. In other words, these coatings enable us to impart some desired properties to the receiver tube surface so that it absorbs maximum possible thermal energy. These coatings have been explored for better adhesion, absorptivity, and thermal stability.

A huge number of materials used as TACs are cited in literature, which show good adhesion properties along with improved absorptivity and thermal stability. It has been reported that *Pyromark 2500* (black) is found to have an absorptivity of about 0.95 when applied to the absorber surface in a PTC system [26]. The coating shows reliable results below 350°C and degrades beyond that. Minor cracks/fractures appear on the surface, which cause energy losses from the cracks. Black paints are also reported to have better absorptive capabilities, but they also offer emissivity at higher degree of temperatures. Common degradations like corrosion and abrasion at high temperatures are found in thermal absorptive coatings. Gupta et al. used the high-velocity oxygen fuel (HVOF) process to develop Al_2O_3-based coatings on stainless-steel substrates [3]. These coatings are actually composite coatings. The coating material chosen comprises Al_2O_3-cerium, while the reinforcement added to it is hexagonal boron nitride. The metallurgic features of the coatings revealed that mechanical properties of the coating material are ascribed to basic structural changes. And similarly, the erosive wear resistance of the hybrid coating. Titanium diboride, along with aluminum oxide coating and nickel–aluminum coating, are suitable TACs with improved absorptivity and thermal stability [27,28]. Some other metal and metal-oxide coatings, including Al/Si/CuO, Cu/Si/Al_2O_3, Al/Si/TiO_2, Mo-Si_3N_4, and $MoSi_2$—Si_3N_4, offer absorption ranging from 0.66 to 0.77 with 0.23 to 0.28 emissivity. Similarly, some rare-earth metals were also studied for their impact on metallurgical properties of coatings when added to metals. In a similar investigation, theoretical and experimental studies were carried out by [29] for an antireflection coating deposited onto silicon wafers. The coating, comprising magnesium fluoride with cerium oxide, was developed using the physical vapor deposition (PVD) method. MgF_2–CeO_2 structures exhibited the reflectivity below 3% in the wavelength window from 0.5 μm (540 nm) to 1.2 μm (1220 nm). The thin film exhibited minimal cracks over the surface as the lateral binding capabilities of cerium bound the particles in a better manner. Still, the surface failure is a major problem in the case of TACs.

The absorptivity, thermal stability, and reliability of a coating depend on its thickness; therefore these coatings require high-quality deposition and controlled coating thickness, along with better adhesion. From this section, it can be clearly noted that using appropriate material is important for better performance of the system. Thermal absorptive coatings need to absorb maximum available heat coming from the heat source. If the coating has antireflective properties, it will enhance the absorptivity and reduce the emissivity of the surface. The antireflective surfaces do offer reduced wear because of their matte finish The stress concentration over a coating particle is equally distributed in all the neighboring particles in all directions. So, it can be concluded that the energy efficiency of a parabolic trough collector (PTC) system highly depends on the operating temperature of the system, and it can be increased by the application of TACs.

The coated surfaces, when exposed to heat for an extended time period, undergo deformations like surface cracks or fractures in the layer. It is obvious to mitigate these kinds of defects while developing a coating by adding some better binding substances that make the surface thermally stable for elevated temperatures, but this cannot be the case every time. So, it becomes necessary to analyze the propagation of these deformations to understand the nature of the break point of a substance. So, enrichment techniques are key to understand these defects and deformations in a precise and accurate way. This study is carried out to analyze the deformations formed on the tungsten carbide, cobalt, and chromium (WC–Co–Cr) coating surface over a stainless-steel substrate.

The upcoming sections explain how the numerical techniques, especially numerical enrichment techniques, can help in predicting the fatigue or crack behavior of the coating.

The upcoming sections also present the materials and methods used for developing the TAC over a stainless-steel substrate, followed by results and discussions, which details about the erosion test carried out for surface degradation. It also explains the scanning electron microscopy (SEM) carried out to understand the surface cracks and discontinuities. The enrichment technique, that is, XIGA, is also well discussed based on past research carried out by various researchers. Further, the conclusive remarks about the study are presented, which prove that XIGA is a helpful tool to predict such types of discontinuities.

15.2 Materials and methods

Since the objective of this study is to analyze the cracks in coating, the coated surface is produced by a thermal spray of coating over the substrate. TAC applicable to a typical PTC receiver tube is developed over the stainless-steel substrates. The coating, as stated earlier, consists of tungsten carbide along with cobalt and chromium (WC–Co–Cr). These coatings are done over $25 \times 25 \times 4\ mm^3$ stainless-steel (SS-304) substrate. First of all, SS-304 has been cut into the said size with the help of a laser cutting machine. Now tungsten carbide, cobalt, and chromium (WC–Co–Cr) are put into the ball milling chamber. The ratio of WC–Co–Cr is 84, 12, and 4 wt.%, respectively. Also, the powder is of a fine grit size, precisely saying less than $50\ \mu m$. Now the powder is ball milled in the ball milling machine so that the mixture becomes homogeneous in nature. The ball milling method causes all the ingredient particles to alloy with tungsten carbide. The alloying process is necessary for the even distribution and morphological symmetry of the coating mixture over the substrate [30]. The alloying of ball-milled powder is visible in Fig. 15.1, which has been snapped through SEM.

Now, the ball-milled powder is fed into a HVOF machine. It pressurizes the feed, followed by heating to form a fluid ready for coating. The SS substrate is cleaned and placed in the target area, and the fluid is sprayed over it using a thermal spray gun. This gun produces a high-velocity flame spray as the coating fluid mixed with oxygen, air, and nitrogen and the kerosene oil ignites the fire and creates a fusible fluid that sticks to the surface of the substrate. Fig. 15.2 shows the coated surface magnified to high resolution using a scanning electron spectroscopic machine.

When the coating was placed in an erosion test setup, it depicted few particles erode from the surface, which can be seen in Fig. 15.3, while Fig. 15.4 shows the crack at the edge of the coated substrate.

15.2.1 Need for enrichment techniques

The reason these discontinuities occurred in the coating is because the coated surface is not able to withstand prolonged thermal stresses impounded over it. These thermal stresses increase with the increasing time for which the thermal load or stress is applied. So, actually, these cracks and discontinuities can occur when the surface of TAC is subjected to a prolonged heating over elevated temperature. Although the advancements like the

FIGURE 15.1 Ball-milled coating powder.

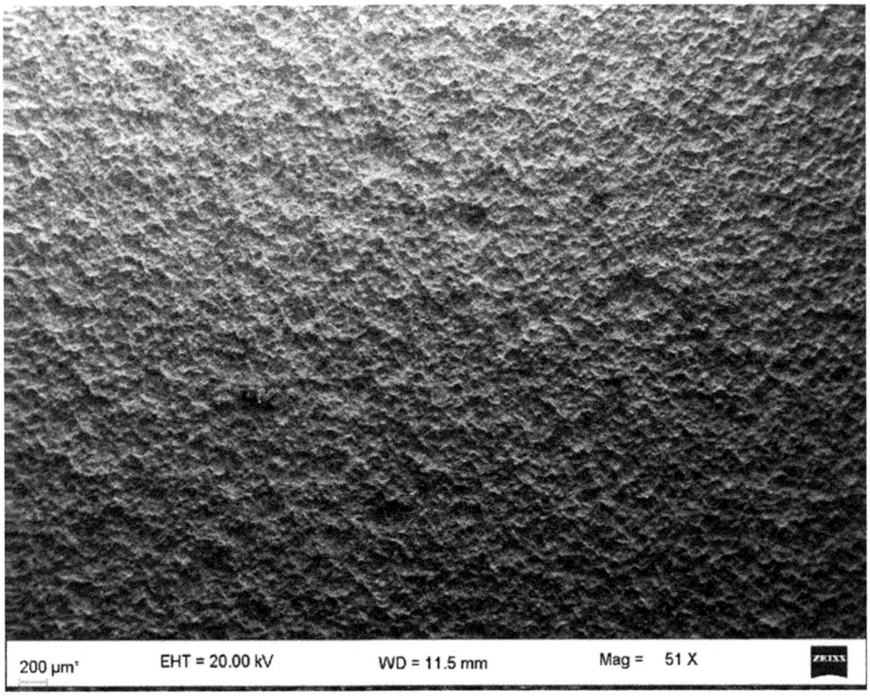

FIGURE 15.2 Coating sprayed over the substrate.

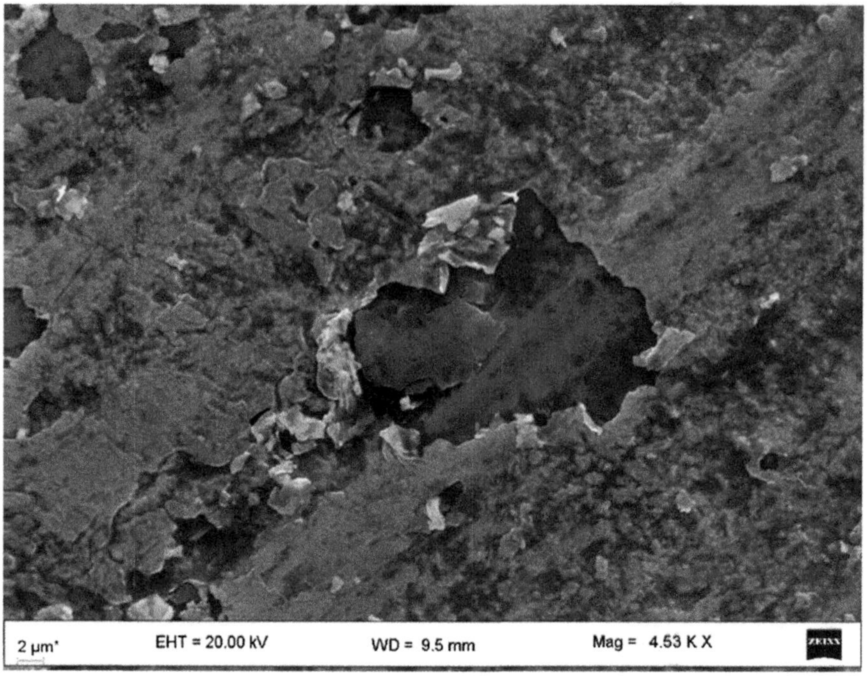

FIGURE 15.3 Eroded coating surface.

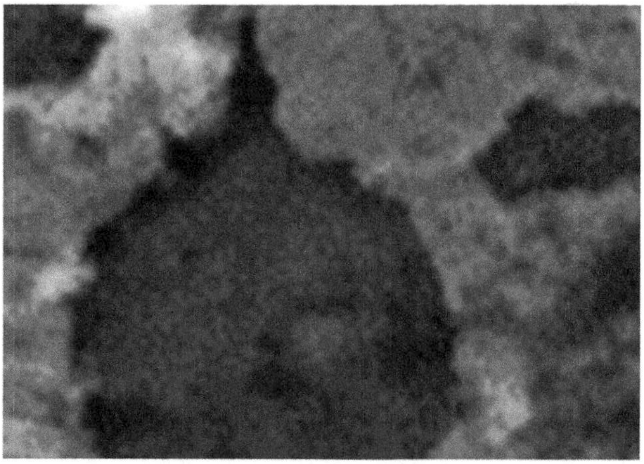

FIGURE 15.4 Crack on the coating surface.

addition of some suitable binding material or substance are already proposed in literature, still the surfaces undergo through such defects and discontinuities. These discontinuities over the surface need to be understood as if the failure of a surface is not known, it may result in some catastrophic situations while working on a system. So, numerical techniques came into picture that are capable of predicting such defects and cracks precisely. Now, the literature suggests that the numerical techniques were reported to be lacking in the computational efforts induced to check a discontinuity of the surface. It was reported that the numerical techniques found it tough to analyze complex geometries, which generated requirement for some advanced enrichment technique that could overcome the shortcomings that numerical techniques possessed. Enrichment techniques are capable of analyzing complex geometries with less computational efforts made. Literature has reported a few techniques in this field that fall under the enrichment technique category, and this study presents the use of one of the enrichment techniques, namely, XIGA, which stands for XIGA. This is an advanced numerical technique, also called an enrichment technique, which can help predict the failure of surface coatings. XIGA is detailed with suitable literature in the next section followed by B-spline curves and then the techniques to evaluate the SIFs.

15.2.2 Isogeometric analysis

The IGA approach utilizes a computer-aided designing (CAD) basis function in order to approximate the geometries and prime variables. It results in the mitigation of geometrical discretization errors [31]. B-spline as well as basis functions of nonuniform rational B-spline (NURBS)-based have found rigorous or intensive application in the IGA for the approximating geometry and unknown field variables because the continuity offered by these functions is of quite higher order. Thus a brief discussion on these basis functions is provided in this study. It was checked for a detailed explanation of these basis functions. Ref. [32] reported that they developed a NURBS-based multipatch IGA technique, and this technique was used to analyze cracks of different geometries. The geometries also included some irregular plane geometries like parabolic shapes, crests and troughs, and circular shapes. The knot insertion technique was used to define exactly the cracked zones present in the area considered for investigation without making any type of changes in the real or initial form of geometry. Similarly, Khademalrsoul [33] made use of the same method to analyze the cracks at different inclinations. Mesh refinement techniques and their different sets were considered in the approach. It reported relatively better results in case when fine meshes were scaled or compared to other coarser mesh forms. Verhoosel et al. [34] suggested a NURBS-based approach to analyze crack growth problems that were cohesive in nature. These cases, comprising discontinuity, are to be inserted directly in the solution using the method of knot insertion by modeling the interface. It is also depicted from the literature that, due to the duplication of knot points in the cracked region, the requirements for the continuity were not met across the crack interfaces. However, the solution to that is to use the higher order basis functions, which will eliminate such issues. Later in the continuity to the research, the extended version of IGA was called XIGA [35]. Various researchers, like [36,37] Shojaee et al. [38], analyzed cracks and static fracture problems using IGA and XFEM for getting accurate and robust solutions. Another research study was reported by [39] in which they simulated the cracks using a NURBS-based XIGA approach. In this investigation, various loading conditions were analyzed on the surface by implementing first-order shear deformation theory, and the reported results were quite promising.

Similarly, [40,41] implemented XIGA to model and analyze cracks in an isotropic and homogeneous plate using Reddy's higher order formulations to analyze the cracks present through the thickness of the plates. Moreover, these studies stated that the analysis in XIGA is NURBS, which has limitations due to its tensor product, which has been employed on the basis function. Similarly, local refinement strategy is another approach that includes Bezier extraction in XIGA, and it was developed by Singh et al. [42]. A coated surface having discontinuities and fractures (cracks), if further provided with thermal load and stress, will cause the crack to propagate. This type of problem causes financial as well as energy losses. So, it becomes important to analyze these fractures to ensure the reliability and life of the TAC under conditions like mechanical, thermal, chemical, and electrical. The direct exposure of these coatings in absorber surface applications leads to failure due to the presence of discontinuities and cracks. These techniques help to evaluate the SIFs, which play a vital role in the reliability and longevity of the coating surface. These methods/techniques provide a detailed monitoring of the discontinuities, which include thermal and mechanical loading conditions, properties of the materials, and the uncertainties that crack geometry inherits, literature suggests that in past few years, researchers have used to simulate the actual geometries under realistic loading conditions. FEA is one of the most famous techniques used to evaluate SIFs of a cracked surface or a structure subjected to thermo-elastic load condition [43,44]. The BEM is another technique used for various investigations carried out to analyze the thermo-mechanical fracture problems in 2D space [45,46]. The fracture problem analysis by using FEA and the BEM has been proven to be a beneficial technique, provided it has limitations like the use of singular tips, conformal mesh, and remeshing while simulating the cracked structures or surfaces. Literature also suggests and details the solution to tackle these limitations, which includes examples like extended finite element, kernel particle reproducing technique, and element-free Galerkin technique. These developed techniques benefited in understanding and handling the propagating and moving discontinuities like crack and fracture propagations. Meanwhile, the geometry approximation explains discretization errors in the solution fields. This is possible because of the condition that basis function used for describing the geometry and the unknown field variables are different. [31] developed a technique named IGA, a novel concept only for resolving such discretization issues. In IGA, the standard basis function is used to describe the geometry and find the unknown field variables. This makes IGA a suitable and precise approach for solving various crack and discontinuity problems. In literature, it has been mentioned that IGA proves to be an efficient, reliable, and precise numerical tool for deducing various types of crack growth problems. Recently, the popularity of immersed boundary techniques has increased, and it gives solutions to the multimaterial problems, which are also characterized above. These methods can accommodate complex domains in terms of interfaces or, we can say, boundaries without need to develop or construct a body-fitted conforming mesh.

It is also utilized for the analysis of discontinuities present in structures more accurately by employing local refinement in the domain. XIGA can be applied and has the potential to solve the complex fracture cases like elasto-plastic growth of the crack and behavior of cracks when put under dynamic loading conditions. In similar kind of research, Montassir et al. [47] and Kaushik and Ghosh [48] applied XIGA in modeling crack growth in composite structures. The literature discussed above is enough to explain the applicability of the numerical techniques in the wide range of fracture problems. However, a limited amount of work has been reported for analyzing thermal absorptive coatings (TACs) under thermal and mechanical loading conditions. The present study is an effort to analyze fracture problems subjected to TACs by implementing XIGA approach.

15.2.3 B-splines

The piecewise approximated polynomials generated by a linear combination of basis function and control points are defined as B-splines. These control points are also referred to as vector-valued coefficients (vectors are broadly classified as uniform and nonuniform). Since the B-spline is formed by joining various knots, the exactness and accuracy of the B-spline curve depend on these knots. This section details the NURBS formulation. B-spline functions show linear independency as well as nonnegative $N_{i,p} \geq \forall \xi$ features.

For $p = 0$

$$N_{i,0}(\xi) = \begin{cases} 1 & if\ \xi_i \leq \xi_{i+1} \\ 0 & otherwise \end{cases} \tag{15.1}$$

$$N_{i,p}(\xi) = \frac{\xi - \xi_i}{\xi_{i+p} - \xi_i} N_{i,p-1}(\xi) + \frac{\xi_{i+p+1} - \xi}{\xi_{i+p+1} - \xi_{i+1}} N_{i+1,p-1}(\xi) \tag{15.2}$$

It is evident from the literature that detailed formulations are already available and reported through many studies [49]. Firstly, a NURBS surface, **S** (ξ, η), in cases of order p in ξ-direction and order q in η-direction, can be depicted as

$$R_i^p(\xi, \eta) = \frac{N_{i,p}(\xi) \times M_{j,q}(\eta) w_{i,j}}{\sum_{i=1}^{n} \sum_{j=1}^{m} N_{i,p}(\xi) \times M_{j,q}(\eta) w_{i,j}} \tag{15.3}$$

where w_i is the weight function associated with the control point vector C_i. Similarly for bivariate basis functions, the function will be given as

$$R_i^p(\xi, \eta) = \frac{N_{i,p}(\xi) \times M_{j,q}(\eta) w_{i,j}}{\sum_{i=1}^{n} \sum_{j=1}^{m} N_{i,p}(\xi) \times M_{j,q}(\eta) w_{i,j}} \tag{15.4}$$

And for trivariate basis functions, the function will be given as:

$$R_i^p(\xi, \eta, \zeta) = \frac{N_{i,p}(\xi) \times M_{j,q}(\eta) \times L_{k,r}(\zeta) w_{i,j,k}}{\sum_{i=1}^{n} \sum_{j=1}^{m} \sum_{k=1}^{l} N_{i,p}(\xi) \times M_{j,q}(\eta) \times L_{k,r}(\zeta) w_{i,j,k}} \tag{15.5}$$

Above-given equations show $N_{i,p}(\xi)$, $M_{j,q}(\eta)$, and $L_{k,r}(\zeta)$. These are the B-spline basis functions, which have orders p, q, and r, respectively. The way B-spline geometries are developed remains the same, and NURBS-based geometries are formulated in the same manner. Following equations provide for one-, two-, and three-dimensional geometries:

Case I: One-dimensional geometries

$$W(\xi) = \sum_{i=1}^{n} R_i^p(\xi) C_i \tag{15.6}$$

Case II: Two-dimensional geometries

$$V(\xi,\eta) = \sum_{i=1}^{n}\sum_{j=1}^{m} R_i^p(\xi,\eta)C_{i,j} \tag{15.7}$$

Case III: Three-dimensional geometries

$$U(\xi,\eta,\zeta) = \sum_{i=1}^{n}\sum_{j=1}^{m}\sum_{k=1}^{l} R_i^p(\xi,\eta,\zeta)C_{i,j,k} \tag{15.8}$$

15.2.4 Crack formulation and extended isogeometric analysis

This section defines the problem formulated based on mixed loading, that is, thermal and mechanical, in the XIGA framework. Since the TACs are subjected to cyclic heating process, this will cause the cracks to propagate (as shown in Fig. 15.5) in multidimensional planes. To understand these cracks, an approximation of temperature and displacement field variables is done. This investigation has led to the analysis of two types **of** cracks one is isothermal while other is adiabatic cracks.

The domain (Ω) when subjected to mechanical loading, the solution needs to satisfy the following equilibrium conditions

$$\nabla \cdot \sigma + f = 0 \tag{15.9}$$

where σ, ∇ and f presents a Cauchy tensor product, divergence operator and body force vector respectively.

The boundary conditions (BC) for the domain under consideration are given as:

$$u = \overline{u} \ on \ \Gamma_u \tag{15.10}$$

$$\sigma.\hat{n} = 0 \ on \ \Gamma_c \tag{15.11}$$

$$\sigma.\hat{n} = \hat{t} \ on \ \Gamma_t \tag{15.12}$$

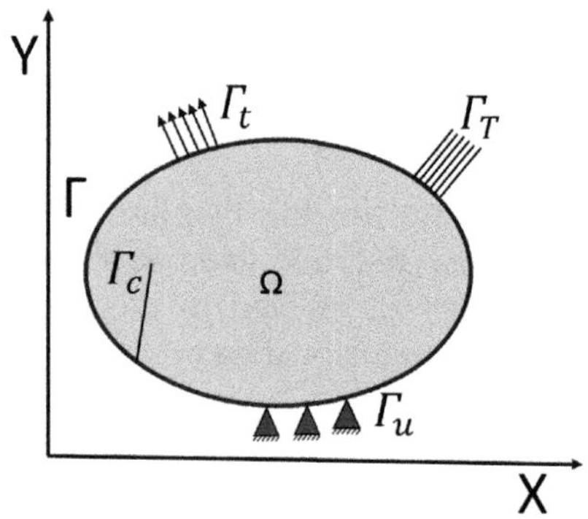

FIGURE 15.5 Problem domain.

Above equations present the mechanical equilibrium and show Hooke's law relations. Thus the constitutive equation in the case of homogeneous elastic material can be derived as:

$$\sigma = C \cdot \epsilon \tag{15.13}$$

where elasticity and strain tensor is presented by C and ϵ respectively.

Based on this, the weak form equation as per [23] can be expressed as:

$$\int_{\Omega} \sigma(u) \colon \epsilon(u) d\Omega = \int_{\Omega} b \colon u d\Omega + \int_{\Gamma} t \colon u d\Gamma \tag{15.14}$$

By putting the test and trial functions and utilizing the arbitrariness of the varied control point values, the discrete system of equations formed as follows:

$$[K]\{d\} = \{f\} \tag{15.15}$$

f, d&K show the external force vector, displacement vector and global stiffness matrix respectively. While the approximation of XIGA is done in a similar manner to XFEM in which a local enrichment approach is utilized for describing cracks independent of the element grid. Thus the approximation of displacement by using the XIGA approach is expressed as follows:

$$u^h(\xi) = \sum_{i=1}^{n_{be}} R_i^p(\xi) u_i + \sum_{j=1}^{n_s} R_j^p(\xi) [W(\xi) - W(\xi_i)] a_j + \sum_{k=1}^{n_T} R_k^p(\xi) \sum_{\alpha=1}^{4} [\beta_\alpha(\xi) - \beta_\alpha(\xi_i)] b_k^\alpha \tag{15.16}$$

where

u_i = specified control points,
n_{be} = shape functions per element;
R_i^p is the NURBS basis function,
n_s = split enriched node elements;
n_T = tip enriched node elements;
a_j = enriched DOF of split elements related to Heaviside function (HF) $W(\xi)$;
b_k^α = enriched degree of freedom of tip elements relates to crack tip enrichment function (CTEF) $\beta_\alpha(\xi)$; $\beta_\alpha(\xi)$ and $W(\xi)$ presents the CTEF and HF respectively. The main purpose of HF $W(\xi)$ is to produce discontinuity across the surface of the crack, and it ranges from -1 to $+1$, which can be understood as

$$W(\xi) = \begin{cases} -1, & | \, otherwise \\ +1, & | \, if \, (\xi - \xi_i) \cdot n_{be} \geq 0 \end{cases} \tag{15.17}$$

however, CTEF is used for the enrichment of tip elements of the surface. This function is defined by [50], and mathematically, it is explained as follows:

$$\beta_\alpha(\xi) \left[\sqrt{r} \cos\frac{\theta}{2}, \sqrt{r} \sin\frac{\theta}{2}, \sqrt{r} \cos\frac{\theta}{2} \sin\theta, \sqrt{r} \sin\frac{\theta}{2} \sin\theta \right] \tag{15.18}$$

where θ and r present polar coordinates at the reference line of the crack and the crack surface. The element matrices of K and f can be obtained by utilizing the approximation function $u^h(\xi)$
is given as:

$$K^{hU} = \begin{bmatrix} K_{ij}^{uu} & K_{ij}^{ua^u} & K_{ij}^{ub^u} \\ K_{ij}^{a^u u} & K_{ij}^{a^u a^u} & K_{ij}^{a^u b^u} \\ K_{ij}^{b^u u} & K_{ij}^{b^u a^u} & K_{ij}^{b^u b^u} \end{bmatrix} \tag{15.19}$$

$$f^{hU} = \begin{bmatrix} f_i^u & f_i^{a^u} & f_i^{b_1^u} & f_i^{b_2^u} & f_i^{b_3^u} & f_i^{b_4^u} \end{bmatrix}^T \tag{15.20}$$

$$K_{ij}^{rs} = \int_{\Omega^e} (B_i^r)^T C B_j^s d\Omega \tag{15.21}$$

$$f_i^u = \int_{\Omega^e} (R_i)^T b d\Omega + \int_{\Gamma_t} (R_i)^T \hat{t} d\Gamma \tag{15.22}$$

$$f_i^{a^u} = \int_{\Omega^e} (R_i)^T H b d\Omega + \int_{\Gamma_t} (R_i)^T H \hat{t} d\Gamma \tag{15.23}$$

$$f_i^{b^u} = \int_{\Omega^e} (R_i)^T \beta_\alpha b d\Omega + \int_{\Gamma_t} (R_i)^T \beta_\alpha \hat{t} d\Gamma \tag{15.24}$$

In these equations, R_i is represented as the NURBS basis function, and B_i^u, $B_i^{a^u}$, $B_i^{b_\alpha^u}$, $B_i^{b^u}$, and $B_i^{c^u}$ are the derivative matrices of R_i that are given as

$$B_i^u = \begin{bmatrix} R_{i,X_1} & 0 \\ 0 & R_{i,X_2} \\ R_{i,X_1} & R_{i,X_2} \end{bmatrix} \tag{15.25}$$

$$B_i^{a^u} = \begin{bmatrix} R_{i,X_1} W & 0 \\ 0 & R_{i,X_2} W \\ R_{i,X_1} W & R_{i,X_2} W \end{bmatrix} \tag{15.26}$$

$$B_i^{b\alpha} = \begin{bmatrix} (R_i\beta_\alpha)_{,X_1} & 0 \\ 0 & (R_i\beta_\alpha)_{,X_2} \\ (R_i\beta_\alpha)_{,X_1} & (R_i\beta_\alpha)_{,X_1} \end{bmatrix} \tag{15.27}$$

$$B_i^b = \begin{bmatrix} B_i^{b1} & B_i^{b2} & B_i^{b3} & B_i^{b4} \end{bmatrix} \tag{15.28}$$

15.2.5 Computation of stress intensity factors

In this section, discussion on SIFs computation with the help of domain-based interaction integral has been considered. In this domain-based approach, only those field variables are chosen that are not near the vicinity of the crack tip, which makes the approach more accurate for the evaluation of SIFs. Also, it is proven to be more favorable during the evaluation of single SIFs from a solution when only auxiliary field is judiciously chosen.

The domain form of interaction integral in the case of mechanical loading only is derived in Nguyen et al. [51] and is given as

$$M^{(1,2)} = \int_A \left[\sigma_{ij}^{(1)} \frac{\partial u_i^{(2)}}{\partial x_1} + \sigma_{ij}^{(2)} \frac{\partial u_i^{(1)}}{\partial x_1} - Z^{(1,2)} \delta_{ij} \right] \frac{\partial s}{\partial x_j} dA \tag{15.29}$$

$$Z^{(1,2)} = \frac{1}{2} \left(\sigma_{ij}^{(1)} \epsilon_{ij}^{(2)} + \sigma_{ij}^{(2)} \epsilon_{ij}^{(1)} \right) \tag{15.30}$$

where "s" presents the scalar weight function, which ranges between 0 and 1; $Z^{(1,2)}$ is mutual strain energy; ϵ and σ show the strains and stresses, respectively, whereas "1" and "2" help in showing the actual as well as auxiliary states, respectively.

The SIFs and $M^{(1,2)}$ are associated as

$$M^{(1,2)} = \frac{2}{E^*} \left(K_I^{(1)} K_I^{(2)} + K_{II}^{(1)} K_{II}^{(2)} \right) \tag{15.31}$$

where $E^* = E$ and $E^* = \frac{E}{(1 - v^2)}$ shows the case of plane stress and strain correspondingly. Further, single SIFs under mixed loads can be derived by taking the appropriate choice of auxiliary states. In the case of mode-I loading, the SIFs are derived as $K_I^{(2)} = 1$ and $K_{II}^{(2)} = 0$, and for mode-II, we have $K_I^{(2)} = 0$ and $K_{II}^{(2)} = 1$.

15.2.5.1 Enrichment of control points

It is evident from the literature that the selection of enriched control points can be done in a same way the XFEM is done, that is, both are quite similar in terms of control point selection. The only visible difference is that the nodes of XFEM are either associated with the crack tip or the crack face, while in the case of XIGA, the domain under the influence is much larger, that is, control points located far from the crack are also well depicted as enrichment. The nonzero basis function value are chosen for enrichment with Heaviside of CTEF function.

15.2.6 Results and discussions

The surface cracked from the edge in a homogeneous structure is considered under analysis, which is presented in Fig. 15.6.

The stainless-steel substrate is a square-shaped plate, and the dimensions taken for this investigation are considered to be 25×25 mm^2 with a crack of length exactly half to that. The modulus of elasticity of the substrate is assumed to be 215×10^3 MPa, while the loading condition is considered to be monotonic until the value of SIFs reaches the value of fracture toughness, which is 1903.36 MPa$\sqrt{\text{mm}}$. The first-order NURBS-based XIGA is considered in the study since the weight component associated with control points becomes similar in the case of XFEM. Thus NURBS elements become similar to the Lagrange elements, and the comparison becomes viable. In XIGA, the substrate is discretized in a control point distribution of 40×80. Further, the top edge of the structure is loaded with 250 N/mm, while the bottom edge is considered constrained, as shown in Fig. 15.6.

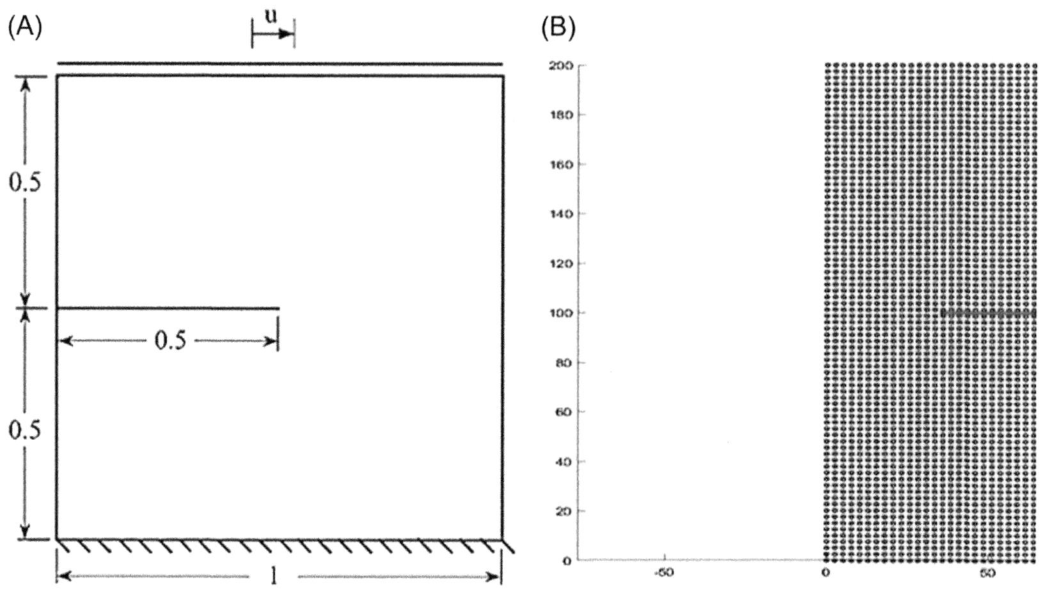

FIGURE 15.6 (A) Geometry under consideration. (B) Mesh representation of crack.

15.2.7 Conclusion

This study has been carried out using XIGA. The objective of the study is to implement XIGA to analyze embedded crack present in the homogeneous stainless-steel substrate. This study can be concluded as follows:

1. XIGA has been found to be quite effective to compute the fatigue life of edge crack problems on a substrate.
2. During the thermal cycling, the UNPs can effectively reduce the thermal stress of TAC.
3. The tensile stress and shear stress regions outside the unmelted particles enhance the initiation of cracks.
4. This NURBS-based XIGA technique employed is extremely vital since it resolves issues related to discretization errors and remeshing, which is quite mandatory in some of the numerical techniques like FEA.
5. The cracked plate structure is considered under a monotonic load at the top edge and constrained at the bottom edge for computing mixed-mode SIFs (KI and KII).
6. Orientation angles of crack impact the evaluation of SIFs greatly.
7. The compressive stress inside the UNPs can effectively prevent the cracks propagation.
8. Arbitrarily oriented cracks mainly propagated parallel to the x-axis, which means that tensile stress is the main driving force for the spallation failure of TBCs.

This study has proved that a structure under monotonic load condition, when put into a thermal cycle, will result in unidirectional cracks when the tensile stress is being applied to the structure.

References

[1] Thappa S, Chauhan A, Sawhney A, Anand Y, Anand S. Thermal selective coatings and its enhancement characteristics for efficient power generation through parabolic trough collector (PTC). Clean Technologies and Environmental Policy 2020;22(3):557–77.

[2] Zhang L, Wang Y, Fan W, Gao Y, Sun Y, Bai Y. A simulation study on the crack propagation behavior of nanostructured thermal barrier coatings with tailored microstructure. Coatings 2020;10(8):722.

[3] Gupta A, Kumar D. Development of Al2O3-based hybrid ceramic matrix composite coating to mitigate the erosive wear of advanced steel. Proceedings of the Institution of Mechanical Engineers, Part L: Journal of Materials: Design and Applications 2021;235(4):752–62.

[4] Lone AS, Jameel A, Harmain GA. Enriched element free Galerkin method for solving frictional contact between solid bodies. Mechanics of Advanced Materials and Structures 2022b. Available from: https://doi.org/10.1080/15376494.2022.2092791.

[5] Jameel A, Harmain GA. Fatigue crack growth in presence of material discontinuities by EFGM. International Journal of Fatigue 2015;81:105–16. Available from: https://doi.org/10.1016/j.ijfatigue.2015.07.021.

[6] Kanth SA, Harmain GA, Jameel A. Modeling of nonlinear crack growth in steel and aluminum alloys by the element free Galerkin method. Materials Today: Proceedings 2018;5:18805–14. Available from: https://doi.org/10.1016/j.matpr.2018.06.227.

[7] Sheikh UA, Jameel A. Elasto-plastic large deformation analysis of bi-material components by FEM. Materials Today. Proceedings 2020;26:1795–802. Available from: https://doi.org/10.1016/j.matpr.2020.02.377.

[8] Gupta V, Jameel A, Verma SK, Anand S, Anand Y. Transient isogeometric heat conduction analysis of stationary fluid in a container. Part E: Journal of Process Mechanical Engineering 2022a. Available from: https://doi.org/10.1177/09544089221112571.

[9] Gupta V, Jameel A, Verma SK, Anand S, Anand Y. An insight on NURBS based isogeometric analysis, its current status and involvement in mechanical applications. Archives of Computational Methods in Engineering 2022b. Available from: https://doi.org/10.1007/s11831-022-09838-0.

[10] Kanth SA, Jameel A, Harmain GA. Investigation of fatigue crack growth in engineering components containing different types of material irregularities by XFEM. Mechanics of Advanced Materials and Structures 2022;29(24):3570–87. Available from: https://doi.org/10.1080/15376494.2021.1907003.

[11] Hansbo A, Hansbo P. A finite element method for the simulation of strong and weak discontinuities in solid mechanics. Computer Methods in Applied Mechanics and Engineering 2004;193(33):3523–40. Available from: https://doi.org/10.1016/j.cma.2003.12.041.

[12] Jameel A, Harmain GA. Extended iso-geometric analysis for modeling three dimensional cracks. Mechanics of Advanced Materials and Structures 2019a;26:915–23. Available from: https://doi.org/10.1080/15376494.2018.1430275.

[13] Peskin CS. Flow patterns around heart valves: a numerical method. Journal of Computational Physics 1972;10(2):252–71.

[14] Glowinski, Kuznetsov Y. Distributed lagrange multipliers based on fictitious domain method for second order elliptic problems. Computer Methods in Applied Mechanics and Engineering 2007;196(8):1498–506. Available from: https://doi.org/10.1016/j.cma.2006.05.013.

[15] Burman E. Ghost penalty. Comptes Rendus Mathematique 2010;348(21):1217–20. Available from: https://doi.org/10.1016/j.crma.2010.10.006.

[16] Hansbo A, Hansbo P. An unfitted finite element method, based on Nitsche's method, for elliptic interface problems. Computer Methods in Applied Mechanics and Engineering 2002;191(47):5537–52. Available from: https://doi.org/10.1016/s0045-7825(02)00524-8.

[17] Burman E, Hansbo P. Fictitious domain finite element methods using cut elements: II. A stabilized Nitsche method. Applied Numerical Mathematics 2012;62(4):328–41. Available from: https://doi.org/10.1016/j.apnum.2011.01.008.

[18] Burman E, Claus S, Hansbo P, Larson MG, Massing A. Cutfem: discretizing geometry and partial differential equations. International Journal for Numerical Methods in Engineering 2015;104(7):472–501. Available from: https://doi.org/10.1002/nme.4823.

[19] Dolbow J, Harari I. An efficient finite element method for embedded interface problems. International Journal for Numerical Methods in Engineering 2009;78(2):229–52.

[20] Babuška I, Melenk JM. The partition of unity method. International Journal for Numerical Methods in Engineering 1997;40(4):727–58.
[21] Strouboulis T, Babuška I, Copps K. The design and analysis of the generalized finite element method. Computer Methods in Applied Mechanics and Engineering 2000;181(1–3):43–69.
[22] Strouboulis T, Copps K, Babuška I. The generalized finite element method: an example of its implementation and illustration of its performance. International Journal for Numerical Methods in Engineering 2000;47(8): 1401–17.
[23] Belytschko T, Black T. Elastic crack growth in finite elements with minimal remeshing. International Journal for Numerical Methods in Engineering 1999;45(5):601–20.
[24] Moës N, Dolbow J, Belytschko T. A finite element method for crack growth without remeshing. International Journal for Numerical Methods in Engineering 1999;46(1):131–50.
[25] Soghrati S, Aragón AM, Armando Duarte C, Geubelle PH. An interface-enriched generalized fem for problems with discontinuous gradient fields. International Journal for Numerical Methods in Engineering 2012;89(8):991–1008. Available from: https://doi.org/10.1002/nme.3273.
[26] Ambrosini A, Lambert TN, Boubault A, Hunt A, Davis DJ, Adams D, et al. Thermal stability of oxide-based solar selective coatings for CSP central receivers. In ASME 2015 9th International Conference on Energy Sustainability Collocated with the ASME 2015 Power Conference, the ASME 2015 13th International Conference on Fuel Cell Science, Engineering and Technology, and the ASME 2015 Nuclear Forum, V001T05A022; 2015.
[27] Suriwong T, Bunmephiphit C, Wamae W, Banthuek S. Influence of Ni-Al coating thickness on spectral selectivity and thermal performance of parabolic trough collector. Materials for Renewable and Sustainable Energy 2018;7(3)14.
[28] Qiu X-L, et al. Structure, thermal stability and chromaticity investigation of TiB2 based high temperature solar selective absorbing coatings. Solar Energy 2019;181:88–94.
[29] Wiktorczyk T, Bieganski P, Zielony E. Preparation and optical characterization of e-beam deposited cerium oxide films. Optical Materials 2012;34(12):2101–7.
[30] He C-Y, Gao X-H, Yu D-M, Guo H-X, Zhao S-S, Liu G. Highly enhanced thermal robustness and photothermal conversion efficiency of solar-selective absorbers enabled by high-entropy alloy nitride MoTaTiCrN nanofilms. ACS Applied Materials & Interfaces 2021;13(14):16987–96.
[31] Hughes TJR, Cottrell JA, Bazilevs Y. Isogeometric analysis: CAD, finite elements, NURBS, exact geometry and mesh refinement. Computer Methods in Applied Mechanics and Engineering 2005;194(39):4135–95. Available from: https://doi.org/10.1016/j.cma.2004.10.008.
[32] Choi M-J, Cho S. Isogeometric analysis of stress intensity factors for curved crack problems. Theoretical and Applied Fracture Mechanics 2015;75:89–103. Available from: https://doi.org/10.1016/j.tafmec.2014.11.003.
[33] Khademalrasoul A. NURBS-based isogeometric analysis method application to mixed-mode computational fracture mechanics. Journal of Applied and Computational Mechanics 2019;5(2):217–30.
[34] Verhoosel CV, et al. An isogeometric approach to cohesive zone modeling. International Journal for Numerical Methods in Engineering 2011;87(1–5):336–60.
[35] Ghorashi SS, Valizadeh N, Mohammadi S. Extended isogeometric analysis for simulation of stationary and propagating cracks. International Journal for Numerical Methods in Engineering 2012;89(9):1069–101. Available from: https://doi.org/10.1002/nme.3277.
[36] Bui TQ. Extended isogeometric dynamic and static fracture analysis for cracks in piezoelectric materials using NURBS. Computer Methods in Applied Mechanics and Engineering 2015;295:470–509. Available from: https://doi.org/10.1016/j.cma.2015.07.005.
[37] De Luycker E, Benson DJ, Belytschko T, Bazilevs Y, Hsu MC. XFEM in isogeometric analysis for linear fracture mechanics. International Journal for Numerical Methods in Engineering 2011;87(6):541–65. Available from: https://doi.org/10.1002/nme.3121.
[38] Shojaee S, Asgharzadeh M, Haeri A. Crack analysis in orthotropic media using combination of isogeometric analysis and extended finite element. International Journal of Applied Mechanics 2014;6(6):1450068.
[39] Bhardwaj G, Singh IV, Mishra BK, Bui TQ. Numerical simulation of functionally graded cracked plates using NURBS based XIGA under different loads and boundary conditions. Composite Structures 2015;126:347–59. Available from: https://doi.org/10.1016/j.compstruct.2015.02.066.

[40] Singh AK, Jameel A, Harmain GA. Investigations on crack tip plastic zones by the extended iso-geometric analysis. Materials Today: Proceedings 2018;5:19284–93. Available from: https://doi.org/10.1016/j.matpr.2018.06.287.

[41] Singh SK, Singh IV, Mishra BK, Bhardwaj G, Singh SK. Analysis of cracked plate using higher-order shear deformation theory: Asymptotic crack-tip fields and XIGA implementation. Computer Methods in Applied Mechanics and Engineering 2018;336:594–639. Available from: https://doi.org/10.1016/j.cma.2018.03.009.

[42] Singh SK, et al. A simple, efficient and accurate Bézier extraction based T-spline XIGA for crack simulations. Theoretical and Applied Fracture Mechanics 2017;88:74–96.

[43] Wilson JL, Townley LR, SaDaCosta A. Mathematical development and verification of a finite element aquifer flow model AQUIFEM-1. Ralph M. Parsons Laboratory for Water Resources and Hydrodynamics. Massachusetts: Institute of Technology; 1979.

[44] Shih CF, Moran B, Nakamura T. Energy release rate along a three-dimensional crack front in a thermally stressed body. International Journal of Fracture 1986;30:79–102.

[45] Prasad NNV, Aliabadi MH, Rooke DP. The dual boundary element method for thermoelastic crack problems. International Journal of Fracture 1994;66:255–72.

[46] Raveendra ST, Banerjee PK, Dargush GF. Three-dimensional analysis of thermally loaded cracks. International Journal for Numerical Methods in Engineering 1993;36(11):1909–26.

[47] Montassir S, et al. Crack propagation modeling using the extended isogeometric. In: Analysis Technique. Advances in Integrated Design and Production: Proceedings of the 11th International Conference on Integrated Design and Production, CPI 2019, October 14–16, 2019, Fez, Morocco, Springer International Publishing; 2021.

[48] Kaushik V, Ghosh A. Experimental and XIGA-CZM based Mode-II and mixed-mode interlaminar fracture model for unidirectional aerospace-grade composites. Mechanics of Materials 2021;154:103722.

[49] Nguyen-Thanh N, et al. An extended isogeometric thin shell analysis based on Kirchhoff–Love theory. Computer Methods in Applied Mechanics and Engineering 2015;284:265–91.

[50] Chopp DL, Sukumar N. Fatigue crack propagation of multiple coplanar cracks with the coupled extended finite element/fast marching method. International Journal of Engineering Science 2003;41(8):845–69.

[51] Nguyen MN, et al. Simulation of dynamic and static thermoelastic fracture problems by extended nodal gradient finite elements. International Journal of Mechanical Sciences 2017;134:370–86.

CHAPTER

16

Enriched techniques for investigating the behavior of structural wood

Ummer Amin Sheikh[1], *Azher Jameel*[2], *Mehnaz Rasool*[1] *and Mohd Junaid Mir*[1]

[1]Department of Mechanical Engineering, Islamic University of Science and Technology, Awantipora, Jammu and Kashmir, India [2]Department of Mechanical Engineering, National Institute of Technology Srinagar, Hazratbal, Srinagar, Jammu and Kashmir, India

16.1 Introduction

Wood is an important natural resource and has many advantages, due to which it is preferred over many other building materials used in construction. It is easily available and easy to transport and handle, provides good thermal insulation, is sound absorbent, and provides good electrical insulation in comparison to structural steel and concrete that have traditionally been employed as potential construction materials in engineering structures. People living at higher altitudes, especially in hilly areas, do not afford concrete-based structures to build their homes and other necessary structures. They are completely dependent on natural resources, such as wood and plants, for their survival. In addition to that, it is also very difficult or sometimes impossible for government organizations to transport raw materials, heavy machinery, and other equipment to these places to construct different structures, such as houses, bridges, and offices. Natural wood available in such areas provides a good structural material and can serve as a potential alternative to other constructional materials because of its mechanical properties. Another advantage of structural wood is the ease of processing, repairing, and replacements, which makes it suitable for areas that are difficult to reach. Wooden structures may sometimes provide a better alternative to concrete structures, if properly designed and fabricated. Proper and accurate mechanical characterization of different trees is very important so that safe and reliable wooden structures are constructed, especially for the people living in hilly and mountainous regions where the transportation of other constructional materials is very difficult and, in some cases, impossible.

Enriched Numerical Techniques
DOI: https://doi.org/10.1016/B978-0-443-15362-4.00004-8

Natural wood itself is one of the oldest building materials and has always remained the material of choice for engineers because of its availability, ease of handling and manufacturing, and excellent relationship weight-to-strength ratios. There is a need to study the mechanical properties, such as strength and hardness, of structural wood so that their mechanical properties are accurately explored and safe and reliable structures are designed. There is also a need to investigate the effect of weather and climatic conditions on the strength-related properties of different trees and plants. Many regions of the world experience severe drop in temperatures during winters, and the temperatures may dip as low as minus 20 degrees, which has a huge impact on the strength of structural wood and hence needs an extensive study. The Himalayan regions also experience a huge amount of rain and snowfall during the winter. The wooden structures are covered with snow for longer durations of time, due to which the wood absorbs moisture, which ultimately affects the strength of the whole structure. The effect of moisture and extremely low temperatures needs investigation. To summarize, it is extremely important and takes an hour to carry out the mechanical characterization of different trees and plants, so that safe and reliable structures are constructed.

Physical and mechanical properties of structural wood are closely related to the moisture content present in wood. Wood present in the tree is of two types: sap wood and heart wood. Water content in sap wood is higher than the moisture present in heart wood. One of the important properties of wood is its fiber saturation point, which represents the state of wood when all moisture content present in it gets removed from the cell cavities. Heart wood is much more durable in compression because it offers significant resistance to the penetration of fluids. Wood is hydroscopic in nature, as it presents a tendency to absorb moisture when it is placed in a humid and moist environment. On the contrary, structural wood loses moisture content when it is placed in a dry environment. Under normal operating conditions, the moisture content of wood cannot be controlled, but there are the chances of reducing the absorption rates by making use of coatings and other heat treatment processes. For structural engineers, the development of appropriate mathematical models, for establishing relationships between the strength and moisture content, has always been a challenge. It is very important to take care of the dimensional stability of wood, such as shrinkage and swelling behavior, when it is exposed to humid environment or in direct contact with different types of fluids. It has been found that due to temperature, the dimensional stability of wood changes, that is, when wood is heated, it expands, contracts when cooled, and swells when exposed to liquids. The most important parameter of good timber is proper seasoning. Seasoning of timber can be done in different ways, which include natural or air seasoning, water seasoning, boiling, kiln seasoning, chemical or salt seasoning, and electric seasoning. Because of excellent resistance offered by wood against different types of chemicals, it has always proved to be a potential structural material under corrosive chemical environments. In many applications, wood provides an excellent alternative to conventional structural materials, such as concrete and steel. This is due to the resistance offered by wood against mild acids having pH values exceeding 2, acidic salt solutions, and various other corrosive agents. Classification of wood is based on the basis of position, grading (structural grading, commercial grading), modulus of elasticity, availability, durability, seasoning characteristics, and on the basis of treatability.

The usability of structural wood is defined by its density, which has always served as an important physical property of structural wood. The density of wood is closely related to its strength under different applications. The density of wood and its strength-related properties exhibit a strong relationship, which has been used by engineering to investigate the behavior of structural wood under different applications. Therefore estimation of the basic density of structural wood is an important step in the design process. All tree species allocate a certain portion of their material to support different types of loads and impart strength to the structural wood. Specific gravity is used for such allocated material. It has been observed that the structures made of wood usually do not withstand extreme environmental conditions, such as high-speed winds, earthquakes, and hurricanes. Such environmental conditions induce cyclic stresses in the structural components, which invokes the need to investigate the fatigue strength of structural wood. During normal operating conditions, damage accumulation occurs in engineering structures made of wood, which demands the detailed investigation of viscoelastic properties of structural wood. The damage process can be defined in terms of low-cycle fatigue conditions. While designing wooden beams, columns, and other end connections, it is very important to understand the compressive behavior of structural wood in the directions parallel and perpendicular to grain boundaries. It has been observed that failure of softwood under direct compressive loads, occurs parallel to grain boundaries, which can be detected in the form of kinks in the cell walls.

16.2 Extended finite element method

Different numerical techniques are available for modeling and simulating different engineering problems, such as crack growth, in structural materials. Out of all numerical techniques available, the conventional finite element method has always proved to be a dominant computational framework for modeling the behavior of different engineering structures. Some of these important and widely used computational methods include the boundary element method [1–3], the extended finite element method (XFEM) [4–6], mesh-free techniques [7–11], models based on peridynamics [12–14], phase field methods [15–17], and conventional FEM [18–21]. Classical FEM has the constraints of conformal meshing for representing the material irregularities present on the domain, such as cracks, inclusions, or holes. It is well known that the development of a conformal mesh for irregular and complex-shaped discontinuities is computationally very demanding and costly, which puts a limit on the conventional FEM for modeling and simulating different types of material irregularities present in structural components. The problems get more complicated and expensive if such discontinuities evolve with time and change their positions, as is the case with propagating cracks and large deformations in engineering structures. Finite element method also suffers from extreme mesh distortion while modeling large deformation problems, which creates a need for the re-meshing of the domain throughout the process of simulation. Re-meshing complex-shaped three-dimensional domains has always remained a challenge for engineers, as it is computationally more costly and consumes a huge amount of resources.

The drawbacks of conformal meshing, refinement of meshes near stress concentrations, and other mesh-related problems are eliminated by XFEM, which is an extended version of the conventional FEM. This updated finite element framework models all types of material irregularities, including cracks, inclusions, and holes, irrespective of the nodal distribution selected for investigation [22–24]. In XFEM, suitable enrichment terms are incorporated into classical displacement-based approximations to model the effect of material irregularities present in structural components. Drawbacks of re-meshing, mesh adaption, and mesh conformation do not arise in XFEM, which makes it computationally more effective in comparison to conventional FEM [25–34]. The enrichment functions are included in the formulations by employing partition of unity approaches [35–40]. Level set (LS) algorithms play a vital role in the XFEM for representing and keeping track of different material irregularities present in the domain. The LS method represents the material irregularities with a function that vanishes over the material interfaces. Such functions are negative on one side and positive on the other side of interface. The coupling of the LS method with XFEM has always provided a strong and efficient computational tool for representing and tracking different types of material irregularities present in the domain.

16.3 Element-free Galerkin method

Meshless methods were developed to eliminate the drawbacks associated with classical FE-based techniques. The element-free Galerkin method (EFGM) is a mesh-free numerical framework that was developed to eliminate the limitations associated with the dependence of the FE mesh to construct the approximate solution. EFGM represents the entire domain as an array of nodes only, and displacement approximation derives from the knowledge of these nodes only. There is no requirement for generating the FE mesh in meshless techniques, which leads to the elimination of all the mesh-related drawbacks like conformal meshing and mesh distortion. The smooth particle hydrodynamics was one of the first meshless computational techniques that was developed to model and simulate various astrophysical phenomena involving dust clouds and exploding stars. Smooth particle hydrodynamics was later applied to investigate different types of solid mechanics problems. The standard smooth particle hydrodynamics method suffered various limitations of instabilities and inconsistencies while modeling solid mechanics problems, due to which many modifications and improvements were carried out to make this numerical framework suitable for computational solid mechanics. The smooth particle hydrodynamics model and its updated versions were based on the strong form of governing differential equations. Later, several meshless methods were developed that were based on the weak form of the governing differential equation. Such meshless methods proved to be very effective in modeling different types of solid mechanics problems. EFGM is one of the meshless techniques that is based on the weak form of the governing differential equation. Other important meshless techniques that have found extensive applications in engineering include the reproducing kernel particle method and the meshless local Petrov–Galerkin method.

EFGM constructs variable approximations in terms of nodal distributions only, and moving least squares methodology has been employed for approximating the displacement field across the domain of interest. Moving least square approximations can be subdivided into polynomial basis function, weight function with a compact support, and coefficients that vary with the position of nodes. If weight function used in the displacement approximations and its first *kth* derivatives maintain their continuity, then EFGM shape functions and their first *kth* derivatives also remain continuous through the domain. Selection of weight functions plays an important role in determining behavior of meshless techniques. Weight functions are defined in such a way that they should always be positive within their compact support and decrease monotonically as we move farther and farther from that particular node. Weight function, thus defined, should maintain the continuity and differentiability over the domain. Various weight functions have been proposed for use in EFGM, out of which the most important ones include the exponential, cubic spline, and quartic spline forms of weight functions. The most commonly used form of weight function in EFGM is the quartic spline.

16.4 Extended isogeometric analysis

The extended isogeometric analysis (XIGA) evolved as a new numerical enrichment technique and gained tremendous importance in the past few decades for the analysis of different engineering problems. Isogeometric analysis is a novel computational tool that was developed to resolve the geometrical discretization issues. In isogeometric analysis, a standard basis function is employed for describing the geometry as well as unknown field variables, which makes it a strong numerical framework for solving various engineering problems. In the case of crack growth problems, isogeometric analysis has proven to be an efficient and potential computational tool for investigation. Isogeometric analysis proves to be an effective numerical framework for modeling complex-shaped domains, as the geometrical discretization errors are minimized. The concept of isogeometric analysis was extended by introducing the partition of unity approach, which was later known as XIGA. In XIGA, the analysis of discontinuous problems can be done efficiently and accurately by coupling different enrichment strategies with standard isogeometric analysis. The domain of control points that are bisected by the face of the crack is enriched by the Heaviside jump function. On the contrary, control points lying near crack tips are enriched by crack tip enrichment functions, which are available in open literature.

16.5 Mechanical properties of wood

Wood has always played an important role in construction. Determination of the mechanical properties of wood is very important for designing safe and reliable engineering structures. Strength of wood varies from species to species, loading conditions and load duration, and environmental factors. Since wood is anisotropic in nature, its mechanical properties vary in all three directions, such as in radial, tangential, and longitudinal directions. Therefore the estimation of mechanical properties of wood in different directions and orientations is very important. It has been found that the strength of structural wood parallel to grain boundaries is 20–30 times greater than that measured in the direction perpendicular to grains. The strength of the structural wood can be obtained by investigating the behavior of wood in tension, compression, shear, torsion, bending, and shock loads. In addition to variations in the mechanical behavior of wood, it is also important to investigate the impact of several factors, such as temperature changes and moisture, on structural wood. This section provides the detailed discussion on the estimation of various strength-related properties of structural wood, such as tensile, shear strength, bending, and impact strength.

16.5.1 Tensile strength of wood

The tensile strength of wood establishes the behavior of structural wood subjected to pure axial loads. As already discussed, the tensile strength of wood along parallel and perpendicular directions to grain boundaries shows huge variations and hence needs to be investigated properly. It has been observed that the strength of wood along the grain boundaries is largely dependent on the strength of fibers. The strength of wood parallel to grain boundaries gets influenced by the arrangement of wood fibers. The dimensions and type of wood fibers also affect the tensile strength of wood. It is well known that the mechanical behavior of structural wood entirely depends on the direction of loading. There is a significant difference between the strengths of wood measured parallel and perpendicular to grain boundaries. From the experimental investigations conducted so far, it has been well established that wood is stronger in tension parallel to grain compared to its tensile strength perpendicular to grain. From the experimental observations, it has been concluded that the tensile strength of compressed softwood, measured parallel to the grain, is almost two times when compared to that of normal spruce. From time to time, several modifications of natural wood have been proposed to improve the strength of structural wood, and a fair amount of success has been achieved in this direction. One such method of improving the strength of wood is the compression of wood samples beyond their strength. It has been found that the compressed wood has higher tensile strength than the normal wood for the same species of structural wood. Compressed wood also shows a higher modulus of elasticity and stiffness as compared to natural wood. It should be noted that the tensile tests for determining the strength of structural wood are performed in both parallel and perpendicular directions as wood is an anisotropic material. Samples collected parallel to grain boundaries are used for estimating the tensile strength parallel to grain boundaries, whereas the tensile strength perpendicular to grain boundaries is obtained by collecting the samples perpendicular to grain boundaries. To get a more detailed insight into the strength of the structural wood, test samples in tangential directions can also be obtained to determine the tensile strength of wood along tangential directions.

16.5.2 Compressive strength of wood

Under normal service conditions, wooden structural elements are usually subjected to compressive loads, which invokes the need for investigation of the compressive strength of wood. It is well known that structural wood offers more resistance in compression as compared to tension. Wood is relatively stronger in compression. Just like tensile testing, structural wood can be exposed to compressive loads either parallel to grain boundaries, perpendicular to grains, or at any inclination to grain boundaries. Because of the anisotropic behavior of structural wood, it is extremely important to estimate the compressive strength of structural wood along different orientations. When the compressive loads are applied parallel to the grain boundaries, strains are produced, which deform the wood cells along the longitudinal direction. On the contrary, when the compressive loads are applied perpendicular to the grain boundaries, strains are developed that deform the wood cells perpendicular to their direction. The compressive loads applied at different inclinations produce stresses that can be resolved in parallel and perpendicular directions to grain boundaries. From a microscopic point of view, structural wood behaves as a natural composite material. The advantage of structural wood is that it has an excellent ratio of mechanical strength to its density. The density of structural wood can be improved by employing various densification techniques. Densification of wood can be obtained by compressing the structural wood, which leads to improvement in the mechanical properties. Compressed wood presents various desirable mechanical properties that make it a suitable constructional material that has the potential to replace classical structural materials. Compression of structural wood in radial direction causes the compression of the early wood cells only. Because of this advantage, compression of structural wood in radial direction is very common, and such types of loadings result in the most desirable characteristics as compared to compressive loads in other directions. Several engineering processes have been designed from time to time to obtain the desired compression of wood to convert it into a more useful product. Compression of wood typically takes place in a hot press with a constant displacement rate at a high temperature, which is usually 120°C–160°C. After obtaining the desired compression, the cooling of the hot-pressed wood is controlled properly to eliminate the spring back effects, which may bring back the processed wood to its initial configuration.

16.5.3 Bending strength of wood

Determination of the bending strength of structural wood is extremely important in the design of sate structures, as the wooden elements are exposed to bending loads under normal service conditions. The bending test is one of the most important tests that has been universally adopted to determine the strength of structural wood, irrespective of the species of wood chosen for investigation. When bending loads are applied to a sample of structural wood or timber, the part of the wood sample lying on one side of the neutral axis experiences tensile forces parallel to grain, whereas the remaining portion experiences compressive loads parallel to grain. The portion of structural wood lying at the neutral axis is not stressed either in compression or tension. Bending loads on wood samples also result in horizontal shear loads parallel to grain boundaries and compressive forces perpendicular to grain boundaries at the supports. Among all bending test configurations available for performing the bending tests, wood is most commonly tested with a four-point bending test, which requires a deflectometer that can accurately and precisely monitor and measure the flexural displacement of the wood specimen subjected to bending loads on the test machine. Bending tests on structural wood provide the flexural strength and flexural modulus of the wood, which are highly useful for the designers during the building of engineering structures. Because of the anisotropic behavior of structural wood, the determination of the bending strength of timber also becomes an area of concern for engineers. Densification of natural wood has a great impact on the bending strength of structural wood. It has been found that the bending strength of compressed softwood gets enhanced by 2.45 times as compared to that of natural uncompressed softwood. Similar impacts of densification can be seen for hardwoods. Experimental investigations show that the bending strength of compressed hardwood is almost twice as high as that of normal uncompressed hardwood. It has also been concluded that the bending strength of compressed hardwood is about 30 MPa higher than that of compressed softwood. Properly compressed wood also displays a better and higher flexure modulus as compared to that of natural uncompressed wood. The flexural modulus of elasticity of the compressed softwood is about 1.8 times the flexure modulus of the natural uncompressed softwood. Experimental investigations also show that the flexural modulus of elasticity of the compressed hardwood is about 2.5 times the flexure modulus of the natural uncompressed hardwood. It has also been found that the mean flexural modulus of elasticity of compressed hardwood is about 2.5 GPa higher than that of compressed softwood. From the application point of view, the columns of a deck must be strong in compression parallel to the grain, whereas the wooden beams running horizontally beneath a deck must be stronger in bending.

16.5.4 Shear strength of wood

Determination of the shear strength of structural wood is also an area of interest for structural engineers. Shear loads can be imposed on the wood samples along the horizontal, vertical, and rolling directions. Determination of shear strength along the vertical direction is normally taken into consideration because the wooden samples are more prone to fail under vertical shear. Horizontal shear also plays an important role and is considered in wooden samples when horizontal shear is imposed in the directions parallel to grain boundaries. In addition to horizontal and vertical shear, rolling shear may also develop in structural wood during its operational stage. Rolling shear is generally caused due to the loads acting normally on cell lengths in a plane parallel to grain boundaries. Stresses generated in the structural wood produce the tendency of the wood cell to rolling shear and in such cases, failure is usually produced by large deformations occurring in cell cross sections.

16.5.5 Applications of enriched techniques

In addition to the experimental investigations performed on structural wood to determine its properties, a number of computational techniques have been proposed in the past few decades for estimating the behavior of wood subjected to different types of loadings. Although the conventional FEM has always remained the widely accepted and impactful numerical tool for modeling the behavior of engineering structures, it faces several issues while modeling the domains containing different types of material irregularities such as cracks, holes, and inclusions. Conventional FEM requires mesh conformation for investigating different types of discontinuities produced by cracks, inclusions, or holes. It is well established that the development of a conformal mesh for complex and irregularly shaped discontinuities is computationally more demanding and costly, which limits the implementation of the classical FEM for modeling such types of material discontinuities. The analysis becomes more complicated if the discontinuity changes with time, as is the case with advancing cracks and large deformation problems where FEM suffers from extreme mesh distortion, which demands re-meshing of the domain throughout the process of simulation. Re-meshing of different engineering domains has always proved to be computationally more costly, especially for three-dimensional engineering problems where re-meshing, conformal meshing, and mesh refinements are very cumbersome.

The issues of conformal meshing and other mesh-related issues do not arise from the use of enriched techniques that model all material interfaces present in the domain, such as cracks, inclusions, or holes, irrespective of the nodal distribution selected for investigation. Enriched techniques model the discontinuities present in the domain by adding the suitable enrichment functions to classical displacement-based approximations. Drawbacks of re-meshing, mesh adaption, and conformal meshing do not occur in enriched techniques, which makes them computationally more effective as compared to conventional FEM. Enrichment functions are incorporated in the mathematical formulations by employing the partition of unity approaches. LS algorithms play a vital role in XFEM for representing and keeping track of different material irregularities present in the domain. The LS method represents the material irregularities with a function that vanishes over the material interfaces. Such functions are negative on one side and positive on the other side of the interface. The coupling of the LS method with XFEM has always provided a strong and efficient computational tool for representing and tracking different types of material irregularities present in the domain. The introduction of enrichment strategies is considered a major breakthrough in modeling different types of material irregularities present in structural components. The basic principle of enriched techniques is to modify the classical displacement approximations with additional enrichment functions, which adds additional degrees of freedom to the system. One of first enrichment techniques, known as XFEM, has found extensive applications in different areas of engineering, which include the investigation of microstructure effects experienced in brittle fracture and delamination mechanisms occurring in thin-layered composite structures [41]. XFEM can be used accurately to investigate the mechanical behavior of wood at different scales of observation.

Wood can be considered an important constructional material that has been widely used for designing engineering structures. A huge number of defects, manufacturing flaws, and in-service damage can be seen in natural wood, and all of these defects can be treated as initial discontinuities produced present in the domain. Pearson estimated the effect of various flaws present in wood by invoking the use of basic fracture mechanics principles [42]. The development of appropriate fracture criteria is presented in Ref. [43], to investigate the influence of end splits on the shear load capacity of cracked wood beams. Linear elastic fracture mechanics principles were developed to investigate the behavior of butted joints in glued laminated timber [44]. A huge amount of research has been dedicated to understand crack growth mechanisms in various kinds of structural wood and on the development of different test methods to determine the mechanical properties of structural wood [45]. Although the conventional finite element method has been invoked to investigate crack tip singularities, SIFs, and the strain energy release rates, significant work has not been reported on the numerical simulation of crack growth in structural wood and other related composites. In view of the presence of flaws in natural wood and timber, XFEM-based models were developed to investigate crack growth behavior in wood subjected to different types of loading conditions [46,47]. Investigation of mixed mode crack growth in engineering structures has always been a domain of interest for structural engineers [48]. To model and simulate crack propagation in structural wood and related composites, appropriate crack initiation and crack propagation theories should be properly determined. Crack growth theories based on energy release rates have been used to define the directions of crack growth [48]. It has been found that the wooden engineering components are subjected to mixed mode loadings during their operational stage. Pure mode-I loading describes the opening mode, whereas mode-II loading involves the shearing action between the crack surfaces. It has always been well established that crack growth in wooden structures is of brittle nature. The study also shows that the main longitudinal crack first develops in the middle depth of the wooden beam, which may be due to the relatively weak tensile strength of structural wood perpendicular to grain boundaries. The crack advanced from the mid-span to both ends of the beam undergoing an immediate brittle fracture.

16.6 Conclusion

The book chapter reports the implementation and applications of the enriched numerical techniques in modeling the behavior of structural wood subjected to different types of loadings. Wood is one of the most commonly used structural materials for constructional purposes, and therefore, proper evaluation of the mechanical properties becomes extremely important to design safe and reliable engineering structures. In the past few decades, enriched numerical techniques such as XFEM, EFGM, and XIGA have gained tremendous importance in investigating the behavior of structural wood under different types of loadings. Enriched numerical techniques model all types of discontinuities, irrespective of the mesh or the grid selected for investigation. Wood geometry is very complex and contains a large number of internal discontinuities. Therefore enriched computational techniques can prove to be very effective and efficient in evaluating the mechanical properties of structural wood.

The tensile strength of wood along parallel and perpendicular directions to grain boundaries shows huge variations and hence needs to be investigated properly. From the experimental and numerical investigations conducted so far, it has been well established that wood is relatively stronger in tension parallel to grain compared to its tensile strength perpendicular to grain. Compressive strength of wood is one of the most important mechanical properties, as structural wood offers more resistance to compression as compared to tension. When the compressive loads are applied parallel to the grain boundaries, strains are produced, which deform the wood cells along the longitudinal direction. On the contrary, when the compressive loads are applied perpendicular to the grain boundaries, strains are developed that deform the wood cells perpendicular to their direction.

Densification of wood can be obtained by compressing the structural wood, which leads to improvement in the mechanical properties. Determination of the bending strength of structural wood is extremely important to ensure the safety and reliability of engineering structures. Among all bending test configurations available for performing the bending tests, wood is most commonly tested with a four-point bending test. Bending tests on structural wood provide the flexural strength and flexural modulus of the wood, which are highly useful for the designers during the building of engineering structures. Densification of natural wood has a great impact on the bending strength of structural wood. Similar impact of densification can be seen for hardwoods. Determination of shear strength of wood is also an area of interest for structural engineers. Shear loads can be imposed on the wood samples along the horizontal, vertical, and rolling directions. Determination of shear strength along the vertical direction is normally taken into consideration because the wood samples are more prone to fail under vertical shear. Horizontal shear also plays an important role and is considered in wood when horizontal shear loads are imposed parallel to the grain boundaries. Such loads introduce a tendency for the upper segment of the wooden sample to slide over the lower portion by breaking the intercellular bonds and causing deformations in the entire cell structure of wood.

References

[1] Aliabadi MH, Brebbia CA. Boundary element formulations in fracture mechanics: a review. Transactions on Engineering Sciences 1998;17:589–98.

[2] Wen PH, Aliabadi MH, Young A. Dual boundary element methods for three-dimensional dynamic crack problems. Journal of Strain Analysis for Engineering Design 1999;34:373–94. Available from: https://doi.org/10.1177/030932479903400601.

[3] Leonel ED, Venturini WS. Non-linear boundary element formulation applied to contact analysis using tangent operator. Engineering Analysis with Boundary Elements 2011;35:1237–47. Available from: https://doi.org/10.1016/j.enganabound.2011.06.005.

[4] Spangenberger AG, Lados DA. Extended finite element modeling of fatigue crack growth microstructural mechanisms in alloys with secondary/reinforcing phases: model development and validation. Computational Mechanics 2020;. Available from: https://doi.org/10.1007/s00466-020-01921-2.

[5] Lone AS, Kanth SA, Jameel A, Harmain GA. A state of art review on the modeling of contact type nonlinearities by extended finite element method. Materials Today: Proceedings 2019;18:3462–71. Available from: https://doi.org/10.1016/j.matpr.2019.07.274.

[6] Sukumar N, Chopp DL, Moës N, Belytschko T. Modeling holes and inclusions by level sets in the extended finite-element method. Computer Methods in Applied Mechanics and Engineering 2001;190:6183–200. Available from: https://doi.org/10.1016/S0045-7825(01)00215-8.

[7] Jameel A, Harmain GA. Fatigue crack growth in presence of material discontinuities by EFGM. International Journal of Fatigue 2015;81:105–16. Available from: https://doi.org/10.1016/j.ijfatigue.2015.07.021.

[8] Lone AS, Jameel A, Harmain GA. A coupled finite element-element free Galerkin approach for modeling frictional contact in engineering components. Materials Today: Proceedings 2018;5:18745–54. Available from: https://doi.org/10.1016/j.matpr.2018.06.221.

[9] Kanth SA, Harmain GA, Jameel A. Modeling of nonlinear crack growth in steel and aluminum alloys by the element free Galerkin method. Materials Today: Proceedings 2018;5:18805–14. Available from: https://doi.org/10.1016/j.matpr.2018.06.227.

[10] Jameel A, Harmain GA. Fatigue crack growth analysis of cracked specimens by the coupled finite element-element free Galerkin method. Mechanics of Advanced Materials and Structures 2019;26:1343–56. Available from: https://doi.org/10.1080/15376494.2018.1432800.

[11] Lone AS, Jameel A, Harmain GA. Enriched element free Galerkin method for solving frictional contact between solid bodies. Mechanics of Advanced Materials and Structures 2022. Available from: https://doi.org/10.1080/15376494.2022.2092791.

[12] Lone AS, Jameel A, Harmain GA. "Modelling of contact interfaces by penalty based enriched finite element method. Mechanics of Advanced Materials and Structures 2022. Available from: https://doi.org/10.1080/15376494.2022.2034075.

[13] Jameel A, Harmain GA. Large deformation in bi-material components by XIGA and coupled FE-IGA techniques. Mechanics of Advanced Materials and Structures 2022;29:850–72. Available from: https://doi.org/10.1080/15376494.2020.1799120.

[14] Kanth SA, Jameel A, Harmain GA. Investigation of fatigue crack growth in engineering components containing different types of material irregularities by XFEM. Mechanics of Advanced Materials and Structures 2022;29(24):3570–87. Available from: https://doi.org/10.1080/15376494.2021.1907003.

[15] Gupta V, Jameel A, Anand S, Anand Y. Analysis of composite plates using isogeometric analysis: a discussion. Materials Today: Proceedings 2021;4:1190–4. Available from: https://doi.org/10.1016/j.matpr.2020.11.238.

[16] Jameel A, Harmain GA. Effect of material irregularities on fatigue crack growth by enriched techniques. International Journal for Computational Methods in Engineering Science and Mechanics 2020;21:109–33. Available from: https://doi.org/10.1080/15502287.2020.1772902.

[17] Lone AS, Kanth SA, Jameel A, Harmain GA. "XFEM modelling of frictional contact between elliptical inclusions and solid bodies. Materials Today: Proceedings 2020;26:819–24. Available from: https://doi.org/10.1016/j.matpr.2019.12.424.

[18] Sheikh UA, Jameel A. Elasto-plastic large deformation analysis of bi-material components by FEM. Materials Today: Proceedings 2020;26:1795–802. Available from: https://doi.org/10.1016/j.matpr.2020.02.377.

[19] Simpson R, Trevelyan J. A partition of unity enriched dual boundary element method for accurate computations in fracture mechanics. Computer Methods in Applied Mechanics and Engineering 2011;200:1–10. Available from: https://doi.org/10.1016/j.cma.2010.06.015.

[20] Noda NA, Oda K. Numerical solutions of the singular integral equations in the crack analysis using the body force method. International Journal of Fracture 1992;58:285–304. Available from: https://doi.org/10.1007/BF00048950.

[21] Rao BN, Rahman S. A coupled meshless-finite element method for fracture analysis of cracks. International Journal of Pressure Vessels and Piping 2001;78:647–57. Available from: https://doi.org/10.1016/S0308-0161(01)00076-X.

[22] Kanth SA, Lone AS, Harmain GA, Jameel A. Modelling of embedded and edge cracks in steel alloys by XFEM,". Materials Today: Proceedings 2020;26:814–18. Available from: https://doi.org/10.1016/j.matpr.2019.12.423.

[23] Kanth SA, Lone AS, Harmain GA, Jameel A. Elasto plastic crack growth by XFEM: a review. Materials Today: Proceedings 2019;18:3472–81. Available from: https://doi.org/10.1016/j.matpr.2019.07.275.

[24] Jameel A, Harmain GA. Modeling and numerical simulation of fatigue crack growth in cracked specimens containing material discontinuities. Strength of Materials 2016;48(2):294–307. Available from: https://doi.org/10.1007/s11223-016-9765-0.

[25] Eberhard P, Gaugele T. Simulation of cutting processes using mesh-free Lagrangian particle methods. Computational Mechanics 2013;51:261–78. Available from: https://doi.org/10.1007/s00466-012-0720-z.

[26] De Lorenzis L, Wriggers P, Zavarise G. A mortar formulation for 3D large deformation contact using NURBS-based isogeometric analysis and the augmented Lagrangian method. Computational Mechanics 2012;49:1–20. Available from: https://doi.org/10.1007/s00466-011-0623-4.

[27] Bhardwaj G, Singh IV, Mishra BK, Bui TQ. Numerical simulation of functionally graded cracked plates using NURBS based XIGA under different loads and boundary conditions. Composite Structures 2015;126:347–59. Available from: https://doi.org/10.1016/j.compstruct.2015.02.066.

[28] Gu YT, Wang QX, Lam KY. A meshless local Kriging method for large deformation analyses. Computer Methods in Applied Mechanics and Engineering 2007;196:1673–84. Available from: https://doi.org/10.1016/j.cma.2006.09.017.

[29] Belytschko T, Krongauz Y, Organ D, Fleming M, Krysl P. Meshless methods: an overview and recent developments. Computer Methods in Applied Mechanics and Engineering 1996;139:3–47. Available from: https://doi.org/10.1016/S0045-7825(96)01078-X.

[30] Rao BN, Rahman S. An enriched meshless method for non-linear fracture mechanics. International Journal for Numerical Methods in Engineering 2004;59:197–223. Available from: https://doi.org/10.1002/nme.868.
[31] Duflot M, Nguyen-Dang H. A meshless method with enriched weight functions for fatigue crack growth. International Journal for Numerical Methods in Engineering 2004;59:1945–61. Available from: https://doi.org/10.1002/nme.948.
[32] Jameel A, Harmain GA. Extended iso-geometric analysis for modeling three-dimensional cracks. Mechanics of Advanced Materials and Structures 2019;26:915–23. Available from: https://doi.org/10.1080/15376494.2018.1430275.
[33] Singh IV, Bhardwaj G, Mishra BK. A new criterion for modeling multiple discontinuities passing through an element using XIGA. Journal of Mechanical Science and Technology 2015;29:1131–43. Available from: https://doi.org/10.1007/s12206-015-0225-8.
[34] Bhardwaj G, Singh IV, Mishra BK. Numerical simulation of plane crack problems using extended isogeometric analysis. Procedia Engineering 2013;64:661–70. Available from: https://doi.org/10.1016/j.proeng.2013.09.141.
[35] Jameel A, Harmain GA. A coupled FE-IGA technique for modeling fatigue crack growth in engineering materials. Mechanics of Advanced Materials and Structures 2019;26:1764–75. Available from: https://doi.org/10.1080/15376494.2018.1446571.
[36] Gupta V, Jameel A, Verma SK, Anand S, Anand Y. "Transient isogeometric heat conduction analysis of stationary fluid in a container,". Part E: Journal of Process Mechanical Engineering 2022;. Available from: https://doi.org/10.1177/0954408922112.
[37] Gupta V, Jameel A, Verma SK, Anand S, Anand Y. An insight on NURBS based isogeometric analysis, its current status and involvement in mechanical applications. Archives of Computational Methods in Engineering 2022;. Available from: https://doi.org/10.1007/s11831-022-09838-0.
[38] Singh AK, Jameel A, Harmain GA. Investigations on crack tip plastic zones by the extended iso-geometric analysis. Materials Today: Proceedings 2018;5:19284–93. Available from: https://doi.org/10.1016/j.matpr.2018.06.287.
[39] Melenk JM, Babuška I. The partition of unity finite element method: Basic theory and applications. Computer Methods in Applied Mechanics and Engineering 1996;139:289–314. Available from: https://doi.org/10.1016/S0045-7825(96)01087-0.
[40] Jameel A, Harmain GA. Extended iso-geometric analysis for modeling three dimensional cracks. Mechanics of Advanced Materials and Structures 2019;26:915–23. Available from: https://doi.org/10.1080/15376494.2018.1430275.
[41] Remmers JJ, Wells GN, de Borst R. A solid-like shell element allowing for arbitrary delaminations. International Journal for Numerical Methods in Engineering 2003;58:2013–40.
[42] Pearson RG. Application of fracture mechanics to the study of the tensile strength of structural lumber. International Journal of the Biology, Chemistry, Physics and Technology of Wood 1974;28:11–19.
[43] Barrett JD, Foschi RO. Mode II stress-intensity factors for cracked wood beams. Engineering Fracture Mechanics 1977;9(2):371–8.
[44] Leicester RH, Bunker PC. Fracture at butt joints in laminated pine. Forest Products Journal 1969;18:59–60.
[45] Mindess S, Nadeau JS, Barret JD. Slow crack growth in Douglas-fir. Wood Science 1975;7:389–96.
[46] Qiu LP, Zhu EC, van de Kuilen JWG. Modeling crack propagation in wood by extended finite element method. European Journal of Wood and Wood Products 2014;72:273–83.
[47] Walsh PF. Linear fracture mechanics in orthotropic materials. Engineering Fracture Mechanics 1972;4:533–41.
[48] Cramer S, Pugel A. Compact shear specimen for wood mode II fracture investigations. International Journal of Fracture 1987;35:163–74.

CHAPTER

17

Analysis of fracture in a receiver tube of parabolic trough collectors by extended finite element method

Amit Kumar[1], Vibhushit Gupta[1], Sanjay Sharma[2], Yatheshth Anand[1] and Kapil Chopra[1]

[1]School of Mechanical Engineering, Shri Mata Vaishno Devi University, Kakryal, Katra, Jammu and Kashmir, India [2]School of Energy Management, Shri Mata Vaishno Devi University, Kakryal, Katra, Jammu and Kashmir, India

17.1 Introduction

Renewable sources of energy are the naturally replenished resources that can be consumed indefinitely without being depleted. They are considered sustainable alternatives to fossil fuels and have a significantly less impact on the environment in terms of greenhouse gas emissions and pollution [1]. A parabolic trough collector (PTC) is a solar thermal technology that concentrates sunlight onto a receiver tube using a parabolic mirror. The heated fluid inside the tube generates steam, which drives a turbine to produce electricity. This process links renewable energy by using abundant and naturally replenished sunlight as the energy source, making it a sustainable and eco-friendly way to generate power [2,3]. The working of a PTC involves concentrating sunlight onto a receiver tube using a parabolic mirror. Parabolic mirror, receiver tube, and support structure are the main component of the parabolic trough collector. These components work together to harness solar energy, heat the fluid in the receiver tube, and generate steam for electricity production in solar thermal power plants [4]. The PTC utilizes an arched shape mirror to concentrate the sunrays onto a receiver tube. The receiver tube holds a heat transfer fluid (HTF) that absorbs the sunlight's heat and gets heated. The heated fluid then produces steam, which drives a turbine connected to an electricity generator, generating renewable electricity. This technology is efficient and environmentally friendly, making it a viable option for clean energy production. The receiver tube in a PTC is subject to several impacts due to the intense solar concentration

Enriched Numerical Techniques
DOI: https://doi.org/10.1016/B978-0-443-15362-4.00001-2

and high temperatures involved in the process. These impacts can affect the performance, durability, and maintenance requirements of the receiver tube. The receiver tube in a PTC plays a critical role in the conversion of sunlight into usable heat for generating electricity [5]. However, it is subject to various impacts due to the harsh conditions it faces. One significant impact is thermal stress caused by the concentrated sunlight, which leads to extreme temperature variations. To withstand this, the receiver tube must be made of highly heat-resistant materials capable of handling such thermal fluctuations without structural damage [6]. Moreover, the high-temperature HTF flowing through the tube can cause chemical reactions, leading to corrosion and erosion over time. Additionally, the receiver tube's inner surface is coated with selective coatings to enhance sunlight absorption and minimize heat loss. However, prolonged exposure to concentrated sunlight can cause the coating to degrade, affecting the tube's overall efficiency. To maintain optimal performance, regular maintenance and cleaning of the receiver tubes are essential, considering the accumulation of dust and debris over time. Proper material selection and engineering are also crucial to manage expansion and contraction during heating and cooling cycles, preventing mechanical failure. Despite these challenges, ongoing research aims to improve receiver tube materials, coatings, and design to enhance durability, efficiency, and lifespan. By addressing these impacts and ensuring proper maintenance, PTC systems can continue to be a reliable and effective means of harnessing solar energy for sustainable electricity generation [5].

Due to various factors, including thermal stress, material fatigue, and mechanical surface cracks, they are developed in receiver tubes. Analyzing cracks in the receiver tube of a PTC typically involves a combination of analytical and experimental methods [5,7]. Analytical methods disadvantages in crack analysis of receiver tubes include reliance on theoretical assumptions, potential inaccuracies, limitations in considering complex interactions, and challenges in handling heterogeneous materials or irregular crack shapes. Complementing analytical techniques with experimental methods can provide a more reliable and comprehensive understanding of cracks in receiver tubes, facilitating effective repair and maintenance strategies for PTC systems. Numerical strategies, such as the finite element method (FEM), are commonly used to analyze receiver tubes of PTCs [8,9]. FEM is a powerful numerical simulation technique that can model the tube's structural behavior and stress distribution, providing valuable insights into crack initiation and propagation. By dividing the tube into small elements and solving the governing equations of stress and deformation, FEM can predict stress concentrations around the crack and analyze its potential impact on the tube's integrity. It allows engineers to simulate different loading conditions, temperature distributions, and material properties to study how these factors influence crack growth. Additionally, FEM can be used to evaluate the effectiveness of potential repair strategies or design modifications to mitigate crack development. While FEM is a valuable tool, it also has some limitations, such as the need for accurate material properties, meshing complexities, and computational resources required for large-scale models. Therefore its successful implementation relies on proper calibration and validation with experimental data to ensure accurate predictions of crack behavior in receiver tubes. Complementing FEM with experimental data and other methods improves accuracy and reliability in assessing crack development for optimal design and maintenance.

The concept of extended finite element method (XFEM) is utilized for dealing with cases involving complex cracks, discontinuities, or singularities in the geometry of the structure. XFEM extends the capabilities of the traditional FEM by allowing for the representation of discontinuities, such as cracks, directly within the finite element mesh [10]. It introduces special enrichment functions that incorporate the presence of the crack and its characteristics into the analysis. This enrichment allows XFEM to model the crack propagation and interaction with other structures more accurately than traditional FEM, which typically requires remeshing to simulate crack growth [11]. By using XFEM, engineers can perform crack analysis and predict crack propagation without the need for remeshing, making it particularly useful for problems where the path of crack is not available a priori or at the position where the crack may evolve over time. Also, the geometry of the crack is defined using level set or Heaviside functions, providing a smooth representation of the discontinuity. Integration is performed over the enriched elements to evaluate the effects of the crack on the system's behavior, such as stress intensity factors (SIFs) and crack opening displacements. XFEM can predict the propagation of cracks and their paths using criteria like the maximum circumferential stress or energy release rate, allowing engineers to determine the growth of crack and its impact on the overall structure. XFEM's versatility extends beyond crack analysis, as it can handle various types of discontinuities, interfaces, and damage simulations in structures with complex geometries. Furthermore, XFEM can be coupled with other numerical strategies to accurately model crack interactions with other cracks or structures. However, validation and verification of XFEM results against analytical solutions or experimental data are crucial to ensure the accuracy and reliability of the method. Overall, XFEM's methodology provides a robust and efficient approach to analyze problems involving discontinuities, making it a valuable tool for engineers in diverse fields such as fracture mechanics, structural integrity assessments, and damage analysis of engineering components and systems. In addition, XFEM is considered a specialized extension of the FEA that allows for accurate modeling and analysis of complex cracks and discontinuities without the need for remeshing. It enhances the capabilities of traditional FEM in scenarios where singularities or evolving cracks need to be accurately captured and simulated. XFEM has applications in various fields, including fracture mechanics, damage analysis, and problems involving contact or interface problem [12].

In the case of mixed loadings like thermo-mechanical (TM) loadings, XFEM has also demonstrated its ability to solve these problems very efficiently and effectively [13–15]. Considering this efficient methodology, it offers significant advantages for assessing surface cracks in the case of receiver tubes of PTC as well. In this case, XFEM enhances the accuracy of crack representation by directly incorporating complex cracks into the finite element mesh without the need for remeshing. This feature allows precise modeling of the crack's shape, orientation, and propagation, which is crucial for understanding its behavior and impact on the receiver tube. Moreover, XFEM excels at predicting crack propagation paths, even when the crack's growth direction is not known beforehand. This capability provides valuable insights into the tube's structural integrity and aids in planning appropriate maintenance strategies. XFEM eliminates the effort and time required for remeshing, simplifying the modeling process and enabling engineers to focus on understanding the crack's behavior [16–18]. Additionally, it efficiently handles complex geometries often present in receiver tubes, ensuring a more realistic representation of the tube and its cracks. The method's versatility allows for capturing multiple cracks within a single finite element, facilitating the study of crack interactions in the receiver tube. With improved accuracy in handling singularities and discontinuities, XFEM enhances the reliability of the analysis, aiding in informed decisions about the receiver tube's repair or replacement. Overall, XFEM's advantages make it a powerful and valuable tool for assessing cracks in receiver tubes, contributing to the safe and efficient operation of PTC systems. In this study, the methodology for the analysis of crack generation in receiver tubes due to various TM loads using XFEM is presented. The variational formulation for the TM cases is shown in Section 17.2, followed by the representation of the approximation of unknown fields, numerical integration, and updation of enriched zone after crack growth is presented in Sections 17.3, 17.4, and 17.5, respectively. The methodology for evaluating SIF for cracks is explained in Section 17.6, and at last, this chapter ends with the conclusion of the study.

17.2 Variational formulation

For the evaluation of surface cracks in PTC receiver tubes, let us assume that the tubes are comprised of homogeneous, isotropic, and linear elastic material. The tube containing cracks on the surface is resented by a domain Ω that is bounded by a boundary "Γ" that is presented in Fig. 17.1. The "Γ" comprises Γ_u (displacement boundary), Γ_T (temperature boundary), Γ_q (heat flux boundary), Γ_t (traction boundary), and Γ_c (crack boundary). Based on this, the relations for TM problem, by assuming small strains and small displacements on Ω Γ_c, are given as

$$\rho c \frac{\partial T}{\partial t}(x,t) + \nabla.q(x,t) = \overline{Q}(x,t) \tag{17.1}$$

$$q(x,t) = -k\nabla T(x,t) \tag{17.2}$$

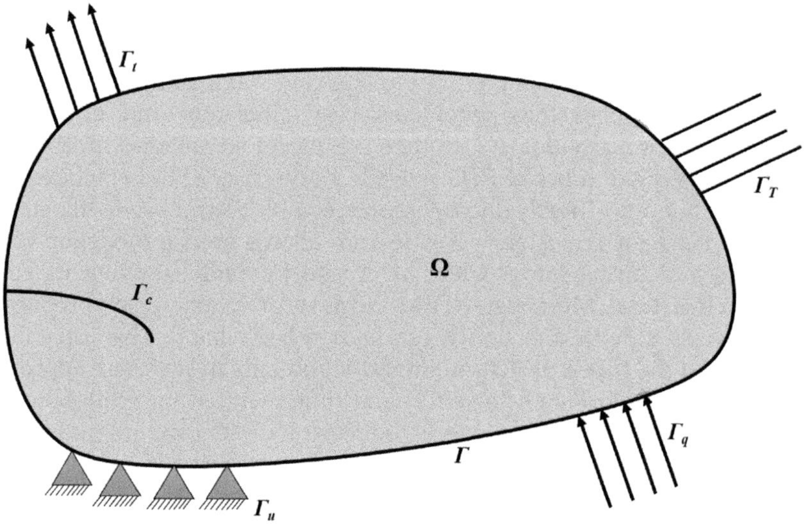

FIGURE 17.1 Problem domain.

$$\nabla.\sigma(x,t) + \overline{b}(x) = 0 \tag{17.3}$$

$$\sigma(x,t) = P{:}(\varepsilon - \varepsilon_T)(x,t) \tag{17.4}$$

$$\varepsilon_T(x,t) = \alpha(T(x,t) - T_0(x))I \tag{17.5}$$

$$\varepsilon(x,t) = \nabla_s u(x,t) \tag{17.6}$$

where the main objective is to evaluate the value of kinematic admissible $u(x,t)$, $T(x,t)$, $\sigma(x,t)$, and $q(x,t)$ for any $(x,t) \in (\Omega\ \Gamma_c) \times [0, \overline{T_f}]$, where $\overline{T_f}$ is the final time. The various fields like temperature vector (T), displacement vector (u), thermal strain tensor (ε_T), strain tensor (ε), and stress tensor (σ) are evaluated w.r.t. a reference temperature T_0. The heat flux vector is represented by (q), k shows the thermal conductivity with ρ, α, and c as density, thermal expansion coefficient, and specific heat, respectively. The fourth-order isotropic Hooke tensor is shown with "P," whereas $\overline{b}(x)$ and $\overline{Q}(x)$ present the imposed body force and heat source on the domain $\Omega\ \Gamma_c$. I and ∇_s are the identity tensor and gradient operator, respectively. The associated boundary conditions prescribed by $\overline{u}$ and $\overline{T}$ are imposed, respectively, on Γ_u and Γ_T, while heat flux $\overline{q}$ and traction t_0 are imposed on Γ_q and Γ_t as

$$\sigma.\hat{n} = \hat{t}\ on\ \Gamma_t \tag{17.7}$$

$$\sigma.\hat{n} = 0\ on\ \Gamma_c \tag{17.8}$$

$$u = \overline{u}\ on\ \Gamma_u \tag{17.9}$$

$$T = \overline{T}\ on\ \Gamma_T \tag{17.10}$$

$$q.n = \overline{q}\ on\ \Gamma_q \tag{17.11}$$

$$u = \overline{u}\ on\ \Gamma_u \tag{17.12}$$

where it is assumed that the surface of the crack Γ_c is free of traction. The problem is well stated with the relations $\Gamma_u \cup \Gamma_t = \Gamma_T \cup \Gamma_q = \Gamma$ and $\Gamma_u \cap \Gamma_t = \Gamma_T \cap \Gamma_q = \varnothing$. Furthermore, two cases have been considered in the upcoming section, that is, isothermal crack ($\Gamma_c \subset \Gamma_T, \overline{T} = \overline{T}_c$ on Γ_c) and adiabatic crack ($\Gamma_c \subset \Gamma_q$, $\overline{q} = 0$ on Γ_c).

Furthermore, the case under consideration involves time-dependent aspects, while the mechanical considerations are treated as quasistatic, neglecting inertial effects [19]. The pseudotime (PT) denoted as $\sigma(.,t)$ and $\varepsilon(.,t)$ is introduced due to the real time (RT) in the time space $[0, \overline{T_f}]$. It is important to note that the present work does not address the staggered TM cases. In the current approach, firstly, the case of transient thermal is solved, which accounts for the time-dependent behavior. Subsequently, the field of temperature at the RT t is computed, which is then used as input in the quasistatic mechanical case described by Eq. (17.3). This allows for the evaluation of the mechanical behavior at a given time $T(t)$, coupling it with the resolved thermal case. The mechanical case advances in PT "marching," which remains constantly conditional in RT due to the resolution of the thermal case. To handle a traction-free crack, the mechanical stress $\sigma(.,t)$ is split into a global stress $\sigma^g(.,t)$ and "thermal stress" $\sigma^{th}(.,t)$, enabling the definition of a continuous space of mechanical pseudotimes:

$$p\mathrm{T}_t = \left\{\tilde{t}{:}\nabla.\sigma^g(x,t) + \overline{b} = F(t); u|_{\Gamma_u} = \overline{u};\ \sigma.n|_{\Gamma_t} = \overline{t}\ \text{such}\ F(t) = {:}\nabla.\sigma^{th}(x,t)\right\} \tag{17.13}$$

where a pseudotime space, represented by $p\mathrm{T}_t$, is introduced alongside the real-time space, denoted as t. This implies that for each RT point, a corresponding PT is naturally and implicitly defined, capturing the proper temporal evolution of the system. In this case, we can simply consider $\tilde{t} = t$, indicating a direct relationship between real-time and pseudotime. From a discretization perspective, the study allows for the consideration of a steady-state case to avoid the need for pseudotime stepping, which is required in nonsteady-state scenarios with minor time steps. However, it's important to note that the current study involves strong coupling, and Eqs. (17.1)–(17.6) encompass the complete dynamics of the system, relying on the transient principle. Consequently, the temporal scenarios in RT reflect the dynamic behavior of the mechanical problem, which evolves into a PT-dependent form. To manage this strong coupling and dynamic behavior, the entire case is treated as a monolithic object, wherein all subcases progress simultaneously. This approach ensures that the interactions and dependencies between the thermal and mechanical aspects are properly accounted for throughout the analysis. By integrating both real-time and pseudotime considerations, the study achieves a comprehensive understanding of the system's behavior during transient dynamics.

The acceptable space for temperature and displacement is $\mathcal{F} \times \mathrm{T}, X = (u, T) \in \mathcal{F} \times \mathrm{T}$, where on Sobolev space $H^1(\Omega)$ the two variational spaces $\mathcal{F}$ and T are defined:

$$\mathcal{F} = \left\{u \in H^1(\Omega)^2 : u = \bar{u} \text{ on } \Gamma_u \text{ and u is discontinuous on } \Gamma_c\right\} \tag{17.14}$$

$$\mathrm{T} = \left\{T \in H^1(\Omega) : T = \overline{T} \text{ on } \Gamma_T \text{ and } T \text{ is discontinuous on } \Gamma_c\right\} \tag{17.15}$$

The homogenous essential conditions spaces can be written as

$$\mathcal{F}_0 = \left\{u \in H^1(\Omega)^2 : u = 0 \text{ on } \Gamma_u\right\} \subset H_0^1(\Omega)^2 \tag{17.16}$$

$$\mathrm{T} = \{T \in H^1(\Omega) : T = 0 \text{ on } \Gamma_T\} \subset H_0^1(\Omega)^2 \tag{17.17}$$

Thus the weak form for this TM case, utilizing the test functions as v and w can be described as follows:

$$W_t^u(u, v) = \int_\Omega \varepsilon(v):C:\varepsilon(u)d\Omega - \int_\Omega v.\bar{b}d\Omega - \int_{\Gamma_t} v.\bar{t}d\Gamma - \int_\Omega \varepsilon(v):C:\varepsilon_T(T)d\Omega = 0, \forall v \in \mathcal{F}_0 \tag{17.18}$$

$$W^T(T, w;t) = \int_\Omega w\left(\rho c\frac{\partial T}{\partial t}\right)d\Omega - \int_\Omega \nabla w.(k\nabla T)d\Omega - \int_\Omega \mathrm{w}.\overline{Q}d\Omega \int_{\Gamma_q} w.\bar{q}d\Gamma, \forall w \in \mathrm{T}_0 \tag{17.19}$$

17.3 Extended finite element method approximation

In XFEM approximation, firstly, field of displacement in the case of cracks is formulated. Then the two cases, that is, isothermal and adiabatic cracks, are reviewed individually for evaluation of temperature field. The approximations are given as follows:

17.3.1 Displacement approximation

By taking an FEM mesh and considering the crack independent of this mesh by using enrichment functions. The XFEM-based approximation of displacement field is given as follows:

$$u^h(x,y) = \sum_{i=1}^{n} N_i(x,y)\mathbf{u_i} + \sum_{j=1}^{n_s} N_j(x,y)[W(x,y) - W(x_j,y_j)]\mathbf{a_j} + \sum_{k=1}^{n_T} N_k(x,y)\sum_{\alpha=1}^{4}[\beta_\alpha(x,y) - \beta_\alpha(x_k,y_k)]\mathbf{b_k^\alpha} \tag{17.20}$$

where $\mathbf{u_i}$ presents the parameters relating to the specified control point, n are the number of shape functions; N_i shows shape function; n_s is the split-enriched node elements; n_T is tip-enriched node elements; $\mathbf{a_j}$ is enriched degree of freedom of split elements related to Heaviside function $W(\xi)$; $\mathbf{b_k^\alpha}$ is enriched degree of freedom of tip elements relates to CTEF $\beta_\alpha(x,y)$; c_l is extra enriched degree of freedoms pertaining to $\chi(x,y)$ that takes 0 for the inner control points of the void and +1 for the outside control points; $\beta_\alpha(x,y)$ and $W(x,y)$ presents the CTEF and Heaviside enrichment function, respectively.

The main purpose of the Heaviside function $W(x,y)$ is to produce discontinuity across the surface of the crack, which ranges from −1 to +1 and is presented as

$$W(x) = \begin{cases} -1, & otherwise \\ +1, & if\ (x - x_\Gamma).n \geq 0 \end{cases} \tag{17.21}$$

The Heaviside function is efficient when an element is divided by a crack interface. However, in the case of tip elements, a portion of the element is only cut, and the Heaviside function is not able to enrich that tip element. Thus CTEF is used for the enrichment of crack tip element. This enrichment function is defined by Chopp and Sukumar [20] and is explained as

$$\beta_\alpha(x,y) = \left[\sqrt{r}\cos\frac{\theta}{2}, \sqrt{r}\sin\frac{\theta}{2}, \sqrt{r}\cos\frac{\theta}{2}\sin\theta, \sqrt{r}\sin\frac{\theta}{2}\sin\theta\right] \tag{17.22}$$

These functions become the basis of the fields of the displacement close and around the tip of the crack. In the case of linear elastic fractures, these functions are utilized for enriching the displacement close to the tip of the crack. As four functions have been added to the conventional displacements, four additional nodes have to be placed on the standard nodes, which gives rise to eight extra degrees of freedom at each node. Hence, the number of DOF is different for split as well as tip nodes.

17.3.2 Temperature approximation

The approximation for thermal field, that is, for the temperature, is done in an analogous manner to the displacement approximations. In this, the approximation of temperature is presented for two cases of cracks, that is, adiabatic and isothermal. In adiabatic case, across the surface of crack, the temperature and displacement field are discontinuous, and the flux of heat is singular near the crack tip [21]. Similarly, the approximation equation for discontinuous temperature field in the case of cracks can be expressed as

$$\mathrm{T^h}(x,y) = \sum_{\mathrm{i}=1}^{n} N_i(x,y)\mathbf{T_i} + \sum_{\mathrm{j}=1}^{n_s} N_j(x,y)\left[\mathrm{W}(x,y) - \mathrm{W}(x_j,y_j)\right]\mathbf{a_j}^T$$

$$+ \sum_{\mathrm{k}=1}^{n_T} N_k(x,y) \sum_{\alpha=1}^{4} [\beta_\alpha(x,y) - \beta_\alpha(x_k,y_k)]\mathbf{b_k^{\alpha T}} \tag{17.23}$$

At the crack surface, exhibits a discontinuous behavior, while the temperature profile remains continuous. The key distinction between the isothermal and adiabatic cases lies in the heat flux and temperature angular variation. For the first crack case, the temperature varies radially from the crack's surface, creating a distribution that extends along the radial direction. On the other hand, in the case of an adiabatic crack, the temperature values vary perpendicular to the crack's surface, resulting in a distribution that extends perpendicularly to the surface. The distribution of temperature for this particular case can be expressed as follows [22]:

$$T = \frac{K_T}{k}\sqrt{\frac{2r}{\pi}}\sin\frac{\theta}{2} \tag{17.24}$$

Thus for capturing crack tip singularity $\sqrt{r}\cos\frac{\theta}{2}$, is the enrichment function used for enriching the approximation function of temperature. The generalized approximation function for the isothermal crack case is explained as

$$\mathrm{T^h}(x,y) = \sum_{\mathrm{i}=1}^{n} N_i(x,y)\mathbf{T_i} + \sum_{\mathrm{j}=1}^{n_s} N_j(x,y)\left[\sqrt{r}\cos\frac{\theta}{2}\right]\mathbf{a_j}^T \tag{17.25}$$

The time discretization in this TM analysis can be derived by considering the two sets of displacement–temperature values, denoted as$\{X_i\}$ at time t_i and $\{X_{i+1}\}$ at time $t_{i+1}, t_{i+1} = t_i + \Delta t$. These two sets of values are related using the general trapezoidal rule, which involves a parameter β:

$$\{X_{i+1}\} = \{X_i\} + \{(1-\beta\}\{X_i\} + \beta\{\dot{X}_{i+1}\}\Delta t \tag{17.26}$$

Also, the relation for a given time-dependent linear system can be defined as

$$[C]\{\dot{X}\} + K\{X\} = \{F\} \tag{17.27}$$

Using this, we can modify Eq. (17.26) for time t_i and t_{i+1} by multiplying with $(1-\beta)$ and then with β. On conducting the addition of the two resulting equations and after performing the elimination of the time derivative term for t_{i+1} the relation for computation of $\{X_{i+1}\}$ at the actual time is given as

$$\begin{cases} \left(\beta[K] + \dfrac{1}{\Delta t}[C]\right)\{X_{i+1}\} = \beta[F_{i+1}] + [C]\left(\dfrac{1}{\Delta t}\{X_i\} + (1-\beta)\{\dot{X}_{i+1}\}\right) \\ \{\dot{X}_{i+1}\} = [C]^{-1}(\{F_i\} - [K]\{X_i\}) \\ \{X_0\} = X_0 \end{cases} \tag{17.28}$$

For improving the stability of the solution, Crank Nicolson of $\beta = \frac{1}{2}$ has been chosen, which is highly stable and results with nullified numeric dissipations into the approximations. Further, let us assume that the global variable is presented by X for a full TM coupled system, such that $\{X\}^T = \{\{U_{std}\}^T, \{U_{enr}\}^T, \{T_{std}\}\{T_{enr}\}\}$, where $\{U_{enr}\}^T$ and $\{U_{std}\}^T$ present the enriched and standard parts of displacement, respectively; that goes the same for the case of temperature. The coupled form of the XFEM matrix can be expressed as

$$\left[K_{global}^{XFEM}\right] = \begin{bmatrix} [K_{UU}] & [K_{UT}] \\ [0] & [K_{TT}] \end{bmatrix}; \left[C_{global}^{XFEM}\right] = \begin{bmatrix} [0] & [0] \\ [0] & [C_{TT}] \end{bmatrix}; \tag{17.29}$$

$$\left[F_{global}^{XFEM}\right]^T = \left[\{K_U\}^T \quad \{F_T\}^T \right]$$

Here in this equation, $[K_{UU}]$ and $[K_{TT}]$ present the purely mechanical and thermal part, respectively, whereas $[K_{UT}]$ represents the coupled part, illustrating the effect of the thermal part on the mechanical part. This coupling term indicates how changes in temperature can affect the mechanical behavior of the system. Conversely, the zero term denotes that there is no direct coupling between the mechanical and the thermal parts in the TM analysis. This implies that in this specific analysis, the mechanical behavior does not influence the thermal behavior of the system.

17.4 Numerical integration

The XFEM is an extended version of the traditional FEM that allows for more accurate modeling of problems involving cracks or discontinuities within the domain. In XFEM, standard numerical integration methods, like the Gauss quadrature, are not sufficient for elements crossed by cracks due to the complexity introduced by these discontinuities. To address this issue, XFEM introduces a concept called "enrichment." The enrichment involves locally enhancing the standard FEM approximation functions in the vicinity of the crack region, which is referred to as the enriched zone. Within this enriched zone, special enrichment functions are used to better capture the crack's behavior and singularities. For elements that are entirely crossed by the crack (split or vertex elements), the domain is divided into two parts, that is, nonconvex domain and convex domain. To handle the various cases that can arise, a subdivision process is applied to create convex disjoint partitions. These partitions form subconvex elements that accurately represent the behavior around the crack region. Similarly, for elements partially crossed by the crack or near the crack tip, further attention is required due to the presence of nonpolynomial nature of approximation functions. These elements are also subdivided into subconvex elements to effectively handle the complexities introduced by the crack [23]. The overall subdivision process at the local element level can be visualized as a spider web pattern centered around the crack tip. This approach enables XFEM to accurately model the crack behavior and obtain more reliable solutions compared to traditional finite element methods [24]. It's important to note that specific enrichment functions and implementation details may vary depending on the problem and the characteristics of the crack. Various research papers and works in the field have supported the development of XFEM for crack problems.

17.5 Updating of enriched zone for crack growth

In the starting stage of crack propagation, crack's position is predetermined as Γ_c^0, and the mesh remains fixed throughout the growth process without any alterations. Instead of relying on the crack's known Cartesian global position, a level-set function is employed to identify the crack independently of the mesh definition. This method uses signed-distance functions to determine the crack's relative position with respect to its neighboring nodes. To formulate the discrete XFEM for the displacement and temperature fields, two types of enriched nodes are selected: Heaviside enriched nodes and tip-enriched nodes. Initially, all mesh nodes, including those fully enriched, are approximated using standard approximation functions. The selection of enriched nodes is performed incrementally. During each increment, the crack propagates in the desired direction according to a predefined crack growth criterion. As a result, a new configuration of the crack geometry emerges, leading to an associated enriched zone called EZ^k. This iterative process continues, yielding different configurations with varying types of enriched nodes. To transfer nodal fields, such as temperature and displacements, from the $(k-1)^{th}$ increment to the kth increment, an L^2 projection is employed. This projection method effectively maps the nodal quantities from the previous configuration to the new configuration, ensuring computational stability and efficiency. As the crack advances, changes occur in the geometrical, topological, and numerical aspects of the TM problem. This adaptability leads to a flexible set of degrees of freedom that accommodates the crack's evolution, resulting in the linear system of discrete equations changing dimensionally. The selection of EZ^k nodes is solely based on the crack's relative position within its nodal environment during the current increment, independent of the previous configuration. The iterative process continues until the estimated or evaluated endpoint of the crack propagation process is reached. This approach provides an accurate representation of crack growth, effectively considering the changing crack geometry and its influence on the numerical solution at each step of the analysis.

17.6 Crack growth criterion and computation of stress intensity factors

17.6.1 Updating of crack and propagation criterion

The level-set function plays a crucial role in implicitly representing cracks and evaluating enriched fields in both mechanical Moës et al. [25] and thermal analyses. In this approach, the crack is represented as the zero-level set of a designated function. To locate the crack tip positions, the intersection between the zero-level contour and a second orthogonal level-set function is considered, as proposed by Stolarska et al. [26]. This involves using the signed-distance function, which is approximated through finite element interpolation on the same mesh used for the mechanical and thermal problems. This interpolation simplifies the evaluation of the level-set function at the element level, allowing the computation of its derivative using the derivative of shape functions θ_c. The maximum hoop (circumferential) tensile stress theory, introduced by Erdogan and Sih [27], is applied for monitoring the growth of the crack. While considering the mixed-mode loading, information near the crack tip is extracted by evaluating the stress state in polar coordinates. The crack extension is assumed to begin radially from its tip in the plane perpendicular to the direction of maximum tension, represented by the critical angle θ_c. Crack propagation initiates when the hoop stress component $\sigma_{\theta\theta}$ reaches a specified critical value. The critical angle θ_c can vary depending on the mode of crack growth. For pure mode-I crack propagation (when $K_{II} = 0$), θ_c is also 0. However, for cases with $K_{II} < 0$, $\theta_c > 0$, and for $K_{II} > 0$, $\theta_c < 0$. Sukumar and Prevost [28] provided a convenient expression for the critical crack growth angle θ_c. This angle governs the direction and initiation of crack growth under different loading conditions:

$$\theta_c = 2\tan^{-1}\left[\frac{-2\left(\frac{K_{II}}{K_I}\right)}{1 + \sqrt{1 + 8\left(\frac{K_{II}}{K_I}\right)^2}}\right] \tag{17.30}$$

The crack path extension can be evaluated by employing constant increment growth, which is considered to be a reliable approach. However, choosing the appropriate value for the crack increment Δa is crucial for achieving accurate results while modeling crack propagation. The various factors that influence the value of the paths of crack propagation is extensively studied by Huang et al. [29], who demonstrated the impact of these choices through various examples. Three main parameters significantly affect the excellence of the crack path. The first parameter is the magnitude of crack growth, representing the length of the incremental crack segment, which should be chosen in limits of $l_e \leq \Delta a \leq \frac{3}{2} l_e$, where l_e is the element size, the value of which is equal to the square root of the average element area A_e. Secondly, the mesh size plays a vital role in providing a finer approximation of the field close to the crack. Lastly, the choice of the *J*-integral domain is crucial for accurately evaluating the *J*-integral, which is essential for determining SIF in different modes (I, II, and mixed mode) and subsequently calculating the orientation of the crack growth θ_c. In the literature, a number of crack extension criteria occur that ensure proper crack progression, particularly in cases of cyclic loading. These criteria are utilized for checking the crack's progress under fatigue conditions and incorporating the rate of crack growth w.r.t. the loading cycle. A popular model is the classical Paris law, a generalized form of fatigue law, which represents the relation between the crack growth rate to the SIF range with the indulgence of two constants known as the Paris law constants. However, this law has limitations, that is, it requires a minimum SIF for the propagation of the crack and does not account for the stress ratio. An improved form of the Paris law was later proposed by Xiaoping et al. [30], which overcomes the drawbacks of the traditional Paris law but requires three extra material-specific parameters. These models offer a better representation of crack propagation and adapt the crack increment for each progression based on fatigue characteristics, but they require additional material parameters determined through fatigue tests. This aspect might pose a drawback for their implementation in some cases. Alternatively, the fixed crack increment method offers advantages, as it requires fewer material parameters in comparison to the aforementioned laws. While its convergence may be lower in certain cases, selecting an appropriate increment for a crack that totally depends on the FEA grid and other relevant constraints for yielding of satisfactory results. Researchers, such as Baydoun et al. [31], have investigated the capability of the fixed crack increment method to get crack paths that line up well with reference solutions available in the open literature.

17.6.2 Evaluation of stress intensity factor

The computation of SIF using the domain-based interaction integral (DBII) method. This approach selectively considers field variables that are away from the immediate vicinity of the crack tip, leading to enhanced accuracy in evaluating the SIFs. By judiciously choosing an auxiliary field, this method proves to be advantageous in efficiently calculating individual SIFs from a solution. The DBII method is particularly useful for mechanical loading scenarios. The derivation of the domain form of the interaction integral for mechanical loading is presented by Baydoun and Fries [32]. This derived form provides a practical expression to determine the SIFs accurately. By utilizing this domain-based approach, we can exclude field variables that are influenced by the crack tip region, resulting in a more reliable evaluation of the SIFs. Moreover, this method allows us to efficiently extract individual SIFs when only a specific auxiliary field is chosen, which is a notable advantage in certain applications. Overall, the DBII offers improved accuracy and efficiency in computing SIFs for mechanical loading, making it a valuable tool in fracture mechanics analyses. The formulation provided in Ref. [32] serves as a practical guide for applying this method in various engineering scenarios that can be expressed as

$$M^{(1,2)} = \int_A \left[\sigma_{ij}^{(1)} \frac{\partial u_i^{(2)}}{\partial x_1} + \sigma_{ij}^{(2)} \frac{\partial u_i^{(1)}}{\partial x_1} - Z^{(1,2)} \delta_{ij} \right] \frac{\partial v}{\partial x_j} dA \tag{17.31}$$

$$Z^{(1,2)} = \frac{1}{2}\left(\sigma_{ij}^{(1)} \epsilon_{ij}^{(2)} + \sigma_{ij}^{(2)} \epsilon_{ij}^{(1)} \right) = \sigma_{ij}^{(1)} \epsilon_{ij}^{(2)} = \sigma_{ij}^{(2)} \epsilon_{ij}^{(1)} \tag{17.32}$$

where "v" shows the scalar weight function (that ranges between 0 and 1); $Z^{(1,2)}$ is mutual strain energy; σ and ϵ represent the stresses and strains, respectively; "1" and "2" represent the actual state and the auxiliary states, respectively.

Analogously, interaction integral for TML can be written as per Ref. [33]:

$$M^{(1,2)} = \int_A \left[\sigma_{ij}^{(1)} \frac{\partial u_i^{(2)}}{\partial x_1} + \sigma_{ij}^{(2)} \frac{\partial u_i^{(1)}}{\partial x_1} - Z^{(1,2)} \delta_{ij} \right] \frac{\partial v}{\partial x_j} dA + \alpha \int_a \frac{\partial T}{\partial x_1} \sigma_{kk}^{(2)} q dA \tag{17.33}$$

Eq. (17.33) can further be modified for linear elastic problems as

$$M^{(1,2)} = \frac{2}{E^*}\left(K_I^{(1)} K_I^{(2)} + K_{II}^{(1)} K_I^{(2)} \right) \tag{17.34}$$

where $E^* = E$ and $E^* = \frac{E}{(1-v^2)}$ show the case of plane stress and strain correspondingly. Further, single SIFs under mixed loads can be derived by taking the appropriate choice of auxiliary states. In the case of mode-I loading, the SIFs are derived as $K_I^{(2)} = 1$ and $K_{II}^{(2)} = 0$ and for mode-II as $K_I^{(2)} = 0$ and $K_{II}^{(2)} = 1$.

17.7 Conclusion

Renewable resources offer a promising path toward sustainable energy solutions, and PTCs stand out as a vital technology in this endeavor. However, the development of surface cracks in receiver tubes due to mixed TM loads poses significant operational challenges. To address this issue, our study presented a methodology based on the XFEM to comprehensively analyze these cracks in receiver tube of a PTC. Additionally, we discussed the concepts of SIF and the updating of crack propagation criteria, enhancing the accuracy of our crack analysis. By incorporating XFEM and SIF concepts into our analysis, we were able to precisely evaluate the SIF, offering a deeper understanding of crack initiation and propagation. This comprehensive evaluation allowed us to predict crack growth under diverse loading conditions, including both thermal and mechanical aspects. By considering mixed TM loads, our approach provided a more realistic representation of the real-world operating conditions faced by PTCs. Also, by studying the combined effects of thermal gradients and mechanical stresses, we were able to identify critical areas of stress concentration and predict crack propagation paths with enhanced precision. This in-depth understanding of crack development can facilitate the implementation of preventive measures, informed maintenance strategies, and targeted material enhancements, ultimately prolonging the operational life and optimizing the efficiency of PTCs. Our study emphasizes the importance of continuous monitoring and thorough analysis of PTC performance to address the cracks' potential detrimental effects. By timely identifying and assessing cracks resulting from mixed TM loads, operators and engineers can implement necessary maintenance and repair actions to prevent catastrophic failures and ensure optimal energy output.

References

[1] Kumar A, Kumar K, Kaushik N, Sharma S, Mishra S. Renewable energy in India: current status and future potentials. Renewable and Sustainable Energy Reviews 2010;14(8):2434–42. Available from: https://doi.org/10.1016/j.rser.2010.04.003.

[2] Murtuza SA, Byregowda HV, H MMA, Imran M. Experimental and simulation studies of parabolic trough collector design for obtaining solar energy. Resource-Efficient Technologies 2017;3(4):414–21. Available from: https://doi.org/10.1016/j.reffit.2017.03.003.

[3] Singh RK, Chandra P. Parabolic trough solar collector: a review on geometrical interpretation, mathematical model, and thermal performance augmentation. Engineering Research Express 2023;5(1):012003. Available from: https://doi.org/10.1088/2631-8695/acc00a.

[4] Jebasingh VK, Herbert GMJ. A review of solar parabolic trough collector. Renewable and Sustainable Energy Reviews 2016;54:1085–91. Available from: https://doi.org/10.1016/j.rser.2015.10.043.

[5] Liu Q, Wang Y, Gao Z, Sui J, Jin H, Li H. Experimental investigation on a parabolic trough solar collector for thermal power generation. Science in China Series E: Technological Sciences 2010;53(1):52–6. Available from: https://doi.org/10.1007/s11431-010-0021-8.

[6] Thappa S, Chauhan A, Sawhney A, Anand Y, Anand S. Thermal selective coatings and its enhancement characteristics for efficient power generation through parabolic trough collector (PTC). Clean Technologies and Environmental Policy 2020;22(3):557–77. Available from: https://doi.org/10.1007/s10098-020-01820-3.

[7] Amein H, Akoush BM, El-Bakry MM, Abubakr M, Hassan MA. Enhancing the energy utilization in parabolic trough concentrators with cracked heat collection elements using a cost-effective rotation mechanism. Renewable Energy 2022;181:250–66. Available from: https://doi.org/10.1016/j.renene.2021.09.044.

[8] Wang Y, Xu J, Liu Q, Chen Y, Liu H. Performance analysis of a parabolic trough solar collector using Al_2O_3/synthetic oil nanofluid. Applied Thermal Engineering 2016;107:469–78. Available from: https://doi.org/10.1016/j.applthermaleng.2016.06.170.

[9] Alkathiri AA, Jamshed W, Uma Devi S S, Eid MR, Bouazizi ML. Galerkin finite element inspection of thermal distribution of renewable solar energy in presence of binary nanofluid in parabolic trough solar collector. Alexandria Engineering Journal 2022;61(12):11063–76. Available from: https://doi.org/10.1016/j.aej.2022.04.036.

[10] Afshar A, Daneshyar A, Mohammadi S. XFEM analysis of fiber bridging in mixed-mode crack propagation in composites. Composite Structures 2015;125:314–27. Available from: https://doi.org/10.1016/j.compstruct.2015.02.002.
[11] Jameel A, Harmain GA. Fatigue crack growth in presence of material discontinuities by EFGM. International Journal of Fatigue 2015;81:105–16. Available from: https://doi.org/10.1016/j.ijfatigue.2015.07.021.
[12] Kanth SA, Shafi Lone A, Harmain GA, Jameel A. Elasto plastic crack growth by XFEM: a review. Materials Today: Proceedings 2019;18:3472–81. Available from: https://doi.org/10.1016/j.matpr.2019.07.275.
[13] Gill P, Davey K. Analysis of thermo-mechanical behaviour of a crack using XFEM for Leak-before-Break assessments. International Journal of Solids and Structures 2014;51(11):2062–72. Available from: https://doi.org/10.1016/j.ijsolstr.2014.02.007.
[14] Habib F, Sorelli L, Fafard M. Full thermo-mechanical coupling using eXtended finite element method in quasi-transient crack propagation. Advanced Modeling and Simulation in Engineering Sciences 2018;5(1)18. Available from: 100.1186/s40323-018-0112-9.
[15] Hosseini SS, Bayesteh H, Mohammadi S. Thermo-mechanical XFEM crack propagation analysis of functionally graded materials. Materials Science and Engineering: A 2013;561:285–302. Available from: https://doi.org/10.1016/j.msea.2012.10.043.
[16] Jameel A, Harmain GA. Modeling and numerical simulation of fatigue crack growth in cracked specimens containing material discontinuities. Strength of Materials 2016;48(2):294–307. Available from: https://doi.org/10.1007/s11223-016-9765-0.
[17] Kanth SA, Harmain GA, Jameel A. Modeling of nonlinear crack growth in steel and aluminum alloys by the element free Galerkin method. Materials Today: Proceedings 2018;5(9, Part 3):18805–14. Available from: https://doi.org/10.1016/j.matpr.2018.06.227.
[18] Kanth SA, Lone AS, Harmain GA, Jameel A. Modeling of embedded and edge cracks in steel alloys by XFEM. Materials Today: Proceedings 2020;26:814–18. Available from: https://doi.org/10.1016/j.matpr.2019.12.423.
[19] Khoei AR, Bahmani B. Application of an enriched FEM technique in thermo-mechanical contact problems. Computational Mechanics 2018;62(5):1127–54. Available from: https://doi.org/10.1007/s00466-018-1555-z.
[20] Chopp DL, Sukumar N. Fatigue crack propagation of multiple coplanar cracks with the coupled extended finite element/fast marching method. International Journal of Engineering Science 2003;41(8):845–69. Available from: https://doi.org/10.1016/S0020-7225(02)00322-1.
[21] Sih GC. On the singular character of thermal stresses near a crack tip. Journal of Applied Mechanics 1962;29:587. Available from: https://doi.org/10.1115/1.3640612.
[22] Duflot M. The extended finite element method in thermoelastic fracture mechanics. International Journal for Numerical Methods in Engineering 2008;74(5):827–47. Available from: https://doi.org/10.1002/nme.2197.
[23] Dolbow J, Moës N, Belytschko T. Discontinuous enrichment in finite elements with a partition of unity method. Finite Elements in Analysis and Design 2000;36(3):235–60. Available from: https://doi.org/10.1016/S0168-874X(00)00035-4.
[24] Laborde P, Pommier J, Renard Y, Salaün M. High-order extended finite element method for cracked domains. International Journal for Numerical Methods in Engineering 2005;64:354–81. Available from: https://hal.science/hal-00815711.
[25] Moës N, Gravouil A, Belytschko T. Non-planar 3D crack growth by the extended finite element and level sets—Part I: Mechanical model. International Journal for Numerical Methods in Engineering 2002;53(11):2549–68. Available from: https://doi.org/10.1002/nme.429.
[26] Stolarska M, Chopp DL, Moës N, Belytschko T. Modelling crack growth by level sets in the extended finite element method. International Journal for Numerical Methods in Engineering 2001;51(8):943–60. Available from: https://doi.org/10.1002/nme.201.
[27] Erdogan F, Sih GC. On the crack extension in plates under plane loading and transverse shear. Journal of Basic Engineering 1963;85(4):519–25. Available from: https://doi.org/10.1115/1.3656897.
[28] Sukumar N, Prévost JH. Modeling quasi-static crack growth with the extended finite element method Part I: Computer implementation. International Journal of Solids and Structures 2003;40(26):7513–37. Available from: https://doi.org/10.1016/j.ijsolstr.2003.08.002.
[29] Huang R, Sukumar N, Prévost JH. Modeling quasi-static crack growth with the extended finite element method Part II: Numerical applications. International Journal of Solids and Structures 2003;40(26):7539–52. Available from: https://doi.org/10.1016/j.ijsolstr.2003.08.001.

[30] Huang X, Torgeir M, Cui W. An engineering model of fatigue crack growth under variable amplitude loading. International Journal of Fatigue 2008;30(1):2–10. Available from: https://doi.org/10.1016/j.ijfatigue.2007.03.004.

[31] Baydoun M, Fries TP. Crack propagation criteria in three dimensions using the XFEM and an explicit–implicit crack description. International Journal of Fracture 2012;178(1):51–70. Available from: https://doi.org/10.1007/s10704-012-9762-7.

[32] Singh IV, Mishra BK, Bhattacharya S, Patil RU. The numerical simulation of fatigue crack growth using extended finite element method. International Journal of Fatigue 2012;36(1):109–19. Available from: https://doi.org/10.1016/j.ijfatigue.2011.08.010.

[33] Banks-Sills L, Dolev O. The conservative M-integral for thermal-elastic problems. International Journal of Fracture 2004;125(1):149–70. Available from: https://doi.org/10.1023/B:FRAC.0000021065.46630.4d.

CHAPTER

18

Extended isogeometric analysis for fracture behavior of isothermally cracked single-walled carbon nanotube-reinforced composite exposed to thermo-mechanical load

Aanchal Yadav, Gagandeep Bhardwaj and R.K. Godara

Mechanical Engineering Department, Thapar Institute of Engineering and Technology, Patiala, Punjab, India

18.1 Introduction

Two or more constituents that are unique, have distinctly dissimilar physical and chemical properties, and are also different at macroscopic or microscopic level make up composite materials. Subsequently, the materials syndicate the beneficial properties of the parental materials and occasionally display novel properties that are not present in the parent compounds. Composite materials that feature at least two different components give an efficient and adaptable method to achieve demand-driven material properties [1]. New design possibilities made possible by composites may be exceedingly difficult to realize with conventional materials. For instance, the majority of structural materials used in engineering must possess both strength and toughness; nevertheless, strength and toughness cannot coexist in the same material [2].

Due to their capability to increase the stiffness and strength of composite materials, CNTs have been utilized as a reinforcing element in the past stiffness [3–7]. Carbon nanotube (CNT) is one of the most intriguing choices of reinforcement materials because of its exceptional electrical, mechanical, and thermal capabilities out of the myriad of reinforcement elements in composites. Carbon nanotube-reinforced composites (CNTRCs) have an

Enriched Numerical Techniques
DOI: https://doi.org/10.1016/B978-0-443-15362-4.00022-X

advantage over other nanocomposites due to their improved fracture toughness, which also leads to superior fatigue and mechanical performance. Since previous studies have demonstrated the strength-amplification of CNTs, CNTRCs have grown in favor not merely in the aerospace sector, which is a weight-sensitive industry, but also in many other disciplines of engineering (including gas turbines, nuclear reactors, automobiles, and locomotives), their high specific stiffness qualities being the reason. In their lifetimes, the majority of engineering components built of epoxy-matrix reinforced with CNT are exposed to thermal and mechanical loads. This makes them prone to temperature-dependent failures. The engineering components collapse catastrophically when there are cracks present under combined loading situations. The analysis of engineering fracture problems typically involves a variety of sophisticated techniques that are compatible with the traditional fracture mechanics framework. Estimating stress intensity factors (SIFs) during thermo-mechanical loading is crucial for the safety assessment. Numerous researchers have in the past used numerical techniques to analyze cracked structures that had undergone thermal loading. For instance, one of the approaches utilized to determine the SIFs of the structure with cracks under thermo-mechanical load was the finite element method (FEM) [8]. The issues of fracture analysis in thermo-mechanical load have also been successfully solved using this method, characterized by a static crack in two dimensions (2D) [9], and next by [10] in three dimensions (3D). When simulating the crack analysis problem, FEMs and boundary element methods each have a unique set of restrictions, for example, conformal mesh, remeshing, and usage of singular elements at the tip of the crack. Numerous techniques, such as the element-free Galerkin method [11], the kernel particle reproduction approach [12], and the extended finite element (XFEM) approach [13–15], were created to address these issues. These techniques were found to be more effective in handling problems related to moving discontinuities, such as the propagation of crack, without the requirement for remeshing or moving nodal points. To resolve the solid cracked body exposed to thermo-mechanical loading [16], used the XFEM method [17]. Expanded the XFEM formulations to simulate a domain of a functionally graded material with many cracks subject to mechanical and thermal loads. Diverse basis functions are used to determine the geometry and the solution. In each of these aforementioned methods, the approximation of geometry brings some error into the solution. By adopting the idea of isogeometric analysis (IGA), this problem is overcome by creating a strong and significant relationship among meshing, analysis, computer-aided design (CAD), and geometry [18]. IGA has the benefit over other approaches since it defines both the geometry and the solution utilizing a common basis function, nonuniform B-splines (NURBS). IGA is hence accurate and exact when used to solve technical fracture problems. Also known as extended isogeometric analysis (XIGA), the IGA was expanded to address issues involving discontinuities utilizing partition of unity enrichment. Many researchers had used XIGA to address the issues with crack formation in diverse structures [19–25].

Mirzaei and Kiani [26] presented research that used IGA to examine how the composite laminated plates consisting of graphene sheets as reinforcement responded to thermal buckling. The first-order shear deformation plate theory is utilized to compute the prebuckling forces produced by thermal work and the plate's strain energy. Then the thermal buckling behavior of the grapheme-reinforced composite plates is examined by the means of a NURBS-based isogeometric FEM. The analysis demonstrates how the plate's critical buckling temperature is influenced by the aspect ratio, lamination scheme, pattern of functionally gradation, side-to-thickness ratio, and conditions at the boundary. Farzam et al. [27] used IGA based on modified couple stress theory to examine the buckling of functionally graded CNTRC (FG-CNTRC) plates (MCST). The analysis of thermal and mechanical buckling uses an improved hyperbolic shear deformation theory. The IGA approach, which utilizes B-spline or NURBS functions, is utilized for numerical analysis. The findings showed that isogeometric analysis (IGA) is an effective numerical technique that may be utilized to address a variety of issues, including size-dependent analysis. In another study, Kiani [28] performs an investigation of the thermal postbuckling of composite laminated graphene sheet–reinforced plates. It was presumed that the thermo-mechanical properties of the composite media in their entirety rely on temperature. It is demonstrated that the lowest postbuckling deflection and greatest critical buckling temperature are produced by the FG-X design of graphene reinforcement. The examination was carried out on modified polymer nanocomposite (PNC) by Negi et al. [8] to examine the impact of discontinuities like inclusions and holes. For the study, the extended FEM (XFEM) was used. Van Do et al. [29] used a Bezier extraction-based IGA to investigate the dynamic transient and free vibration response of plates made of CNTRC. The projected IGA approach coupled with the HSDT is demonstrated to be more efficient, effective, and accurate in the aforementioned study when equated to the traditional FEM, which is the first-order shear deformation theory-based method. In another study, Van Do et al. [30] used IG-FEM integrated with a new hybrid-type higher order shear deformation theory (HSDT) that is a Bezier extraction-based method to examine the buckling and bending of the FG-CNTRC structure. Isogeometric analysis (IGA) can be applied in the traditional framework of the FEM by transferring NURBS basis functions to Bezier elements in the form of Bernstein polynomial basis using the Bezier extraction operator. As the polynomial order rises, the isogeometric technique proposed by Van Do et al. [30] is thought to be more precise and successful and produces higher convergence. Zhang et al. [31] solved the free vibration of CNT-reinforced and FGM sector shells of cylindrical shape using an isogeometric numerical approach. Consideration is given to the consequences of shear deformation, rotary inertia, and axis extensibility. The mixture rule and effective CNT characteristics are used to estimate the material properties. The virtual principle is used to determine the weak form. According to the convergence study, the IGA with higher order basis functions is proven to be converged more quickly. The findings demonstrate that the boundary condition impacts the mode forms significantly, whereas the CNT volume percentage and CNT distribution type have little effect. An LR NURBS-based multipatch local mesh refinement XIGA is suggested by Yuan et al. [32] for complex fractured Mindlin–Reissner plates. Numerous numerical cases are provided to show the effectiveness of the adaptive multipatch XIGA for modeling the growth of cracks in complex structures. An adaptive XIGA approach was provided by Li et al. [33] for shakedown analysis of hole problems. When solving larger scale optimization issues, the second-order cone is used. Fang et al. [34] developed an adaptive XIGA model for the investigation of the thermal buckling behavior of FG composite plates consisting of flaws. It also discusses how several numerical factors, for example, gradient index, fracture direction and position, and hole size, might affect the critical thermal buckling temperature. The results of their study concluded that, in contrast to uniform global refinement, adaptive local refinement has a quick convergence rate. Additionally, it was shown that cracks have a greater refinement priority than holes and that small holes have a higher refinement priority than large holes. To forecast the behavior of fatigue crack formation in CNT-reinforced polymer composites exposed to thermo-mechanical loading conditions, Dwivedi et al. [35] presented a dual-scale modeling method. In meso-scale modeling, the effective orthotropic characteristic of a polymer composite reinforced with CNT is calculated by the means of the mean field homogenization method. Additionally, the higher XFEM method is employed by Dwivedi et al. [35] to examine the macro-scale evolution of fatigue cracks utilizing the equivalent composite property. Enrichment terms of higher order in XFEM are used in macro-scale modeling to improve the crack tip's solution accuracy. Using the XIGA technique, Yadav et al. [36] explored the CNT-reinforced FGM's fracture behavior when exposed to thermo-mechanical loading in the existence of multiple discontinuities. The gradation of the properties of the FG structures, which are composed of metal (Ti−6Al−4V) and ceramic (ZrO_2), is subject to the exponential rule. The equivalent thermal and mechanical properties of CNTs are evaluated by utilizing various micromechanics models. The computational modeling for the loading of thermal nature considers adiabatic cracks. SIFs are extracted by the means of the technique of interaction integral. The findings of their research demonstrate that holes have a greater impact on SIFs than inclusions.

The authors are using XIGA in the current work to resolve the crack problem in a CNT-reinforced epoxy-matrix under thermo-mechanical loading conditions.

The organization of this chapter is as trails: Section 18.2 provides a quick outline of the CNT-reinforced epoxy-matrix equivalent properties. The 2D plane strain isothermal and thermo-mechanical problem's mathematical formulation is shown in Section 18.3. Calculating the equivalent properties of the CNT-reinforced epoxy is the focus of Section 18.4. Section 18.5 goes into great detail about the numerical results. The detailed conclusion of this investigation is provided in Section 18.6.

18.2 Modeling of carbon nanotube-reinforced composite

One of the crucial components of the crack simulations is that, using this research, the anticipated material properties are derived. For an epoxy-matrix reinforced with CNT, it is essential to evaluate equivalent thermal and mechanical properties. Aimed at computing the equivalent properties, determining the CNT and matrix properties is necessary. The equivalent properties are computed using a number of micromechanics models, which include the mechanical as well as thermal properties that consist of: elastic modulus, thermal conductivity, fracture energy, fracture toughness, CTE, and Poisson's ratio. Out of many micromechanics models to calculate the CNT-reinforced nanocomposite's elastic modulus, in this chapter, the Halpin−Tsai (HT) approach is utilized [37]. This semiempirical approach was developed to forecast the stiffness in a composite, that is, unidirectional, which is regarded as the function of the aspect ratio [38]. The rule of mixtures, which is a rough forecast for Poisson's ratio, is produced via the mechanics of materials method. The energy that a crack in a body needs to proliferate is the measure of the fracture energy of the material. The processes referred to as toughening mechanisms include pull-out and CNT debonding [39]. The relationship between energy release rate (G) and mode-I SIF (K_I) is used to calculate the fracture toughness of the linearly elastic modified epoxy material. The crack tip stress field magnitude at a cracked structure under load is denoted by the constant K. Only if the K value hits the critical level, that is, K_c will the crack begin to propagate. The modulus of elasticity in the orientation of fiber can be utilized to calculate the CTE (E_l) in the fiber's direction of CNT, which is mostly controlled by the fibers. For thermal control, it is essential to evaluate the CNT nanocomposite's effective thermal conductivity. The rule of mixing, which can be used to estimate the effective heat conductivity in the CNT fiber's direction, is a good source for additional information.

18.3 Formulation of the problem

The problem of a CNT-reinforced nanocomposite body in the presence of crack subjected to dual-natured thermo-mechanical loadings is addressed in this investigation. For this coupled field case, the mechanical (displacement) and thermal (isothermal condition) approximations necessitate a separate approximation for each. The structure with crack is modeled utilizing the XIGA, and it is modeled utilizing a particular assortment of enrichment functions that are distinct from the mesh of finite element. In this section, governing equations, the XIGA approximation for the cracks, and lastly, the estimation of values for SIF will all be briefly discussed.

18.3.1 Governing equations

In the problem space, the relevant linear elastic solid body is taken into account as a 2-D domain (Ω). Additionally, the domain that the border's contour Γ delineates is divided into four categories: temperature boundary (Γ_T), traction boundary (Γ_t), displacement boundary (Γ_u), and traction-free boundary (Γ_{tf}). This solid structure in the relevant domain is believed to be made up of arbitrary cracks (Γ_c). The body under concern has governing electrostatic equations, which are bounded by Γ with a small displacement and are articulated as follows:

$$-\nabla \mathbf{q} + \overline{Q} = 0 \tag{18.1}$$

$$\mathbf{q} = -k\nabla T \tag{18.2}$$

$$\nabla.\boldsymbol{\sigma} + \mathbf{f} = 0 \quad \text{in } \Omega \tag{18.3}$$

$$\boldsymbol{\sigma} = \mathbf{C}:(\varepsilon - \varepsilon_T) \tag{18.4}$$

Eqs. (18.1) and (18.3) correspondingly depict the thermal and mechanical equilibria. Eqs. (18.2) and (18.4), which represent Fourier's law of heat conduction in Eq. (18.2) and Hooke's law in Eq. (18.4), respectively, are used to express the constitutive relations. One definition of the thermo-elastic strain tensor is

$$\varepsilon = \nabla_s \mathbf{u} \tag{18.5}$$

$$\varepsilon_T = \alpha(T - T_0)\,\mathbf{I} \tag{18.6}$$

where T is the temperature, $\mathbf{u}$ signifies displacement, $\mathbf{q}$ symbolizes heat flux, $\boldsymbol{\sigma}$ stress tensor, ε is for strain tensor, and thermal strain is denoted by ε_T that is determined in respect with the T_0, that is, reference temperature. In this case, the constitutive matrix is denoted by $\mathbf{C}$, the expansion coefficient is denoted by α, and k is the thermal diffusivity; the obligatory values are heat source $\overline{Q}$ and body force $\mathbf{f}$.

18.3.2 Extended isogeometric analysis approximation for crack

For cracks, the XIGA approximation is constructed, and furthermore, the thermal loading case is constructed by taking the isothermal crack into consideration. It is feasible to generalize the displacement approximation for cracks at a particular control point, which corresponds to ξ_i in parametric space as in the following equation:

$$\mathbf{u}^{\mathrm{h}}(\boldsymbol{\xi}) = \sum_{i=1}^{n_{en}} R_i(\xi)\mathbf{u}_i + \left\{\sum_{j=1}^{n_{cf}} R_j(\xi)\left[H(\xi) - H(\xi_i)\right]\mathbf{a}_j\right\} + \left\{\sum_{k=1}^{n_{ct}} R_k(\xi)\left\{\sum_{\alpha=1}^{4}\left[\beta_\alpha(\xi) - \beta_\alpha(\xi_i)\right]\ \mathbf{b}_k^\alpha\right\}\right\} \tag{18.7}$$

where $H(\xi)$ is the Heaviside function; β_α is crack tip enrichment functions.

In Eq. (18.7), $H(\xi)$ will equal $+1$ if the parametric coordinate ξ corresponding to a particular Gauss point is above the crack face, and $H(\xi)$ will equal -1 if ξ is below the crack face.

Singular stress fields can be added to the fields close to the crack tip to improve the accuracy of the solution. This can be done by the means of crack tip enrichment functions, which are expressed as

$$\beta_\alpha(\xi) = \left[\sqrt{r}\cos\frac{\theta}{2},\ \sqrt{r}\sin\frac{\theta}{2},\ \ \sqrt{r}\cos\frac{\theta}{2}\cos\theta,\ \sqrt{r}\sin\frac{\theta}{2}\cos\theta\ \right] \tag{18.8}$$

In this case, r and θ stand for polar coordinates with an origin at the tip of the crack and a orientation line $\theta = 0$ that runs parallel to the crack.

18.3.3 Isothermal crack

In the isothermal crack problem, the crack surface is defined at a specific temperature, and the body is subjected to the required boundary conditions. Here, it is discovered that the temperature profile at the surface of the crack is continuous, but the heat flux is discontinuous elsewhere on the surface of the crack. The following equation describes the temperature distribution over an isothermal crack:

$$T = -\frac{K_T}{k}\sqrt{\frac{2r}{\pi}}\cos\left(\frac{\theta}{2}\right) \tag{18.9}$$

As a result, $\sqrt{r}\cos\frac{\theta}{2}$ is enriched into the temperature field approximation function to capture crack tip singularity. The following is the generalized temperature field for isothermal crack:

$$T^{h}(\xi)=\sum_{i=1}^{n_{en}}R_i(\xi)\mathrm{T}_i+\left\{\sum_{j=1}^{n_{ct}}R_j(\xi)\sqrt{r}\cos\frac{\theta}{2}\mathbf{a}_j\right\} \tag{18.10}$$

Similar to how notations used in Eq. (18.7) were explained, Eq. (18.10) notations have the same meaning.

Using the test and trial functions, the temperature and displacement approximations, and the nodal variation arbitrariness, a collection of discrete equations are stated as follows:

$$\left[\mathbf{K}^{TT}\right]\{T\}=\left\{\mathbf{F}^{T}\right\} \tag{18.11}$$

$$[\mathbf{K}^{uu}]\{\mathbf{u}\}=\left\{\mathbf{F}^{u}\right\} \tag{18.12}$$

where $\mathbf{u}$ and T stand for the nodal unknowns, $\mathbf{K}^{TT}$ and $\mathbf{K}^{uu}$ for temperature and displacement field global stiffness matrices, and $\mathbf{F}^{T}$ and $\mathbf{F}^{u}$ for temperature and displacement field external force vectors, correspondingly. Authors initially create and calculate the matrices on the element level before assembling them to their counterparts with the help of the finite element assembly approach. The thermal and structural difficulties associated with thermo-elastic cracks are separated here. The discrete equation described in Eq. (18.11) is resolved to get the temperature distribution of geometry comprising crack. The elastic discrete equation Eq. (18.12) for displacement field is then solved using this temperature data from Eq. (18.11).

18.3.4 Extended isogeometric analysis discretization

For describing the discontinuity in the XIGA, the primary temperature equation and the displacement field must be obtained. It is possible to discretize the mechanical equation from Eq. (18.7) over the element as follows:

$${}^{u}\mathbf{K}^{h}\mathbf{u}^{h}={}^{u}\mathbf{f}^{h} \tag{18.13}$$

where $\mathbf{u}^{h}$ symbolizes displacement vector made up of the standard degree of freedom (DOF) and the extra DOF brought on by enrichment (crack), which can be defined by

$$\mathbf{u}^{h}=\left\{\mathbf{u}\quad\mathbf{a}^{u}\quad\mathbf{b}_1^{u}\quad\mathbf{b}_2^{u}\quad\mathbf{b}_3^{u}\quad\mathbf{b}_4^{u}\right\} \tag{18.14}$$

As stated, $\mathbf{a}^{u}=\left\{\mathbf{a}_x^{u},\ \mathbf{a}_y^{u}\right\}^{T}$ and $\mathbf{b}_\alpha^{u}=\left\{\mathbf{b}_{\alpha x}^{u},\mathbf{b}_{\alpha y}^{u}\right\}^{T}(\alpha=1,2,3\text{ and }4)$ represent additional DOFs that are used to generate crack faces and crack tip fields, correspondingly. $\mathbf{u}=\left\{\mathbf{u}_x,\ \mathbf{u}_y\right\}^{T}$ stands for the standard DOFs.

The typical XIGA approximation is represented by the use of first term on the right-hand side of Eq. (18.7), whereas the subsequent terms are those that are utilized to simulate the CRACK. Therefore the elemental stiffness matrix and mechanical force vector are constructed as follows:

$$ {}^{u}\mathbf{K}_{ij}^{h} = \begin{bmatrix} \mathbf{K}_{ij}^{uu} & \mathbf{K}_{ij}^{ua^u} & \mathbf{K}_{ij}^{ub^u} & \mathbf{K}_{ij}^{uc^u} \\ \mathbf{K}_{ij}^{a^u u} & \mathbf{K}_{ij}^{a^u a^u} & \mathbf{K}_{ij}^{a^u b^u} & \mathbf{K}_{ij}^{a^u c^u} \\ \mathbf{K}_{ij}^{b^u u} & \mathbf{K}_{ij}^{b^u a^u} & \mathbf{K}_{ij}^{b^u b^u} & \mathbf{K}_{ij}^{b^u c^u} \\ \mathbf{K}_{ij}^{c^u u} & \mathbf{K}_{ij}^{c^u a^u} & \mathbf{K}_{ij}^{c^u b^u} & \mathbf{K}_{ij}^{c^u c^u} \end{bmatrix} \tag{18.15} $$

$$ {}^{u}\mathbf{f}_{i}^{h} = \left\{ \mathbf{f}_{i}^{u} \quad \mathbf{f}_{i}^{a^u} \quad \mathbf{f}_{i}^{b_1^u} \quad \mathbf{f}_{i}^{b_2^u} \quad \mathbf{f}_{i}^{b_3^u} \quad \mathbf{f}_{i}^{b_4^u} \quad \mathbf{f}_{i}^{c^u} \right\}^{T} \tag{18.16} $$

where

$$ \mathbf{K}_{ij}^{rs} = \int_{\Omega^e} (\mathbf{B}_i^r)^T \mathbf{C}\, \mathbf{B}_j^s \; d\Omega \tag{18.17} $$

$$ \mathbf{f}_i^u = \int_{\Omega^e} \mathbf{R}_i^T \mathbf{b}\, d\Omega + \int_{\Gamma_t} \mathbf{R}_i^T \hat{\mathbf{t}}\, d\Gamma \tag{18.18} $$

$$ \mathbf{f}_i^{a^u} = \int_{\Omega^e} \mathbf{R}_i^T H\, \mathbf{b}\, d\Omega + \int_{\Gamma_t} \mathbf{R}_i^T H\; \hat{\mathbf{t}}\, d\Gamma \tag{18.19} $$

$$ \mathbf{f}_i^{b^u\alpha} = \int_{\Omega^e} \mathbf{R}_i^T \beta_\alpha \mathbf{b}\, d\Omega + \int_{\Gamma_t} \mathbf{R}_i^T \beta_\alpha\; \hat{\mathbf{t}}\, d\Gamma \tag{18.20} $$

$$ \mathbf{f}_i^{c^u} = \int_{\Omega^e} \mathbf{R}_i^T \chi\; \mathbf{b}\, d\Omega + \int_{\Gamma_t} \mathbf{R}_i^T \chi\; \hat{\mathbf{t}}\; d\Gamma \tag{18.21} $$

where R_i signifies the NURBS basis function, and $\mathbf{B}_i^u$, $\mathbf{B}_i^{a^u}$, $\mathbf{B}_i^{b^u\alpha}$, and $\mathbf{B}_i^{c^u}$ symbolize matrices of derivatives of NURBS basis function:

$$ \mathbf{B}_i^u = \begin{bmatrix} R_{i,X_1} & 0 \\ 0 & R_{i,X_2} \\ R_{i,X_2} & R_{i,X_1} \end{bmatrix} \tag{18.22} $$

$$ \mathbf{B}_i^{a^u} = \begin{bmatrix} (R_i)_{,X_1} H & 0 \\ 0 & (R_i)_{,X_2} H \\ (R_i)_{,X_2} H & (R_i)_{,X_1} H \end{bmatrix} \tag{18.23} $$

$$ \mathbf{B}_i^{b^u} = \begin{bmatrix} \mathbf{B}_i^{b^u 1} & \mathbf{B}_i^{b^u 2} & \mathbf{B}_i^{b^u 3} & \mathbf{B}_i^{b^u 4} \end{bmatrix} \tag{18.24} $$

$$ \mathbf{B}_i^{b^u\alpha} = \begin{bmatrix} (R_i\beta_\alpha)_{,X_1} & 0 \\ 0 & (R_i\beta_\alpha)_{,X_2} \\ (R_i\beta_\alpha)_{,X_2} & (R_i\beta_\alpha)_{,X_1} \end{bmatrix} \tag{18.25} $$

$$ \mathbf{B}_i^{c^u} = \begin{bmatrix} (R_i)_{,X_1} \chi & 0 \\ 0 & (R_i)_{,X_2} \chi \\ (R_i)_{,X_2} \chi & (R_i)_{,X_1} \chi \end{bmatrix} \tag{18.26} $$

where $r, s = u, a^u, b^u, c^u$, and $\alpha = 1, 2, 3, 4$.

For the analysis of isothermal fracture, the elements in Eq. (18.5) thermal equation are discretized as

$$ {}^{iT}\mathbf{K}^h \quad {}^{i}\mathbf{T}^h = {}^{iT}\mathbf{f}^h \tag{18.27} $$

Similar to Eq. (18.14), the temperature vector in respect with an isothermal crack are defined by means of following equation:

$$^{i}\mathbf{T}^{h} = \left\{ \mathbf{T} \quad \mathbf{b}_{1}^{T} \right\} \tag{18.28}$$

As previously established, $\mathbf{T}$ stands for the temperature field's standard DOF, whereas $\mathbf{b}_1^T$ is the additional DOF used to replicate the crack tip field. Additionally, the force vector $^{iT}\mathbf{f}^h$ and elemental stiffness matrix $^{iT}\mathbf{K}^h$ for the heat field are generated as follows:

$$^{iT}\mathbf{K}_{ij}^{h} = \begin{bmatrix} \mathbf{K}_{ij}^{TT} & \mathbf{K}_{ij}^{Tb} & \mathbf{K}_{ij}^{Tc} \\ \mathbf{K}_{ij}^{bT} & \mathbf{K}_{ij}^{bb} & \mathbf{K}_{ij}^{bc} \\ \mathbf{K}_{ij}^{cT} & \mathbf{K}_{ij}^{cb} & \mathbf{K}_{ij}^{cc} \end{bmatrix} \tag{18.29}$$

$$^{iT}\mathbf{f}_{i}^{h} = \left\{ \mathbf{f}_{i}^{T} \quad \mathbf{f}_{i}^{b_1^T} \quad \mathbf{f}_{i}^{c^T} \right\}^{T} \tag{18.30}$$

The terms in the above equations can be written similarly to how it is shown in Eqs. (18.17)–(18.21).

18.3.5 Stress intensity factor computation

To find each SIF, a domain-based interaction integral is applied. Applying a path-independent integral yield, a sufficiently precise method for determining SIFs. This is the case because only field variables being far from the tip of the crack are being utilized for assessing the SIFs using domain-based integration integral. Only when the auxiliary field is carefully selected, interaction integral proves promising in terms of extracting the different SIFs from a solution. This domain form interaction integral for mechanical loading is shown as [40–49]

$$M^{(1,2)} = \int_A \left[\sigma_{ij}^{(1)} \frac{\partial u_i^{(2)}}{\partial x_1} + \sigma_{ij}^{(2)} \frac{\partial u_i^{(1)}}{\partial x_1} - W^{(1,2)} \delta_{ij} \right] \frac{\partial q}{\partial x_j} dA \tag{18.31}$$

$$W^{(1,2)} = \frac{1}{2} \left(\sigma_{ij}^{(1)} \varepsilon_{ij}^{(2)} + \sigma_{ij}^{(2)} \varepsilon_{ij}^{(1)} \right) \tag{18.32}$$

For this problem, at the crack's tip, the scalar weight function q has a value of 1, 0 at its contour, and arbitrary between those points. While $\varepsilon^{(1)}$ and $\varepsilon^{(2)}$ are the engineering strain tensor, $\sigma^{(1)}$ and $\sigma^{(2)}$ stand for the actual and auxiliary Cauchy stress tensor. $W^{(1,2)}$ stands for the energy density of the strain in both the primary and secondary states. Similar to the thermo-mechanical interaction integral, it has the following notation:

$$M^{(1,2)} = \int_A \left[\sigma_{ij}^{(1)} \frac{\partial u_i^{(2)}}{\partial x_1} + \sigma_{ij}^{(2)} \frac{\partial u_i^{(1)}}{\partial x_1} - W^{(1,2)} \delta_{ij} \right] \frac{\partial q}{\partial x_j} dA + \alpha \int_A \frac{\partial T}{\partial x_1} \sigma_{kk}^{(2)} q dA \tag{18.33}$$

which is to be utilized to find the interaction integral under mixed-mode load in the case related to linear elasticity:

$$M^{(1,2)} = \frac{2}{H}\left(K_I^{(1)}K_I^{(2)} + K_{II}^{(1)}K_{II}^{(2)}\right) \tag{18.34}$$

where $H = E$ and $H = E/(1-\nu^2)$ stand for the homogeneous material's plane stress and strain conditions, respectively.

It is possible to present the direction of crack propagation as

$$\theta_c = 2\tan^{-1}\frac{1}{4}\left(\frac{K_I}{K_{II}} \pm \sqrt{\left(\frac{K_I}{K_{II}}\right)^2 + 8}\right) \tag{18.35}$$

The equivalent SIF value is expressed as follows:

$$K_{Ieq} = K_I\cos^3\left(\frac{\theta_c}{2}\right) - 3K_{II}\cos^2\left(\frac{\theta_c}{2}\right)\sin\left(\frac{\theta_c}{2}\right) \tag{18.36}$$

18.4 Calculation of equivalent properties

The equivalent properties (such as fracture toughness, fracture energy elastic modulus, and Poisson's ratio) are to be calculated by fluctuating the single-walled carbon nanotube (SWCNT) content in the epoxy-matrix. Table 18.1 displays the geometrical and material features that were implemented to assess the equivalent attributes. The elastic modulus of composite material fabricated by reinforcing SWCNTs in the epoxy composite was calculated with the help of modified HT model equation. The fracture energy was calculated using CNT debonding and CNT pull-out as the mechanisms for toughening the fracture energy.

TABLE 18.1 Evaluation of equivalent properties using following material parameters.

Parameter	SWCNT	Epoxy
Density (kg/m^3)	1.8×10^3	1.2×10^3
Poisson's ratio	0.16	0.35
Elastic modulus (Pa)	1002×10^9	2.9×10^3
Thermal conductivity $(W/mm^\circ C)$	3	0.25×10^{-3}
CTE $(^\circ C^{-1})$	19.6×10^{-6}	11.7×10^{-6}
Outer diameter (m)	1.36×10^{-9}	–
Inner diameter (m)	0.68×10^{-9}	–
Length (m)	306×10^{-9}	–
Strength (Pa)	30×10^9	–
Fracture energy of unmodified epoxy (J/m^2)	–	133
Interfacial fracture energy (J/m^2)	–	25

TABLE 18.2 Equivalent material properties (mechanical and thermal) of epoxy reinforced by single-walled carbon nanotube (SWCNT).

CNT	Vol %	Young's modulus (N/mm^2)	Fracture energy (J/mm^2)	Fracture toughness ($MPa\sqrt{mm}$)	Thermal conductivity ($W/mm°C$)	Coeff. of thermal expansion ($°C^{-1}$)
SWCNT	0%	2.9×10^3	133×10^{-6}	19.638	0.25	1.17×10^{-5}
	1%	5.088×10^3	215.19×10^{-6}	33.087	0.2775	1.784×10^{-5}
	2%	7.295×10^3	297.38×10^{-6}	46.577	0.305	1.862×10^{-5}
	3%	9.523×10^3	379.57×10^{-6}	60.121	0.3325	1.892×10^{-5}
	4%	11.771×10^3	461.76×10^{-6}	73.725	0.36	1.909×10^{-5}

The fracture energy was calculated using CNT debonding and CNT pull-out as mechanisms for toughening the fracture energy. The findings demonstrate that as the CNT's volume percentage in matrix upsurges, the fracture energy rises as well (Table 18.2).

The predicted mechanical and thermal characteristics of the epoxy reinforced with SWCNT at various SWCNT volume percentages are enlisted in Table 18.2. The corresponding mechanical and thermal properties of SWCNT, reinforced in the epoxy-matrix, consisting of volume fractions ranging from 0% to 4%, are shown in Table 18.2. As Table 18.2 suggests, the value of Young's modulus is rising with respect to the percentage volume of CNT content in the epoxy-matrix. A similar form of hiking in the values of other mechanical properties (fracture energy and fracture toughness) and thermal properties (CTE and thermal conductivity) with the surge in CNT volume percentage can be noted from Table 18.2.

18.5 Numerical results and discussion

This segment includes two different cases of rectangular bodies with horizontal and inclined isothermal cracks exposed to thermo-mechanical load. XIGA is being used in the current study to investigate how SWCNT reinforcement affects the fracture pattern of CNTRCs. The structure being investigated for cracks was modeled using a polymer composite reinforced with SWCNTs. The corresponding equivalent properties of the body have been calculated by increasing the SWCNT content of the polymer matrix from 0% to 4%. To make the comparison of results easier, the geometric dimensions are maintained the same (i.e., 100 × 200 mm) throughout both cases. This numerical problem is resolved considering the plane strain condition for both cases. The integration across the complete body is done using the Gauss quadrature rule. Though the exactness of integration is declined because of the presence of discontinuities like cracks, hence, the discontinuous elements are integrated using polar integration scheme and the subtriangulation technique. The current work uses an isothermal fracture model to derive the equivalent SIFs of thermally generated cracks. The obtained normalized SIF values are verified against the published results. The solid rectangular body is discretized using the third order NURBS basis function and a control net of 30×50. Next subsections include the problems for analyzing fracture behavior of SWCNT-reinforced composite body.

18.5.1 Thermo-mechanical loading on an single-walled carbon nanotube-reinforced composite body with a horizontal center crack

An SWCNT-reinforced epoxy body dimensioned as 100×200 mm consisting of a horizontal isothermal crack exposed with thermo-mechanical load is neatly illustrated in Fig. 18.1. A mechanical stress of (tensile nature) 4.5 MPa is applied to the area under investigation in a perpendicular direction with respect to crack. The material properties taken into account are listed in Table 18.1. The domain is also exposed to 0°C at the crack face and 10°C at all four edges of the rectangular solid body. The bottom edge of the body's extreme left is confined in both the x- and y-directions, whereas the bottom edge is completely constrained in the y-direction.

The temperature distribution in the SWCNT-reinforced composite body under the abovementioned boundary condition is displayed in Fig. 18.2. The horizontal crack size is 40 mm, and hence, the $a/W = 0.2$. The displacement in direction x- and y-axes is represented in Figs. 18.3 and 18.4. On the other hand, Figs. 18.5 and 18.6 depict the stress distributions σ_{xx} and σ_{yy} in the identical rectangular structure exposed to same loading conditions.

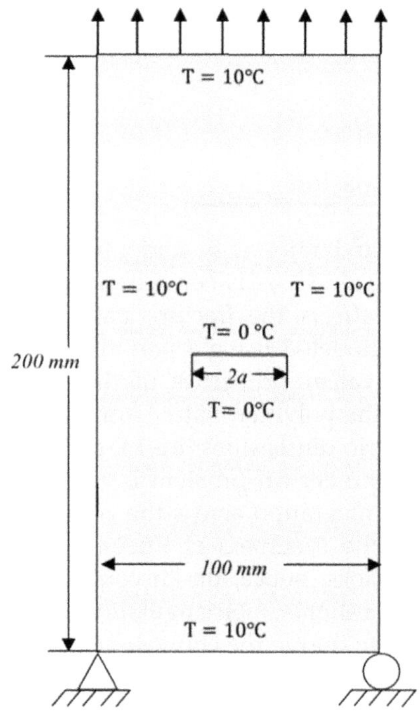

FIGURE 18.1 Boundary conditions showing a composite reinforced by SWCNT structure subjected to thermo-mechanical load with isothermal horizontal center crack. *SWCNT*, Single-walled carbon nanotube.

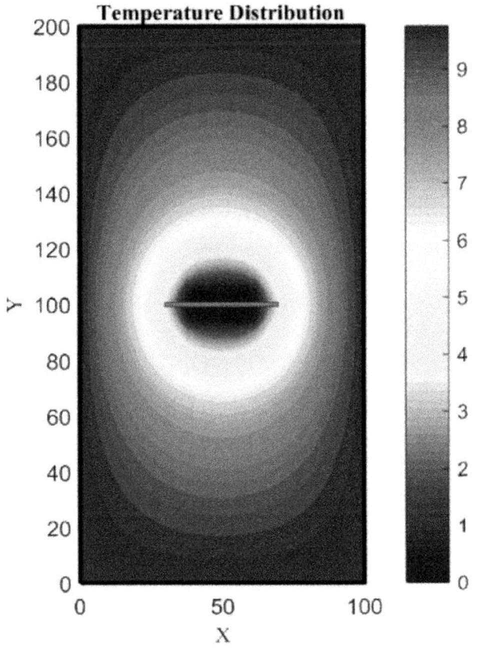

FIGURE 18.2 Profile of temperature distribution in rectangular structure subjected to thermo-mechanical load consisting isothermal center crack ($a/W = 0.4$).

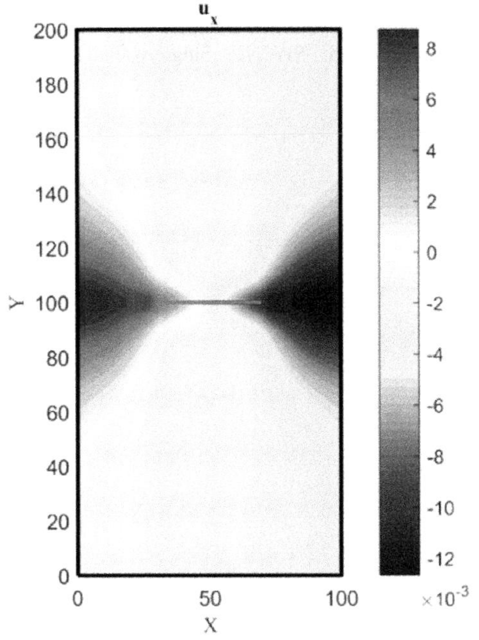

FIGURE 18.3 Displacement represented by contours in x-direction for a composite structure (reinforced by 2% SWCNT) consisting a horizontal isothermal crack for $a/W = 0.2$ under thermo-mechanical load. *SWCNT*, Single-walled carbon nanotube.

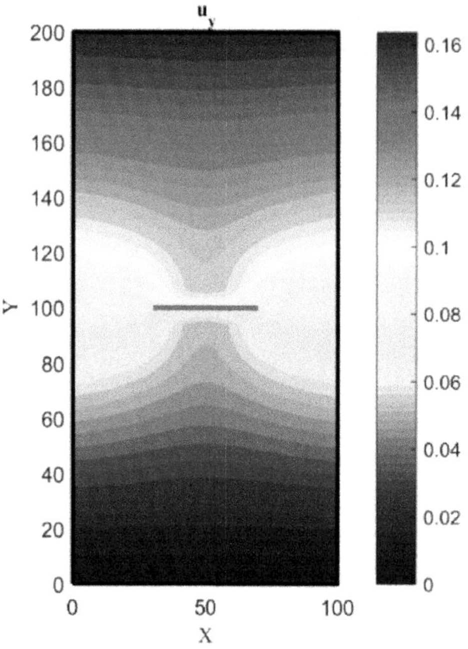

FIGURE 18.4 Displacement represented by contours in the *y*-direction for a composite structure (reinforced by 2% SWCNT) consisting a horizontal isothermal crack for $a/W = 0.2$ under thermo-mechanical load. *SWCNT*, Single-walled carbon nanotube.

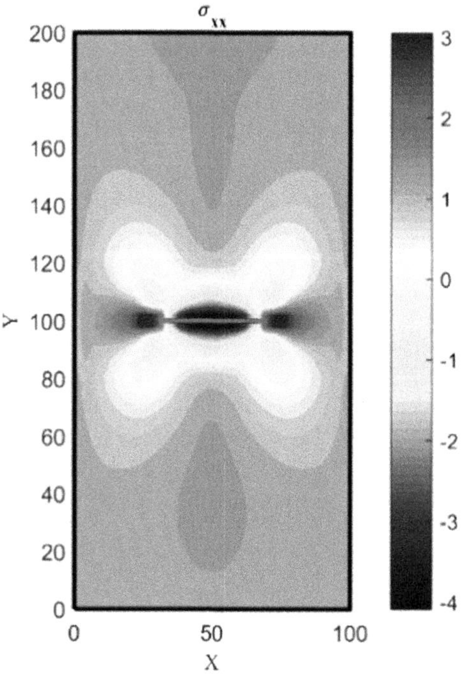

FIGURE 18.5 Stress distribution σ_{xx} represented by contours for a composite structure (reinforced by 2% SWCNT) consisting a horizontal isothermal crack for $a/W = 0.2$ under thermo-mechanical load. *SWCNT*, Single-walled carbon nanotube.

For a better comprehension of the fracture pattern of the SWCNT-reinforced composite exposed to thermo-mechanical load, the value of equivalent SIF is extracted at various crack lengths. The length of the mentioned crack is increased equally (5 mm) at each step. Hence, the horizontal crack length is increased by 2.5 mm at the crack tip (left and right crack tip) after every step. The temperatures at all the edges are set at 10°C and are set to 0°C at the crack face. For different SWCNT content percentages (0%–4%) in the composite structure, the corresponding equivalent SIFs are shown in Fig. 18.7. The graph presented in Fig. 18.7 is drawn between equivalent SIFs and crack tip extensions on the right-hand side. Figs. 18.7 and 18.8 in the text are throwing some light on the pattern of failure of FG structures reinforced with different volume percentages of CNT. After evaluating the equivalent SIF value at varying crack lengths using interaction integral, a curve is plotted between equivalent SIF (y-axis at left side) and right crack tip extension (x-axis). The horizontal dotted lines denote the fracture toughness of the structure for a distinct volume percentage of SWCNT at the right crack tip. These fracture toughness values for varying SWCNT volume percentages (0%–4%) are indicated on the right side of the y-axis. The color of the corresponding curve of similar SIF corresponds to the horizontal dotted lines. The black horizontal line at the top indicates the fracture toughness of a structure with 5% SWCNT. It intersects the black curve on the graph that represents the corresponding SIFs at different right crack tips of the same structure with 5% SWCNT. For every volume percentage of CNT, a vertical dotted line is drawn at the junction of the horizontal line and the curve; this vertical line on the x-axis represents the corresponding critical fracture length where the structure fails. Since the structure is

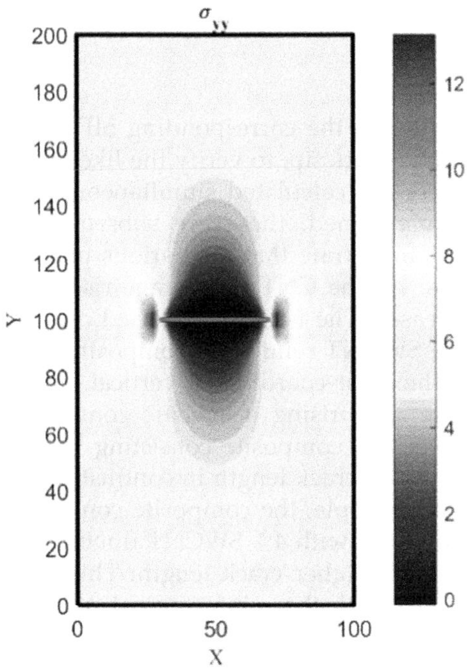

FIGURE 18.6 Stress distribution σ_{yy} represented by contours for a composite structure (reinforced by 2% SWCNT) consisting a horizontal isothermal crack for $a/W = 0.2$ under thermo-mechanical load. *SWCNT*, Single-walled carbon nanotube.

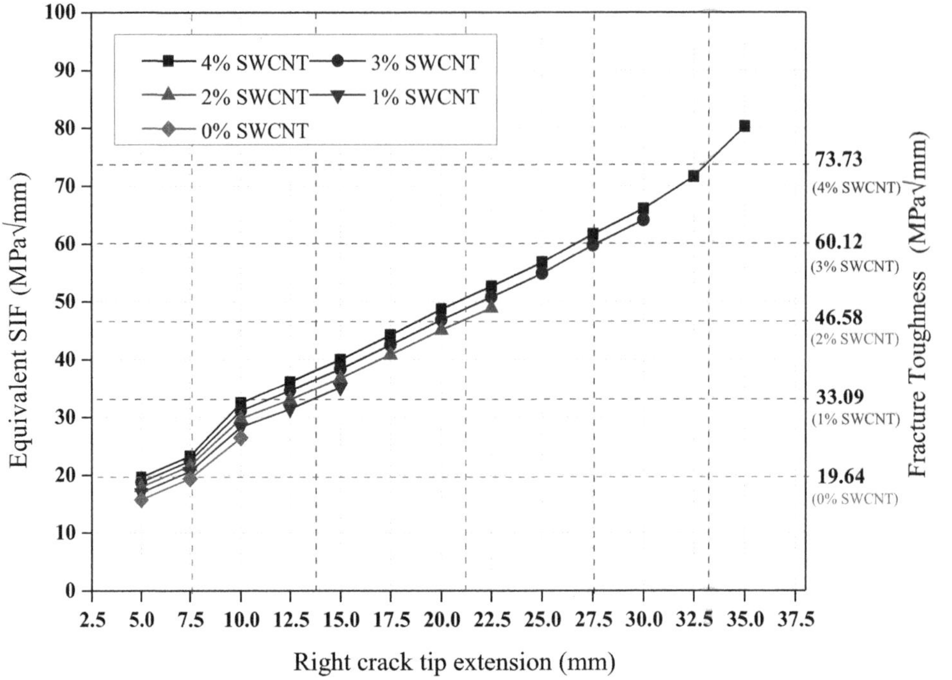

FIGURE 18.7 Variation in the equivalent SIF (K_{eq}) along extension of right crack tip in composite structure reinforced by SWCNT comprising a horizontal isothermal center crack subjected to thermo-mechanical load. *SWCNT*, Single-walled carbon nanotube.

already known to collapse at a critical crack length when the corresponding SIF value exceeds the fracture toughness value at the appropriate crack tip, to verify the likelihood of failure, the equivalent SIFs and fracture toughness are calculated simultaneously at each crack length and compared. As previously mentioned, these are subsequently represented in the graph. Thus the graph's results demonstrate that the various percentages of SWCNT in FG have failed at various times. As the CNT percentage rises, the critical crack length or failure-crack length also increases. The horizontal dotted colored lines denote the respective fracture toughness of the SWCNT-reinforced composite with a specific percentage of SWCNT. In a similar way, the color-coordinated vertical dotted lines denote the crack length where the composite comprising a specific content of SWCNT fractures. As the graph in Fig. 18.7 suggests, the composite consisting with a lower percentage volume of SWCNT content fails at lesser crack length in contrast to the composites with a higher percentage of SWCNT. For example, the composite consisting of no SWCNT content fails at 15.02 mm and the composite with 4% SWCNT fractures at 66.35 mm of length of the crack, that is, at least 77.36% higher crack length. This phenomenon may indicate the strengthening of composite with the reinforcement of more volume fraction of SWCNT.

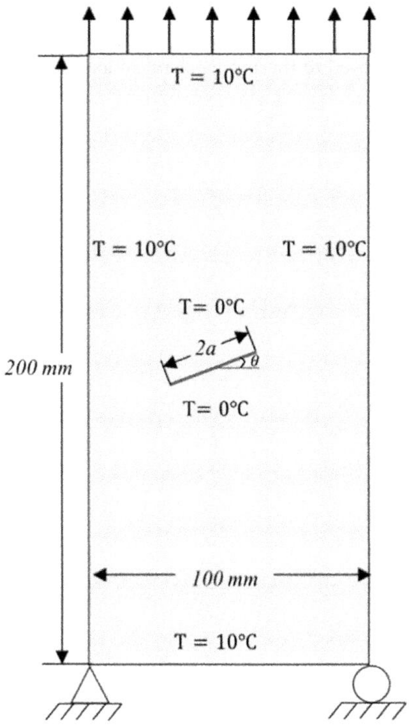

FIGURE 18.8 Boundary conditions showing a composite reinforced by SWCNT structure subjected to thermo-mechanical load with isothermal inclined center crack. *SWCNT*, Single-walled carbon nanotube.

18.5.2 Thermo-mechanical loading on an single-walled carbon nanotube-reinforced composite body with a inclined center crack

An SWCNT-reinforced composite structure dimensioned as 100×200 mm consisting of a horizontal isothermal crack under thermo-mechanical load is neatly illustrated in Fig. 18.8. A mechanical load of 4.5 MPa of tensile nature is applied to the area under investigation in a direction perpendicular to the top and bottom edges. The crack is considered to be inclined at 30° from the bottom edge. The material properties taken into account are listed in Table 18.1. In Addition to the mechanical load, the domain is also exposed to 0°C at the face of the crack and 10°C at all four edges of the rectangular solid body. The bottom edge of the body's extreme left is confined in both the x- and y-directions, whereas the bottom edge is completely constrained in the y-direction.

The temperature distribution in the SWCNT-reinforced composite body under the abovementioned boundary condition is displayed in Fig. 18.9. The inclined crack length is considered to be 40 mm and hence the $a/W = 0.2$. The displacement in direction x- and y-axes is represented Figs. 18.10 and 18.11. On the other hand, Figs. 18.12 and 18.13 depict the stress distributions σ_{xx} and σ_{yy} in the identical rectangular structure under same loading conditions.

For studying the fracture pattern of the SWCNT-reinforced composite consisting of inclined crack, the value of equivalent SIF is extracted at various crack lengths. The crack is increased equally (5 mm) at each step. The temperatures at all the edges are set at 10°C

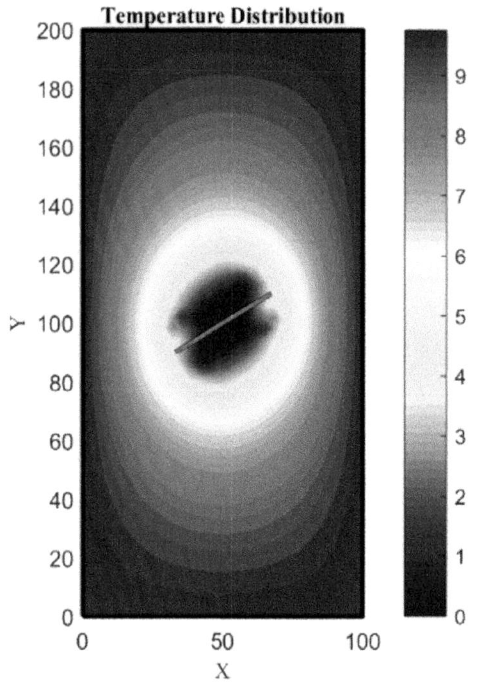

FIGURE 18.9 Profile of temperature distribution in rectangular structure consisting an inclined isothermal center crack ($a/W = 0.4$) exposed to thermo-mechanical loading.

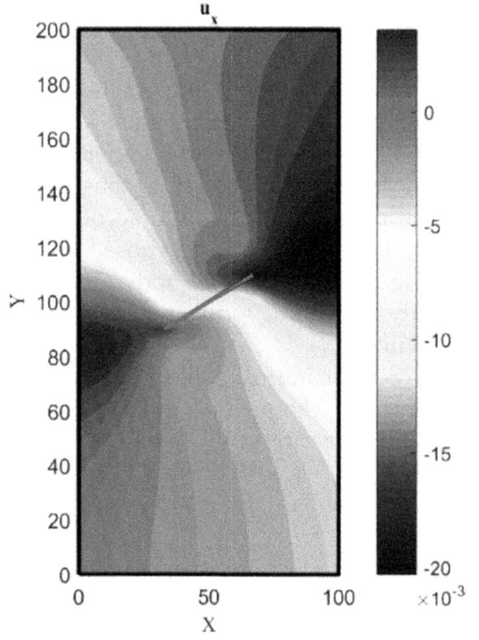

FIGURE 18.10 Displacement represented by contours in the x-direction for a composite body (reinforced by 2% SWCNT) consisting an inclined isothermal crack for $a/W = 0.2$ under thermo-mechanical load. *SWCNT*, Single-walled carbon nanotube.

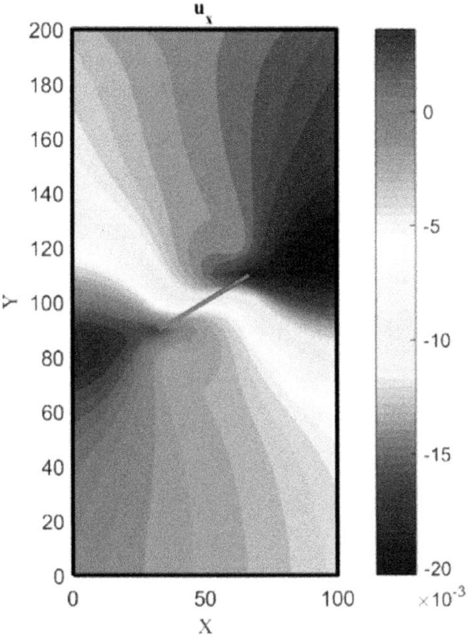

FIGURE 18.11 Displacement represented by contours in the y-direction for a composite body (reinforced by 2% SWCNT) consisting an inclined isothermal crack for $a/W = 0.2$ under thermo-mechanical load. *SWCNT*, Single-walled carbon nanotube.

and are set to 0°C at the crack face. The crack is considered to be inclined at 30° from the bottom edge. For different SWCNT content percentages (0%–4%) in the composite structure, the corresponding equivalent SIFs are shown in the plot in Fig. 18.14 plotted with respect to the extension of crack tip at right-hand side. The horizontal dotted colored lines denote the respective fracture toughness of the SWCNT-reinforced composite with specific percentage of SWCNT. In a similar way, the color-coordinated vertical dotted lines denote the crack length where the epoxy composite comprising a specific percentage fraction of SWCNT fractures. As the graph in Fig. 18.14 suggests, the epoxy-composite consisting of a lower content of SWCNT content fractures at lesser crack length in contrast to the composites with a higher percentage of SWCNT. For example, the composite consisting of 1% SWCNT content fails at 33.47 mm, and the composite with 4% SWCNT fails at 74.62 mm of crack length, that is, at least 55.15% higher crack length. This phenomenon may indicate the strengthening of composite with the reinforcement of more volume fraction of SWCNT.

18.5.3 Comparison of horizontal and inclined center crack

An SWCNT-reinforced composite body of identical dimension (100 × 200 mm), as stated in Section 18.5.1, has a different volume content of SWCNTs, which is taken into account for this problem. A thermo-mechanical stress is considered to be employed at the top edge of the rectangular structure. Fig. 18.15 shows a plot comparing the two equally sized composite bodies with 2% SWCNT reinforcement, one with horizontal crack and other with inclined crack (30 degrees). From the plot, it can clearly be deduced that the composite

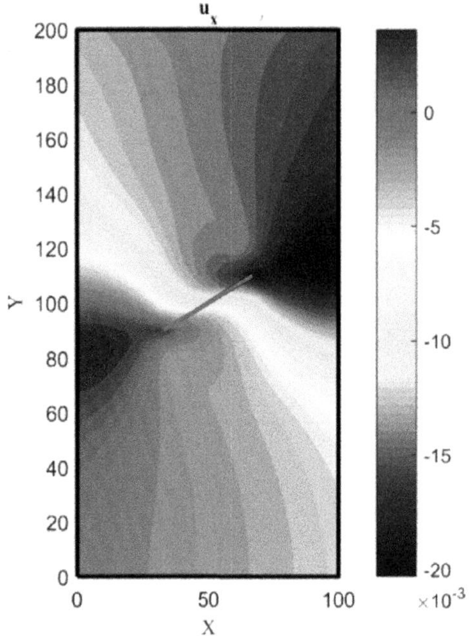

FIGURE 18.12 Stress distribution σ_{xx} represented by contours for a composite body (reinforced by 2% SWCNT) consisting an inclined isothermal crack for $a/W = 0.2$ under thermo-mechanical load. *SWCNT*, Single-walled carbon nanotube.

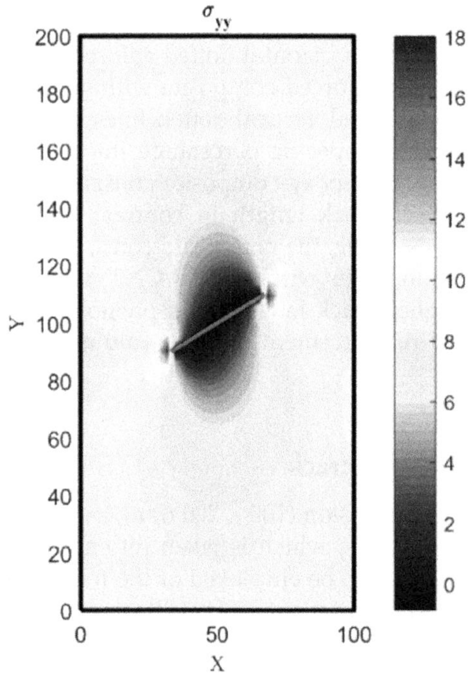

FIGURE 18.13 Stress distribution σ_{yy} represented by contours for a composite body (reinforced by 2% SWCNT) consisting an inclined isothermal crack for $a/W = 0.2$ under thermo-mechanical load. *SWCNT*, Single-walled carbon nanotube.

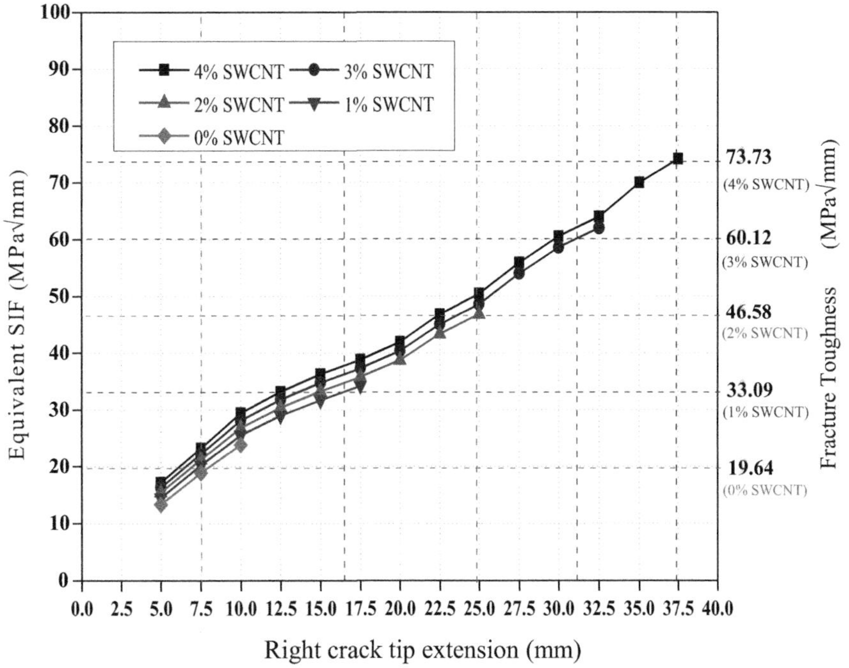

FIGURE 18.14 Variation of the equivalent SIF (K_{eq}) along the extension of right crack tip in composite structure reinforced by SWCNT comprising an inclined isothermal center crack exposed with thermo-mechanical load. *SIFs*, Stress intensity factors; *SWCNT*, single-walled carbon nanotube.

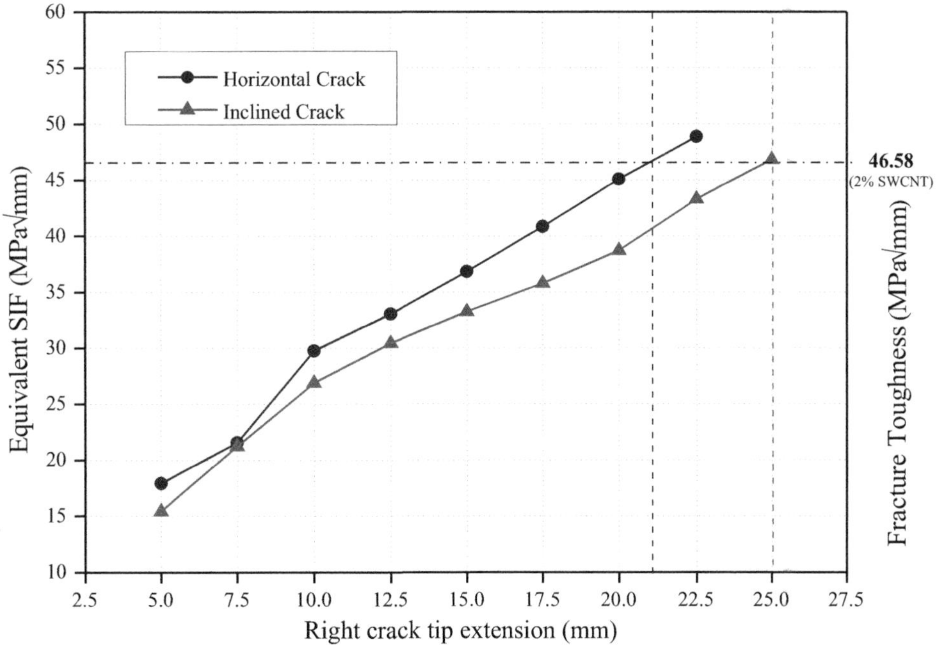

FIGURE 18.15 Comparison of horizontal and inclined crack in composite with 2% SWCNT reinforcement by means of variation of equivalent SIF (K_{eq}) with respect to various crack lengths. *SWCNT*, Single-walled carbon nanotube.

rectangular body with horizontal crack fails earlier at 41.53 mm crack length, whereas the body with inclined crack fails at 49.93 mm. So, the body with horizontal crack fails at least 15.34% earlier crack length. This concludes that out of two SWCNT-reinforced composite bodies of identical dimensions, the one with horizontal crack fails at smaller crack length when compared with the one with inclined crack. Hence, a rectangular body under thermo-mechanical load with isothermal crack is more susceptible to failure when the crack is horizontal in comparison to when the crack is inclined at an angle.

18.6 Conclusion

In this investigation, XIGA is expanded for the modeling of horizontal and inclined isothermal cracks in the SWCNT-reinforced composite structure under thermo-mechanical loading. To calculate the equivalent properties of the CNT-embedded composite structure, few micromechanics models are used. An in-depth investigation is performed to predict the influence of the amount of reinforcement (by changing the volume content of SWCNT between 0% and 4%). The present simulations have led to the following conclusions:

- The plot of equivalent SIFs against the crack extension for all cases considered here shows a positive slope, which indicates that as the crack extends, there is an increment in the value of the SIFs.
- The elastic modulus of the nanocomposite continues to rise as the volume of SWCNT material is raised from 0% to 4%.
- By the increase in SWCNT volume fraction in the epoxy-matrix, the properties like fracture toughness and fracture energy also improve.
- This current work indicates to the strengthening of composite with the reinforcement of more volume fraction of SWCNT.
- The composite rectangular body with horizontal crack failed earlier at 41.93 mm of crack length; on the other hand, the body with inclined crack failed at 49.53 mm. So, the body with horizontal crack fails at least 15.34% earlier crack length.

Acknowledgment

The Science and Engineering Research Board (SERB), Department of Science and Technology (DST), New Delhi is to be thanked for funding this work with an Early Career Research Award under grant number **ECR/2018/000592**.

References

[1] Gu J, Yu T, Lich LV, Tanaka S, Qiu L, Bui TQ. Adaptive orthotropic XIGA for fracture analysis of composites. Composites Part B: Engineering 2019;. Available from: https://doi.org/10.1016/j.compositesb.2019.107259.

[2] Ritchie RO. The conflicts between strength and toughness. Nature Materials 2011;10(11):817–22.

[3] Launey ME, Ritchie RO. On the fracture toughness of advanced materials. Advanced Materials 2009;21 (20):2103–10.

[4] Kuronuma Y, Shindo Y, Takeda T, Narita F. Fracture behaviour of cracked carbon nanotube-based polymer composites: experiments and finite element simulations. Fatigue & Fracture of Engineering Materials & Structures 2010;33(2):87–93.

[5] Jameel A, Harmain GA. Extended iso-geometric analysis for modeling three dimensional cracks. Mechanics of Advanced Materials and Structures 2019;26:915–23. Available from: https://doi.org/10.1080/15376494.2018.1430275.

[6] Yamamoto G, Shirasu K, Hashida T, Takagi T, Suk JW, An J, et al. Nanotube fracture during the failure of carbon nanotube/alumina composites. Carbon 2011;49(12):3709–16.

[7] Jameel A, Harmain GA. Modeling and numerical simulation of fatigue crack growth in cracked specimens containing material discontinuities. Strength of Materials 2016;48(2):294–307. Available from: https://doi.org/10.1007/s11223-016-9765-0.

[8] Negi A, Bhardwaj G, Saini JS, Khanna K, Godara RK. Analysis of CNT reinforced polymer nanocomposite plate in the presence of discontinuities using XFEM. Theoretical and Applied Fracture Mechanics 2019;103102292.

[9] Jameel A, Harmain GA. Fatigue Crack Growth in Presence of Material Discontinuities by EFGM. International Journal of Fatigue 2015;81:105–16. Available from: https://doi.org/10.1016/j.ijfatigue.2015.07.021.

[10] Karami J, Ayatollahi MR, Saboori B. Experimental fracture investigation of CNT/epoxy nanocomposite under mixed mode II/III loading conditions. Fracture of Engineering Materials & Structures 2020;43(5):879–92.

[11] Henshell RD, Shaw KG. Crack tip finite elements are unnecessary. International Journal for Numerical Methods in Engineering 1975;9(3):495–507.

[12] Wilson WK, Yu IW. The use of the J-integral in thermal stress crack problems. International Journal of Fracture 1979;15(4):377–87.

[13] Gupta V, Jameel A, Verma SK, Anand S, Anand Y. An insight on NURBS based isogeometric analysis, its current status and involvement in mechanical applications. Archives of Computational Methods in Engineering 2022;. Available from: https://doi.org/10.1007/s11831-022-09838-0.

[14] Shih CF, Moran B, Nakamura T. Energy release rate along a three-dimensional crack front in a thermally stressed body. International Journal of Fracture 1986;30(2):79–102.

[15] Belytschko T, Lu YY, Gu L. Element-free Galerkin methods. International Journal for Numerical Methods in Engineering 1994;37(2):229–56.

[16] Lone AS, Jameel A, Harmain GA. Enriched element free Galerkin method for solving frictional contact between solid bodies. Mechanics of Advanced Materials and Structures 2022. Available from: https://doi.org/10.1080/15376494.2022.2092791.

[17] Liu WK, Jun S, Zhang YF. Reproducing kernel particle methods. International Journal for Numerical Methods in Fluids 1995;20(8–9):1081–106.

[18] Belytschko T, Black T. Elastic crack growth in finite elements with minimal remeshing. International Journal for Numerical Methods in Engineering 1999;45(5):601–20.

[19] Singh IV, Mishra BK, Bhattacharya S, Patil RU. The numerical simulation of fatigue crack growth using extended finite element method. International Journal of Fatigue 2012;36(1):109–19.

[20] Patil RU, Mishra BK, Singh IV. A new multiscale XFEM for the elastic properties evaluation of heterogeneous materials. International Journal of Mechanical Sciences 2017;122:277–87.

[21] Jameel A, Harmain GA. Effect of Material Irregularities on Fatigue Crack Growth by Enriched Techniques. International Journal for Computational Methods in Engineering Science and Mechanics 2020;21:109–33. Available from: https://doi.org/10.1080/15502287.2020.1772902.

[22] Kanth SA, Lone AS, Harmain GA, Jameel A. Elasto plastic crack growth by XFEM: a review. Materials Today: Proceedings 2019;18:3472–81. Available from: https://doi.org/10.1016/j.matpr.2019.07.275.

[23] Pathak H. Crack interaction study in functionally graded materials (FGMs) using XFEM under thermal and mechanical loading environment. Mechanics of Advanced Materials and Structures 2018;27(11):903–26.

[24] Hughes TJR, Cottrell JA, Bazilevs Y. Isogeometric analysis: CAD, finite elements, NURBS, exact geometry and mesh refinement. Computer Methods in Applied Mechanics and Engineering 2005;194:4135–95.

[25] Gupta V, Jameel A, Verma SK, Anand S, Anand Y. Transient isogeometric heat conduction analysis of stationary fluid in a container. Part E: Journal of Process Mechanical Engineering 2022. Available from: https://doi.org/10.1177/0954408922112.

[26] Rabczuk T, Gracie R, Song JH, Belytschko T. Immersed particle method for fluid–structure interaction. International Journal for Numerical Methods in Engineering 2010;81(1):48–71.

[27] Ghorashi SS, Valizadeh N, Mohammadi S. Extended isogeometric analysis for simulation of stationary and propagating cracks. International Journal for Numerical Methods in Engineering 2012;89 (9):1069–101.

[28] Bhardwaj G, Singh IV, Mishra BK. Numerical simulation of plane crack problems using extended isogeometric analysis. Procedia Engineering 2013;64:661–70.

[29] Van Do VN, Lee CH. Bending analyses of FG-CNTRC plates using the modified mesh-free radial point interpolation method based on the higher-order shear deformation theory. Composite Structures 2017;168:485497.

[30] Van Do VN, Lee YK, Lee, CH. Isogeometric analysis of FG-CNTRC plates in combination with hybrid type higher-order shear deformation theory. Thin-Walled Structures 2020;148:106565.

[31] Zhang Y, Jin G, Chen M, Ye T, Liu Z. Isogeometric free vibration of sector cylindrical shells with carbon nanotubes reinforced and functionally graded materials. Results in Physics 2020;16:102889.

[32] Yuan H, Yu T, Bui TQ. Multi-patch local mesh refinement XIGA based on LR NURBS and Nitsche's method for crack growth in complex cracked plates. Engineering Fracture Mechanics 2021;250:107780.

[33] Li K, Yu T, Bui TQ. Adaptive XIGA shakedown analysis for problems with holes. European Journal of Mechanics-A/Solids 2022;93:104502.

[34] Fang W, Zhang J, Yu T, Bui TQ. Analysis of thermal effect on buckling of imperfect FG composite plates by adaptive XIGA. omposite Structures 2021;275:114450.

[35] Dwivedi K, Arora G, Pathak H. Fatigue crack growth in CNT-reinforced polymer composite. Journal of Micromechanics and Molecular Physics 2022;7(03n04):173–84.

[36] Yadav A, Bhardwaj G, Godara RK. Thermally induced fracture analysis of CNT reinforced FG structures with multiple discontinuities using XIGA. Engineering Fracture Mechanics 2022;275:108822.

[37] Shoo S, Chitturi VR, Agarwal R, Jiang JW, Katiyar RS. Thermal conductivity of freestanding single wall carbon nanotube sheet by Raman spectroscopy. ACS Applied Materials & Interfaces 2014;6(22):19958–65.

[38] Yu MF, Files BS, Arepalli S, Ruoff RS. Tensile loading of ropes of single wall carbon nanotubes and their mechanical properties. Physical Review Letters 2000;84(24):5552.

[39] Samani MK, Khosravian N, Chen GCK, Shakerzadeh M, Baillargeat D, Tay BK. Thermal conductivity of individual multiwalled carbon nanotubes. International Journal of Thermal Sciences 2012;62:40–3.

[40] Jameel A, Harmain GA. A coupled FE-IGA technique for modeling fatigue crack growth in engineering materials. Mechanics of Advanced Materials and Structures 2019;26:1764–75. Available from: https://doi.org/10.1080/15376494.2018.1446571.

[41] Singh AK, Jameel A, Harmain GA. Investigations on crack tip plastic zones by the extended iso-geometric analysis. Materials Today: Proceedings 2018;5:19284–93. Available from: https://doi.org/10.1016/j.matpr.2018.06.287.

[42] Lone AS, Kanth SA, Jameel A, Harmain GA. A state of art review on the modeling of contact type nonlinearities by extended finite element method. Materials Today: Proceedings 2019;18:3462–71. Available from: https://doi.org/10.1016/j.matpr.2019.07.274.

[43] Kanth SA, Lone AS, Harmain GA, Jameel A. Modelling of embedded and edge cracks in steel alloys by XFEM. Materials Today: Proceedings 2020;26:814–18. Available from: https://doi.org/10.1016/j.matpr.2019.12.423.

[44] Lone AS, Kanth SA, Jameel A, Harmain GA. "XFEM modelling of frictional contact between elliptical inclusions and solid bodies. Materials Today: Proceedings 2020;26:819–24. Available from: https://doi.org/10.1016/j.matpr.2019.12.424.

[45] Gupta V, Jameel A, Anand S, Anand Y. Analysis of composite plates using isogeometric analysis: a discussion. Materials Today: Proceedings 2021;44:1190–4. Available from: https://doi.org/10.1016/j.matpr.2020.11.238.

[46] Kanth SA, Jameel A, Harmain GA. Investigation of fatigue crack growth in engineering components containing different types of material irregularities by XFEM. Mechanics of Advanced Materials and Structures 2022;29(24):3570–87. Available from: https://doi.org/10.1080/15376494.2021.1907003.

[47] Jameel A, Harmain GA. Large deformation in bi-material components by XIGA and coupled FE-IGA techniques. Mechanics of Advanced Materials and Structures 2022;29:850–72. Available from: https://doi.org/10.1080/15376494.2020.1799120.

[48] Jameel A, Harmain GA. Fatigue crack growth analysis of cracked specimens by the coupled finite element-element free Galerkin method. Mechanics of Advanced Materials and Structures 2019;26:1343–56. Available from: https://doi.org/10.1080/15376494.2018.1432800.

[49] Kanth SA, Harmain GA, Jameel A. Modeling of nonlinear crack growth in steel and aluminum alloys by the element free Galerkin method. Materials Today: Proceedings 2018;5:18805–14. Available from: https://doi.org/10.1016/j.matpr.2018.06.227.

CHAPTER

19

Extended finite element method for the simulation of edge dislocations

Neha Duhan, B.K. Mishra and I.V. Singh

Department of Mechanical and Industrial Engineering, Indian Institute of Technology Roorkee, Roorkee, Uttarakhand, India

19.1 Introduction

The solid materials found in the earth's crust can be of crystalline, semicrystalline, or amorphous structures. These structures in solids are decided by the arrangement of atoms, that is, regular packaging of atoms will result in crystalline form, and a completely irregular packing of atoms forms amorphous solids. Most metals, ceramics, semiconductors, and a few polymers are naturally crystalline materials. The crystalline materials are found to have a particular 3D pattern of the atoms, and the arrangement is generally known as a lattice. The lattices are generally made of some standard crystal systems, such as cubic, tetragonal, orthorhombic, hexagonal, rhombohedral, monoclinic, or triclinic. The lattices can be simple, such as body- or face-centered cubic or some complex-shaped. The materials, which are compounds of two different materials, have zincblende and wurtzite crystal structures. The zincblende structure is a cubic close packing (ccp), and wurtzite structure is a hexagonal close packing (hcp). The directions and planes of the crystal lattices are defined by the Millers indices.

In reality, crystalline materials do not have a perfect arrangement of atoms; however, there can be some local disturbances, that is, imperfections. The imperfections can be in the form of 0D, 1D, 2D, or 3D defects, such as point, line, surface, or volume defects, respectively. Point defects can occur from missing atoms or self-interstitial atoms. A number of point defects can combine together to form line defects. These line defects are also known as dislocations, classified as edge, screw, or mixed dislocations depending on the geometry of atomic arrangement [1]. The dislocation in the crystalline material can be defined with the help of the Burgers circuit, which is a closed loop around dislocation in the crystal. When the same atom-to-atom distance is moved on the parallel sides of the Burgers circuit, the vector closing the circuit is the Burgers vector. The angle between the

Enriched Numerical Techniques
DOI: https://doi.org/10.1016/B978-0-443-15362-4.00007-3

dislocation line tangent and Burgers vector decides the type of dislocation in an imperfect crystalline material. An angle of 90° indicates the presence of edge dislocation, and of 0° means screw dislocation. Any other angle will result in a mixed dislocation (a combination of edge and screw dislocation). When the Burgers vector of dislocation is the lattice vector (a vector between two adjacent lattice points), the dislocation is perfect; otherwise, it is a partial dislocation.

The continuum concept of dislocations was first given by Volterra while explaining the deformations in an elastic, homogeneous, and isotropic media [2]. He considered a cylinder having the Cartesian coordinates such that the length was in the z-direction and the cross section was in the $x-y$ plane. When the cylinder was cut at one location with a plane parallel to the $x-z$ plane, the two cut surfaces were generated. The displacement of these cut surfaces with respect to each other resulted in different types of dislocations. The edge-type dislocations formed when the movement between the two cut surfaces was either in the x- or y-directions. This movement resulted in a dislocation line that had the dislocation line vector in the z-direction and the Burgers vector perpendicular to it. Hence, the angle between the dislocation line and the Burgers vector was 90° for the edge dislocations. Further, the screw-type dislocations were formed when the cut surfaces were moved within the cut plane in the z-direction. The dislocation line vector was in the direction of the Burgers vector; hence, the angle between the dislocation line and the Burgers vector was 0° for the screw dislocations. The mixed-type dislocations were formed by combining different movements of the pure edge and screw dislocations.

The dislocations present in a domain will be under the influence of different forces. The forces exerted on the closed line dislocations are called material/configurational forces, which were initially discussed by Ref. [3] and hence given the name the Peach–Koehler force. Head (1953) analytically obtained the stress fields around dislocations in different bimetallic mediums when the Burgers vector is parallel or perpendicular to the free surface [4]. Different boundary conditions, such as free surface, welded boundary, and slipping boundary between the bi-materials, were considered. Also, the simplified expressions of stresses for a dislocation near the bi-material interface were given by Ref. [5]. The expression of the Peach–Koehler force tending to move the dislocation near the material interface was also given. Other than this, the analytical studies of dislocation pileups near the material interfaces were done by Lubarda in Refs. [6–8]. He considered the straight edge, screw, and mixed dislocations near the interfaces. The individual dislocations in the dislocation arrays were placed at the equal distance from each other. The discussions on strain energies, stress fields, and dislocation forces were present in these research articles. The cases of edge dislocation arrays with the Burgers vector normal and parallel to the interface were studied. The dislocations near the circular inhomogeneity were also examined with the mathematical expressions.

Other than the analytical studies of dislocations, the numerical modeling of dislocations was also done by different researchers. The finite element method (FEM) was initially used to model the dislocations by thermal analog. By introducing dummy thermal strains equivalent to the extra half plane of atoms, the straight dislocations were incorporated in the FEM domain [9,10]. The Burgers vector was defined in terms of the Poisson's ratio, thermal expansion coefficient, and temperature difference to avoid the formation of disclinations. The Peach–Koehler force was also evaluated with the energy-based method given by Parks. However, the complex modeling of dislocations in the domains was not always easy with FEM. Therefore to capture the discontinuity and singularity of the dislocation glide plane and core, the extended finite element method (XFEM) was proposed, where the FEM shape functions were modified by adding the enrichment functions.

Gracie et al. used the concept of extrinsic enrichment, where the analytical solution behavior is modeled by enriching the standard shape functions according to the concept of partition of unity [11]. The glide plane discontinuity of the Volterra dislocations was considered with the help of the Heaviside enrichment. To apply the Heaviside enrichment, the nodes above and below the glide planes were recognized by defining the level set functions. The singularity in the displacement field at the dislocation core was captured with the help of core enrichment functions given in Ref. [12]. The core enrichment functions were the analytical displacement expressions for the dislocation in an infinite domain. The discrete equation obtained from the XFEM was solved to get the approximate solution of primary variables. The primary field results were then used to compute the Peach–Koehler force of dislocations by obtaining the secondary variables as strains and stresses. Further, the modifications in the XFEM for modeling dislocations in the thin films, epitaxial islands, and strained layers were proposed in Refs. [13,14]. Other than the elastic modeling of dislocations, the dislocations in piezoelectric materials have also been studied [15]. The analytical expressions of electro-elastic fields of dislocations (edge and screw) were illustrated by Mishra and Pak [16]. They gave the exact form of the displacements, electric potential, strains, electric fields, stresses, and electric displacements. The numerical modeling of the electro-elastic fields was also done by Ref. [17] for the problem of an edge dislocation dipole. Furthermore, the thermo-elastic modeling of dislocations for linear and nonlinear material properties was done by Refs. [18,19]. They analyzed the variation of the Peach–Koehler force with regard to the temperature. The Joule heat effect in the semiconductor materials due to the presence of electric fields around dislocations was also considered to obtain the Peach–Koehler force.

By following the available XFEM studies of the dislocations present in literature, the present chapter is organized to give a basic idea of using XFEM to model the discontinuity and singularity around the dislocations. The enrichment functions required for modeling the primary fields near edge dislocations present in the crystalline materials are discussed. Two problems of edge dislocations are also solved to obtain the Peach–Koehler force of dislocations from the numerical results. The concept of *J*-integral defined for the computation of Peach–Koehler force is considered, where the Gauss points in an annular region near the dislocation core are considered for performing the numerical integration.

This chapter is outlined in terms of various sections, such as Introduction, Mathematical fundamentals, XFEM formulation, Numerical examples, Discussions, and Conclusions, respectively. Section 19.1 gives a basic introduction to the material dislocations in crystalline materials and their continuum mechanics–based studies present in the literature. Section 19.2 contains a list of fundamental mathematical equations (governing differential equations, constitutive relations, and boundary conditions) required for defining any boundary value problem. Section 19.3 illustrates the enrichment-based XFEM formulation of the dislocation problem for obtaining the discrete form equations from the partial differential equations. Section 19.4 explains the numerical examples of the simulations performed to obtain the Peach–Koehler force of dislocations. Finally, in Section 19.5, a discussion about the entire procedure and the results is presented, along with the conclusions.

19.2 Mathematical fundamentals

This section includes the fundamental equations of a problem in the form of governing partial differential equations, constitutive relations, and boundary conditions. The expression of the Peach–Koehler force in terms of the integral equation is also discussed here. The focus of the current section is to elaborate the mathematical foundation of dislocations. The entire mathematical formulation is within the concept of continuum mechanics. Here, the basic equations are considered only for the elastic simulations of edge-type dislocations in the crystalline materials.

19.2.1 Governing differential equations

A physical phenomenon can be generally illustrated with the help of some differential equations. In continuum linear elastic materials, the elastic behavior is explained in terms of the force equilibrium equations, which are the partial differential equations defined in the entire domain. These force equilibrium equations are the governing differential equations where the inertia force term is neglected for the steady state condition. The governing force equilibrium equation for a linear elastic solid under static conditions (in the indicial form) is as follows:

$$\sigma_{ji,i} + f_j = 0 \tag{19.1}$$

where σ_{ji} is the stress tensor of order "2," and f_j is the body force in the domain. The differentiation in the above equation is defined with the help of the indices following the comma.

Eq. (19.1) is defined in the domain Ω bounded with Γ that encloses edge dislocation, as shown in Fig. 19.1. The line defects or dislocations in the domain change the nature of primary variables (displacements) in the governing equations by generating the discontinuity and singularity. The displacements are coming into the picture of force equilibrium equations with the definition of constitutive equations. The constitutive equations are material-dependent relations that relate the strains with the stresses. The strains indicate the change in the geometry due to the application of forces, that is, the kinematic equations relate the strains with the infinitesimal deformations. An elaborated discussion of the constitutive relations and kinematics equations is discussed in the next subsection.

19.2.2 Constitutive relations

The material behavior is governed with the help of some characteristic equations, which depend on the material type. For the linear elastic materials, the constitutive relations between the strains and the stresses are defined with the help of Hooke's law given as follows:

$$\sigma_{ji} = c_{jilk}\,\varepsilon_{lk} \tag{19.2}$$

where ε_{lk} is the strain tensor of order "2," and c_{jilk} is the elastic tensor of order "4."

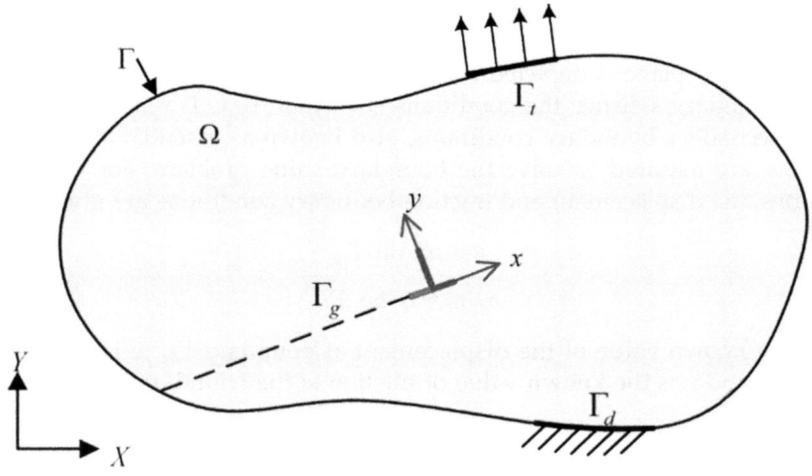

FIGURE 19.1 An arbitrary domain showing the glide plane and core of edge dislocation along with displacement and traction boundary conditions.

The kinematic equations relate the strains to the deformations when considered small strains and infinitesimal deformations. The expression for the same in the indical notation is given as follows:

$$\varepsilon_{lk} = \frac{1}{2}(u_{l,k} + u_{k,l}) \tag{19.3}$$

where $u_{l,k}$ is the derivative of the displacement u_l with respect to indices k.

For the 2D plane problems, some of the stresses and strains in one of the directions are neglected according to the plane conditions. For the plane strain condition, the strain is zero in the direction perpendicular to the 2D plane of analysis. This plane strain condition corresponds to the high thickness of plates. For 2D analysis, Hooke's tensor is also reduced, which relates the stresses and strains present only in the plane of analysis. The matrix form of reduced Hooke's tensor c in terms of the elastic moduli (E, ν) is given as follows:

$$c = \frac{E}{(1+\nu)}\begin{bmatrix} \frac{(1-\nu)}{(1-2\nu)} & \frac{\nu}{(1-2\nu)} & 0 \\ \frac{\nu}{(1-2\nu)} & \frac{(1-\nu)}{(1-2\nu)} & 0 \\ 0 & 0 & \frac{1}{2} \end{bmatrix} \tag{19.4}$$

where E is Young's modulus, and ν is Poisson's ratio of the considered material.

19.2.3 Boundary conditions

The boundary of any arbitrary domain Ω is defined by Γ as depicted in Fig. 19.1, at which different boundary conditions are applied to solve a boundary value problem.

The parts of the boundary Γ represented by Γ_d correspond to the location of displacement boundary condition and Γ_t corresponds to the location of traction boundary condition. The dislocation glide plane is depicted as Γ_g, at the end of which the core of dislocation is present. The boundary satisfies the conditions such that, $\Gamma_d \cup \Gamma_t = 0$ and $\Gamma_d \cap \Gamma_t = 0$. The primary and secondary boundary conditions, also known as essential and natural boundary conditions, are required to solve the boundary value problem. For the force equilibrium equations, the displacement and traction boundary conditions are given as follows:

$$u = u^0 \quad \text{on } \Gamma_d \tag{19.5}$$

$$\sigma_{ji}\, n_i = t_j \quad \text{on } \Gamma_t \tag{19.6}$$

where u^0 is the known value of the displacement at boundary Γ_d, n_i is the vector normal to boundary Γ_t, and t_j is the known value of traction at the boundary.

19.2.4 Peach–Koehler force

The Peach–Koehler force is the force on the dislocation line due to the stress field near the dislocation line. This force is defined as the ratio of the change in strain energy to the change in dislocation position. The strain energy change corresponds to the work required to be done to move the dislocation line. There are different ways available for the computation of the Peach–Koehler force. Directly by knowing the equivalent shear stress or using the potential energy concept. Out of a number of existing methods for the computation of the Peach–Koehler force, the J-integral method is easy to use with the numerical results. The direct use of the path-independent form of the Peach–Koehler force is not very convenient. Hence, the domain form is available, which is expressed in Eq. (19.7). The Peach–Koehler force given in Eq. (19.7) is a function of the stress tensor σ_{ji}, displacement gradients $\frac{\partial u_j}{\partial x_k}$, strain energy W, and weight function w. The integration in Eq. (19.7) is evaluated on an area A_n defined around the dislocation core. For 2D analysis, the area near the dislocation core is taken in the form of an annular disk, as explained in Refs. [18,19]. The weight function w varies linearly in the annular area with a value of 1 at the inner radius and 0 at the outer radius of the annular disk taken into consideration:

$$J_k = \int_{A_n} \left[\sigma_{ji} \frac{\partial u_j}{\partial x_k} - W\delta_{ki} \right] \frac{\partial w}{\partial x_i}\, dA_n \tag{19.7}$$

19.3 Extended finite element method formulation

In this section, the overall XFEM formulation is elaborated to obtain the discrete algebraic equations from the strong form of the governing partial differential equations of the problem. The first step is to obtain the weak form of the governing differential equations, and then the approximate discrete form is obtained with the help of approximate equations of the field variables. The approximate field equations are expressed in terms of the interpolation (shape) functions by following the concept of partition of unity. Further modifications in the interpolation functions are involved for incorporating the dislocation discontinuity and singularity. These modifications are defined in terms of the enrichment functions in XFEM. The explanation of all these equations is now discussed in the following subsections.

19.3.1 Weak form of governing differential equations

The weak form equation is a modified governing equation where the integration is used to reduce the highest order derivative in the differential equation by order one. The weak form equation is a variational equation obtained by applying the virtual work principle or the Galerkin method. The Galerkin method is a weighted residual method defined to minimize the error. The virtual work is a multiplication of virtual displacement with the force equilibrium equation [20]. The virtual work principle considers all the virtual works of a domain in the form of an integral equation that has to be approximated to zero. Here, the weak form of the governing equation in the absence of body force in Eq. (19.1) is obtained by multiplying the virtual displacement δu_j with the forces as

$$\int_{\Omega} \delta u_j \, \sigma_{ji,i} \, d\Omega = 0 \tag{19.8}$$

Now, in Eq. (19.8), after using the concept of integration by parts and Hooke's relation given by Eq. (19.2), the weak form equation is as follows:

$$\int_{\Omega} \delta u_j \left(c_{jilk} \, \varepsilon_{lk}\right)_{,i} \, d\Omega = 0 \tag{19.9}$$

After using the concept of divergence theorem, Eq. (19.9) can be simplified as given in the following equation:

$$\int_{\Omega} \delta \varepsilon_{ji} \, c_{jilk} \, \varepsilon_{lk} \, d\Omega = \int_{\Gamma_t} \delta u_j \, t_j \, d\Gamma \tag{19.10}$$

Eq. (19.10) is organized in such a way that the force term due to traction is separated and placed on the right-hand side of the equation.

19.3.2 Approximate field equations

The approximate field equations are the discrete equations of the field variables (displacements) involving the interpolation functions. The interpolation functions are the shape functions that are nothing but the polynomials having the piecewise continuity over the elements in a domain [20]. Commonly, the shape functions are the Lagrange interpolation polynomials defined in terms of the local element coordinates. A simple form of the Lagrange shape function (N_j) of order n in terms of the isoparametric coordinate (ξ) is as

$$N_j(\xi) = \prod_{\substack{i=1 \\ i \neq j}}^{n+1} \frac{\xi - \xi_i}{\xi_j - \xi_i} \tag{19.11}$$

The mapping of the local isoparametric coordinates ($\boldsymbol{\xi}$) to the physical coordinates ($\mathbf{x}$) of an element is done through the definition of Jacobian.

The displacements over a standard finite element with p nodes can be defined with the help of the shape functions $N_j(\mathbf{x})$ and the nodal displacements u_j when the partition of unity concept is fulfilled as given in the following equation:

$$u(\mathbf{x}) = \sum_{j=1}^{p} N_j(\mathbf{x})\, u_j \tag{19.12}$$

The approximate displacement equation in Eq. (19.12), for the enriched elements in XFEM, is required to be modified to capture the local behavior of the element.

19.3.3 Enrichments

A discontinuous or singular behavior of a problem is difficult to capture with the help of standard FEM elements. The XFEM overcomes the difficulties of FEM by adding the enrichment terms to the approximate field equations given in Eq. (19.12). These enrichment terms can be added in the form of extrinsic or intrinsic enrichments. In the extrinsic type enrichment, the enrichment term has a standard shape function multiplied by some enrichment functions. However, in the intrinsic type enrichment, the shape function itself is modified according to the enrichment functions. Extrinsic enrichment has some extra unknowns, while this is not the case for intrinsic enrichment. Here, for modeling a Volterra-type dislocation, the extrinsic enrichment is done. In the XFEM formulation, to capture a strong discontinuity corresponding to the dislocation glide plane and singularity of the dislocation core, the local enrichment functions are incorporated as follows:

$$u(\mathbf{x}) = \sum_{j=1}^{p} N_j(\mathbf{x})\, u_j + Enrich^1 + Enrich^2 \tag{19.13}$$

$$Enrich^1 = \sum_{j=1}^{n_h} N_j(\mathbf{x})\, H_j\big(f_1(\mathbf{x})\big)\, a_j \tag{19.14}$$

$$Enrich^2 = \sum_{j=1}^{n_c} N_j(\mathbf{x})\, \varsigma_j(\mathbf{x})\, b_j \tag{19.15}$$

where p is the number of nodes of an element, $Enrich^1$ and $Enrich^2$ are the enrichment terms added to capture the displacement discontinuity and singularity of a dislocation, respectively. a_j and b_j are extra degrees of freedom in Eqs. (19.14) and (19.15), respectively. $N_j(\mathbf{x})$ are the standard shape functions, $H_j\big(f_1(\mathbf{x})\big)$ are the Heaviside functions when $f_1(\mathbf{x})$ is the level set function, and $\varsigma_j(\mathbf{x})$ are the core enrichment functions. n_h are the number of nodes where the Heaviside enrichment functions are added, and n_c are the number of nodes where the core enrichment functions are added.

The enriched elements are identified with the help of the level set method. Usually, The two level set functions are defined, such as $f_1(\mathbf{x})$ is along the glide plane and $f_2(\mathbf{x})$ is perpendicular to the glide plane at the core location. The glide plane elements are identified as the elements at which the $f_1(\mathbf{x})$ will have zero value. The core element is the one where $f_1(\mathbf{x})$ and $f_2(\mathbf{x})$ both will have a value of zero. The Heaviside enrichment function is used for the elements through which the glide plane passes, and the core enrichment function is for the element where the core is located. The Heaviside function has a different value for a positive or negative value of $f_1(\mathbf{x})$, such as

$$\begin{aligned} H\big(f_1(\mathbf{x})\big) &= \frac{1}{2} \qquad \text{for } +\text{ve } f_1(\mathbf{x}) \\ H\big(f_1(\mathbf{x})\big) &= \frac{-1}{2} \qquad \text{for } -\text{ ve } f_1(\mathbf{x}) \end{aligned} \tag{19.16}$$

The core enrichment function has two parts for adding the singularity in the x- and y-directions of the displacement. The core enrichment as a function of the local coordinates (x, y) at the core, material property (ν), and the basis vectors (e_x and e_y) is given as follows:

$$\varsigma_j(\mathbf{x}) = \varsigma_x e_x + \varsigma_y e_y \tag{19.17}$$

$$\begin{aligned} \varsigma_x &= \frac{e_t}{2\pi}\left[\left(tan^{-1}\left(\frac{y}{x}\right) + \frac{xy}{2(1-\nu)(x^2+y^2)}\right)\right] \\ \varsigma_y &= -\frac{e_n}{2\pi}\left(\frac{(1-2\nu)}{4(1-\nu)}\ln(x^2+y^2) + \frac{(x^2-y^2)}{4(1-\nu)(x^2+y^2)}\right) \end{aligned} \tag{19.18}$$

where $e_t$and e_n are the vectors in the tangent and normal direction to the glide plane, respectively. The vectors in Eq. (19.18) ensure that the correct enrichment expression is added to capture the singularity in the x- and y-directions. First expression on the right-hand side of Eq. (19.17) is active when the tangent vector is in the direction of the basis vector e_x, and the second expression is active when the normal vector is in the direction of the basis vector e_y.

So, now the approximate displacement field after considering the shifted enrichment functions in Eq. (19.13) is given as follows:

$$u(\mathbf{x}) = \sum_{j=1}^{p} N_j(\mathbf{x})\, u_j + \sum_{j=1}^{n_h} N_j(\mathbf{x})\,\big(H\big(f_1(\mathbf{x})\big) - H\big(f_1(\mathbf{x}_j)\big)\big)\, a_j + \sum_{j=1}^{n_c} N_j(\mathbf{x})\,\big(\varsigma(\mathbf{x}) - \varsigma(\mathbf{x}_j)\big)\, b_j \tag{19.19}$$

The shifted enrichment functions are defined to avoid the use of blending elements in the XFEM. The blending elements are the elements other than the standard and enriched ones with one or more enriched nodes. A simplified form of Eq. (19.19) is

$$u(\mathbf{x}) = \sum_{j=1}^{p} N_j(\mathbf{x})\, u_j + \sum_{j=1}^{n_h} N_j^{1e}(\mathbf{x})\; a_j + \sum_{j=1}^{n_c} N_j^{2e}(\mathbf{x})\, b_j \tag{19.20}$$

19.3.4 Discrete form

The weak form equation given as Eq. (19.10), is converted into an algebraic equation with the substitution of the approximate displacement field equation given in Eq. (19.20) and its derivatives. So, the discrete form of the governing equation is written as follows,

$$\left[\int_{\Omega} [B]\, c_{jilk}\, [B]\, d\Omega\right]\{u\} = \int_{\Gamma_t} [N]\; t_j\; d\Gamma \tag{19.21}$$

where $[N]$ is the shape function matrix and $[B]$ is derivatives of the shape functions matrix, given as follows:

$$[N] = \begin{bmatrix} N_j & 0 & N_j^{1e} & N_{jx}^{2e} \\ 0 & N_j & N_j^{1e} & N_{jy}^{2e} \end{bmatrix} \tag{19.22}$$

$$[B] = \begin{bmatrix} (N_j)_{,x} & 0 & \left(N_j^{1e}\right)_{,x} & \left(N_{jx}^{2e}\right)_{,x} \\ 0 & (N_j)_{,y} & \left(N_j^{1e}\right)_{,y} & \left(N_{jy}^{2e}\right)_{,y} \\ (N_j)_{,y} & (N_j)_{,x} & \left(N_j^{1e}\right)_{,y} + \left(N_j^{1e}\right)_{,x} & \left(N_{jx}^{2e}\right)_{,y} + \left(N_{jy}^{2e}\right)_{,x} \end{bmatrix} \tag{19.23}$$

where N_j is the standard FEM shape function, N_j^{1e} is the shape function with the Heaviside enrichment as in Eq. (19.20), N_{jx}^{2e} is the shape function with core enrichment due to ς_x and N_{jy}^{2e} is the shape function with core enrichment due to ς_y.

19.3.5 Numerical integration

The integration of the terms in the discrete equation (Eq. 19.21) needs to be evaluated to get the solution of the displacements. Direct integration is not always possible; hence, numerical integration is used [19]. The Gauss quadrature is a numerical integration procedure that defines the definite integration of a continuous function for limits -1 to 1, in the form of a summation equation where the function is multiplied with some weights. The number of weights will depend on the order of the Gauss quadrature used. The Gauss quadrature is indirectly applicable to the enriched elements as the shape functions are discontinuous there. So, to apply the Gauss quadrature in the enriched elements, the elements are subdivided into smaller elements such that discontinuity does not exist inside the subelements. For example, a number of triangles can be used to divide a quadrilateral element where the Gauss quadrature is used on each triangle. The integration points, also known as the Gauss points, and weights are selected within each subelement according to the order of polynomial function required to be integrated. The sum of the integration done at all the Gauss points within an element is the total integration over the area of an element. After performing the integration at all the Gauss points, the overall values at the nodes can be obtained with the help of interpolation functions.

19.3.6 Numerical procedure

The overall numerical procedure to implement the XFEM incorporates some preprocessing, processing, and postprocessing. Preprocessing includes the input of data to define the geometrical aspects of the domain and dislocation. Then the discretization of the domain is done to get the elements, and the material properties are defined in each element. In the meshed domain, the enriched nodes corresponding to the elements cut by dislocation glide plane and core are identified for the enrichment application. For the global conditions, the stiffness matrix, force vector, and the vector of unknown displacements are set up. The processing includes the solution of discrete governing equations on the application of the load and boundary conditions. The solution results in obtaining all the nodal values of the vector of unknown primary variables. Then, in postprocessing, the secondary variables, which are the derivatives of primary variables such as strains and stresses, are evaluated. Finally, the Peach–Koehler force on dislocations as in Eq. (19.7) is also computed after knowing the primary and secondary variables. A related flowchart diagram showing the XFEM procedure for modeling the dislocations can be found in Refs. [18,19].

19.4 Numerical examples

The present section considers the XFEM simulations of the problems of edge dislocations present in crystalline materials. Here, the quadrilateral elements "Q4" with four nodes are considered to discretize the problem domain in 2D plain strain conditions. The discrete equations are converted into algebraic equations with the help of Lagrange shape functions and their derivatives. The numerical integrations in the discrete equations are performed with the help of the Gauss quadrature and subtriangulation techniques. The primary and secondary fields of the domain are obtained from the solution of the algebraic equations, and the Peach–Koehler force is also computed from the results.

19.4.1 Edge dislocation with Burgers vector parallel to bi-material interface

Here, edge dislocation at a distance h from the interface is considered with the Burgers vector b parallel to the interface. The location of dislocation from the bottom surface is at half of the height of the domain considered. The two crystalline materials in the bi-material are semiconductors named Germanium—Ge and Silicon—Si denoted with numbers (1) and (2), respectively. A vertical material interface is considered that separates Ge on the right side from the Si on the left side. Edge dislocation is considered on the right-hand side of the interface in Ge. The domain depicted in Fig. 19.2 is a square of length $L = 100b$ when the magnitude of b is taken as $1\,\mathring{A}$ $(10^{-10}\,m)$ and distance $h = L/8$. The elastic properties of Ge and Si are defined with the help of elastic constants E and ν, such that the values are $E_{Ge} = 103$ GPa, $E_{Si} = 130$ GPa, $\nu_{Ge} = 0.26$, and $\nu_{Si} = 0.28$ [19].

The boundary conditions are implemented on the external boundaries of the problem domain shown in Fig. 19.2 by using the traction boundary conditions. The tractions are obtained from the stress fields of edge dislocation near a bi-material interface with the Burgers vector parallel to the interface [4,6]. The stresses are the functions of Poisson's ratios (ν_1, ν_2) and the modulus of rigidity (G_1, G_2) of both the materials along with the Cartesian coordinates (x, y) defined at the material interface. The stresses in the plane condition, that is, σ_{xx}, σ_{yy}, and σ_{xy} in materials "1" and "2" of the bi-material, are separately represented with the help of superscripts. Such that Eqs. (19.24), (19.25), and (19.26) represent the stresses in material (1) and Eqs. (19.27), (19.28), and (19.29) represent the stresses in material (2) as given in the following equation:

$$\sigma_{xx}^{(1)} = k_1 b\left\{\frac{(x-h)\left[(x-h)^2-y^2\right]}{\left((x-h)^2+y^2\right)^2} + k_2\frac{(x+h)\left[(x+h)^2-y^2\right]}{\left((x+h)^2+y^2\right)^2} - k_3\beta\frac{(x+h)}{\left((x+h)^2+y^2\right)}\right\} - 2k_1k_2bh\frac{\left[(x+h)^4+2x(x+h)^3-6x(x+h)y^2-y^4\right]}{\left((x+h)^2+y^2\right)^3} \tag{19.24}$$

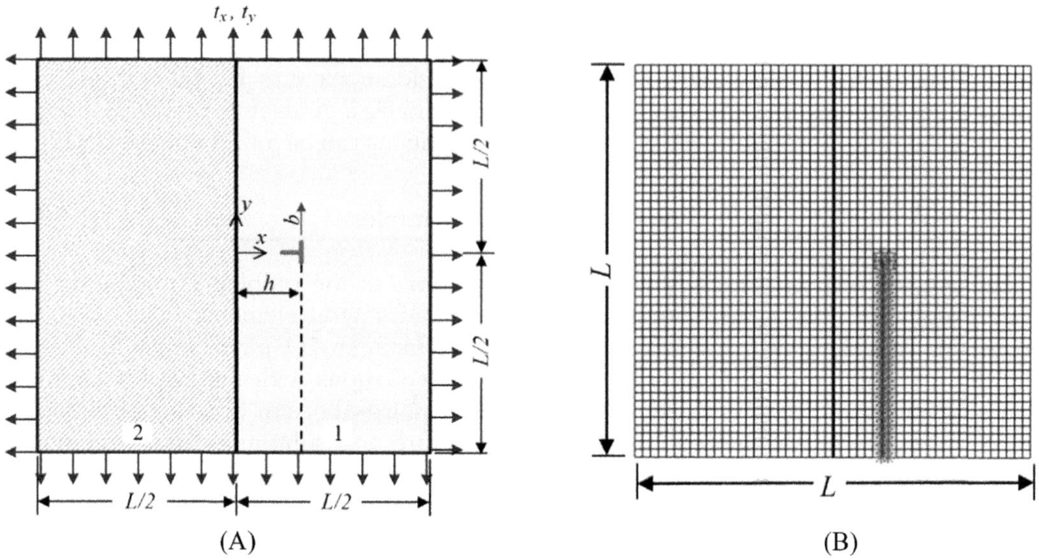

FIGURE 19.2 (A) A schematic diagram showing edge dislocation with the Burgers vector b parallel to the bi-material ("1" is Ge, "2" is Si) interface at a distance h away from the interface. At the external boundary of the domain, the traction is applied in terms of the boundary conditions. (B) The discretized domain of part (A) is shown with 50 elements in the x-direction (25 on each side of bi-material) and 49 elements in the y-direction with total 2450 elements.

$$\sigma_{yy}^{(1)} = k_1 b\left\{\frac{(x-h)\left[(x-h)^2+3y^2\right]}{\left((x-h)^2+y^2\right)^2} + k_2\frac{(x+h)\left[(x+h)^2+3y^2\right]}{\left((x+h)^2+y^2\right)^2} + k_3\beta\frac{(x+h)}{\left((x+h)^2+y^2\right)}\right\} - 2k_1k_2bh\frac{\left[(x+h)^4-2x(x+h)^3+6x(x+h)y^2-y^4\right]}{\left((x+h)^2+y^2\right)^3} \tag{19.25}$$

$$\sigma_{xy}^{(1)} = k_1 b\left\{\frac{y\left[(x-h)^2-y^2\right]}{\left((x-h)^2+y^2\right)^2} + k_2\frac{y\left[(x+h)^2-y^2\right]}{\left((x+h)^2+y^2\right)^2} - k_3\beta\frac{y}{\left((x+h)^2+y^2\right)}\right\} - 4k_1k_2bhxy\frac{\left[3(x+h)^2-y^2\right]}{\left((x+h)^2+y^2\right)^3} \tag{19.26}$$

and

$$\sigma_{xx}^{(2)} = k_1k_3b\left\{\frac{(x-h)\left[(x-h)^2-y^2\right]}{\left((x-h)^2+y^2\right)^2} + 2\beta\frac{x(x-h)^2-hy^2}{\left((x-h)^2+y^2\right)^2}\right\} \tag{19.27}$$

$$\sigma_{yy}^{(2)} = k_1k_3b\left\{\frac{(x-h)\left[(x-h)^2+3y^2\right]}{\left((x-h)^2+y^2\right)^2} - 2\beta\frac{h(x-h)^2-xy^2}{\left((x-h)^2+y^2\right)^2}\right\} \tag{19.28}$$

$$\sigma_{xy}^{(2)} = k_1 k_3 b \left\{ \frac{y\left[(x-h)^2 - y^2\right]}{\left((x-h)^2+y^2\right)^2} + 2\beta y \frac{x^2 - h^2}{\left((x-h)^2+y^2\right)^2} \right\} \tag{19.29}$$

where different parameters are defined as $\alpha = \frac{(1-\nu_1)G_2 - (1-\nu_2)G_1}{(1-\nu_1)G_2 + (1-\nu_2)G_1}$, $\beta = \frac{1}{2}\left[\frac{(1-2\nu_1)G_2 - (1-2\nu_2)G_1}{(1-\nu_1)G_2 + (1-\nu_2)G_1}\right]$, $k_1 = \frac{G_1}{2\pi(1-\nu_1)}$, $k_2 = \frac{\alpha-\beta}{1+\beta}$, and $k_3 = \frac{1+\alpha}{1-\beta^2}$.

Fig. 19.2B shows the discretization of the problem domain shown in Fig. 19.2A, where the 2D plane strain problem is subdivided with the help of bi-linear elements denoted as "Q4." The "Q4" elements are the Lagrange interpolation functions, which are combinations of the polynomials in bi-directions (x and y). Here, the selected mesh has 25×49 elements for individual materials of the bi-material domain. With a perfect interface, the bi-material mesh is composed of 2450 elements, which is capable of capturing the dislocation behavior within an acceptable error limit. The nodes of enriched glide elements are shown with red * symbols.

After knowing the geometrical data, material properties, and mesh of an XFEM problem, we will set up the discrete equations in the algebraic form as $KU = F$. The stiffness matrix K is an organized collection of the elemental stiffness matrices of all elements. The matrix of unknowns is U, which denotes the unknown degrees of freedom. The matrix of forces is F, which incorporates the forces acting from the traction boundary conditions. The displacement unknowns are obtained by multiplying the force vector (F) with the inverse of the stiffness matrix (K^{-1}) and imposing the simply supported boundary conditions. Fig. 19.3A shows the displacement field in the y-direction for edge dislocation with the Burgers vector parallel to the interface. The contour plot shows the values as a ratio of u_y/b to compare the displacements with the magnitude of the Burgers vector. The values are positive on the left of the glide plane because the Heaviside enrichment function is

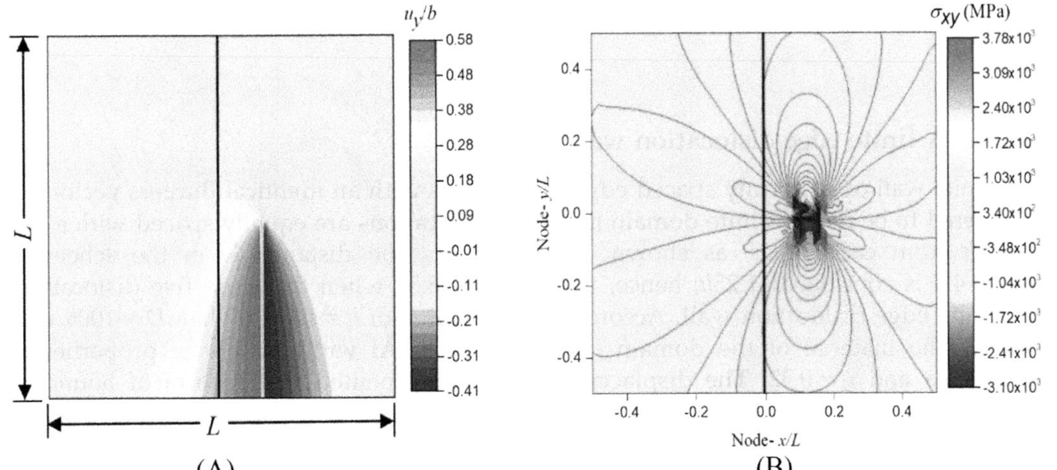

FIGURE 19.3 (A) The displacement field in the y-direction for problem defined in Fig. 19.2 is illustrated in the form of a contour plot of normalized value when u_y is divided with the Burgers vector magnitude b. (B) The shear stresses σ_{xy} (MPa) obtained at the nodes are represented with the help of a contour plot.

positive on this side of the domain and negative on the right of the glide plane due to the negative value of the Heaviside enrichment function.

In Fig. 19.3B, the shear stress σ_{xy} contour plot is shown, where the maximum and minimum stress values are located at the core. The shear stress has negative values above the core and positive values below the core along the glide plane. A slightly high magnitude of stress is generated in tension (+ve) than in compression (−ve). These plots when compared with the analytical stress field are found to be in accordance with the considered mesh. The analytical shear stresses σ_{xy} are computed with the help of Eqs. (19.26) and (19.29) for the right-hand side and the left-hand side material of the bi-material domain, respectively.

Further, after obtaining the displacements and stresses from the simulation results, the Peach–Koehler force on the edge dislocation is evaluated. The expression of the Peach–Koehler force for edge dislocation with the Burgers vector parallel to the material interface [5] is a function of the magnitude of the Burgers vector b, the perpendicular distance of the dislocation from the interface h, the parameters $A' = \frac{G_1 - G_2}{G_1 + G_2(3 - 4\nu_1)}$, and $B' = \frac{(3 - 4\nu_2)G_1 - G_2(3 - 4\nu_1)}{(3 - 4\nu_2)G_1 + G_2}$ as follows:

$$F_{PK} = \frac{-(B' + A')G_1 b^2}{2\pi((3 - 4\nu_1) + 1)h} \tag{19.30}$$

For the discretized domain shown in Fig. 19.2B, the J-integral-based Peach–Koehler force given in Eq. (19.7) is evaluated. The numerical integration is done on the area "A_n," which is the shape of an annular disk with an inner diameter of "$2r$" and outer diameter of "2 R." The weight function "w" is such that it has a value of "1" at radius "r" and "0" at "R." To compute the Peach–Koehler force from the numerical results, the area "A_n" is enclosed between radius, $r = 2.125 \times$ Element length and $R = 2 \times r$. The ratio of the numerically obtained Peach–Koehler force and the analytically obtained Peach–Koehler force given in Eq. (19.30) is 1.001. This verifies the numerical procedure with the analytical results.

19.4.2 A finite edge dislocation wall

A finite wall of uniformly spaced edge dislocations with an identical Burgers vector b is considered to be in an infinite domain [21]. The dislocations are equally spaced with a distance h from each other, as shown in Fig. 19.4A. The distance lh in the schematic Fig. 19.4A is considered 1.25h; hence, the total $D = 6.5h$ when there are five dislocations in a finite edge dislocation wall. Accordingly, the value of $h = D/6.5$ when $D = 100b$ with $b = 1\,\mathring{A}$. The material of the domain is considered Ti_3Al with the elastic properties as $G = 44$ GPa and $\nu = 0.32$. The displacements are implemented in the form of boundary conditions at the external surface of the domain. The applied displacements are equal to the superposition of displacements of each dislocation in the domain. Analytical displacement field of edge dislocation in an infinite domain is given as [22]

$$u_x = \frac{b}{2\pi}\left[\left(\tan^{-1}\left(\frac{y}{x}\right) + \frac{xy}{2(1-\nu)(x^2+y^2)}\right)\right] \tag{19.31}$$

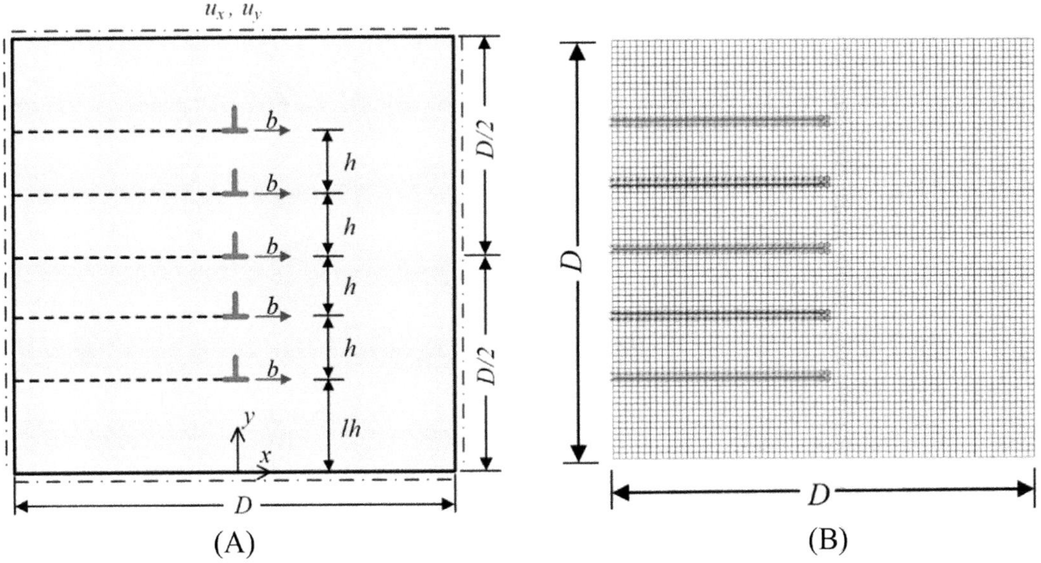

FIGURE 19.4 (A) A schematic of an infinite body showing a finite wall of edge dislocations with the identical Burgers vector b and at an equal spacing of h between them. The displacement boundary conditions are applied on the outer surface of the domain as a superposition of the displacement of individual dislocations. (B) Mesh of part (A) domain when discretized with 69 elements in the x- and y-directions each contributing to a total of 4761 elements.

$$u_y = -\frac{b}{2\pi}\left(\frac{(1-2\nu)}{4(1-\nu)}\ln(x^2+y^2) + \frac{(x^2-y^2)}{4(1-\nu)(x^2+y^2)}\right) \tag{19.32}$$

where x and y are the local coordinates measured at the dislocation core. The coordinates shown in Fig. 19.4A at the middle of the bottom surface are used for superimposing the displacement fields in Eqs. (19.31) and (19.32) for all the five dislocations at the boundary.

Fig. 19.4B is the illustration of the type and size of elements considered in the edge dislocation wall problem of Fig. 19.4A. The dislocation lines are believed to be extended in the direction perpendicular to the 2D domain considered here. The core of the dislocation is present at the end of the glide plane, which is represented by enriched nodes on the mesh. The entire domain has 69 × 69 elements, so a sufficient number of elements are present in the gap between any two dislocations of the wall. All the total 4761 elements are of the same 4-noded plane elements of type, that is, "Q4." The red and blue * are used to specify the enriched nodes in the domain.

The numerical problem set up given in Fig. 19.4 is solved to obtain the elastic field solution. The primary unknowns are displacements, which are obtained for the x- and y-directions. The displacements in the y-direction will have an interaction with the nearby dislocations present in the edge dislocation wall. The ratio of displacements in the y-direction, that is, u_y and the Burgers vector magnitude, that is, b, is plotted in the form of contours in Fig. 19.5A. The maximum values are present at the core locations, as seen by red loops. The results correspond to the equivalent displacements applied as boundary conditions on the extremities of the domain. On postprocessing the displacement results,

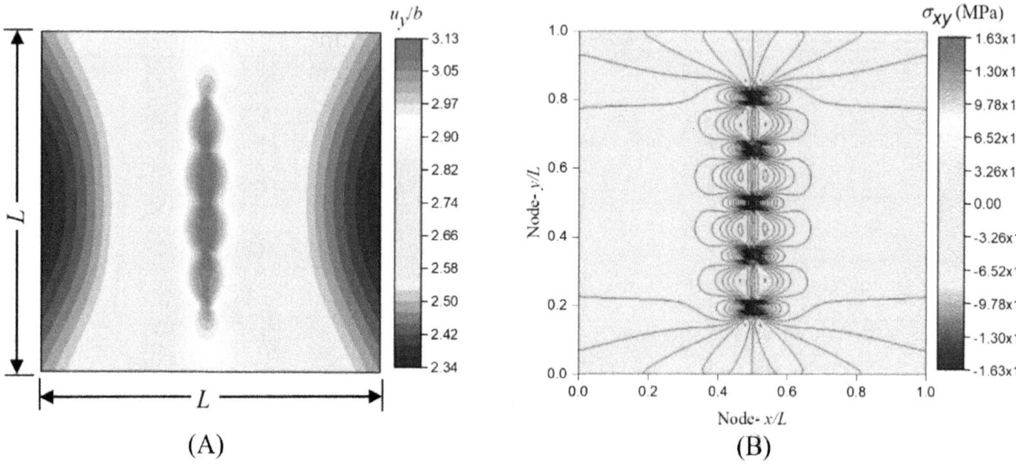

FIGURE 19.5 (A) A contour plot of the displacement field in the y-direction for the problem defined in Fig. 19.4 for normalized values of u_y with respect to the magnitude of the Burgers vector b. (B) A contour plot of the nodal values of the shear stresses σ_{xy} (MPa) is represented in the domain with an edge dislocation wall.

the stresses are computed in the problem domain. Usually, the shear force component is the main factor affecting the Peach–Koehler forces on dislocations; hence, the contour plot of σ_{xy} is depicted in Fig. 19.5B. The nodal stress values are plotted to obtain the contour for the stress field, so both the axes show a normalized value of the nodal coordinates as "Node - x/L" and "Node - y/L." The loops on the stress diagram have positive values ahead of the dislocation core (red) and negative values behind the core along the glide plane direction (blue). This type of result is in accordance with the analytical shear stress distribution of a single-edge dislocation in an infinite domain. Each dislocation of the finite wall surrounded by other dislocations on both sides has the interaction between the individual stresses that result in the formation of loops on the contour plot.

The displacements and stress fields of the edge dislocations in the finite dislocation wall are used to evaluate the Peach–Koehler force on the individual dislocations. The procedure of the numerical evaluation of the Peach–Koehler force is the same as that explained in Section 19.4.1. Here also the annular area around the dislocation is having the inner ($w = 1$) and outer ($w = 0$) radius as $r = 2.125 \times$ element length and $R = 2 \times r$. The maximum value of the Peach–Koehler force is found to be at the core of the central dislocation in the edge dislocation wall. The dislocations above and below the middle dislocations have approximately the same amount of force on them, but those are approximately 0.6 times the force on the central dislocation. The other two dislocations near the bottom and top surfaces of the domain in the dislocation wall have the Peach–Koehler force more than their adjacent dislocations but less than that of the central dislocation.

19.5 Conclusion

The present chapter discusses the XFEM-based modeling of dislocations in crystalline materials. Here we considered the continuum model of the dislocations given by Volterra.

The edge-type dislocations are modeled with the help of enrichment functions to capture the discontinuities and singularities. The Heaviside enrichment function is used to capture the discontinuity at the glide plane, and the core enrichment function is used to capture the singularity due to the dislocation line. The standard quadrilateral elements in the FEM defined with the Lagrangian shape functions are used with the current XFEM formulation. One problem of edge dislocation present near the interface of the semiconductor bi-material (Ge and Si) is considered when the Burgers vector is directed parallel to the interface. This problem is available in the literature to evaluate the analytical solution. The numerical results are obtained in the form of displacements that are used to compute the stresses and the Peach–Koehler force. The numerical results are then compared with the available analytical results to verify the numerical procedure. The contour plots of the fields were in accordance with the analytical results, and the ratio of the numerical value to the analytical value of the Peach–Koehler force was obtained as 1.001. Another problem of a finite edge dislocation wall with five dislocations is considered in an infinite body made of Ti_3Al. The edge dislocations in the wall have the same magnitude of Burgers vector that are parallel to each other. The vertical spacing between the dislocations is constant; however, the distance of two dislocations from the extremities of the top and bottom domain surfaces is more than dislocation spacing. The Peach–Koehler force is found to be maximum for the central dislocation and then for the dislocations at the top and bottom of the wall and minimum for the remaining two dislocations.

References

[1] Hull D, Bacon DJ. Introduction to dislocations, 37. Elsevier; 2011.

[2] Volterra V. Sur l'équilibre des corps élastiques multiplement connexes. In Annales scientifiques de l'École normale supérieure 1907;24:401–517.

[3] Peach M, Koehler JS. The forces exerted on dislocations and the stress fields produced by them. Physical Review 1950;80(3):436.

[4] Head AK. Edge dislocations in inhomogeneous media. Proceedings of the Physical Society. Section B 1953;66 (9):793.

[5] Dundurs J, Sendeckyj GP. Behavior of an edge dislocation near a bimetallic interface. Journal of Applied Physics 1965;36(10):3353–4.

[6] Lubarda VA. Energy analysis of dislocation arrays near bi-material interfaces. International Journal of Solids and Structures 1997;34(9):1053–73.

[7] Lubarda VA. An analysis of edge dislocation pileups against a circular inhomogeneity or a bimetallic interface. International Journal of Solids and Structures 2017;129:146–55.

[8] Lubarda VA. A pileup of edge dislocations against an inclined bimetallic interface. Mechanics of Materials 2018;117:32–40.

[9] Giannakopoulos AE, Baxevanakis KP, Gouldstone A. Finite element analysis of Volterra dislocations in anisotropic crystals: a thermal analogue. Archive of Applied Mechanics 2007;77(2–3):113–22.

[10] Baxevanakis KP, Giannakopoulos AE. Finite element analysis of discrete edge dislocations: configurational forces and conserved integrals. International Journal of Solids and Structures 2015;62:52–65.

[11] Gracie R, Ventura G, Belytschko T. A new fast finite element method for dislocations based on interior discontinuities. International Journal for Numerical Methods in Engineering 2007;69(2):423–41.

[12] Gracie R, Oswald J, Belytschko T. On a new extended finite element method for dislocations: core enrichment and nonlinear formulation. Journal of the Mechanics and Physics of Solids 2008;56(1):200–14.

[13] Oswald J, Gracie R, Khare R, Belytschko T. An extended finite element method for dislocations in complex geometries: Thin films and nanotubes. Computer Methods in Applied Mechanics and Engineering 2009;198 (21–26):1872–86.

[14] Oswald J, Wintersberger E, Bauer G, Belytschko T. A higher-order extended finite element method for dislocation energetics in strained layers and epitaxial islands. International Journal for Numerical Methods in Engineering 2011;85(7):920–38.
[15] Skiba O, Gracie R, Potapenko S. Electro-mechanical simulations of dislocations. Modelling and Simulation in Materials Science and Engineering 2013;21(3):035003.
[16] Mishra D, Pak YE. Electroelastic fields for a piezoelectric threading dislocation in various growth orientations of gallium nitride. European Journal of Mechanics-A/Solids 2017;61:279–92.
[17] Duhan N, Mishra BK, Singh IV. Electro-elastic analysis of edge dislocation dipole in GaN using XFEM. Recent Advances in Computational and Experimental Mechanics, I. Singapore: Springer; 2022. p. 141–51.
[18] Duhan N, Patil RU, Mishra BK, Singh IV, Pak YE. Nonlinear thermo-elastic analysis of edge dislocations with Internal Heat Generation in Semiconductor Materials. Mechanics of Materials 2022;169104322.
[19] Duhan N, Patil RU, Mishra BK, Singh IV, Pak YE. Thermo-elastic analysis of edge dislocation using extended finite element method. International Journal of Mechanical Sciences 2021;192106109.
[20] Zienkiewicz OC, Taylor RL. The finite element method for solid and structural mechanics. Elsevier; 2005.
[21] Lubarda VA, Kouris DA. Stress fields due to dislocation walls in infinite and semi-infinite bodies. Mechanics of Materials 1996;23(3):169–89.
[22] Hirth JP, Lothe J. Theory of dislocations. Krieger Publishing Company; 1982.

Index

Note: Page numbers followed by "*f*" and "*t*" refer to figures and tables, respectively.

D

F

G

H

I

J

K

L

M

Q

R

S

T

9780443153624